P. C. MICHAELIS 12/84

Handbooks and Tables in Science and Technology

Second Edition

Edited by Russell H. Powell

Contents

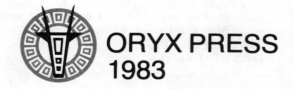

ORYX PRESS
1983

The rare Arabian Oryx is believed to have inspired the myth of the unicorn. This desert antelope became virtually extinct in the early 1960s. At that time several groups of international conservationists arranged to have 9 animals sent to the Phoenix Zoo to be the nucleus of a captive breeding herd. Today the Oryx population is nearing 300 and herds have been returned to reserves in Israel, Jordan, and Oman.

Copyright © 1983 by The Oryx Press
2214 N Central at Encanto
Phoenix, AZ 85004

Published simultaneously in Canada

Printed and Bound in the United States of America

Library of Congress Cataloging in Publication Data

Powell, Russell H., 1943–
   Handbooks and tables in science and technology.

   Includes indexes.
   1. Science—Handbooks, manuals, etc.—Bibliography.
2. Technology—Handbooks, manuals, etc.—Bibliography.
3. Medicine—Handbooks, manuals, etc.—Bibliography.
4. Business—Handbooks, manuals, etc.—Bibliography.
5. Lawyers—Handbooks, manuals, etc.—Bibliography.
6. Military art and science—Handbooks, manuals,
etc.—Bibliography.   I. Title.
Z7405.T3P68   [Q199]   016.5′02′02   82-19842
ISBN 0-89774-039-4

Preface

This book is a compilation of 3,403 handbooks and tables in science, technology, and medicine, completely indexed by subject, keyword, author/editor, and title. All entries have complete bibliographic citations and most have annotations which briefly describe the intellectual content of the handbooks and tables.

What is a handbook? It would be useful to review the characteristics of most handbooks for the convenience of the user. Perhaps we should consider the following in our definition:

1. A handbook is a ready-reference book.
2. Most handbooks are in one volume, although many are multivolume.
3. A handbook is generally prepared by a team of specialists in their respective fields.
4. A handbook is a condensed treatment of a subject and arranged in systematic order.
5. A handbook is a compilation of data and information on a specialized subject.
6. A handbook usually has an extensive table of contents and a comprehensive index, and it often contains numerous references to the literature.
7. A handbook is a secondary source; i.e., a compilation of data that have originally been reported in the research literature.
8. A handbook is frequently in tabular form or it includes a considerable number of tables.

9. A handbook is usually updated by an occasional new edition to ensure reliability and timeliness.

The usage of the term "handbook" may be somewhat misleading because it does not really describe what is included in this bibliography. The emphasis of this book is on data tabulations or compilations of physical and chemical values whether they are in a manual, databook, handbook, sourcebook, table, or guide. The type of information that is included in these tabulations ranges from such topics as properties of materials, physical constants, equations, mathematical functions, conversion factors, formulas, and chemical composition data to circuits, reactions, testing information, instruments, and numerous condensed treatments of extensive subjects.

This book is designed to help the user select the appropriate handbook or table by listing the handbooks in title sequence and by providing the necessary indexes to make it easier to find the information that is needed. Grogan[1] has mentioned that a library with no more than a sound collection of handbooks can answer 90 percent of the ready-reference questions that it receives. Certainly, this book should assist libraries and individuals in the process of finding answers to the many questions that arise in scientific and technical situations.

---

[1]Grogan, Denis. *Science and Technology: An Introduction to the Literature*. 3rd edition, London: Clive Bingley. Hamden, CT: Linnet Books, 1976, p. 68.

Acknowledgements

A book of this magnitude cannot be prepared by one person. Consequently, I would like to take this opportunity to formally thank Andrea Bryant for her kind encouragement in the preparation of this manuscript and her assistance in proofreading the computer printout. I am also deeply grateful to Linda Ferrell for her accurate typing and to Melany Brite for transcribing annotations and keywords during the preparation of this publication. Many other people have played a key role in the preparation of this publication, including Mary Anne Stout, Carol Brinkman, Donna Aubrey, Robin Barnard, David Shippy, Jr., and many others who have helped to alphabetize, sort, resort, search bibliographic entries, and proofread this publication in its various stages of preparation.

Introduction

This book is divided into the following sections:

1. Main Entry Section
2. Subject Index
3. Author/Editor Index
4. Publisher's Index

Main Entry Section

Each item in the Main Entry Section is listed alphabetically by title and may contain any or all of the following information: title, author or editor, publisher, edition statement, copyright date, pages, number of volumes if more than one, Library of Congress catalog card number, International Standard Book Number, and cost. In the event that the handbook is a government document, such information as Superintendent of Documents number, technical report number, and accession number are provided. In many cases, the date of the first edition is listed if the handbook has been published in more than one edition. Most of the handbooks are followed by a brief annotation describing the nature of the handbook and the type of information it contains.

It was decided to include the cost of each handbook primarily as a guide to determine the relative worth of the publication and to indicate how substantial a particular book might be. The prices can certainly be used for comparative purposes when deciding which handbook to purchase. However, it would be wise to check a current pricing guide before ordering any particular handbook because chances are very good that the price has changed. Some of the books do not have prices included. This may be an indication that the book is out of print. Always check with the publisher before concluding that a particular title is out of print.

Each item in the Main Entry Section is preceded by a reference number. This number is used as a locating device from the author and subject indexes to the title listing. The title of each entry is printed in boldface type so that the section will be easier to use.

Subject Index

The Subject Index is primarily an enriched keyword index. The keywords have been taken from the titles or the annotations of each entry. Each book has also been indexed under subject if the keywords do not describe the intellectual content of the book. Keywords have a tendency to scatter a subject, whereas a subject index has a tendency to bring like subjects together under one heading. The "enriched" aspect of this index is an attempt to bring the subjects together by selecting the appropriate terms that do not appear in the title or the annotation. Every effort has been made to provide consistency. Many *see* references have been added to direct the user from an unused term to a used term. In addition, numerous *see also* references have been added to provide interrelationships between subject headings.

In order to make the index easier to use, it should be pointed out that the majority of terms are nouns, although entries were occasionally made under adjectives. In addition, if an entry has parentheses in it, it would be advisable to also look under the term in parentheses. For example, if an entry is "Design (Electronics)," then the user should also look under the term "Electronics" in order to get better coverage of a subject. A word of caution should be mentioned with regard to combining medical, engineering, and science terminology in one index. Frequently, the same word may have several different meanings depending upon the subject. The user should be alert to this phenomenon and approach the index with this in mind.

Each subject term is followed by a reference number which indicates the book on that subject.

Author/Editor Index

This is a combination author and editor index, and no distinction has been made between the two. Corporate authors are included as well as personal authors. When a book does not have an author listed, the publisher is considered to be the author. Translators are also frequently listed in this index. Entries are provided under co-authors. Each author entry is followed by a reference number which refers to the main entry listing.

Publisher's Index

The Publisher's Index is provided so that the user can easily obtain purchasing information and availability.

Types of Handbooks Included and Excluded

The emphasis in this bibliography is on the "hard" sciences. For example, handbooks and tables were included if they were on such subjects as physics, chemistry, engineering, mathematics, astronomy, biology, geology, agriculture, medicine, dentistry, etc. Numerous medical handbooks have been included because of the interdisciplinary nature of the physical and biological sciences. A substantial number of medical atlases also have been included in the Second Edition. Occasionally, books on the following subjects are listed in this handbook: psychology, nursing, statistics, physical therapy, business, management, etc.

The popular "how to" literature is generally excluded from this book. For example, such titles as *Appliance Service Handbook, TV Troubleshooter's Handbook, Handbook of Practical CB Service,* and *Motor Auto Repair Manual* have not been included in this bibliography.

Dictionaries, encyclopedias, field guides, maps, directories, biographical sources, indexing, abstracting, and current awareness sources are generally omitted from this book.

Many data sources appear in journal format or in a continuing series. For example, the following titles are good sources for technical data: *Bulletin of Chemical Thermodynamics; Journal of Chemical and Engineering Data; Journal of Physical and Chemical Reference Data; Nuclear Data* (Journal), Academic Press; American Society for Testing and Materials, *Data Series.* Occasionally, individual publications in these series have been included if they represent a substantial compilation of data. The reader is encouraged to refer to the indexes of these series for further sources of technical data.

Although "buyers' guides" are generally excluded, some are included if they contain such things as basic properties, tables, tabulated data, and testing information on various products. For example, *Metal Finishing Guide Book Directory* was included because it provides the aforementioned information.

Although there are many translations of foreign language handbooks included in this book, there are actually very few handbooks included that are in languages other than English, except for several German handbooks (treatises). Some of the major German language treatises were included because they contain a considerable amount of tabulated data, in spite of the fact that they are not really handbooks in the traditional sense. The following are examples of German language treatises that were included: *Beilstein's Handbuch der Organischen Chemie, Gemlins Handbuch der Anorganischen Chemie, Landolt-Böernstein Numerical Data and Functional Relationships in Science and Technology, New Series.*

Generally speaking, standards have been omitted from this book. However, many engineering manuals and handbooks include standards as a part of the design or specification data. For example, the *ACI Manual of Concrete Practice* and the *SAE Handbook* include many standards and recommended practice. Another example is the *ASTM Annual Book of Standards* which was included because it is used like a handbook.

The *Engineering Design Handbooks* of the U.S. Army Materiel Command are included in this bibliography because they represent a substantial collection of data and specifications on many different subjects that are not just limited to military projects. A considerable number of environmental topics are discussed in addition to such subjects as pyrotechnics, properties of materials, and hydraulics.

For NASA handbooks, please refer to the following publication: *A Guide to NASA Data Handbooks,* Scientific and Technical Information Division, published by the National Aeronautics and Space Administration, 1967. This publication is a bibliography including approximately 1,000 handbooks that have been sponsored by NASA.

There are many sources of critical data. The reader's attention should be drawn to the publication entitled: *Continuing Numerical Data Projects, A Survey and Analysis,* published by the Office of Critical Tables, National Academy of Sciences, National Research Council, 2nd edition, 1966, Publication No. 1463, 203 pages. This publication describes over 50 projects that produce and compile critical data. An additional listing of sources of critical data can be found in the *CRC Handbook of Chemistry and Physics,* 63rd edition, 1982–83, pages F-349 through F-362.

Many of these critical data projects are sponsored by an information analysis center. A good example of one is the Center for Information and Numerical Data Analysis and Synthesis (CINDAS) at Purdue University. CINDAS has produced such compilations as the *Thermophysical Properties of Matter,* and they are currently working on the development of a new 80-volume CINDAS handbook series on material properties and special projects.

For further reading on the subject of data compilations, please refer to the following:

1. International Council of Scientific Unions. Committee on Data for Science and Technology. *International Compendium on Numerical Data Projects: A Survey and Analysis.* Springer-Verlag, 1969.
2. N.B. Gove and K. Way. "The Data Compilation as Part of the Information Cycle," *Journal of Chemical Documentation,* v. 2, 1962, pp. 179–81.
3. U.S. National Bureau of Standards. Office of Standard Reference Data. *Critical Evaluation of Data in the Physical Sciences—A Status Report on the National Standard Reference Data System, January 1977.* U.S. Government Printing Office, Washington, D.C., 1977. (*USNBS Technical Note* 947).
4. Brenda Mountstephens *et al. Quantitative Data in Science and Technology* (ASLIB, 1971).
5. National Academy of Sciences. National Research Council. Office of Critical Tables. *Continuing Numerical Data Projects: A Survey and Analysis.* 2nd edition, Washington, D.C. National Academy of Sciences, 1966 (*NAS-NRC Publication* No. 1463).

Every effort has been made to make this book as complete as possible. However, it is recognized that omissions or errors may have occurred. Consequently, the editor is requesting that any user who identifies an omission or error please call it to his attention, so that corrections can be made for future editions. All corrections, additions, and deletions should be sent to:

Russell H. Powell
Engineering Librarian
University of Kentucky
Lexington, Kentucky 40506

Main Entry

1. Ab Initio Valence Calculations in Chemistry. D.B. Cook. Halsted Press, 1974, 271 pp. LC: 73-15144. ISBN: 0-470-17000-X. Cost: $32.95.

2. Abbreviated Tables of Thermodynamic Properties to 85km for the US Standard Atmosphere, 1974. A.J. Kantor; A.E. Cole. NTIS, 1974, 23 pp. Accession: N74-21008; AD-774404. Tech Report: AFCRL-AFSIG-278; AFCRL-TR-73-0687.

This set of tables was prepared under contract with the U.S. Air Force. Provides background information and abbreviated tables of thermodynamic properties of U.S. standard atmosphere, 1974. These tables provide temperature, pressure, and density from the surface to 85km altitude.

3. Abridged Thermodynamic and Thermochemical Tables in SI Units. F.D. Hamblin. Pergamon Press, 1971, 88 pp. ISBN: 0-08-016456-0. Cost: $9.75.

4. Absolute Configurations of 6000 Selected Compounds with One Asymmetric Carbon Atom. J.J.C. Gros; S. Bourcier. Heyden and Sons Inc, 1977, 622 pp. ISBN: 0-9940010-3-7. Cost: $110.00. Series: *Stereochemistry Fundamentals and Methods Series*, Vol. 4.

This volume comprises a set of tables as a compilation of organic compounds whose absolute configuration is known. The compounds can be located and identified either by formula, by classification, or by family. This book includes a formula index, a subject index, and an extensive bibliography providing source information for the tables.

5. Accident Prevention Manual for Industrial Operations. Engineering and Technology Volume. Frank E. McElroy (ed). National Safety Council, 1980, 768 pp. LC: 80-81376. ISBN: 0-87912-026-6. Cost: $35.00.

6. ACI Detailing Manual-1980. American Concrete Institute. American Concrete Institute, 1980, 206 pp. LC: 80-68476. Series: ACISP 66.

Collection of standards recommended by the American Concrete Institute. It includes a considerable amount of supporting reference data. The manual also includes data on highway structures as well as nonhighway structures. Information is also provided on reinforcing bars, welded wire fabric, and bar supports.

7. ACI Manual of Concrete Inspection. American Concrete Institute, 6th ed., 1975, 275 pp. LC: 74-83156. Cost: $9.50 members; $13.75 nonmembers. Series: No. SP-2.

8. ACI Manual of Concrete Practice. American Concrete Institute, 1981, 5 parts, various paging. Cost: $139.75 per set; $35.95 per part.

The 5 parts of this manual is entitled as follows: Part 1, *Materials and General Properties of Concrete*; Part 2, *Construction Practices and Inspection—Pavements*; Part 3, *Use of Concrete in Buildings-Design, Specifications, Related Topics*; Part 4, *Bridges, Structures, Sanitary and other Special Structures*; Part 5, *Masonry-Precast Concrete-Special Processes*. The *Manual* contains such diverse subjects as painting concrete, protection of concrete against chemical attack, low-pressure steam curing, precast concrete panels, joints and connections in precast structural concrete, practice for shotcreting, shear and diagonal tension, torsion, reinforced concrete nuclear power structures, building codes, reinforced concrete pavements for highways, concrete formwork, hot and cold weather concreting, mass concreting for dams, and use of aggregates for concrete.

9. Activation Analysis Handbook. R.C. Koch. Academic Press, 1960, 219 pp. LC: 60-14271. ISBN: 0-12-417550-3. Cost: $33.50.

10. Activation and Decay Tables of Radioisotopes. E. Bujdoso; I. Feher; G. Kardos. Elsevier-North Holland Publishing Co, 1973, 575 pp. LC: 79-135492. ISBN: 0-444-99937-X. Cost: $68.50.

11. Active Filter Design: Handbook for Use with Programmable Pocket Calculators and Mini Computers. G.S. Moschytz; P. Horn. John Wiley and Sons Inc, 1981, 296 pp. LC: 80-40845. ISBN: 0-471-27850-5. Cost: $49.50.

Provides information on 20 different active filters. The filters selected have been chosen for low resistance-capacitance component count, minimum power, low sensitivies to passive and active components, easy tunability, and good stability.

12. Active Filter Handbook. Frank P. Tedeschi. TAB Books, 1979, 280 pp. LC: 79-12530. ISBN: 0-8306-9788-8. Cost: $9.95.

This handbook provides information on active filter circuits, low-pass active filter design, high-pass active filter design, band-pass active filter design as well as information on operational amplifiers. Butterworth and Chebyshev transfer functions are also covered in this handbook.

13. Acupuncture Handbook. D. Lawson-Wood; J. Lawson-Wood. British Book Centre, 2d ed., 1973. ISBN: 0-8277-1427-0. Cost: $10.95.

14. Adhesive and Sealant Compounds and Their Formulations. E.W. Flick. Noyes Data Corp, 1978, 402 pp. LC: 78-56014. ISBN: 0-8155-0713-5. Cost: $36.00.

Over 500 adhesives and sealant formulations are included. This book covers such topics as coatings, hot melts, industrial adhesives, and adhesives for paper and packaging.

15. Adhesive Bonded Aerospace Structures Standardized Repair Handbook. J.E. McCarty; R.E. Horton. Boeing Commercial Airplane Co (available from NTIS), phase 2 of a 5 phase program, Dec. 1974, 90 pp. Accession: AD-A017 779. Tech Report: AFML-TR-75-158. Cost: $2.25 microfiche; $5.00 paper copy.

This handbook is primarily concerned with standardized methods of repairing bonded aircraft structures. It includes such instructions as component rebuilding and the selection of standard procedures for small area repairs. The handbook also covers such areas as surface preparation methods and evaluating adhesives.

16. Adhesives Handbook. J. Shields. Newnes-Butterworths, 2d rev. ed., March 1976, 370 pp. LC: 77-861540. ISBN: 408-00210-7. Date of First Edition: 1970. Cost: $51.95.

This handbook is primarily concerned with adhesive bonding in connection with assembly processes. Adhesive materials are covered and, particularly, synthetic polymers. Such broad areas as adhesive selection, surface preparation and the bonding process receive treatment, as well as joint design, nondestructive testing, fatigue and standard test methods. The *Adhesives Handbook* has a number of tables indicating physical properties of adhesives, a substantial bibliography and a trade names list with manufacturers and sources of additional information. In addition, specific information is given on costs of adhesive selection, rubber-resin blends, wet bonding, reactivation bonding, curing, ultrasonic activation, durability of adhesives and strength properties.

17. Adhesives. International Plastics Selector, Inc, 1978/79, 2 Vols. Cost: $49.95. Series: *Desk-top Data Bank.*

This 2-volume set includes information on over 4,500 adhesives, sealants, films, and primers. Numerous properties are listed for these adhesives including density, flashpoint, drying time, percent solids, service temperatures, and lap shear. Such information as commercial names, manufacturers, and addresses are provided. One of the most useful aspects of this 2-volume set is the paired substrate index which includes over 240 substrate combinations. For example: it provides information on recommended adhesives for such combinations as asbestos—wood, asphalt—concrete, ceramic—glass, clay—paper, concrete—metal, leather—metal, etc.

18. Adhesives: Guidebook and Directory. Noyes Data Corp, annual, 1976.

19. Advanced Composites: Mechanical Properties and Hardware Programs for Selected Resin Matrix Materials . . . Considering Space Shuttle Applications. E.K. Welhart. NTIS, 1976, 38 pp. Accession: N76-24364. Tech Report: NASA-13970; NASA-CR-147705; REPT-1.2-DN-B0104-1. Cost: $4.00.

Tabulated data are provided for 6 epoxies, 8 polyimides, and one polyquinoxaline. Boron and graphite are the fiber reinforcements.

20. Advanced Oscilloscope Handbook for Technicians and Engineers. Derek Cameron. Reston Publishing Co, 1977, 228 pp. LC: 76-25044. ISBN: 0-87909-008-1. Cost: $18.95.

21. Aerospace Fluid Component Designers' Handbook. Glen W. Howell et al. TRW Systems Group (available from NTIS), rev. D, 1970, 2 Vols. Accession: Vol. 1, AD-874542; Vol. 2, AD-874543. Tech Report: AFRBL-TDR-64-25.

22. Aerospace Structural Metals Handbook. W.F. Brown Jr et al. Mechanical Properties Data Center, 1978, 5-Vol. set, loose-leaf service with annual updating. Cost: $325.00 plus $50.00 for the annual update service.

This handbook provides information on mechanical properties of over 200 metals and alloys which have been selected primarily for their high efficiency structural applications in the aerospace industry.

23. AGMA Gear Handbook. American Gear Manufacturers Association, 1972. Cost: $5.00.

Vol. 1 is entitled: *Gear Classification, Materials and Measuring Methods for Unassembled Gears.*

24. Agricultural Engineer's Handbook. C.B. Richey; P. Jacobson; C. Hall. McGraw-Hill, 1961, 880 pp. LC: 59-13942. ISBN: 0-07-052617-6. Cost: $55.00.

25. Agricultural Statistics: A Handbook for Developing Countries. N.M. Idaikkadar. Pergamon Press Inc, 1979, 100 pp. LC: 78-41193. ISBN: 0-08-023388-0 hardback; 0-08-023387-2 flexicover. Cost: $20.00 hardback; $10.00 flexicover.

This book is a statistical analysis of various agricultural topics for developing countries. Included are such topics as production of crops and livestock, crop estimation, national food balance sheets, crop yields, as well as price statistics.

26. Agricultural V-Belt Drive Design. L.R. Oliver; C.O. Johnson; W.F. Breig. Dayco Corp, 1977, 168 pp. LC: 77-079837. Cost: $50.00.

This book represents a compilation of data on V-Belt drives with such design calculations as tension of belt drives, axial force, torsional frequency, and vibration.

27. Air and Noise Pollution Control. Lawrence Wang; Norman C. Pereira. Humana Press, 1979, 484 pp. LC: 78-78033. ISBN: 0-89603-001-6. Cost: $49.50.

This is the first in a 5-volume series entitled *Handbook of Environmental Engineering.* The major methods of air pollution control are discussed in this handbook including fabric filtration, cyclones, electrostatic precipitation, wet scrubbing, and atmospheric dilution as well as mechanical particulate collectors and absorption methods. Although the major portion of this book is devoted to air pollution control, there are approximately 75 pages devoted to the control of noise pollution.

28. Air Conditioning and Refrigeration Troubleshooting Handbook. Billy C. Langley. Reston Publishing Co, Inc, 1980, 848 pp. ISBN: 0-8359-0204-8. Cost: $29.95.

This book provides practical service procedures, recommended service operations, and lists common problems that are encountered while troubleshooting air conditioning and refrigeration equipment. Charts are provided to help isolate problems and localize the cause of malfunctions. Wiring diagrams are included and the book is well illustrated. Such difficult service procedures as stack control check-out, oil test procedures, compressor replacement, and systems charging are included.

29. Air Conditioning for Building Engineers and Managers: Operations and Maintenance. Seymour G. Price. Industrial Press, 1970, 136 pp. ISBN: 0-8311-3001-6. Cost: $23.00.

This handbook is designed primarily for the operation and maintenance of air conditioning equipment and commercial buildings. Such topics as refrigeration, central chilled water, water cooling towers, water treatment, air supply systems, and filters are discussed. In addition, there is a section on preventive maintenance, safety, and a section on reference material for further reading.

30. Air Pollution Arthur C. Stern. Academic Press, 3rd ed., Vol.1, 1976; Vol.2, 1977; Vol.3, 1976; Vol.4, 1977; Vol.5, 1977, Vol.1, 752 pp.; Vol.3, 824 pp.; Vol.4, 976 pp.; Vol.5, 672 pp. LC: 61-18293 (rev. 1st ed.). ISBN: Vol.1, 0-12-666601-6; Vol.2, 0-12-666602-4; Vol.3, 0-12-666603-2; Vol.4, 0-12-666604-0; Vol.5, 0-12-666605-9. Date of First Edition: 1962. Cost: Vol.1, $39.50; Vol.2, $58.50; Vol.3, $42.50; Vol.4, $49.50; Vol.5, $50.50.

This is a well-known and respected standard work among environmental engineers. The first edition had 2 vols., the 2d 3 vols., and the 3d is in 5 vols., which is a reflection of how the knowledge of air pollution has expanded. This 5-Vol set covers such areas as pollution from automobiles and incinerates, sampling, analysis of air pollution on animals, humans, and plants, data handling, and air pollutant dispersal by meteorological factors. In addition, Stern's comprehensive work also includes legal aspects of air pollution, regulations, standards, emission inventory, mathematical modeling, control techniques, and equipment. For related works, refer to *Air Pollution Engineers Manual*, *CRC Handbook of Environmental Control*, and the *Environmental Engineers Handbook*.

31. Air Pollution Control and Design Handbook. Paul H. Cheremisinoff (ed); Richard A. Young (ed). Marcel Dekker, Inc, 1977, 2 Vols.; Part 1, 624 pp; Part 2, 404 pp. LC: 76-588. ISBN: Part 1, 0-8247-6444-7; Part 2, 0-8247-6448-X. Cost: Part 1, $49.75; Part 2, $39.75.

Primarily covers air pollution problems associated with stationary sources.

32. Air Pollution Engineering Manual. John A. Danielson. U.S. Public Health Service, U.S. Government Printing Office, 1967, 892 pp. SuDoc: FS2.300:AP-40. Cost: $5.75. Series: *Public Health Service Publication*, No. 999-AP-40.

33. Air Pollution Handbook. P. Magill; F. Holden; C. Ackley. McGraw-Hill, 1956, 645 pp. LC: 55-11934. ISBN: 0-07-039490-3. Cost: $44.50.

34. Air Pollution Sampling and Analysis Deskbook. Paul N. Cheremisinoff; A.C. Morresi. Ann Arbor Science Publishers, Inc, 1978, 490 pp. LC: 77-93385. ISBN: 0-250-40234-3. Cost: $29.50.

Sampling and analytical methods are discussed for point sources and fugitive emissions. Suggested methods of measuring all gases, particles, anions, and the various advantages, disadvantages, and limitations of each method are presented.

35. Air Tables. Duane P. Jordan; Michael D. Mintz. McGraw-Hill, 1965, 797 pp. LC: 64-66045.

The subtitle of this book is *Tables of the Compressible Flow Functions for One-Dimensional Flow of a Perfect Gas and of Real Air*. Part 1 includes isentropic flow, frictionless constant—area flow, adiabatic constant—area flow, isothermal constant—area flow and normal—shock flow.

36. Air Weapons Materials Application Handbook—Metals and Alloys. R.F. Pray III; G. Sachs. NTIS, 1959, 805 pp. Accession: N70-79082. Tech Report: ARDC-TR-59-66.

This applications handbook includes information on chemical properties of metals and alloys, mechanical properties, and thermodynamic properties. Ferrous as well as nonferrous metals are included. Missiles and airborn equipment and weapons are given coverage in this handbook.

37. Aircraft and Missile Design and Maintenance Handbook. Charles A. Overbey. Macmillan, 1960, 369 pp. LC: 59-14331.

38. Airport Capacity Handbook. AIL (division of Cutler-Hammer). NTIS, 2d ed., 1969, various paging. Accession: AD-690470. Tech Report: RD-68-14.

This handbook deals with the traffic at airports and the number of airplanes that can be handled in terms of the number of runways the airport has. Various types of airport designs are considered, ranging from single runway airports to parallel runways, 3 parallel runways, intersection, and open-V runways. This handbook was prepared for the Department of Transportation, Federal Aviation Administration, Systems Research and Development Service.

39. Alkali Metal Halides in Organic Solvents—Solubilities of Solids. B. Scrosati. Pergamon Press Inc, 1980, 350 pp. ISBN: 0-08-023917-X. Cost: $100.00. Series: *Solubility Data Series*, Vol. 11.

40. Alkali Metal, Alkaline-earth Metal and Ammonium Halides in Amide Solvents. A.S. Kertes. Pergamon Press Inc, 1980, 374 pp. LC: 79-41710. ISBN: 0-08-023917-X. Cost: $100.00. Series: *Solubility Data Series*, Vol. 11.

This International Union of Applied and Pure Chemistry (IAUPC) publication is a collection of solubility data of the halides and pseudohalides of the alkali and alkali-earth and ammonium ions in formamide, N-methylformamide, and various other amide solvents. A registry number index is provided in addition to a system index.

41. Alkali-and Alkaline-earth Metal Oxides and Hydroxides in Water—Solubilities of Solids. J.E. Bauman. Pergamon Press Inc, 1979, 350 pp. ISBN: 0-08-023920-X. Cost: $100.00. Series: *Solubility Data Series*, Vol. 10.

42. Alkali-Metal Chlorides (Binary Systems)—Solubilities of Solids. R. Cohen-Adad. Pergamon Press Inc, 1979, 350 p. ISBN: 0-08-023918-8. Cost: $100.00. Series: *Solubility Data Series*, Vol. 9.

43. Alkaline-earth Metal Sulfates—Solubilities of Solids. J.W. Lorimer. Pergamon Press Inc, 1979, 350 pp. ISBN: 0-08-023916-1. Cost: $100.00. Series: *Solubility Data Series*, Vol. 6.

Alkaline-earth Sulfates in all Solvents—Solubilities of Solids. *See Solubility Data Series*, Vol. 6.

44. Alloy Cross Index, 1980 Edition. Battelle, Columbus Laboratories, Mechanical Properties Data Center, 1980. Cost: $75.00.

The purpose of this index is to facilitate the identification of equivalent materials with a common designation so that information can be more efficiently stored and retrieved. This index lists over 22,000 designations in alphanumeric sequence with a materials listing and common code number.

45. Alloy Digest. Engineering Alloys Digest, Inc, monthly loose-leaf subscription service.

Monthly data sheets are provided on approximately 500 different types of alloys, both ferrous and nonferrous, giving such information as properties, heat treatment, composition, corrosion resistance, and fabrication.

46. Alloying of Structural Steel B.B. Vinokur; B.N. Beinisovich; A.L. Geller; M.E. Natanson. Izdatel'stno Metallurgia, Moscow (available from NTIS), 1977, 200 pp. Accession: A77-46503.

This book, in Russian, provides information on chromium steels containing one, 2, and 3 percent chromium, and also alloyed with molybdenum, tungsten, vanadium, titanium, and their compounds. Microstructure is studied as a function of heat treatment.

47. Alloys for Cryogenics. P.F. Koshelev.. Scientific Information Consultants Ltd., 374 estimated pp.

Translation edited by Dr. R. Kerr, Scientific Consultant, British Oxygen Co. Ltd., London. Data presented on mechanical properties, notch sensitivity, the strength of welded joints, and the impact strength of numerous metals and alloys at temperatures down to 20°K. There are over 385 graphs which show variation of the mechanical properties with temperature from 300° to 20°K. Included are such metals as steels and ferrous alloys, titanium alloys, aluminum alloys, magnesium alloys, copper alloys, and various other special metals such as nickel and brazing alloys.

48. Alphabetical List of Compound Names, Formulae, and References to Published Infrared Spectra; An Index to 92,000 Published Infrared Spectra, ASTM-AMD-34. American Society for Testing and Materials, 1969. Cost: $50.00. Series: *ASTM Publication Code* No.: 10-034000-39.

An additional 10,000 compounds are listed in: *Alphabetical List of Compoound Names, Formulae, and References to Published Infrared Spectra, AMD-34-S14,* 1972. Publication Code No. 10-034014-39.

49. Aluminum Alloys: Structure and Properties. L.F. Mondolfo. Butterworths, 1976, 971 pp. ISBN: 0-408-70680-5. Cost: $89.50.

This work has an extensive bibliography from which the data have been compiled. Many properties of aluminum are provided including thermal, mechanical, and electrical properties.

50. Aluminum Construction Manual. J.W. Clark. Aluminum Association Inc, 3d ed., 1976, 222 pp. (in 3 sections). Date of First Edition: 1967.

There are 3 sections to this manual. The first section is entitled *Specifications for Aluminum Structures,* copyright April, 1976, 67 pp. The second section entitled *Illustrative Examples of Design: Based on Specifications for Aluminum Structures,* copyright April, 1978, 51 pp. The third section, entitled *Engineering Data for Aluminum Structures,* copyright January, 1975, 104 pp.

51. Aluminum Standards and Data. Aluminum Association, 1972-73, 200 pp.

52. Aluminum Wire Tables. National Bureau of Standards. NTIS, Feb., 1972. Accession: COM 72-50183. Cost: $4.50. Series: NBS Handbook 109.

53. Aluminum. Kent Robertson Van Horn. Aluminum Company of America, American Society for Metals (also Chapman), 1967, 3-Vol. set. LC: 66-16222.

54. AMA Drug Evaluations. American Medical Association, Department of Drugs; American Society for Clinical Pharmacology and Therapeutics. Wiley Medical, 4th ed., 1980, 1,580 pp. ISBN: 0-471-08125-6. Cost: $48.00.

Contains evaluations of over 1,300 drugs. Gives information on dosage, contraindications, adverse reactions, and provides chemical information such as formulas.

55. Amateur Astronomer's Handbook. John Benson Sidgwick. Enslow Publishers, 4th ed., 1980, 568 pp. LC: 80-20596. Cost: $24.95.

This book is a companion volume to *Observational Astronomy for Amateurs.*

56. Amateur Astronomer's Handbook. James Muirden. Thomas Y. Crowell (subs of Harper-Row), 2d ed., 1974, 404 pp. LC: 74-5411. ISBN: 0-690-00505-9. Cost: $9.95.

57. American Civil Engineering Practice. R.W. Abbott. John Wiley and Sons, 1956-57, 3-Vol. set.

Successor to Merriam's *American Civil Engineers' Handbook.*

American Concrete Institute. *See* **ACI.**

58. American Cotton Handbook. D.S. Hamby. John Wiley and Sons, 3d ed., Vol.1, 1965; Vol.2, 1966. LC: 65-21455. ISBN: Vol.1, 0-470-34640-X; Vol.2, 0-470-34644-2. Cost: Vol.1, $42.50; Vol.2, $55.00.

59. American Drug Index 1977 Norman F. Billups. Lippincott, 1977, 735 pp. ISBN: 0-397-50372-5. Cost: $14.50.

60. American Electricians' Handbook: A Reference Book for the Practical Electrical Man. John H. Watt; Terrell Croft; Clifford C. Carr. McGraw-Hill, 10th ed, 1981, Paging by section numbers (no consecutive paging), 1,648 pp. LC: 80-14757. ISBN: 0-07-013931-8. Date of First Edition: 1913. Cost: $39.50.

This is a standard reference work that has been published in 10 different editions since 1913. This handbook has 11 different sections covering such subjects as wiring, lighting, outside distribution systems, conductors, circuits, batteries, solid-state devices, transformers, generators, and motors. There are many tables including such information as data on copper conductors, aluminum wire, metals and alloys for resistance wire, sheet metal gages, interior-illumination design, and resistance of conductors. There is a detailed table of contents and a good index to help locate information of interest. In addition, information is supplied for such objects as switches, nickel-cadmium batteries, cathode-ray tubes, solid-state diodes, thyristors, thermionic vacuum tubes, A-C and D-C motors, residential and farm wiring, infrared heating lamps, and fluorescent and incandescent lighting. Also refer to the *National Electrical Code; Handbook of Wiring, Cabling, and Interconnecting for Electronics,* the *NFPA Handbook of the National Fire Code,* and the *IES Lighting Handbook.*

61. American Handbook of Psychiatry. Silvano Arieti et al. Basic Books, 2d ed., 1974-75, multivolume set, 6 Vols. LC: 72-89185. ISBN: 0-465-00156-4. Cost: $179.50.

62. American Institute of Physics Handbook. Dwight E. Gray. McGraw-Hill, 3d ed., 1972, paging by section numbers (no consecutive paging), 2,200 pp. LC: 72-3248. ISBN: 07-001485-X. Date of First Edition: 1957. Cost: $87.50.

This handbook has generally been recognized as one of the major compilations of data in the physical sciences and has been sponsored by the American Institute of Physics. Each section is written by a different authority and includes such subjects as solid-state physics, atomic and molecular physics, nuclear physics, heat, optics, electricity and magnetism, mechanics, and acoustics. An excellent bibliography on SI units as it applies to mathematics is presented at the beginning of the handbook. In addition, a great deal of emphasis is placed on properties of materials such as semiconductors, ionic crystals, nuclides, superconductors, metals, transducer materials, and dielectrics. Physical data such as thermal conductivity, heat capacities, vapor pressure, index of refraction, wavelengths, atomic spectra, density, melting points, surface tension, and viscosities also receive treatment.

63. American Metric Handbook. Robert Lytle. McGraw-Hill, 1976, 312 pp. ISBN: 0-07-039277-3. Cost: $14.95 paper.

Provides information, charts, and tables on conversion of numerous measuring systems to the metric system. Primarily designed for the construction industry, architects, builders, and contractors.

American Physiological Society Handbook of Physiology. *See* Handbook of Physiology.

American Society for Testing and Materials. *See* ASTM.

American Society of Mechanical Engineers. *See* ASME.

American Society of Plumbing Engineers. *See* ASPE.

American Society of Tool and Manufacturing Engineers. *See* ASTME.

64. Amplifier Handbook. R. Shea. McGraw-Hill, 1966, 1,496 pp. LC: 64-66296. ISBN: 0-07-056503-1. Cost: $54.50.

65. Analysis of Contingency Tables. B.S. Everitt. Chapman and Hall, 1977, 128 pp. LC: 77-4244. ISBN: 0-412-14970-2. Cost: $12.95.

This book is primarily for those interested in statistics, medical statistics, psychiatry, and psychology. The book includes information on the chi-square test, multidimensional tables, and various different contingency tables.

66. Anemonefishes of the World: Handbook for Aquarists, Divers, and Scientists. Hans A. Baensch. Available from Otto Harrassowitz, 1979, 104 pp. ISBN: 3-88244-005-8.

67. Angiography of the Human Brain Cortex: Atlas of Vascular Patterns and Stereotactic Cortical Localization. G. Szikla; G. Bouvier; T. Hori. Available from Otto Harrassowitz, 1977, 273 pp. ISBN: 3-540-08285-9. Cost: $149.60.

This atlas has 199 plates.

68. Animal Identification: A Reference Guide. R.W. Sims. John Wiley and Sons, Inc, 1980, Vol.1 and Vol.2, 108 pp. ISBN: 0-471-27768-1. Cost: $86.50.

The 3 volumes of this set are entitled as follows: *Animal Identification Vol. 1: Marine and Brackish Water Animals.* R.W. Sims. LC: 80-40006, ISBN: 0-471-27765-7, $25.00; *Animal Identification Vol. 2: Land and Freshwater Animals.* R.W. Sims. LC: 80-40006, ISBN: 0-471-27766-5, $25.00; *Animal Identification Vol. 3: Insects.* D. Hollis. LC: 80-40006, ISBN: 0-471-27767-3, $36.50.

69. Animals for Research: A Directory of Sources. Institute of Laboratory Animal Resources. Institute of Laboratory Animal Resources, 10th ed., 1979. ISBN: 0-309-02920-1. Cost: $6.25 paper.

This directory provides information with regard to the availability of laboratory animals, and helps to locate these animals for research purposes. Numerous types of animals are listed including ducks, chickens, turkeys, geese, rabbits, mice, rats, hamsters, gerbils, guinea pigs, dogs, cats, cattle, goats, swine, sheep, and horses.

70. Annual Book of ASTM Standards, 1982. American Society for Testing and Materials, 1982, 48-volume set (over 52,000 pages). LC: 40-10712. Date of First Edition: 1940. Cost: complete set $1,925.00. Each volume ranges in price from $19.00 to $54.00.

This multivolume set is revised annually and consists of over 6,600 standards, test methods, specifications, and recommended practices that have been approved by ASTM. The 48th volume is an index to these standards, providing a subject approach as well as an alphanumeric approach. Volumes 1 through 47 include standards and test methods on such subjects as nuclear materials, electrical insulation, plastics, textiles, paint, wood, coke and coal, concrete, and metals. In addition, standards and test methods are also provided for such subjects as road and paving materials, petroleum products, electronics, rubber, soap and polishes, stone, and leather.

71. Antenna Engineering Handbook. Henry Jasik. McGraw-Hill, 1961, paging by section numbers (no consecutive paging), 1,013 pp. LC: 59-14455. ISBN: 07-032290-2. Cost: $49.50.

Although this handbook is somewhat dated, it is still one of the best sources of information on antenna design. Each of the 35 chapters is written by an expert on his/her respective area of specialization with emphasis placed upon antenna types and structures. Many different types of antennas are covered, including linear antennas; loop antennas; helical, slot, long-wire, and horn antennas; corner-reflector antennas; scanning antennas; and low-, medium-, and high-frequency antennas. Attention is also given to TV and FM antennas, VHF and UHF antennas, aircraft and radar antennas, as well as transmission lines, ionospheric transmission, and properties of antennas such as impedance, gain, polarization and bandwidth. This handbook has an excellent table of contents and index, and each chapter is preceded by an even more detailed contents for that chapter to help in locating needed information.

72. Aphasia Handbook: For Adults and Children. Aleen Agranowitz; Milfred R. McKeown. C.C. Thomas, 1975. ISBN: 0-398-00017-4. Cost: $13.75.

73. Apparel Manufacturing Handbook; Analysis Principles and Practice. Jacob Solinger. Van Nostrand Reinhold

Co (division of Litton Educational Publishing, Inc), 1980, 992 pp. LC: 79-21951. ISBN: 0-442-26154-3. Cost: $60.00.

This comprehensive handbook covers product design, selection and purchase of raw materials as well as processing, methods engineering, production control, packaging, and warehousing. In addition, plant layout, sales engineering, marketing, and personnel management are also treated in this handbook. Tables and charts have been provided. Also covered in this handbook is a thorough analysis of cutting, sewing, pressing, and molding production equipment and techniques. An excellent section on pattern-making has been included, as well as a chapter on computer construction and operations and uses in the apparel manufacturing field.

74. Appendices to Survey to Select a Limited Number of Hazardous Materials to Define Amelioration Requirements. Vol. 2. W. Lyman; L. Nelson; L. Partridge; A. Kalelkar; J. Everett. NTIS, 1974, 438 pp. Accession: N75-26380; AD-A004312. Tech Report: USCG-D-46-75-VOL-2..

This volume covers the physico-chemical categories of the CHRIS chemicals. Includes risk indices for 400 chemicals.

75. Applications Handbook for Electrical Connectors. John D. Lenk. Sams, 1966, 160 pp. LC: 66-25346.

76. Applied Ergonomics Handbook. I.P.C. Science and Technology Press Ltd, 1974, 126 pp. ISBN: 0-902852-38-8. Cost: $15.00.

This volume has been reprinted 4 times but earlier dates are not available. Covers such topics as ergonomics in industry, thermal comfort in industry, lighting of work places, layout of panels and machines, displays, and ergonomics versus accidents.

77. Aqueous Solutions: Data for Inorganic and Organic Compounds. Rolf K. Freier. Walter De Gruyter, Inc, Vol.1, 1976, Vol.2, 1978, Vol.1, 440 pp.; Vol.2, 447 pp. ISBN: Vol. 1, 3-11-003627-3; Vol. 2, 3-11-006537-1. Cost: Vol.1, $92.50; Vol.2, $110.00.

This 2-volume set is available through Stechert Macmillan, Inc, 866 3rd Avenue, New York, New York 10022. It is a good source of information on physical and chemical properties of water with special concentration on water as a solvent. Vol.1 contains the solubilities and redox potentials of various inorganic and organic compounds. Vol.2 contains information on reaction equations, ion-exchangers, and supplements to the inorganic and organic compounds listed in Vol.1.

78. Architect's Data: The Handbook of Building Types. Ernst Neufert. Halsted Press (division of John Wiley and Sons), 2d (International) English ed., 1980, 420 estimated pp. ISBN: 0-470-26947-2. Cost: $69.95 (tentative).

This handbook includes substantial design considerations, such as requirements for the elderly and disabled, site selection, basic dimensions, and general planning criteria. Also covered are flow-charting techniques, along with numerous illustrations of plans, sections, and site layouts. Over 3,500 different plans are included for various building types.

79. Architects' and Builders' Handbook. Frank E. Kidder; H. Parker. Wiley-Interscience, 18th ed., 2,315 pp. ISBN: 0-471-47421-5. Date of First Edition: 1931. Cost: $65.00.

80. Architectural Graphic Standards Charles G. Ramsey; Harold R. Sleeper. Wiley-Interscience, 6th ed., 1970, 695 pp. LC: 79-136970. ISBN: 0-471-70780-5. Cost: $47.25.

81. Architectural Handbook: Environmental Analysis, Programming Design, Technology and Construction Administration. T.D. Kemper. John Wiley and Sons, Inc, 1978, 944 pp. LC: 78-8202. ISBN: 0-471-02697-2. Cost: $35.00.

Covers such topics as building materials, human behavior, social structures, environmental impact, and environmental law. Various examples of environmental impact reports and feasibility studies are provided. Energy design suggestions are made on such topics as solar energy, earthquakes, and wind power applications.

82. Argon—Gas Solubilities. H.L. Clever et al. Pergamon Press Inc, 1980, 348 pp. LC: 79-41555. ISBN: 0-08-022353-2. Cost: $100.00. Series: *Solubility Data Series*, Vol. 4.

This compilation provides solubility data of the noble gases and liquids up to and through 1978. Over 2,000 literature citations have been used to provide data for this work. A registry number index is provided as well as a system index.

83. Army Medical Department Handbook of Basic Nursing. Superintendent of Documents, U.S. Government Printing Office, 1970, 632 pp. SuDoc: D 101.11:80230/3. S/N: 008-020-00336-1. Cost: $7.55.

This handbook is used in training medical corpsmen, medical specialists, and clinical specialists in nursing. Includes information on anatomy and physiology, injury, pharmacology and drug administration, as well as obstetrical care and emergency service.

84. ARRL Antenna Book. American Radio Relay League, Inc. American Radio Relay League, Inc, 13th ed., 4th printing, 1977, 336 pp. LC: 55-8966. ISBN: 0-87259-413-0. Cost: $5.00. Series: *Radio Amateur's Library Series*, No. 15.

This book covers such information as fundamentals of antennas, VHF and UHF antennas, different types of antennas such as rotatable antennas, mobile antennas, and wire antennas. Also covered in this book are HF antennas, antennas for various ranges of frequency, multiband antennas, and transmission lines.

85. ASBC Methods of Analysis. American Society of Brewing Chemists. American Society of Brewing Chemists, 7th ed., 1976. Cost: $110.00.

This is a loose-leaf service for the laboratory bench person involved with brewing. The book covers such topics as barley, malt, cereals, sugars, syrups, brewers' grains, hops, beer, microbiology, and packaging materials.

86. Asbestos Particle Atlas. Walter C. McCrone. Ann Arbor Science Publishers Inc, 1980, 122 pp. LC: 80-66474. ISBN: 0-250-40372-2. Cost: $85.00.

This book was designed to help identify the presence of asbestos fibers. The book includes over 400 full-color light photomicrographs. The book also covers background information on asbestos, including definition, minerology, composition, and properties. Helps to identify such asbestos fibers as chrysotile, tremolite, anthophyllite, amosite, actinolite, and crocidolite. Also covers crystal morphology, crystal optics, and dispersion staining.

87. Asbestos Vol. 1: Properties, Applications and Hazards. L. Michaels; S.S. Chissick. John Wiley and Sons, Inc, 1979, 553 pp. LC: 78-16535. Cost: $25.00.

This is Vol.1 of a 2-volume set. It is a well-written volume consisting of 16 chapters written by different authors. It is a factual account of the properties, applications, and hazardous nature of asbestos. Provides essential information for this timely topic.

88. ASDRI Oxygen Technology Survey. Vol. 1: Thermophysical Properties (Handbook of Thermophysical Properties of Liquid and Gaseous Oxygen.). H.M. Roder; L.A. Weber. National Bureau of Standards (available from NTIS), 1972, 432 pp. Accession: N73-13952. Tech Report: NASA-SP3071. Cost: $6.00.

This handbook presents information on the thermodynamic functions, transport properties, and physical properties of both liquid and gaseous oxygen.

89. ASHRAE Handbook and Product Directory 1978: Applications. American Society of Heating, Refrigerating and Air Conditioning Engineers, Inc, 1978, paging by section numbers (no consecutive paging), 1st of 4 volumes. Date of First Edition: 1961. Cost: $40.00. Series: Published as part of the *ASHRAE Guide and Data Book.*

As mentioned in the preface, this volume, as well as the other volumes in the *ASHRAE Handbook and Product Directory* series, will be revised on a 4-year cycle. This volume provides applications data on air conditioning, heating, ventilation and refrigeration for a wide variety of processes and building structures. For example, refrigeration information is tabulated for such applications as meat products, poultry, citrus fruits, vegetables, frozen foods, beverages, and bakery products. Air conditioning and heating information is provided for such applications as printing plants, computer areas, drying and storing farm crops, and textile and photographic processing. Applications data are also provided for such buildings as skating rinks, concrete dams, health spas, swimming pools, ships, airplanes, churches, theaters, hotels, and nursing homes. For related works, please refer to *IHVE Guide Book; Handbook of Heating, Ventilating and Air Conditioning; Handbook of Ventilation for Contaminant Control;* and *Handbook of Refrigerating Engineering.*

90. ASHRAE Handbook and Product Directory 1979: Equipment. American Society of Heating, Refrigerating and Air Conditioning Engineers, Inc, 1979, paging by section numbers (no consecutive paging), 1 of 4 Vols. Date of First Edition: 1961. Cost: $40.00. Series: Published as part of the *ASHRAE Guide and Data Book.*

The *Equipment* volume of the *ASHRAE Handbook and Product Directory* includes a Product Directory and Catalog Section listing the manufacturers and the latest equipment available, as well as an index to all the volumes in the ASHRAE series. The emphasis in this volume is naturally on air conditioning, heating, ventilation and refrigeration equipment data. Components such as pumps, motors, fittings and turbine drives receive coverage in this handbook, as well as data on humidifiers, air cleaners, dehumidifying coils, fans, cooling towers, and air diffusing equipment. Unitary equipment such as drinking water coolers, freezers, refrigerators, refrigerated vending machines, heat pumps, and room air conditioners are thoroughly discussed in this reference source. In addition, data are provided for such heating equipment as boilers, furnaces and space heaters, radiators, fireplace systems and high-intensity infrared heaters. For related works, please refer to *IHVE Guide Book; Handbook of Heating, Ventilating and Air Conditioning; Handbook of Ventilation for Contaminant Control;* and *Handbook of Refrigerating Engineering.*

91. ASHRAE Handbook and Product Directory 1980: Systems. American Society of Heating, Refrigerating and Air Conditioning Engineers, Inc, 1980, paging by section numbers (no consecutive paging), 1 of 4 Vols. Date of First Edition: 1961. Cost: $40.00. Series: Published as part of the *ASHRAE Guide and Data Book.*

This handbook is the recognized reference source for engineering data on air conditioning, heating, ventilation, and refrigeration systems. It includes 46 chapters plus a Product Directory and a Catalog Data section. An index is also included which indexes this volume plus the other volumes in the *ASHRAE Handbook and Product Directory* series. The *Systems* volume contains such information as heat recovery systems, heat pump systems, forced air heating and cooling systems, steam heating systems, and water and electric heating systems. Of interest also will be the sections on owning and operating costs, odor control, sound and vibration control, evaporative air cooling, corrosion and water treatment, and the system practices for the various refrigerants like ammonia and halocarbon. For related works, please refer to *IHVE Guide Book; Handbook of Heating, Ventilating and Air Conditioning; Handbook of Ventilation for Contaminant Control;* and *Handbook of Refrigerating Engineering.*

92. ASHRAE Handbook of Fundamentals. American Society of Heating, Refrigerating, and Air Conditioning Engineers, Inc, 1981, 760 pp., 1 of 4 Vols. LC: 72-186171. Date of First Edition: 1961. Cost: $40.00. Series: Published as part of the *ASHRAE Guide and Data Book.*

Along with the other 3 volumes (*Systems, Equipment, and Applications*), the *Handbook of Fundamentals* completes the *ASHRAE Handbook and Product Directory* set. This handbook concentrates on the basic data and tables needed to calculate problems dealing with air conditioning, heating, ventilating or refrigerating. For example, it covers such information as fluid flow, combustion, secondary coolants, pipe sizing, load calculations, psychrometric tables, thermodynamics and refrigeration cycles. In addition, weather and design conditions, refrigerant tables, sorbents and desiccants, thermal insulation and water vapor barriers, and heat transmission coefficients are tabulated and presented in this comprehensive reference manual. For related works, please refer to *IHVE Guide Book; Handbook of Heating, Ventilating and Air Conditioning; Handbook of Ventilation for Contaminant Control;* and *Handbook of Refrigerating Engineering.*

93. ASI Tables of Diurnal Planetary Motion. Arthur D. Libin. ASI Publications Inc, 1975, 190 pp. LC: 75-22432. ISBN: 0-88231-025-9. Cost: $4.25 paper.

94. ASME Boiler and Pressure Vessel Code. American Society of Mechanical Engineers. American Society of Mechanical Engineers, 1977. Cost: $1,850.00.

This is a collection of standards in loose-leaf format. Includes information on nuclear power plants, welding rods, material specifications, ferrous materials, nonferrous materials, electrodes, filler metals, codes for concrete reactor vessels, heating boilers, information on nondestructive examination, welding and brazing qualifications, and fiberglass-reinforced plastic pressure vessels.

95. ASME Guide for Gas Transmission and Distribution Piping Systems - 1976. American Society of Mechanical Engineers, 4th ed., 1976, 378 pp. LC: 72-96733. ISBN: 0-685-67491-6. Date of First Edition: 1970. Cost: $100.00.

This guide provides information, design recommendations and recommended practices of the ASME Gas Piping Standards Committee as well as Federal Gas Pipeline Safety Standards.

ASME Hanbook of Industrial Metrology. *See* **Handbook of Industrial Metrology.**

96. ASME Handbook: Engineering Tables. Jesse Huckert. McGraw-Hill, 1956, 692 pp. LC: 52-5326. ISBN: 07-001516-3. Cost: $44.50.

Engineering Tables is the third volume of the 4-volume series of the *ASME Handbook*. It includes engineering tables that are not readily found elsewhere on standards of shape and dimension of various components. For example, data on bearings, helical gears, worm gears, herringbone gears, screw threads, bolts, nuts, washers, wrench openings, and springs are included in this volume. Although mathematical tables have generally been excluded from this volume, information has been included on such devices as gaskets, electric motors, pipe threads and fittings, aircraft tubing, and bar stock. The data are nicely tabulated, and a bibliography is included at the end of the text for further information and sources. For related works, please refer to *Metals Handbook* and *Woldman's Engineering Alloys*.

97. ASME Handbook: Metals Engineering Design. Oscar J Horger. McGraw-Hill, 2d ed., 1965, 605 pp, Vol. 1 of 4. LC: 63-13133. ISBN: 07-001518-X. Date of First Edition: 1953. Cost: $55.00.

This handbook deals primarily with the physical properties of metals and the design considerations that would result in the selection of one metal over another for a specific function. Each section is written by an authority. Tabulations of data are made for such physical properties of metals as plasticity, fatigue, corrosion, radiation, and electrical properties. Processing of metals is included, such as cold working, case carburizing, nitriding, different joints, and decarburization. In addition, several nondestructive test methods are discussed including such methods as ultrasonic testing, X-ray, magnetics and electromagnetics. Design considerations are also made based upon such information as friction, wear, vibration, coatings, and stress range, which receive coverage in this handbook. For related works please refer to *Metals Handbook* and *Woldman's Engineering Alloys*.

98. ASME Handbook: Metals Engineering—Processes. Roger W. Bolz. McGraw-Hill, 1958, 448 pp, Vol. 4 of 4. LC: 52-5326. ISBN: 07-001514-7. Cost: $38.50.

Metals Engineering-Processes is the 4th volume in the 4-volume series of the *ASME Handbook*. This volume concentrates on the different processes that metals undergo in order to be converted into a finished product. It provides information on the advantages and disadvantages of these processes and indicates whether certain metals are suitable for each process, based upon the metals' natural tolerances and the desired or obtainable surface finish. Each section is written by an authority in his/her specialization and includes such processes as cold working, stretch forming, wire forming, stamping and drawing, furnace brazing, barrel tumbling, and electroforming. In addition, such processes as sand casting, die casting, centrifugal casting, extension, forging and hot forming are also discussed in the volume. For related works, please refer to *Metals Handbook* and *Woldman's Engineering Alloys*.

99. ASME Handbook: Metals Properties. Samuel L. Hoyt. McGraw-Hill, 1954, 445 pp, Vol. 2 of 4. LC: 52-5326. ISBN: 07-001513-9. Cost: $38.50.

The *Metals Properties* handbook is the second volume of the *ASME Handbook* series and contains tabulated data on the mechanical, physical, fabrication, and metallurgical properties of various metals. Similar metals and alloys are grouped together including such metals as steels and iron, and nonferrous metals such as copper, magnesium, tin, lead, nickel, zinc, aluminum, etc. Properties such as specific compositions, static and tensile properties, fatigue, low and high temperature properties, corrosion and magnetic properties and various treatments are included in this handbook. Although not all available steels are included, a generous listing of both ASTM steels and AISI steels receive coverage. For related works, please refer to *Metals Handbook* and *Woldman's Engineering Alloys*.

100. ASME Steam Charts: Thermodynamic Properties of Steam, H-V, in Graphical Form for the Superheated, Vapor and Liquid Conditions in Both SI (Metric) and U.S. Customary Units. James H. Potter. American Society of Mechanical Engineers, 1975, 128 pp. Cost: $12.50 members; $25.00 nonmembers. Series: ASME No. E00090.

101. ASME Steam Tables in SI (Metric) Units for Instructional Use. American Society of Mechanical Engineers, (based on 3d ed. of *ASME Steam Tables*), 1977, 19 pp. Cost: $1.50 members; $3.00 nonmembers. Series: ASME Order No. H00093.

Abbreviated tables in SI based on the 1976 IFC Formulation for Industrial Use.

102. ASME Steam Tables: Thermodynamics and Transport Properties of Steam. Charles A. Meyer et al. American Society of Mechanical Engineers, 4th ed, 1979, 330 pp. LC: 77-367546. Date of First Edition: 1st and 2d ed., 1967 and 1968. Cost: $22.50. Series: Bk. No. G-00038.

The *ASME Steam Tables* represents an effort to tabulate the thermodynamics and transport properties of steam in a one-volume reference source. This book includes an excellent collection of tables and charts, such as properties of superheated steam and compressed water; enthalpy and entropy of superheated steam and water-steam mixture; specific heat, viscosity, thermal conductivity of steam and water; and choking velocity for water-steam mixture. In addition, this handbook also includes such data as conversion factors, units, notation, and constants. The appendixes include values and tables established at various conferences including the 1976 IFC Formulation for Industrial Use.

103. ASPE 1971 Data Book. American Society of Plumbing Engineers. American Society of Plumbing Engineers, 1971, various paging.

This handbook is a collection of data and information which will be helpful for the design engineer on the subject of plumbing systems. This book has 28 chapters of information ranging from such subjects as private sewage disposal systems, storm drainage systems, and demineralized distilled water systems, to irrigation systems, reflection pools, and fountains. In addition, information on pumps, swimming pools, valves, piping, fire protection, wiring, insulation, and drinking water coolers is also included. Compressed air distribution, fuel gas systems, gasoline and oil dispensing, and steam piping also receive coverage in the *ASPE Data Book*. For related works, please refer to the *Pump Handbook*; *ASHRAE Handbook and Product Directory*; and the *Piping Handbook*.

104. Asphalt Handbook. Asphalt Institute, rev. ed., 1960, 398 pp. plus index. Series: *Manual Series*, No. 4.

105. Assessment of a Prototype Human Resources Data Handbook for Systems Engineering. David Metster. NTIS, 1976, 106 pp. Accession: AD-A039-269.

This handbook was developed for the U.S. Air Force and it covers such areas as human factors engineering, job analysis, human performance, aviation personnel, jet fighters, and systems engineering. A considerable amount of human resources data are provided.

106. Assignments for Vibrational Spectra of Seven Hundred Benzene Derivatives. Gyorgy Varsanyi. John Wiley and Sons Inc, rev., 1974, 668 pp. (2 Vols.). LC: 74-8113. ISBN: 0-470-90330-9. Cost: $71.95 per set.

107. ASTM Metric Practice Guide American National Standards Institute. Cost: $4.00. Series: ANSI Z210.1-1976.

NOTE: IEEE Std 268-1976

108. ASTM Standards in Building Codes. American Society for Testing and Materials, 13th ed, 1975, 2 Vols. LC: 75-330781. ISBN: 0-300-13751-0. Date of First Edition: 1955. Cost: $89.50.

This book represents a useful compilation of standards, test methods, and tentatives on building materials and construction methods approved by ASTM. The standards are arranged in alphanumeric sequence and include such subjects as steel pipe and tubing, steel bolting and riveting material, wood preservatives, sealants, pipe and nonmetallic drain tile, bituminous roofing materials, concrete, and brick. In addition, standards are also included for such areas as acoustical material, fire testing, waterproofing, thermal insulation, storm drains, and sewer pipes. For related works, please refer to *Building Construction Handbook*; *Construction Specifications Handbook*; and the *Annual Book of ASTM Standards*.

109. ASTM Viscosity Tables for Kinematic Viscosity Conversions and Viscosity Index Calculations. American Society for Testing and Materials. American Society for Testing and Materials, 1972, 35 pp. LC: 70-180913. Cost: $3.00. Series: *ASTM Special Technical Publication* 43 C.

This publication provides a set of tables for the 2 most frequently used ASTM viscosity methods. The titles of the 2 methods are as follows: ASTM D 2270-64 (1968)—*ASTM standard methods for calculating viscosity index from kinematic viscosity*, and ASTM D 2161-66 (1971)—*ASTM standard method for conversion of kinematic viscosity to Saybolt Universal viscosity or to Saybolt furol viscosity*.

110. ASTME Die Design Handbook. F.W. Wilson (ed) (American Society of Tool and Manufacturing Engineers). McGraw-Hill, 2d ed., 1965, 788 pp. LC: 64-21625. ISBN: 0-07-001523-6. Cost: $41.50.

Includes cold pressworking, stamping design, press feeding and die setting.

111. Astrionics System Designers Handbook. International Business Machines (available from NTIS), 1973, 2 Vols. Accession: N74-27365. Tech Report: NASA-CR-120229.

112. Astronomy Data Book. J. Hedley Robinson. John Wiley and Sons, 2d ed., 1979, 271 pp. LC: 78-21698. ISBN: 0-470-26594-9. Cost: $16.95.

Includes information on such topics as telescopes, planets, comets, asteroids, stars, and meteors.

113. Astronomy: A Handbook. Gunter D. Roth ed; A. Beer trans. Springer-Verlag, 1975. LC: 74-11408. ISBN: 0-387-06503-2. Cost: $26.00.

Observation techniques are discussed in this compendium, along with physical principles associated with the stars, planets, and galaxies.

114. Astrophysical Formulae: A Compendium for the Physicist and Astrophysicist. K.R. Lang. Springer-Verlag, 1974. LC: 73-20809. ISBN: 0-387-06605-5. Cost: $95.20.

115. Astrophysical Quantities. C.W. Allen. Athlone Press, 3d ed., 1973, 320 pp. ISBN: 0-485-11150-0. Cost: $37.50.

116. Atlas and Manual of Plant Pathology. Ervin H. Barnes. Plenum Publishing Corp, 2d ed., 1979, 343 pp. LC: 79-10575. ISBN: 0-306-40168-1. Cost: $19.50.

Covers such topics as bacterial diseases, root nodules of legumes, virus diseases, tobacco mosaic, potato latent mosaic, clubroot of cabbage, the Downey mildews, damping-off, black leaf spot of elm, and sycamore anthracnose. In addition, this book covers such rusts as stem rust of wheat, hollyhock rust, cedar-apple rust, white pine blister rust, as well as numerous wood rots.

117. Atlas and Sourcebook of the Lesser Bushbaby, Galago Senegalensis. James L. Stevens. Chemical Rubber Co, 1981, 208 pp. ISBN: 0-8493-6320-9. Cost: $79.95.

The Lesser Bushbaby is presented as an excellent model for biomedical research and provides information on morphology, physiology, and experimentation. Twenty-two tables are included, as well as 102 plates.

118. Atlas for Stereotaxy of the Human Brain. G. Saltenbrand; W. Wahren. Georg Thieme (distributed in USA by Year Book Medical Publishers), 2d ed., 1977. ISBN: 3-13-393702-2. Cost: DM 600.

Comes with an accompanying guide entitled, *Architectonic Organization of the Thalamic Nuclei*, by R. Hassler. This work includes 69 plates, 50 overlays, and 25 drawings in color.

119. Atlas Intrakardialer Druckkurven—Atlas of Intracardiac Pressure Curves—Atlas de Curvas Tensionales Intracardiacas. O. Bayer; H.H. Wolter. Georg Thieme (distributed in USA by Year Book Medical Publishers), 1959, 185 pp. ISBN: 3-13-304201-7. Cost: DM 84.

Text in German, English, and Spanish. Includes 55 illustrations and 42 plates.

120. Atlas of Acupuncture: Points and Meridians in Relation to Surface Anatomy. Felix Mann. International Ideas Inc, 1970. Cost: $14.95.

121. Atlas of Acute Hand Injuries. Sigurd C. Sandzen. MaGraw-Hill, 1980, 441 pp. LC: 79-21848. Cost: $95.00.

Contains over 1,100 illustrations covering acute and chronic hand problems.

122. Atlas of Angiography. K.E. Loose; R.J.A.M. van Dongen. Georg Thieme (distributed in USA by PSG Publishing, 656 Great Rd, Littleton, MA 01460), 1976, 435 pp. ISBN: 3-13-514401-1. Cost: DM 190.

Contains 707 illustrations.

123. Atlas of Animal Bones. E. Schmid. Elsevier, 1972, 159 pp. ISBN: 0-444-40831-2. Cost: $68.25.

124. Atlas of Arteriography in Occlusive Cerebrovascular Disease. J. Weibel; W.S. Fields. Georg Thieme (distributed in USA by G.B. Saunders), 1969, 190 pp. ISBN: 3-13-455101-2. Cost: DM 120.

Contains 315 illustrations.

125. Atlas of Arthroscopy. M. Watanabe et al. Springer-Verlag, 3d ed., 1979, x + 156 pp. ISBN: 3-540-07674-3. Cost: $81.40.

Contains information on disorders and surgery of the knee joint as well as a new chapter on arthroscopy of smaller joints.

126. Atlas of Binary Alloys: A Periodic Index. K.P. Staudhammer; L.E. Murr. Marcel Dekker Inc, 1973, 112 pp. ISBN: 0-8247-6081-6. Cost: $28.75. Series: *Monographs and Textbooks in Material Science Series*, Vol. 5.

127. Atlas of Carbon-13 NMR Data. E. Breitmaier; G. Haas; M. Voelter (comp). Heyden, 1979, 2 Vols., 712 pp. LC: 76-2126. ISBN: 0-85501-480-6. Cost: $165.00.

Includes NMR data for over 3,000 compounds.

128. Atlas of Cerebrospinal Fluid Cells. H.W. Kolmel. Available from Otto Harrassowitz, 2d ed., 1977, 142 pp. ISBN: 3-540-08186-0. Cost: $43.20.

This atlas contains 251 figures, 139 in color.

129. Atlas of Clinical Conditions in Pedodontics. Robert Rapp; G.B. Winter. Year Book Medical Publishers, 1979, 142 pp. LC: 79-25413.

This atlas is concerned with oral pathology and pathologic conditions that involve the face and cranium. Contains an excellent collection of clinical photographs and roentgenographic illustrations on oral pathology and pedodontics.

130. Atlas of Continuous Cooling Transformation Diagrams for Engineering Steels. M. Atkins. American Society for Metals, 1980, 260 pp. LC: 79-28514. ISBN: 0-87170-093-X. Date of First Edition: 1977. Cost: $74.00.

First published in 1977 by the British Steel Corporation, this atlas represents a comprehensive and authoritative collection of continuous cooling transformation diagrams that illustrate typical patterns of transformation response of numerous standard steels when cooled in air, oil, or water. This is a practical guide for heat treatment and hot processing of engineering steels. The AISI-SAE designations are provided for all the applicable transformation diagrams. The atlas contains 172 CCT (continuous cooling transformation) diagrams.

131. Atlas of Cross-Sectional Anatomy. Stephen A. Kieffer; E.R. Heitzman. Harper and Row, 1979, 299 pp. LC: 78-18859. Cost: $42.50.

This atlas correlates cross-sectional human anatomy of specimen sections with computerized tomography scans and X-ray studies.

132. Atlas of Diagnostic Cytology. C. Gompel. John Wiley and Sons, Inc, 1978, 237 pp. LC: 77-27068. ISBN: 0-471-02278-0. Cost: $42.00.

133. Atlas of Diseases of the Breast: Synopsis of Clinical, Morphological and Radiological Findings, with a Consideration of Special Investigative Methods. Volker Barth. Georg Thieme (available from Otto Harrasowitz, distributed in USA by Year Book Medical Publishers), 1979, approx. 208 pp. LC: 78-28552. ISBN: 3-13-542401-4. Cost: DM 120.

134. Atlas of Diseases of the Eye. R. Thiel. Georg Thieme, 1963. ISBN: 3-13-409806-7. Cost: DM 490.

Contains 1,881 illustrations and 21 tables.

135. Atlas of Electron Micrographs Functional Morphology of Endocrine Glands. K. Kurosumi; H. Fujita. Georg Thieme, 1975, 377 pp. ISBN: 3-13-522401-5. Cost: DM 188.

Contains 236 illustrations.

136. Atlas of Electron Microscopy of Clay Minerals and Their Admixtures. H. Beutelspacher; H.W. Van der Marel. Elsevier-North Holland Publishing Co, 1968, 333 pp. ISBN: 0-444-40041-9. Cost: $102.50.

137. Atlas of Electron Spin Resonance Spectra. B.H.J. Bielski; J.M. Gebicki. Academic Press, 1967, 665 pp. ISBN: 0-12-096650-6. Cost: $70.00.

138. Atlas of Emergency Medicine. Peter Rosen; G.L. Sternbach. Williams and Wilkins, 1979, 165 pp. LC: 78-9651. Cost: $17.95.

139. Atlas of Equine Anatomy. Chris J. Pasquini. 1978.

140. Atlas of Flexible Bronchofiberoscopy. S. Ikeda. Georg Thieme, 1974, 230 pp. ISBN: 3-13-507001-8. Cost: DM 198.

Contains 373 plates in color and 247 illustrations.

141. Atlas of Gastrointestinal Cytodiagnosis. N. Henning; S. Witte. Georg Thieme (distributed in USA by Thieme-Stratton Inc, 381 Park Ave S, New York, NY 10016), 2d ed., 1970, 132 pp. ISBN: 3-13-459502-8. Cost: DM 78.

This book contains 125 illustrations in color, 184 in black and white, and 11 tables.

142. Atlas of Haematology. George A. McDonald; T.C. Dodds; Bruce Cruickshank. Churchill Livingstone, 4th ed., 1978, 312 pp. LC: 77-30131. ISBN: 0-443-01376-4. Cost: $38.00.

This book contains illustrations of normal and abnormal blood cells. A new chapter on ultrastructural appearances of blood and bone marrow cells is provided.

143. Atlas of Histology. Everett E. Dodd. McGraw-Hill, 1979, 384 pp. LC: 78-14449. ISBN: 0-07-017230-7. Cost: $32.50.

This book includes over 304 4-color photomicrographs as well as 46 line drawings. Illustrations are provided for muscular tissues, as well as the respiratory, reproductive, and the nervous systems.

144. Atlas of Hot Working Properties of Nonferrous Metals: Vol.1: Aluminum and Aluminum Alloys; Vol.2: Copper and Copper Alloys. Deutsche Gesellschaft fur Metallkunde. Available from The Metals Society, 1 Carlton House Terrace, London SW1Y 5DB, England, 1978, Vol.1, 245 pp.; Vol.2, 480 pp. ISBN: Vol.1, 3-88355-000-0; Vol.2, 3-88355-001-0. Cost: Vol.1, $72.00/DM 145; Vol.2, $84.00/DM 170.

This bilingual (German-English) 2-volume set is basically a collection of stress-strain curves for hot-working. The aluminum volume contains approximately 700 stress-strain curves and the copper volume contains about 1,200 stress-strain curves. These diagrams are accompanied with references to the literature.

145. Atlas of Infrared Spectroscopy of Clay Minerals and Their Admixtures. H.W. Van der Marel; H. Beutelspacher. Elsevier-North Holland Publishing Co, 1976, 396 pp. ISBN: 0-444-41187-9. Cost: $107.25.

146. Atlas of Insect Morphology. Henrik Steinmann; Lajos Zombori. Heyden and Sons, Inc, 1981, 248 pp. Cost: $26.00.

Over 756 line drawings are provided illustrating external morphology of insects. The main insect body regions that are covered are the homologized skeletal elements, appendages, superficial appertures of the body, and the copulating organs. A comprehensive appendix containing over 100 illustrations is also included. Such illustrations as auditory organs, the form of segmentation and articulation, openings of various glands, and wing patterns are presented.

147. Atlas of Insects Harmful to Forest Trees, Vol. 1. V. Novak; F. Hrozinka; B. Stary. Elsevier-North Holland Publishing Co, 1976, 126 pp. ISBN: 0-444-99874-8. Cost: $39.00.

148. Atlas of Interference Layer Metallography. Hans-Eugen Buhler; Hans Paul Hougardy. American Society for Metals, 1980, 237 pp. Cost: $148.00.

This atlas is beautifully illustrated in color, presenting the techniques of interference layer metallography as a method of studying the microstructure of metals by a purely optical technique without the use of chemical etching.

Atlas of Intracardiac Pressure Curves. *See* **Atlas Intrakardialer Druckkurven.**

149. Atlas of Isothermal Transformation and Cooling Transformation Diagrams. American Society for Metals, 1977, 432 pp. Cost: $32.00.

This atlas includes over 396 transformation diagrams. It covers 27 carbon steels, 223 alloy steels, 37 tool steels, 7 stainless steels, and 15 cast steels.

150. Atlas of Lymphography. M. Viamonte; A. Ruettimann. Georg Thieme, 1979, 435 pp. ISBN: 3-13-547801-7. Cost: $165.00; DM 300.

This atlas includes over 893 figures, many of them in color, as well as 16 tables.

151. Atlas of Mammalian Chromosomes. T.C. Hsu; K. Bernirschke. Springer-Verlag, 1968-77, 10 Vols. (looseleaf). LC: 67-19307. ISBN: Vol.1, 0-387-03878-7; Vol.2, 0-387-04198-2; Vol.3, 0-387-04563-2; Vol.4, 0-387-04882-0; Vol.5, 0-387-05280-1; Vol.6, 0-387-05590-8; Vol.8, 3-540-06755-8; Vol.9, 0-387-07365-5; Vol.10, 3-540-90273-2; vinyl binder, 0-387-90016-0; ring binder for individual volume, 0-387-90005-5. Cost: Vol.1, $19.70; Vol.2, $19.70; Vol.3, $19.70; Vol.4, $28.30; Vol.5, $28.30; Vol.6, $26.30; Vol.8, $19.80; Vol.9, $19.80; Vol.10, $23.00; vinyl binder, $17.30; ring binder for individual volume, $4.10 each.

152. Atlas of Mammalian Ova. S. Suzuki. Georg Thieme, 1974, 139 pp. ISBN: 3-13-502001-0. Cost: DM 128.

Contains 40 color plates and 90 illustrations.

153. Atlas of Mammalian Reproduction. E.S.E. Hafez. Georg Thieme, 1975, 429 pp. ISBN: 3-13-532501-6. Cost: DM 240.

Contains 358 illustrations.

154. Atlas of Metal-Ligand Equilibria in Aqueous Solution. J. Kragten. Halsted Press (division of John Wiley and Sons Inc), 1977, 781 pp. ISBN: 0-470-99309-X. Cost: $99.50.

This atlas provides information about the behavior of 45 different metals in the presence of 29 common ligands. It contains numerous graphs for separate metal-liquid combinations.

155. Atlas of Metal. J. Kragten. Halsted Press (division of John Wiley and Sons), 1978, 781 pp. LC: 77-12168. ISBN: 0-470-99309-X. Cost: $77.50.

This atlas presents data on 45 different metals in the presence of ligands.

156. Atlas of Models of Crystal Surfaces. J.F. Nicholas. Gordon and Breach, Science Publishers, Inc, 1965, 238 pp. ISBN: 0-677-00580-6. Cost: $72.75.

157. Atlas of Normal Roentgen Variants That May Stimulate Disease. Theodore E. Keats. Year Book Medical Publishers, Inc, 1979, 616 pp. LC: 78-25773. Cost: $69.50.

158. Atlas of Nuclear Medicine. Sheldon Baum. Appleton-Century-Croft, 1980, 400 pp. ISBN: 0-8385-0447-7. Cost: $70.00.

This book includes over 2,800 illustrations.

159. Atlas of Orthopaedic Operations. F.W. Rathke; K.F. Schlegel. Thieme, Stuttgart, 1979, Vol.1, 270 pp.; Vol.2, 360 pp. ISBN: Vol.1, 3-13-550301-1; Vol.2, 3-13-550201-5.

Vol.1 is entitled *Surgery of the Spine*, and Vol.2 is entitled *Lower Leg and Foot*.

160. Atlas of Palaeobiogeography. A. Hallam. Elsevier-North Holland Publishing Co, 1973, 531 pp. ISBN: 0-444-40975-0. Cost: $92.75.

161. Atlas of Pathologic. W. Doerr; G. Schumann; G. Ule. Georg Thieme (available from Otto Harrassowitz), 1978, 319 pp. ISBN: 3-13-509001-9.

Contains 874 illustrations.

162. Atlas of Pediatric Diseases. H. Moll; W. Kleindienst (trans). Georg Thieme (distributed in USA by W.B. Saunders Co), 1976, 275 pp. ISBN: 3-13-546401-6. Cost: DM 98.

Contains 378 illustrations most of which are in color.

163. Atlas of Photogrammetric Instruments. V.J. Cimerman; Z. Tomasegovic. Elsevier-North Holland Publishing Co, 1970, 216 pp. ISBN: 0-444-40700-6. Cost: $107.25.

164. Atlas of Photomicrographs of the Surface Structures of Lunar Regolith Particles. Olga Rode et al. D. Reidel Publishing Co, 1979, 244 pp. LC: 78-12367. ISBN: 90-277-0877-0. Cost: $39.00.

This atlas contains 164 photomicrographs of the surface features of the lunar regolith particles collected by the Soviet Union.

165. Atlas of Polymer and Plastics Analysis. Dieter Hummel; Friedrich Scholl. Verlag Chemie International, 2d ed., Vol.1, 1978; Vol.2, 1981; Vol.3, 1980, Vol.1, 671 pp. ISBN: Vol.1, 3-527-25801-9; Vol.2, 3-527-25802-7; Vol.3, 3-527-25803-5. Cost: DM 398.

The 3 volumes in this work are entitled as follows: Vol.1, *Polymers, Structures, and Spectra*; Vol.2, *Plastics, Fibers, Rubbers, Resins*; Vol.3, *Additives and Processing Aids*. Vol.1 contains 1,903 spectra, and Vol.3 contains 1,000 infrared spectra and 100 ultraviolet spectra. This atlas will be very useful in the identification of polymers and plastics.

166. Atlas of Radiologic Anatomy. L. Wicke. Urban and Schwarzenberg, 1978, 234 pp. ISBN: 3-541-72111-1. Cost: $18.00.

This atlas has over 100 radiographs. Includes information on skeletal radiography and roentgen anatomy.

167. Atlas of Rectoscopy and Colonoscopy. P. Otto; K. Ewe. Springer-Verlag, 1979, 120 pp. LC: 79-673. ISBN: 0-387-09296. Cost: $53.90.

168. Atlas of Rock-Forming Minerals in Thin Section. W.S. Mackenzie; C. Guilford. Longman Inc, 1980, 98 pp. LC: 79-41162; 79-27822. ISBN: 0-470-26921-9. Cost: $22.50.

Approximately 90 common rock-forming minerals in thin sections are covered in this atlas. It is well-illustrated with 220 color photomicrographs. This atlas is basically designed to be used as a laboratory handbook for the purpose of identifying minerals in thin section.

169. Atlas of Scanning Electron Microscopy in Microbiology. Z. Yoshii; J. Tokunaga; J. Tawara. Georg Thieme, 1976, 233 pp. ISBN: 3-13-532701-9. Cost: DM 160.

Contains 207 illustrations.

170. Atlas of Small Animal Surgery; Surgical Techniques for Practitioners. Thomas David. Available from Otto Harrassowitz, 1977, 624 pp. ISBN: 3-87706-053-6.

171. Atlas of Spectral Data and Physical Constants for Organic Compounds. Jeanette G. Grasselli; William M. Ritchey. CRC Press Inc, 2d ed., 1975, 6 Vols., 4,688 pp. LC: 72-2452. ISBN: 0-87819-317-0. Date of First Edition: 1973. Cost: $700.00.

This massive collection of spectral data and physical constants for approximately 21,000 organic compounds, includes a remarkable group of indexes that help to efficiently retrieve this organic chemical information from a multitude of different access points. Such spectral data as Mass Spectra, Raman, Carbon 13, Infrared, Ultraviolet and HNMR are included, as well as such physical constants as melting point, boiling point, density, solubilities, molecular weight, and refractive index. In addition, information is also provided on organic chemical structural diagrams, line formulas, and conversion tables, along with selected bibliographic literature references for sources of further investigation.

172. Atlas of Spectral Interferences in ICP Spectroscopy. M.L. Parsons; Alan Forster; Donn Anderson. Plenum Publishing Corp, 1980, 644 pp. LC: 79-24222. ISBN: 0-306-40334-X. Cost: $59.50.

This book is divided into 4 Tables. Table I includes the atomic and ionic lines for elemental analysis. Table II is divided into 2 sections; the first section including all transitions within $+1$ angstrom of the analytical line; the second section includes all prominent lines within 5 angstroms of the analytical lines. Table III is a compilation of lines below 2,000 angstroms. Table IV contains all the argon transitions.

173. Atlas of Surgery (2 Vols.). H.E. Grewe; K. Kremer. Georg Thieme (distributed in USA by W.B. Saunders Co), 1979, Vol.1, 555 pp.; Vol.2, 435 pp. ISBN: Vol.1, 3-13-569701-0; Vol.2, 3-13-569801-7. Cost: Vol.1, DM 198; Vol.2, DM 148.

Vol.1 contains 1,375 figures, and Vol.2 contains 1,155 figures.

174. Atlas of Surgical Approaches to the Bones of the Dog and Cat. Donald L. Piermattei. 2d ed., 1979.

175. Atlas of Surgical Approaches to the Bones of the Horse. Dennis W. Milne; A.S. Turner. Saunders, 1979, 210 pp. LC: 78-64718. Cost: $30.00.

176. Atlas of Surgical Techniques. Philip Thorek. Lippincott, 1970. Cost: $22.00.

177. Atlas of the Human Brain for Computerized Tomography. Takayoshi Matsui; Asao Hirano. Available from Otto Harrassowitz, 1978, 570 pp. ISBN: 3-437-10564-7.

178. Atlas of the Newborn. Neil O'Doherty. MTP Press, 1979, 412 pp. LC: 78-70614. Cost: £14.95.

179. Atlas of the Ocular Fundus. Hans Sautter; Wolfgang Straub; Herman Rossman. Urban and Schwarzenberg (available from Otto Harrassowitz), 2d ed., 1977, 160 pp. ISBN: 3-541-02462-3.

180. Atlas of the Textural Patterns of Basalts and Their Genetic Significance. S.S. Augustithis. Elsevier-North Holland Publishing Co, 1978, 324 pp. ISBN: 0-444-41566-1. Cost: $92.75.

181. Atlas of the Textural Patterns of Basic and Ultrabasic Rocks and Their Genetic Significance. S.S. Augustithis. Walter de Gruyter, 1979, 384 pp. ISBN: 3-11-006571-0. Cost: $159.40.

This atlas contains the microstructures of basic and ultrabasic rocks, showing the most important mineral phases and their intergrowths. The book contains over 750 photomicrographs showing textural patterns of rocks and their genetic significance.

182. Atlas of the Textural Patterns of Granites, Gneisses and Associated Rock Types. S.S. Augustithis. Elsevier, 1973, 378 pp. ISBN: 0-444-40977-7. Cost: $102.50.

183. Atlas of Thermoanalytical Curves. G. Liptay. Heyden and Sons Inc, 6 Vols. (5 Vols. and Index). ISBN: 0-85501-150-5. Cost: $175.00 set; $47.50 per Vol.

This is a collection of differential thermal analysis curves for organic and inorganic compounds of minerals. Over 350 compounds have been selected for inclusion in this 6-volume set.

184. Atlas of Topographical and Applied Human Anatomy. Eduard Pernkopf; Helmut Ferner; Harry Monsen (trans). Urban and Schwarzenberg, 1980, Vol.1, 308 pp.; Vol.2, 418 pp. Cost: $98.00.

Vol. 1 is entitled, *Head and Neck*, and Vol. 2 is entitled, *Thorax, Abdomen, Extremities*.

185. Atlas of Transbronchial Biopsy. E. Tsuboi. Georg Thieme, 1971, 151 pp. ISBN: 3-13-477-301-5. Cost: DM 139.

Contains 35 color plates and 180 illustrations.

186. Atlas of Ultrasonic Diagnosis in Obstetrics and Gynecology (Atlas der Ultrschalldiagnostik in Geburtshilfe und Gynakologie). K.H. Schlensker; W.B. Schulze (trans). Georg Thieme (distributed in USA by PSG Publishing), 1975, 158 pp. ISBN: 3-13-512501-7. Cost: DM 80.

Contains 359 illustrations.

187. Atlas of Ultrastructure. Howard Tseng. Appleton-Century-Croft, 224 pp. ISBN: 0-8385-0462-0. Cost: $26.50.

188. Atlas of Veterinary Surgery. John Hickman. Lippincott, 2d ed., 1980, 244 pp. LC: 80-130508. Cost: $49.50.

189. Atlas of X-ray Diagnosis of Early Gastric Cancer. H. Shirakabe. Georg Thieme, 1972, 244 pp. ISBN: 3-13-496401-5. Cost: DM 148.

Contains 576 illustrations.

190. Atlas zur Trinkwasserqualitaet der Bundesrepublik Deutschland (BIBIDAT). Rudolf Wolter; Hans Biermann. 1980, 180 pp. ISBN: 3-503-01823-8. Cost: DM 90.

191. Atomic and Nuclear Data Reprints. Katherine Way. Academic Press, 1973, Vol.1, 380 pp.; Vol.2, 574 pp. ISBN: Vol.1, 0-12-738901-6; Vol.2, 0-12-738902-4. Cost: Vol.1, $30.00; Vol.2, $35.50.

This 2-volume set is entitled as follows: Vol.1, *Internal Conversion Coefficients*, and Vol.2, *Reaction List for Charged-Particle-Induced Nuclear Reactions* (F.K. McGowan and W.T. Milner). This is basically a collection of data from the *Journal Nuclear Data Tables*.

192. Atomic and Nuclear Data Tables. National Bureau of Standards. Academic Press. Cost: $69.00 per year by subscription.

Replaces National Bureau of Standards Circular 499, *Nuclear Data*.

193. Atomic Data and Nuclear Data Tables; A Journal Devoted to Compilations of Experimental and Theoretical Results. Katherine Way. Academic Press. ISSN: 0092-640X. Cost: $71.00 per year.

This publication is a bimonthly journal devoted to the compilation of atomic data. This journal evaluates experimental and theoretical data in such areas as energy levels, wave functions, line-broadening parameters, collision processes, interaction cross-sections of atomic and simple molecules, transition probabilities, and penetration through matter of charged particles.

194. Atomic Energy Level and Grotrian Diagrams. S. Bashkin; J.O. Stoner. Elsevier, Vol.1, 1976; Vol.1, Addenda 1978; Vol.2, 1978; Vol.3, 1980, Vol.1, xx + 616 pp.; Vol.1, Addenda viii + 176 pp.; Vol.2, xviii + 650 pp.; Vol.3, in prep. ISBN: Vol.1, 0-7204-0322-7; Vol.1, Addenda 0-444-85236-0; Vol.2, 0-444-85149-6; Vol.3, 0-444-86006-1. Cost: Vol.1, $87.75; Vol.1, Addenda $24.50; Vol.2, $87.75; Vol.3, in prep.

This set is entitled as follows: Vol.1, *Hydrogen* 1- *Phosphorus* XV. AEL, 1.; Vol.1, (Addenda), *Hydrogen* 1- *Phosphorus* XV, Addenda AEL, 1(Addenda); Vol.2, *Sulphur* 1- *Titanium* XXII, AEL, 2; Vol.3, *Vanadium* 1, *Chromium* XXIV. AEL, 3.

195. Atomic Energy Levels as Derived form the Analyses of Optical Spectra, Vol. 1, ^{1}H to ^{23}V. C.E. Moore (National Bureau of Standards). U.S. Government Printing Office, 1971. SuDoc: C13.48:35/Vol.1. Cost: $9.25. Series: NSRDS-NBS-35, Vol.I.

196. Atomic Energy Levels as Derived from the Analyses of Optical Spectra, Vol. II. ^{24}Cr to ^{41}Nb. C.E Moore (National Bureau of Standards). U.S. Government Printing Office, 1971. SuDoc: C13.48:35/Vol.II. Cost: $7.95. Series: NSRDS-NBS-35, Vol.II.

197. Atomic Energy Levels as Derived from the Analyses of Optical Spectra, Vol.III. ^{42}Mo to ^{72}La, ^{72}Hf to ^{89}Ac. C.E. Moore. U.S. Government Printing Office, 1971. SuDoc: C13.48:35/Vol.III. Cost: $8.30. Series: NSRDS-NBS-35, Vol.III.

198. Atomic Energy Levels—The Rare Earth Elements, The Spectra of Lanthanium, Cerium, Praseodymium, Neodymium, Promethium, Samarium, Europium, Gadolinium, Terbium, Dysprosium, Holmium, Erbium, Thulium, Ytterbium, Lutetium. W.C. Martin; Romuald Zalubas; Lucy Hagan. U.S. Government Printing Office. SuDoc: C13.46:60. Series: NSRDS-NBS-60.

199. Atomic Energy Levels: Data for Parametric Calculations. Serafin Fraga; K.M.S. Saxena; Jacek Karwowski. Elsevier-North Holland, 1979, 482 pp. LC: 79-20809. ISBN: 0-444-41838-5. Cost: $92.25. Series: *Physical Sciences Data*; 4.

This volume includes values of atomic parameters for electronic configurations associated with isoelectronic series ranging between 5 and 102 electrons. The values provided are as follows: the average energy of the primary configuration; radical integrals; and spin-orbit constants.

200. Atomic Gas Laser Transition Data: A Critical Evaluation. William R. Bennett Jr. IFI/Plenum, 1979, 300 pp. LC: 79-22073. Cost: $75.00. Series: *The Lebedev Physics Institute Series*, Vol. 91.

This book includes references on 1,000 articles regarding atomic gas lasers. The author has covered the time period between 1964 and 1979 and has scanned over 30,000 abstracts on lasers. The study was requested by the National Bureau of Standards and contains the most accurate list of gas laser wavelengths. All the data has been critically evaluated for the 51 elements presented. Both Paschen and Racah notations are listed for the neutral noble gas transitions. Vacuum and air wavelengths are also listed.

201. Atomic Handbook: Europe, 1965. John W. Shortall. Morgan Brothers, Ltd., 1965. LC: 66-47790.

202. Atomic Spectra and Radiative Transitions. I.I. Sobelman. Otto Harrassowitz, 1979, 330 pp. ISBN: 3-540-09082-7. Cost: $32.50. Series: *Springer Series in Chemistry and Physics*, 1.

Contains 21 figures and 58 tables.

203. Atomic Transition Probabilities, Vol. 1 Hydrogen Through Neon. W.L. Wiese; M.W. Smith; B.M. Glennon (National Bureau of Standards). U.S. Government Printing Office, 1966. SuDoc: C13.48:4/Vol. I. Cost: $6.45. Series: NSRDS-NBS-4, Vol. I.

204. Atomic Transition Probabilities, Vol. II. Sodium through Calcium, A Critical Data Compilation. W.L. Wiese; M.W. Smith; B.M. Miles (National Bureau of Standards). NTIS, 1969. Accession: AD-696884. Cost: $9.75. Series: NSRDS-NBS-22, Vol. II.

205. Audubon Illustrated Handbook of American Birds. Edgar M. Reilly Jr. National Audubon Society, 1968. ISBN: 0-685-48741-5. Cost: $20.00.

206. Auger Electron Spectroscopy Reference Manual; A Book of Standard Spectra for Identification and Interpretation of Auger Electron Spectroscopy Data. G.E. McGuire. Jan Press, 1979, unpaginated. LC: 79-24223. ISBN: 0-306-40333-1. Cost: $39.50.

This spiral bound paper handbook provides graphical data on 43 common elements. Up to 3 scans are provided on each element. Such elements as beryllium, oxygen, aluminum, phosphorus, argon, titanium, chromium, iron, nickel, copper, germanium, arsenic, silver, tin, tellurium, xenon, rhenium, and gold are included.

207. Austenite Transformation Kinetics of Ferrous Alloys. Witold W. Cais. Climax Molybdenum Co, 83 pp.

This volume includes numerous transformation diagrams describing the characteristics of austenite and transformation behavior. Description of various alloying methods are presented and there is an abundance of continuous cooling transformation diagrams, as well as hardenability of hypereutectoid steels.

208. Automatic Control Handbook. G.A.T. Burdett. Newnes, 1962. LC: 64-3791.

209. Automatic Data Processing Handbook. Carl Heyel. McGraw-Hill, 1977, 967 pp. LC: 76-28331. ISBN: 0-07-016807-5. Cost: $38.95.

This work sponsored by the Diebold Group, Inc.

210. Automotive Aerodynamics Handbook. Henry Clyde Landa. Film Instruction Co of America, 2d ed., 1979, various paging. ISBN: 0-931974-07-0. Cost: $14.65.

211. Automotive Handbook. Robert Bosch. Society of Automotive Engineers, 1979, 516 pp. ISBN: 0-89883-510-0. Cost: $13.50.

This durable, vinyl-bound handbook covers a wide range of automotive topics, as well as materials, mechanics, electrical engineering, fuels, and lubricants. For example, aerodymanics, electric vehicles, carburetors, fuel injection, transmissions, cooling systems, heating and air conditioning systems, electromagnetic interference suppression, braking equipment, and international passenger car specifications are among the topics covered.

212. Avalanche Handbook. Public Document Distribution Center, 1976 (rep. 1979), 254 pp. Cost: $5.25.

Contains information for people interested in avalanche control, mountain weather, and snow. Procedures are outlined for avoiding avalanches in ski areas and near roads and settlements.

213. Aviation Electronics Handbook. Ed Safford. TAB Books, 1975, 406 pp. LC: 72-97217. ISBN: 0-8306-4631-0. Cost: $8.95.

214. Aviation Ground Operations Handbook. National Safety Council Staff, 2d ed., 1972. ISBN: 0-87912-029-0. Cost: $10.00.

215. Azeotropic Data-III. Lee H. Horsley. American Chemical Society, 1973, 628 pp. LC: 73-75991. ISBN: 0-8412-0166-8. Cost: $45.50. Series: *Advances in Chemistry*, No. 116.

Data are presented on azeotropes, nonazeotropes, and vapor-liquid equilibria for more than 17,000 sytems. Tables are included for binary, ternary, quaternary, and quinary systems. A formula index, bibliography, and charts are also presented.

216. Azimuths for Plane Coordinates. O.S. Adams. U.S. Dept of Commerce, Coast and Geodetic Survey, (available from National Geodetic Survey C13x4 NOS/NOAA, Rockville MD 20852), 1936, 14 pp. Series: Serial No.: 584.

217. Band Spectrum of Carbon Monoxide. P.H. Krupenie (National Bureau of Standards). NTIS, 1966. Cost: $5.00. Series: NSRDS-NBS-5.

218. Barker Index of Crystals, A Method for the Identification of Crystalline Substances. Barker Index Committee; M. W. Porter et al. W. Heffer and Sons, 1951-64, 3 Vols. in 7 parts.

This 3-volume set has the following titles: Vol. 1, *Crystals of the Tetragmal, Hexagonal, Trigonal and Orthorhombic Systems* (in 2 parts); Vol. 2, *Crystals of the Monoclinic System* (in 3 parts); and Vol. 3, *Crystals of the Anorthic System* (in 2 parts).

219. Barlow's Tables of Squares, Cubes, Square Roots, Cube Roots, and Reciprocals of All Integers Up To 12,500. P. Barlow; L.J. Comrie. Halsted Press, 4th ed., 1969, 258 pp. ISBN: 0-470-04976-6. Cost: $7.00.

220. Basic Atlas of Cross-sectional Anatomy. Walter J. Bo et al. Saunders Press, 1980, 357 pp. LC: 79-66032. Cost: $41.00.

Includes anatomic photographs that correlate with xerograms and computerized tomograms.

221. Basic Data—Thermodynamic Properties/ Thermophysical Properties-Metal Oxides. Advisory

Group for Aerospace Research and Development. NTIS, 1975, 25 pp. Accession: N76-11245.

Tabular data are presented on metal oxides. Such properties as crystal structure, melting point, boiling point, equilibrium pressures, mechanical properties, and reaction kinetics are presented. Phase diagrams are also provided for metal-oxygen systems.

222. Basic Electrocardiography Handbook. Leonard J. Lyon. Van Nostrand Reinhold, 144 pp. ISBN: 0-442-24960-8. Cost: $9.95.

223. Basic Electronic Instrument Handbook. Clyde F. Coombs Jr. McGraw-Hill, 1972, 832 pp (various paging). LC: 72-1394; 78-141917. ISBN: 0-07-012615-1. Cost: $38.50.

This handbook includes 40 chapters of information on electronic instrumentation ranging in function from measuring instruments to signal generators. Each chapter is written by an authority in his/her particular area of specialization. Such instruments as oscillators, pulse generators, microwave signal generators, electronic counters, timekeeping instruments, and DC and AC voltmeters are discussed. In addition, information is provided on such subjects as transducers, ammeters, amplifiers, frequency synthesizers and frequency analyzers, and instrumentation print-out devices and recorders. For related works, please refer to *Standard Handbook for Electrical Engineers*, the *Electrical Engineer's Reference Book*, the *Electronics Engineers' Handbook*, and *Handbook of Modern Electronic Data*.

224. Basic Engineering and Mathematical Tables. L.S. Srinath; K. Sarma; S. Patankar. McGraw-Hill, 1973, 136 pp. ISBN: 0-07-096608-7. Cost: $12.50.

225. Basic Handbook: An Encyclopedia of the Basic Computer Language. David A. Lien. CompuSoft, 1978, 360 pp. LC: 78-64886. ISBN: 0-932760-00-7. Cost: $14.95.

This is a handbook on the computer language, BASIC. Two-hundred fifty of the most frequently used words are listed alphabetically. The handbook tells whether each word is a statement, a command, a function, or an operator.

226. Basic Tables in Chemistry. Roy Alan Keller. McGraw-Hill, 1967, 400 pp. LC: 66-24477.

227. Basic Tables in Electrical Engineering. Granino A. Korn. McGraw-Hill, 1965, 370 pp. LC: 64-66023.

228. Basic Tables in Physics. John Robson. MaGraw-Hill, 1967, 354 pp. LC: 66-26581.

229. Basic Theorems in Matrix Theory. Marvin Marcus (National Bureau of Standards). U.S. Government Printing Office, 1960, 27 pp. SuDoc: C13.32:57. Cost: $0.65. Series: NBS AMS57.

230. Beacon Boiler Reference Book. Industry Publications, Inc, 6th ed., 1973, various paging. Date of First Edition: 1940. Cost: $10.00.

This reference book is a collection of data from boiler manufacturers and includes boilers rated up to 3,000 sq. ft. of steam. Although this book does not include a considerable amount of physical data on each boiler, it does provide basic information, model numbers, ratings and qualifying data such as total load and equivalent direct radiation. The boiler specifications are alphabetically arranged by company, and there is a cross index of boiler trade names. This handbook is quite useful in tracking down information on old or obsolete boilers.

231. Beginner's Handbook in Biological Electron Microscopy. Brenda S. Weakley. Churchill-Livingstone, 1972, 240 pp. ISBN: 0-443-00908-2.

232. Beginner's Handbook of Electronics. George Olsen; Forest Mims III. Prentice-Hall, Inc, 1979, 320 pp. ISBN: 0-13-074211-2 (P). Cost: $17.95.

This cloth book provides information on how to use resistors, transistors, capacitors, and other electronic components. Integrated circuits are also discussed in this handbook.

233. Beilstein's Handbuch Der Organischen Chemie. Beilstein Institute for Literature of Organic Chemistry. Springer-Verlag, 4th ed., 1918-, over 130 Vols. Cost: For price, consult publisher.

A massive comprehensive treatise on the preparation and properties of carbon compounds. Contains research data, reported in the literature, which has been critically evaluated and correlated. Portions of this work are available in microfilm.

234. Benzene: Basic and Hazardous Properties. Paul N. Cheremisinoff; Angelo C. Morresi. Marcel Dekker Inc, 1979, 256 pp. ISBN: 0-8247-6860-4. Cost: $51.00. Series: *Pollution Engineering and Technology: A Series of Reference Books and Textbooks*, Vol. 9.

235. Bergey's Manual of Determinative Bacteriology. R.E. Buchanan; N.E. Gibbons. Williams and Wilkins, 8th ed., 1974, 1,268 pp. LC: 73-20173. ISBN: 0-683-01117-0. Cost: $55.00.

The first 7 editions are entered under American Society of Microbiology.

236. Betz Handbook of Industrial Water Conditioning. Betz Laboratories, Inc, 6th ed., 1962 (rep. in 1967), 427 pp. LC: 62-21097. Cost: $6.00.

Although this handbook is somewhat dated, it is still an excellent source of information on water treatment and analysis. The *Betz Handbook* has 69 chapters which discuss such problems as boiler corrosion control, boiler metal embrittlement, scale control, hydrogen sulfide removal, sodium zeolite softening and many other problems associated with industrial water treatment and processes. It also has a fairly comprehensive section on cooling water treatment as well as information on nitrates, phosphates, sulfates, and slime and algae control. For related works, refer to such titles as *CRC Handbook of Environmental Control, Pollution Engineering Practice Handbook, Environmental Engineers' Handbook*, and *Handbook of Environmental Civil Engineering*.

237. Bibliography and Index of Experimental Range and Stopping Power Data. H.H. Andersen. Pergamon Press Inc, 1977, 214 pp. LC: 77-22415. ISBN: 0-08-021604-8. Cost: $38.50. Series: *The Stopping and Ranges of Ions in Matter*, Vol. 2.

Over 900 references have been cross-indexed under ions. Each ion is listed by its atomic number, and under each ion is a catalog of target materials and energy range of ions.

238. Binomial Reliability Table: Lower Confidence Limits for the Binomial Distribution. James R. Cooke; Mark

T. Lee; John P. Vanderbeck. NTIS, 1964, 331 pp. Accession: AD-444-344. Tech Report: NOTS-TP-314D.

Binomial reliability table is presented for use in the reliability estimation of missiles and electronic components. Includes information on statistical distributions and probability.

239. Bioastronautics Data Book. NASA (available from NTIS), 1964. Series: NASA SP 3006, 1964.

240. Biochemists' Handbook. Cyril Long. Van Nostrand Reinhold. ISBN: 0-442-04869-6. Cost: $28.50.

Biological Data Book. *See* **Biology Data Book.**

241. Biological Laboratory Data. Leslie J. Hale. Wiley (Halsted Press), 2d ed., 1965, 147 pp. ISBN: 0-470-34081-9. Cost: $4.00.

242. Biological Stains: A Handbook on the Nature and Use of the Dyes Employed in the Biological Laboratory. Harold Joel Conn. Williams and Wilkins, 8th ed., 1969, 498 pp. LC: 69-16600.

243. Biology Data Book. Philip L. Altman; Dorothy S. Dittmer. Federation of American Societies for Experimental Biology, 2d ed., 1972-74, 3 Vols., Vol. 1, 606 pp; Vol. 2, 826 pp; Vol. 3, 691 pp. LC: 72-87738. ISBN: 0-913822-09-4. Cost: $50.00 per Vol. or $120.00 per set.

Covers data on cytology, genetics, nutrition, digestion, metabolism, blood respiration, circulation, enzymes, hormones, and reproduction.

244. Blasters' Handbook. Sales Development Section of the Explosive Dept, E.I. du Pont de Nemours and Co, 16th ed., 1977, 494 pp. Date of First Edition: 1942. Cost: $18.00.

Bofors Handbook. *See* **Steel and Its Heat Treatment; Bofors Handbook.**

245. Bond Dissociation Energies in Simple Molecules. B. deB. Darwent (National Bureau of Standards). U.S. Government Printing Office, 1970. SuDoc: C13.48:31. Cost: $0.95. Series: NSRDS-NBS-31.

246. Bone Dysplasias of Infancy: A Radiological Atlas. B.J. Cremin; P. Beighton. Available from Otto Harrassowitz, 1978, 109 pp. ISBN: 3-540-08816-4. Cost: $39.00.

247. Borides. G.V. Samsonov; T.I. Serebriakova; V.A. Neronov. Atomizcat (available from NTIS), 1975, 376 pp. Accession: A76-32271.

This book, in Russian, contains data on the chemical composition and crystal structure of borides along with the characteristics of all the borides, metals, and nonmetals. Phase diagrams of boron compounds are presented along with various chemical properties and physical properties.

248. Brazing Manual. American Welding Society, Inc, 3rd rev. ed., 1976, 309 pp. LC: 76-15811. ISBN: 0-686-51724-5. Cost: $18.00.

249. Broadcast Engineering and Maintenance Handbook. Patrick S. Finnegan. TAB Books, 1976, 532 pp. LC: 76-24783. ISBN: 0-8306-6852-8. Cost: $19.95.

250. BSCC High Temperature Data. British Steelmakers Creep Committee. Iron and Steel Institute, 1973, 735 pp.

This tabulation provides information on British long-term creep rupture and elevated temperature tensile data on steels for high temperature service. Includes carbon and carbon-manganese steels, alloy steels, and austenitic steels.

251. Buchsbaum's Complete Handbook of Practical Electronic Reference Data. Walter H. Buchsbaum. Prentice-Hall, 2d ed., 1978, 645 pp. LC: 78-1055. ISBN: 0-13-084624-4. Date of First Edition: 1973. Cost: $19.95.

Covers properties of materials, wire tables, filters, alternators, antennas circuits, semiconductors, radar, power supply, and transducers.

252. Building Code Requirements for Reinforced Masonry. National Bureau of Standards. American National Standards Institute. Cost: $3.25. Series: NBS Handbook 74.

Available from American National Standards Institute, 1430 Broadway, New York, NY 10018 as A41.2-1960 (R1970).

253. Building Construction Cost Data. Robert S. Godfrey. Means, 39th ed., 1981. LC: 55-20084. ISBN: 0-911950-29-X. Cost: $24.50 paper.

Provides over 18,000 prices for such items as building materials, labor, overhead, and profit. Information such as required equipment and crew sizes is included.

254. Building Construction Handbook. Frederick S. Merritt. McGraw-Hill, 3d ed., 1975, 1,120 pp. (various paging). LC: 75-6553. ISBN: 0-07-041520-X. Date of First Edition: 1958. Cost: $46.50.

This is a very comprehensive handbook that includes a wealth of information on all phases of building construction. Each major section, written by a different authority in his/her respective specialization, includes such subjects as soil mechanics, concrete, structural-steel windows, walls, and wood construction. In addition, information is also provided for such subjects as plumbing, fire protection, electricity, elevators, sound and vibration control, air conditioning, and insulation. There is also an excellent section on the various properties of many different building materials, as well as bibliographies for further information. Please refer to *ASTM Standards in Building Codes* and *Construction Specifications Handbook.*

255. Building Cost Manual. Craftsman Book Co, 1979, 240 pp. Cost: $10.00.

Square feet cost for residential, commercial, or industrial military structures is presented. Over 300 cost tables are included. All types of buildings are discussed including shopping centers, hospitals, warehouses, steel buildings, farm buildings, banks, etc.

256. Building for Energy Conservation. P.W. O'Callaghan. Pergamon Press Inc, 1978, 200 pp. LC: 77-30332. ISBN: 0-08-022120-3. Cost: $25.50.

Includes information on building heat balance and thermal insulating techniques; heat losses from buildings are discussed along with alternative energy sources and waste heat recovery.

257. Bulbs: A Complete Handbook. Roy Genders. Bobbs-Merrill, 1973. LC: 72-88269. ISBN: 0-672-51783-3. Cost: $19.95.

258. Bulletin of Alloy Phase Diagrams. Lawrence H. Bennett. American Society for Metals, Vol.1, 1981. ISBN: 0197-0216. Cost: $75.00.

This publication is actually a journal published on a quarterly basis. Includes approximatley 20 to 30 phase diagrams in each issue. Basically, this journal provides graphic displays of the thermodynamic relationships of 2 or more substances at different temperatures and compositions.

259. Burgess Engineering Manual; Complete Battery Data for the Design Engineer. Burgess Battery Co, 1961.

260. Burnham's Celestial Handbook: An Observer's Guide to the Universe Beyond the Solar System. Robert Burnham Jr. Dover Publications, 1980, 3 Vols. (2,000 pp. total). ISBN: Vol.1 and Vol.2, 0-486-24065-7; Vol.3, 0-486-24065-7. Cost: $20.00 each.

This is basically a star guide and catalog of celestial objects. Hundreds of illustrations, photographs, and charts are included. The arrangement of this 3-volume set is by constellation.

261. Business Systems Handbook: Analysis, Design Documentation Standards. Robert Gilmour. Prentice-Hall Inc, 1979, 272 pp. ISBN: 0-13-107755-4. Cost: $19.95.

Written for systems analysts, comptrollers, and programers. Discusses such analytical techniques as PERT and regression analysis.

262. Business Systems Handbook: Strategies for Administrative Control. J.W. Haslett. McGraw-Hill, 1979, 480 pp. LC: 78-18361. ISBN: 0-07-026980-7. Cost: $19.95.

263. Butterfly and Angelfishes of the World. Vols.1 and 2. Hans A. Baensch; Roger Steene; Gerald R. Allen. Merges, Melle, Vol.1, 1977; Vol.2, 1980, Vol.1, 144 pp.; Vol.2, 208 pp. LC: Vol.1, 78-17351. ISBN: Vol.1, 0-471-04737-6; Vol.2, 0-471-05618-0; German: Vol.1, 3-88244-000-7; Vol.2, 3-88244-003-1. Cost: Vol.1, $22.95; Vol.2, $30.00.

Vol.1 (Steene) covers butterfly and angelfishes in Australia and New Guinea as well as the Indo-Pacific Region. Vol.2 (Allen) covers the Atlantic, Caribbean, and Red Sea.

264. Calculated Liquid Phase Thermodynamic Properties and Liquid Vapor Equilibria for Fluorine-oxygen (FLOX) Mixtures. W.R. Parrish; M.J. Hiza. NTIS, 1973, 30 pp. Accession: N75-10844. Tech Report: NASA-CR-140709.

Experimental results are tabulated for fluorine-oxygen mixtures. Liquid phase thermodynamic properties and liquid-vapor equilibria are presented. Molar volumes, enthalpy, entropy, and constant pressure specific heat are listed.

265. Calculation of Rotational Energy Levels and Rotational Line Intensities in Diatomic Molecules. Jon T. Hougen (National Bureau of Standards). NTIS, 1973. Accession: PB-192874. Cost: $4.50. Series: NBS Monograph 115.

266. Calculation of Thermodynamic Properties of Ideal Gases at High Temperatures: Monatomic Gases. J.R. Downey. NTIS, 1978, 35 pp. Accession: N78-30981; AD-A054854. Tech Report: AFOSR-78-0960TR.

This report presents the problems involved in extending the *JANAF Thermochemical Tables* for monatomic gases to higher temperatures. Stastical mechanics, high temperature gases, free energy, thermodynamic properties, and extrapolation are presented.

267. Cameron Hydraulic Data. Ingersoll-Rand, 14th ed., 1977, 320 pp. Cost: $8.50.

This compilation of data includes friction tables, vapor pressures and viscosities of liquids, and many topics of interest to those who operate hydraulic equipment.

268. Canadian Mines Handbook 1980-81. Northern Miner Press Limited, 1981. Cost: $17.50 paper.

Details are provided on oil reserves, mine development, and exploration. Numerous maps are included.

269. Canadian Oil and Gas Handbook 1980-81. Northern Miner Press Ltd, 1981. Cost: $17.00 paper.

This book is a good source of information for petroleum reserves, land holdings, partnerships, and working capital.

270. Cancer Chemotherapy Handbook. R.T. Dorr; W.L. Fritz. Elsevier, 1980, 786 pp. ISBN: 0-444-00330-4. Cost: $40.00.

271. Cancer Handbook of Epidemiology and Prognosis. J.A. Waterhouse. Longman, 1974. ISBN: 0-443-00938-4. Cost: $21.00.

272. Cane Sugar Handbook: A Manual for Cane Sugar Manufacturers and Their Chemists. George P Meade; James C.P. Chen. Wiley-Interscience, 10th ed., 1977, 976 pp. LC: 76-51046. ISBN: 0-471-58995-0. Cost: $67.50.

This handbook includes a bibliography and index and includes information on processing, refining, and various types of equipment and instruments used in the sugar manufacturing industry. The 9th edition was published under the title *Spencer-Meade Cane Sugar Handbook*.

273. Carbides. Metals and Ceramics Information Center. Metals and Ceramics Information Center, 1980, 135 pp. Cost: $45.00. Series: *MCIC-HB-07V2*.

This is Vol.II of *Engineering Property Data on Selected Ceramics.* This volume replaces section 5.2 entitled *Carbides* of the 1966 *Databook*. The following properties of carbides are listed in the tables: thermal conductivity, compressive strength, bulk modulus, fracture toughness, corrosion, density, thermal expansion, impact strength, hardness, melting point, Young's modules, creep, thermal stress, specific heat, tensile strength, shear modulus, and erosion.

274. Carbon Adsorption Handbook. Paul N. Cheremisinoff; Fred Ellerbusch. Ann Arbor Science Publishers, Inc, 1978, 1,064 pp. LC: 77-093382. ISBN: 0-250-40236-X. Cost: $49.95.

This book discusses carbon systems as a method of pollution control and as purifying agents.

275. Carbon-13 NMR Spectral Data. W. Bremser; L. Ernst; B. Franke; R. Gerhards; A. Hardt. Verlag Chemie International, 3d ed., 1981. ISBN: 3-527-25899-X. Cost: DM 1,980.00.

This comprehensive work on nuclear magnetic resonance spectroscopy has been published on 175 microfiche and contains 30,000 spectra plus 9 indexes. The indexes provide rapid access to the spectral information. Access points are chemical names, molecular formulas, molecular weights, CAS registry number, literature references, structure versus coupling constants, coupling versus constants structure, chemical shifts, and substructure codes.

276. Carbon-Carbon Coupling Constants: A Compilation of Data and a Practical Guide. V. Wray. Pergamon Press Inc, 80 pp. ISBN: 0-08-024891-8. Cost: $15.50.

This compilation includes data from the literature on Carbon-Carbon spin-spin coupling constants.

277. Carpentry Handbook. Public Document Distribution Center, 1962 (rep. 1977), 354 pp. ISBN: 008-050-00149-6. Cost: $6.00.

This is a loose-leaf service, providing time data for different types of carpentry jobs.

278. Cast Metals Handbook. American Foundrymen's Society, 4th ed., 1957, 816 pp.

279. Catalogue of Meteorological Data for Research. World Meteorological Organization. Unipub, Part 1, 1965; Part 2, 1970; Part 3, 1972; Part 4, 1980, 4 Vols. ISBN: Part 1, 0-685-36783-5; Part 2, 0-685-36784-3; Part 3, 0-685-29207-X; Part 4, 0-686-65617-2. Cost: Part 1, $30.00; Part 2, $20.00; Part 3, $50.00; Part 4, $17.00.

280. CDP Review Manual: A Data Processing Handbook. Kenniston W. Lord; James B. Steiner. Van Nostrand Reinhold, 2d ed., 1981, 632 pp. Cost: $19.95.

Provides information on electronic data processing in order to help the user prepare for the certificate in data processing (CDP) examination.

281. Celestial Handbook: An Observers Guide to the Universe Beyond the Solar System. Robert Burnham. Celestial Handbook, 1967, 5 Vols..

This 5-volume set includes numerous illustrations, photographs, and charts of the stars and other celestial phenomena.

282. Cell Biology. Philip L. Altman; Dorothy D. Katz. Federation of American Societies for Experimental Biology, 1976, Vol.1, 454 pp. LC: 75-42787. ISBN: 9-913822-10-8. Cost: $45.00.

This collection includes over 100 tables, which have organized data, on such topics as cell characteristics, Lysomes, Microsomes, Ribosomes, Mitochondria, and Endoplasmic Reticulum.

283. Cement Engineer's Handbook. O. Labahn; W.A. Kaminsky. Bauverlag GmBH, 4th ed. in preparation. ISBN: 3-7625-0975-1.

Provides important schedules, formulas and numerical values for the cement industry.

284. Cement Manufacturer's Handbook. Kurt E. Peray. Chemical Publishing Company, Inc, 1979, 382 pp. ISBN: 0-8206-0245-0. Cost: $32.50.

Contains the most essential engineering formulas used in the cement manufacturing process. Provides information on chemical and physical properties of materials, mix calculations, heat balance, kiln performance, sulfur and alkali balance, and rotary kilns, cooler air balance, heat transfer, and dust collection.

285. Cement-Data-Book: International Process Engineering in the Cement Industry. Walter H. Duda. Bauverlag GmbH (available from Otto Harrassowitz), 2d ed., 1977, 539 pp. ISBN: 3-7625-0834-8; 0-686-56597-5. Cost: $160.00.

This volume contains data on such topics as rotary kilns, dust collection, grinding, roller mills, kiln refractory linings, and air separators. The text is in German and English and has numerous illustrations, tables, and literature references.

286. Cerebral Palsied and Learning Disabled Children: A Handbook Guide to Treatment, Rehabilitation and Education. Nancy C. Marks. C. C. Thomas, 1974. ISBN: 0-398-02911-3. Cost: $17.50.

287. Cerium-144 Data Sheets. E.E. Ketchen; S.J. Rimshaw. NTIS, 1967, 30 pp. Accession: N68-24383. Tech Report: ORNL-4185.

Provides chemical, physical, and mechanical data for Cerium-144 and cerium compounds.

288. Chambers Shorter Six-Figure Mathematical Tables. L.J. Comrie. Halsted Press, 1972, 389 pp. ISBN: 0-470-16725-4. Cost: $7.95.

289. Characteristics of Sodium. A. Pee. NTIS, 1969, 93 pp. Accession: N69-40018. Tech Report: KFK-924 EUR-4168.

This work is a collection of tables on the thermophysical properties of sodium.

290. Characteristics of Urban Transportation Systems. A Handbook for Transportation Planners. DeLeuw, Cather and Co (avaialable from NTIS), 1975, 193 pp. Accession: PB-245809 (supersedes PB-233580). Tech Report: UMTA-IT-06-0049-75-1. Cost: $2.25 microfiche; $7.00 paper copy.

Compares performance measures of rail transit, highway systems, and bus transit.

291. Charles Press Handbook of Current Medical Abbreviations. Charles Press. Charles Press, 1976, 176 pp. ISBN: 0-913486-80-9. Cost: $4.95 paper copy.

292. Chemical and Physical Properties of Human Urine Concentrates. D.F. Putnam. Douglas Aircraft Co (available from NTIS), 1968, 70 pp. Accession: N68-26600. Tech Report: NASA-CR-66612.

293. Chemical Compositions and Rupture Strengths of Superalloys. American Society for Testing and Materials, 1970, 25 pp. ISBN: 0-8031-01367-6. Cost: $3.50 soft. Series: *ASTM Series DS 9E.*

Lists chemical composition and rupture strengths for over 400 alloys; includes ferritic superalloys and age hardening stainless steels.

294. Chemical Compounds in the Atmosphere. T.E. Graedel. Academic Press, 1978, 456 pp. ISBN: 0-12-294480-1. Cost: $29.50.

Presents information on sources, ambient concentrations, reactions, products, and atmospheric lifetimes of more than 1,600 compounds in the earth's atmosphere.

295. Chemical Data Guide for Bulk Shipment by Water. Public Documents Distribution Center, 1976, 317 pp. ISBN: 050-012-00117-1. Cost: $4.25.

This chemical data guide is an alphabetical listing of over 250 chemicals, providing information on explosion, hazards, spill or leak procedures, and fire.

296. Chemical Economics Handbook. Stanford Research Institute, 1967, 19-Vol. set. Date of First Edition: 1950.

The *Chemical Economics Handbook* is a loose-leaf service that is updated monthly and is available from the Stanford Research Institute on a subscription basis. Essentially, this handbook provides information on new and old chemicals and chemical groups that would be helpful to those scientists and engineers who need such data as current production and consumption, national trends, different processes associated with the development of the chemicals and manufacturers of the chemicals. It has an excellent index and the chemicals are listed under generic names, as well as trade names.

297. Chemical Engineering Drawing Symbols. D.G. Austin. Halsted Press (division of John Wiley and Sons, Inc), 1979, 96 pp. ISBN: 0-470-26601-5. Cost: $19.95.

Graphic symbols are listed and standard symbols are recommended. Symbols are provided for such areas as piping systems, materials handling, processing equipment, heat transfer equipment, valve actuation, and automatic control.

298. Chemical Engineers' Handbook. Robert H. Perry; Cecil H. Chilton. McGraw-Hill, 5th ed., 1973, 1,958 pp. various paging. LC: 73-7866. ISBN: 0-07-049478-9. Date of First Edition: 1934. Cost: $51.50.

This is a classical reference work in the chemical engineering field and has been through 5 different editions, each edition updating and retabulating the latest available information. A detailed table of contents is provided at the beginning of each section, which is written by an authority in that particular subject. The fifth edition of *Perry's* (which it is often called) covers such subjects as heat transfer gas absorption, distillation, fluid transport, reaction kinetics, and thermodynamics. This handbook has an excellent section on the physical and chemical properties of various compounds and includes many illustrations and tables. *See also* the *CRC Handbook of Chemistry and Physics* and *Lange's Handbook of Chemistry*.

299. Chemical Equilibria in Carbon-Hydrogen-Oxygen Systems. Robert E. Baron; James H. Porter; Ogden H. Hammond Jr. MIT Press, 1976, 110 pp. LC: 75-44374. ISBN: 0-262-02121-8. Cost: $17.50. Series: *Energy Laboratory Series*.

These tables are designed for calculating equilibria for systems containing solid carbon.

300. Chemical Formulary. H. Bennett. Chemical Publishing Co., Inc, 1933-76, 23 Vols. ISBN: 0-8206-0208-6. Cost: $25.00 each Vol.; cumulative index for Vols. 1-15 $33.00; entire set $575.00.

This multivolume, set, with index volume is a very interesting collection of chemical formulas. The emphasis is upon mixtures, blends and compounds that, when prepared, make useful products, such as paints, cosmetics, adhesives, weed killers, insecticides, fertilizers, vinegar, beer, and various types of ink. The ingredients for each chemical formula are provided, indicating the percentage or volume of each chemical component to be added. In addition, such useful products as toothpaste, soap, textile dyes, varnish, detergents, as well as various foods, such as marshmallow, cheese, sausage, and Philadelphia scrapple are covered in this remarkable compilation.

301. Chemical Handbook. Lefax Publishing Co, loose-leaf binding. ISBN: 0-685-14125-X. Cost: $8.50. Series: *Lefax Technical Manual*, No. 777.

302. Chemical Kinetic and Photochemical Data for Modelling Atmospheric Chemistry. Robert F. Hampson, Jr; David Garvin eds (National Bureau of Standards). U.S. Government Printing Office, 1975. SuDoc: C13.46:866. Cost: $1.85. Series: NBS Tech. Note 866.

303. Chemical Reference Handbook. Public Documents Distribution Center, 1967 (rep. 1975). ISBN: 008-020-00547-9. Cost: $2.55.

This is an Army field manual which is military oriented. It covers such categories as radiological defense and biological defense.

304. Chemical Resistance Data Sheets. L.T. Butt et al. Rubber and Plastics Research Association of Great Britain, 2 Vols., loose-leaf, (various paging).

Provides information on the chemical resistance of rubbers and plastics.

305. Chemical Shift Ranges in ^{13}C-NMR Spectroscopy. W. Bremser. Verlag Chemie International, 1981, 760 pp. ISBN: 3-527-25908-2. Cost: DM 220.

Chemical shifts of ^{13}C-NMR signals are correlated with the structural elements for whcih they are characteristic. The data are compiled in tables based upon evaluation of 30,000 spectra.

306. Chemical Stability of Pharmaceuticals: A Handbook for Pharmacists. K.A. Connors; G.L. Amidon; L. Kennon. John Wiley and Sons, Inc, 1979, 367 pp. LC: 78-1759. ISBN: 0-471-02653-0. Cost: $21.50.

This handbook is designed for the pharmacist and is concerned with drug stability.

307. Chemical Statistics Handbook. Manufacturing Chemists' Association, 6th ed., 1966, 606 pp.

One volume supplemented by *Statistical Summary*, Nos. 5-7, 1967-70. Formerly entitled *Chemical Facts and Figures*.

308. Chemical Tables for Laboratory and Industry. Wolfgang Helbing; Adolf Burkart; R. Vardharajan (trans); Sudhanshu Gupta (trans). John Wiley and Sons, Inc, published in English in 1979, 272 pp. LC: 79-26137. ISBN: 0-470-26910-3. Cost: $19.95.

This book, originally published in German in 1969, is divided into 3 sections. The first is entitled *Matter*. It includes tables of inorganic and organic compounds and their corresponding properties. The second section is entitled *Number* and includes information on various constants and rules of calculation. Certainly, the gas reduction table is one of the important items in this section. The third section is entitled *Process*. It includes various types of methods that are used in research laboratories and industries.

309. Chemical Tables. Lefax Publishing Co, loose-leaf binding. ISBN: 0-685-14126-8. Cost: $3.00. Series: *Lefax Data Books*, No. 605.

310. Chemical Tables. Bela A. Nemeth; Istvan Finlay (trans). Halsted Press (division of John Wiley and Sons), 1976. LC: 75-19223. ISBN: 0-470-63161-9. Cost: $32.95.

311. Chemical Technicians' Ready Reference Handbook. Gershon Shugar. McGraw-Hill, 2d ed., 864 pp. LC: 80-22017. ISBN: 0-07-057176-7. Cost: $39.50.

312. Chemical Technology Handbook. Robert L. Pecsok et al. American Chemical Society, 1975, 215 pp. LC: 75-22497. ISBN: 0-8412-0242-7. Cost: $15.75.

The subtitle of this volume is *Guidebook for Industrial Chemical Technologists and Technicians* and was actually written by the Writing Team for the Chemical Technician Curriculum Project. This book is primarily designed for the laboratory technician who is interested in learning more about building laboratory apparatus and using equipment safely in an experimental setting. In addition, safety is also covered with regard to toxic chemicals, fires, radiation, and electrical hazards. Information is also provided for recording experimental data properly and for finding additional information in the library.

313. Chemical Thermodynamic Properties of Hydrocarbons and Related Substances. Properties of the Alkane Hydrocarbons. C_1 through C_{10} in the Ideal Gas State from 0 to 1500 K. D.W. Scott. Bureau of Mines, Washington, DC (available from NTIS), 1974, 194 pp. Accession: N74-34587. Tech Report: PB-233210. SuDoc: I28.3:666. Cost: $2.25 microfiche; $2.55 hard copy.

These tables include chemical thermodynamic properties for the ideal gas state from 0 to 1500 K for all the alkane hydrocarbons through C_{10}. Such properties as Gibbs energy function, enthalpy function, enthalpy, entropy, heat capacity, and enthalpy of formation are included.

314. Chemical Thermodynamics in Non-Ferrous Metallurgy. J.I. Gerassimov et al.. Metallurgical Publishing House, Moscow, Vols. I and II, 1960; Vol. III, 1963; Vol. IV, 1966, Vol. I, 231 pp.; Vol. II, 231 pp.; Vol. III, 283 pp.; Vol. IV, 428 pp.

The titles of each volume are as follows: Vol. I, *Theoretical Introduction, Thermodynamic Properties of Important Gases, Thermodynamics of Zinc and Its Important Compounds*; Vol. II, *Thermodynamics of Copper, Lead, Zinc, Silver and Their Important Compounds*; Vol. III, *Thermodynamics of Tungsten, Molybdenum, Titanium, Zirconium, Niobium, Tantalum, and Their Important Compounds* (English translation available as NASA-TT-F-285 or NTIS-TT-65-50111); Vol. IV, *Thermodynamics of Aluminum, Antimony, Magnesium, Nickel, Bismuth, Cadmium, and Their Important Compounds.*

315. Chemical Thermodynamics of Organic Compounds. Daniel Richard Stull.. John Wiley and Sons, Inc, 1969, 865 pp. LC: 68-9250. ISBN: 0-471-83490-4. Cost: $55.00.

316. Chemist's Companion: A Handbook of Practical Data, Techniques, and References. Arnold J. Gordon; Richard A. Ford. John Wiley and Sons, 1973, 560 pp. LC: 72-6660. ISBN: 0-471-31590-7. Cost: $27.50.

317. Chemistry Handbook. Donald B. Summers. Willard Grant Press, 2d ed., 1975. ISBN: 0-87150-715-3.

Chemists' Yearbook. *See* **Handbook of Chemical Data.**

318. Choline and Acetylcholine: Handbook of Chemical Assay Methods. Israel Hanin. Raven Press, 1974. LC: 73-79289. ISBN: 0-911216-51-0. Cost: $17.50.

319. CHRIS/HACS Chemical Property File. E. Atkinson. Arthur D Little Inc (available from NTIS), 1976, 905 pp. Accession: N77-24218; AD-A034607. Tech Report: USCG-D124-76. Cost: $3.50 microfiche; contact NTIS for hard copy price.

This file contains the physical and chemical properties of over 900 chemical substances, and numerous properties are recorded for each chemical.

320. CHRIS: A Condensed Guide to Chemical Hazards. U.S. Coast Guard. NTIS, Jan. 1974, 4 Vols., 459 pp. Accession: AD/A-002 390. Tech Report: CG-446-1. Cost: $2.25 microfiche; $12.00 paper copy.

This 4-volume handbook has been compiled by the Chemical Hazards Response Information System (CHRIS). The handbook is primarily about the water transport of hazardous chemicals and what to do in the event of an accident. These 4 volumes contain chemical data and hazard-assessment methods that would be helpful in achieving improved levels of safety in the bulk shipment of dangerous chemicals.

321. Chromatography: A Laboratory Handbook of Chromatographic and Electrophoretic Methods. Erich Heftman. Van Nostrand Reinhold, 3d ed., 1975, 1,002 pp. LC: 75-5804. ISBN: 0-442-23280-2. Cost: $47.50.

322. Circuit Design Idea Handbook. Bill Furlow. Cahners Books, 1974, 207 pp. LC: 73-76440. ISBN: 0-8436-0205-8. Cost: $19.50.

The *Circuit Design Idea Handbook* is a collection of 287 electronic circuits ranging in subject from digital and pulse circuits to power supply circuits. Each circuit has been selected for inclusion in this handbook on the basis that it has already received recognition in *EDN* and *EEE* magazines. Some examples of the circuits and schematics that have been included are a double-balanced diode mixer, a track-and-hold amplifier, a reversible linear counter, a phase shifter for phase-locked loops, a stable low-distortion bridge oscillator, and a sequential bipolar multivibrator. An excellent cross-referenced index should help the user find the circuits s/he desires.

323. Citizens Band Radio Handbook (title varies). David E. Hicks. Sams, 5th ed., 1976. LC: 75-46225. ISBN: 0-672-21336-2. Cost: $5.95.

324. Civil Engineer's Reference Book. L.S. Blake. Newnes-Butterworths (Transatlantic), 3d ed., 1975, approx. 1,800 pp., various paging. LC: 75-309179. ISBN: 0-408-70475-6. Date of First Edition: 1951. Cost: $85.00.

This reference work attempts to cover all branches of civil engineering, and has been written by experienced practitioners in their respective fields of engineering. Emphasis has been placed on the theory and practice of the basic fields of civil engineering as well as the more specialized areas, such as tunnelling, demolition, aluminum design, and coastal and maritime engineering. This handbook is an excellent source of information for quick look-up data on such subjects as surveying, soil and rock mechanics, hydrodynamics, reinforced and prestressed concrete, foundations, bridge structures, highways, airports, railway structures, and sewage disposal. Selective lists of reference and bibliographies are provided as sources of information for further study. For related works, please refer to *Standard Handbook for*

Civil Engineers; *Handbook of Environmental Civil Engineering*, and *Civil Engineering Handbook*.

325. Civil Engineering Handbook. Leonard Church Urquhart. McGraw-Hill, 4th ed., 1959 (copyright renewed 1962), various paging. LC: 58-11195. ISBN: 07-066148-0. Date of First Edition: 1934. Cost: $44.50.

As is common with McGraw-Hill handbooks, each section is written by an authority who has expertise in his/her respective specialization. This handbook includes a wealth of civil engineering information, from airport design to water treatment. A considerable amount of design data are included, particularly as it relates to loads, as in design loads on foundations, reinforced concrete design for flexural loading, oblique loading on beams, and loads on framed structures. Information is provided for a number of subjects, including such areas as hydraulics, properties of soils, sewage disposal, welded girders, surveying, photogrammetric surveying, the interstate highway system, and continuous beams and frames. For related works, please refer to *Civil Engineer's Reference Book*; *Standard Handbook for Civil Engineers*, and *Handbook of Environmental Civil Engineering*.

326. Classification, Standards of Accuracy, and General Specifications of Geodetic Control Surveys. Federal Geodetic Control Committee. National Geodetic Survey, U.S. Dept. of Commerce, 1974, 12 pp.

327. Cleaning, Hygiene and Maintenance Handbook. Terence McLaughlin. Prentice-Hall, 1973, 240 pp. LC: 72-14158. ISBN: 0-13-136608-8. Cost: $18.95.

328. Climates of the States. James A. Ruffner; Frank E. Bair. Gale Research Co, 2d ed., 1980, 2 Vols., 1,175 pp. LC: 76-11672. ISBN: 0-8103-1036-8. Cost: $95.00.

This 2-volume set consists of the following titles: Part 1, *Climates of the States*; Part 2, *Federal and State Public Services in Climate and Weather*. These volumes include many tables, graphs, and maps on such topics as snowfall, sunshine, temperatures and fog.

329. Climatography of United States: Climate of the States. U.S. Environmental Data Service. U.S. Government Printing Office, 1959-. SuDoc: C55.221:Nos. Earlier volumes in C30.71/3:no..

330. Climatological Data (State Sections). Environmental Data Service. U.S. Government Printing Office, Vol.1, 1914-, monthly with annual summary. SuDoc: C55.214/2-C55.214/47.

331. Climatological Data; National Summary. Environmental Data Service. U.S. Government Printing Office, Vol.1, 1950, monthly with annual summary. SuDoc: C55.214:v. and no..

332. Clinical Dosimetry. National Bureau of Standards. NTIS, Aug. 9, 1963. Accession: PB-248-564. Cost: $4.50. Series: NBS Handbook 87.

Supersedes parts of Handbook 78; Handbooks 84 through 89 extend and largely replace Handbook 78.

333. Clinical Drug Index. Nathalie Fedynskyi. C. V. Mosby, approx. 1977, 600 cards. ISBN: 0-8016-1554-2. Cost: Approx. $35.00.

334. Clinical Echocardiography: A Handbook for Analysis of the Echocardiogram. Jacob I. Haft; Michael Horowitz. Futura Publishers, 1977. ISBN: 0-87993-079-9.

335. Clinical Handbook of Psychopharmacology. Alberto Di Mascio; Richard I. Shader (eds). Science House, 1970, 395 pp. LC: 75-118581.

336. Clinical Pharmacy Handbook for Patient Counseling. Brandt Rowles; Scott E. Apelgren. Drug Intelligence Publications, 1975. LC: 75-17157. ISBN: 0-914768-22-0. Cost: $12.00.

337. Clinical Pharmacy Handbook. Hugh K. Kabat. Lea and Febiger, 1969. LC: 74-78539. ISBN: 0-8121-0138-3. Cost: $6.50.

338. Clinical Toxicology of Commercial Products. Lab Safety Supply Company. Lab Safety Supply Company, 4th ed., 1976, 1,783 pp. Cost: $74.25.

This is a monumental work containing 17,500 commercial chemical products that are potentially toxic in nature. Toxicology information is provided along with specific diagnosis and treatment procedures.

339. Closed-Circuit Television Handbook. Leon A. Wortman. Sams, 3d ed., 1974. LC: 71-92465. ISBN: 0-672-21097-5. Cost: $7.95.

340. Clothoid Design and Setting Out Manual. A. Krenz; H. Osterloh. Bauverlag, 12th ed., 1975, 492 pp. ISBN: 3-7625-0576-4. Cost: $13.75 (394 tables).

341. Coblentz Society Evaluated Infrared Reference Spectra. Coblentz Society. Coblentz Society, 1969-74, Vol. 6-11; 1,000 Spectra. Cost: $295.00 per Vol.; cumulative Coblentz Indices $50.00.

Edited and published by the Coblentz Society, sponsored by the Joint Committee on Atomic and Molecular Physical Data (1969-1974). Available from Sadtler Research Laboratories, Philadelphia, Pa., 19104 (Microfilm available from the Coblentz Society.).

342. CODASYL Data Description Language-Journal of Development. National Bureau of Standards. U.S. Government Printing Office, June, 1973. SuDoc: C13.11:113. Cost: $2.15. Series: NBS Handbook 113.

343. Code for Protection Against Lightning. National Bureau of Standards. National Bureau of Standards. Cost: $4.25.

A United States of American Standards Institute Code for Protection Against Lightning (NFPA-78-1975) is available from the American National Standards Institute, 1430 Broadway, New York, NY as C5.1-1975.

344. Colour Index. The Society of Dyers and Colourists (England), 3d ed., 1971, 5 Vols.

345. Commercial Explosives and Initiators—A Handbook. B.D. Rossi. NTIS (translated from the Russian for the Army Foreign Science Technology Center, Charlottesville, VA), 1973, 218 pp. Accession: AD-786785. Tech Report: FSTC-HT-23-587-73.

346. Communication Systems Handbook. Federal Electric Corp (available from NTIS), 1972, 155 pp. Accession: AD-755917. Tech Report: SAMTEC-TR-72-2.

347. Communications System Engineering Handbook. D.H. Hamsher. McGraw-Hill, 1967, 600 pp. LC: 64-8973. ISBN: 0-07-025960-7. Cost: $52.50.

348. Communicative Disorders: A Handbook for Prevention and Early Intervention. Curtis Weiss; Herold Lillywhite. C.V. Mosby, 1976, 500 pp. ISBN: 0-8016-5386-X. Cost: $6.95 paper copy.

349. Compendium of Ab-Initio Calculations of Molecular Energies and Properties. M. Krauss (National Bureau of Standards). NTIS, 1967. Accession: AD-665245. Cost: $6.75. Series: NBS Tech. Note 438.

350. Compendium of Analytical Nomenclature. H.M.N.H. Irving; T.S. West; H. Freiser. Pergamon Press Inc, 1978, 240 pp. ISBN: 0-08-022008-8 hardback; 0-08-22347-8 flexicover. Cost: $25.00 hardback; $15.00 flexicover.

This book is commonly referred to as *The Orange Book*. Provides information of recommended terminology to be used with such subjects as precision balances, scales, contamination phenomena, automatic analysis, thermal analysis, mass spectrometry, and titrimetric analysis.

351. Compendium of Corn Diseases. American Phytopathological Society. American Phytopathological Society, 2d ed., 1980, 72 pp. ISBN: 0-89054-021-7. Cost: $7.00.

This compendium is designed for the plant pathologist and includes over 107 photographs and 97 drawings of examples of corn diseases.

352. Compendium of Cotton Diseases. American Phytopathological Society. American Phytopathological Society, 1981, 87 pp. ISBN: 0-89054-030-4.

353. Compendium of Lake and Reservoir Data Collected by the National Eutrophication Survey in Eastern Northcentral and South-eastern United States. Environmental Protection Agency. National Eutrophication Survey (available from NTIS), 1978, 270 pp. Accession: N79-21666; PB-289820. Tech Report: Working paper 475. Cost: $3.50 microform; $20.00 hard copy.

Morphometric, limnological, and nutrient loading data are summarized for 153 lakes studied between 1973 and 1974 by EPA's National Eutrophication Survey.

354. Compendium of Organic Synthetic Methods. Ian T. Harrison; Shuyan Harrison. Wiley-Interscience, 1971-80, 4 Vols. LC: 71-162800. ISBN: Vol.1, 0-471-35550-X; Vol.2, 0-471-35551-8; Vol.3, 0-471-36752-4 Vol.4, 0-471-04923-9. Cost: Vol.1, $24.50; Vol.2, $19.95; Vol.3, $19.95; Vol.4, $22.50.

Includes over 5,000 examples of preparations of different organic compounds. Examples are provided for such compound groups as alcohols, phenols, alkyls, esters, ethers, olefins, ketones, acetylenes, and halides.

355. Compendium of Registration Laws for the Design Professions. National Council of Engineering Examiners; National Council of Architectural Registration Boards, 1978, loose-leaf for update. LC: 78-61486.

This compendium is designed specifically for the engineer, architect, land surveyor, landscape architect, and planner. Registration laws are presented in the actual text and listed geographically based upon the jurisdiction. Also available from the National Council of Engineering Examiners, PO Box 5000, Seneca, SC 29678.

356. Compendium of Shock Wave Data. M. Van Thiel; J. Shaner; E. Salinas. University of California, Berkeley, Lawrence Livermore Laboratory (available from NTIS), 1977. Tech Report: UCRL-50108. Cost: $3.50 microform.

The titles of the individual volumes of this compendium are as follows: Vol.1, Section A1 *Elements,* 663 pp. (N68-25705); Vol.2, Section A2 *Inorganic Compounds,* Section B *Hydrocarbons,* 353 pp. (N79-70056); Vol.3, Section C *Organic Compounds Excluding Hydrocarbons,* Section D *Mixtures,* Section E *Mixtures and Solutions Without Chemical Characterization, Compendium Index,* 535 pp. This 3-volume set includes thermodynamic data obtained by shock-wave techniques. Inorganic compounds and hydrocarbons are listed as well as alloys and various materials.

357. Compendium of Soil Fungi. K.H. Domsch. Academic Press. LC: 80-41403. ISBN: Vol.1, 0-1222-0401-8; Vol.2, 0-1222-0402-8. Cost: $96.50.

358. Compendium of Soybean Diseases. American Phytopathological Society, 1975, 69 pp. ISBN: 0-89054-022-5. Cost: $7.00.

359. Compendium of Structural Aids. Louis A. Hill. Arizona State University, 1975, 150 pp. ISBN: 700-22462-3.

This compendium includes information on the Uniform Building Code, timber, steel construction, and the American Concrete Institute Code, as well as information on reinforced concrete and various other extracts from major codes.

360. Compendium of the Properties of Materials at Low Temperatures. V.J. Johnson (ed). NTIS, 1960-61, 4 Vols. Tech Report: WADD 60-56, Parts I, II, III, IV.

The volumes are entitled as follows: Phase I, Part I, *Properties of Fluids* (PB-171-618); Phase I, Part II, *Properties of Solids* (PB-171-619); Phase II, Part III, *Bibliography of References* (PB-171-620); Phase II, Part IV, "Untitled" (501 pp. of data), R. B. Stewart and V. J. Johnson, eds. (AD-272-769).

361. Compendium of Wheat Diseases. American Phytopathological Society, 1977. ISBN: 0-89054-023-3. Cost: $10.00.

362. Compilation and Analysis of Helicopter Handling Qualities Data: Volume 2: Data Analysis. Robert K. Heffley. Systems Technology, Inc (available from NTIS), 1979, 176 pp. Accession: N79-31222. Tech Report: NASA CR-3145 (Systems Technology—TR-1087-2-vol.2). Cost: $3.50 microform; $15.00 hard copy.

363. Compilation and Index of Trade Names, Specifications, and Producers of Stainless Alloys and Superalloys. American Society for Testing and Materials, 1972, 57 pp. ISBN: 0-8031-0138-4. Cost: $5.25. Series: *ASTM Data Series,* 45A; ASTM Publication Code No.: 05-045010-02.

Over 300 alloys are listed in this compilation which includes compositions, specifications, trade names, and manufacturers of stainless steels and superalloys.

364. Compilation of Available High-Temperature Creep Characteristics of Metals and Alloys. Creep Data Section of Joint Research Committee on Effect of Temperature on the Properties of Metals. American Society for Testing Materials and American Society of Mechanical Engineers, 1938, 848 pp.

This compilation provides information on wrought steels, ferrous alloys, cast steels, and nonferrous materials. The steels are described with individual logarithmic charts of stress and corresponding creep rate for each material at each temperature. Such information as heat treatment, grain size, and chemical composition is also provided.

365. Compilation of Electron Collision Cross Section Data for Modeling Gas Discharge Lasers. L.J. Kieffer. NTIS, 1969. Accession: COM 74-11661. Cost: $6.75. Series: JILA Report No. 13.

366. Compilation of Level 1 Environmental Assessment Data: TLSP Final Report. N.H. Gaskins; F.W. Sexton. Research Triangle Institute (available from NTIS), 1978, 504 pp. Accession: N79-16439; PB-286924. Tech Report: EPA-600/2-78-211. Cost: $3.50 microform; $35.00 hard copy.

This compilation is organized by industrial type; for example, chemically active fluidized bed combustor, coal-fired boiler, coal-fired power plant, coke production, electric arc furnace, fluidized bed combustor, ocean incinerator, oil burner, and the textile industries.

367. Compilation of Low Energy Electron Collision Cross Section Data, Part I: *Ionization; Dissociation; Vibrational Excitation.* L.J. Kieffer. NTIS, 1969. Accession: PB-189127. Cost: $5.50. Series: JILA Report No. 6.

368. Compilation of Low Energy Electron Collision Cross Section Data, Part II: *Line and Level Excitation.* L.J. Kieffer. NTIS, 1969. Accession: AD-696467. Cost: $6.75. Series: JILA Report No. 7.

369. Compilation of Mass Spectral Data. A. Cornu; R. Massot. Heyden and Sons Inc, 2d ed., 1975, 2 Vols. ISBN: Vol.1, 0-085501-086-X; Vol.2, 0-855-1-087-4; 2-Vol. set, 0-085501-088-6. Cost: Vol.1, $57.00; Vol.2, $105.50; 2-Vol. set, $146.00.

This comprehensive collection of mass spectral data covers 10,000 compounds in 2 vols. Vol.1 provides complete data listed in order of increasing molecular weight. Vol.2 provides a listing by fragment ion values.

370. Compilation of NMR Data. W. Brugel. Heyden and Sons Inc. ISBN: 0-85501-170-X.

A compendium of nuclear magnetic resonance, couplings, shifts, and correlations of over 7,000 chemical compounds.

371. Compilation of Odor and Taste Threshold Values Data. American Society of Testing and Materials, 2d ed., 1978, 497 pp. LC: 77-83047. ISBN: 0-8031-0087-6. Cost: $27.50. Series: *ASTM Data Series*, 48A; ASTM Publication Code No.: 05-048010-36.

This compilation contains odor threshold values and taste threshold values on 2,626 compounds. Literature sources are provided.

372. Compilation of Physical and Chemical Properties of Materials and Streams Encountered in the Chemical Processing Department. R.F. Fleming. Hanford Atomic Products Operation, Richland, Washington, 1958, 135 pp. Accession: N69-75858. Tech Report: HW-57386.

373. Compilation of Selected Thermodynamic and Transport Properties of Binary Electrolytes in Aqueous Solution. T.W. Chapman; J. Newman. University of California, Berkeley, Lawrence Radiation Laboratory (available from NTIS), 1968, 297 pp. Accession: N68-35921. Tech Report: UCRL-17767.

374. Complete Concrete, Masonry, and Brick Handbook. J.T. Adams. McGraw-Hill, 1979, 1,130 pp. LC: 77-14003. ISBN: 0-668-04340-7. Cost: $19.95.

This is a comprehensive handbook on working with concrete, masonry, brick, stucco, and tile.

375. Complete Guide to Reading Schematic Diagrams. John Douglas-Young. Prentice-Hall (a Parker Publication), 2d ed., 1979, 288 pp. ISBN: 0-13-160416-3 (4). Cost: $14.95.

Contains information on interpreting schematic diagrams and troubleshooting modern equipment. Information is provided for logic circuits, amplifiers, radios, and television.

376. Complete Handbook for Professional Ambulance Personnel. S. C. Morris. Year Book Medical Publishers, 1970, 288 pp. ISBN: 0-8151-5948-X. Cost: $12.50.

377. Complete Handbook of Centrifugal Casting. Philip Romanoff. TAB Books, 1981, 256 pp. LC: 80-20591. ISBN: 0-8306-9919-8. Cost: $15.95.

378. Complete Handbook of Electrical and Housing Wiring. S. Blackwell Duncan. TAB Books, 1977, 473 pp. LC: 77-1770. ISBN: 0-8306-7913-8. Cost: $10.95.

379. Complete Handbook of Magnetic Recording. Finn Jorgensen. TAB Books, 1980, 448 pp. LC: 80-14359. ISBN: 0-8306-1059-6. Cost: $15.95.

Contains information on magnetic heads, magnetic tapes and disks, recording, AC-bias, and FM and PCM recording.

380. Complete Handbook of Maintenance Management. John E. Heintzelman. Prentice-Hall, 1976, 335 pp. LC: 76-6490. ISBN: 0-13-160994-7. Cost: $24.95.

This handbook includes valuable information on management principles that can be applied to maintenance operations. The main emphasis of this book is how to apply these principles in order to reduce costs, regardless of what type of facility is being maintained. For example, this handbook discusses planning, organizing, staffing, controlling, and directing maintenance operations. It is well illustrated, as indicated by a 3-page listing of forms, figures, and checklists in the table of contents. In addition, such information as janitorial contracts, recruiting, training, and handling repair work receives treatment.

381. Complete Handbook of Poultry-Keeping. Stuart Banks. Van Nostrand Reinhold, 1979, 216 pp. LC: 79-

14305. ISBN: 0-442-23382-5; 0-442-23383-3 paper. Cost: $14.95; $8.95 paper.

382. Complete Handbook of Power Tools. George Drake. Reston, 1975, 415 pp. LC: 75-14286. ISBN: 0-87909-150-9. Cost: $17.95.

383. Complete Handbook of Radio Receivers. Joseph J. Carr. TAB Books, 1980, 319 pp. LC: 80-14491. ISBN: 0-8306-1182-7. Cost: $7.95.

Contains information on electronic test equipment for radio receivers, safety precautions, and circuit diagrams.

384. Complete Handbook of Radio Transmitters. Joseph J. Carr. TAB Books, 1980, 350 pp. LC: 80-14458. ISBN: 0-8306-1224-6. Cost: $8.95.

385. Complete Handbook of Robotics. Edward L. Safford Jr. TAB Books, 1978, 358 pp. LC: 78-10692. ISBN: 0-8306-9872-8. Cost: $12.95.

Presents information on power sources, robot mobility, servomechanisms, radio control, and interfacing with computers, as well as numerous diagrams, schematics, and photographs.

386. Complete Handbook of Sand Casting. C.W. Ammen. TAB Books, 1979, 238 pp. LC: 78-26495. ISBN: 0-8306-9841-8. Cost: $9.95.

Contains information on sand foundry, molding sand, and various patterns that can be used in sand casting. Designing of sand castings and usage of nonferrous melting equipment is described.

387. Complete Handbook of Slow-Scan TV. Dave Ingram. TAB Books, 1977, 304 pp. LC: 77-3735. ISBN: 0-8306-7859-X; 0-8306-6859-4 paper. Cost: $14.95; $9.95 paper.

A description of slow-scan TV (SSTV) is presented. Information is also provided on satellite communications, SSTV stations, and SSTV transmitters and receivers.

388. Compliance Manual: Methods for Meeting OSM Requirements. Skelly and Loy. McGraw-Hill, 1979, 600 pp. ISBN: 07-606641-X. Cost: $90.00.

Regulations and compliance information is presented for coal mine officials and those involved in surface mining. Such topics as topsoil handling, erosion, sediment control, stream buffer zones, environmental protection, blasting, soil disposal, coal processing waste, revegetation and reclamation, road construction, auger mining, and steep slope mining are discussed.

389. Composition and Properties of Oil Well Drilling Fluids. Walter F. Rogers. Gulf Publishing Co, 4th ed., 1979, 600 pp. ISBN: 0-87201-129-1. Cost: $50.00.

This publication, revised by George Gray and H.C.H. Darley, discusses clay mineralogy and colloid chemistry of drilling fluids along with rheology of drilling fluids and various types of drilling problems. Filtration properties, surface chemistry, hole stability, and packer fluids are also presented.

Composition of Foods. *See* **Handbook of the Nutritional Contents of Food.**

390. Comprehensive Analytical Chemistry. C.L. Wilson; D.W. Wilson. Elsevier, 1959, multivolume set. ISBN: Vol. 1, 0-444-40647-6. Cost: Consult publisher for subscription price.

This is a multivolume set which presents methods for analytical chemistry. Included are such methods as liquid chromatography, gas chromatography, distillation, ion exchangers, inorganic titrimetric analysis, paper chromatography, thin-layer chromatography, and radiochemical methods of analysis. In addition, such methods as coulometric analysis, nuclear magnetic resonance, X-ray spectrometry, vacuum fusion analysis, and emission spectroscopy are discussed.

391. Comprehensive Index of API 44-TRC Selected Data on Thermodynamics and Spectroscopy. Bruno J. Zwolinski; R.C. Wilhoit; C.O. Reed Jr. Thermodynamics Research Center, 2d ed., 1974, 732. LC: 73-86988. Cost: $30.00.

392. Comprehensive Inorganic Chemistry. J.C. Bailar et al. Pergamon Press (distributed in U.S. by Compendium Pubs), 1973, 5 Vols., 6,224 pp. LC: 77-189736. ISBN: 0-08-017275-X. Cost: 155.00 per Vol.; $900.00 set.

Arranged by element as listed in the periodic table. Covers preparation, nuclear properties, physical and chemical properties, and use.

393. Comprehensive Organic Chemistry. Derek Barton. Pergamon Press, 1979, approx. 8,034 pp. ISBN: 0-080213-19-7. Cost: $1,375.00.

Covers the use of over 2,500 reagents and approximately 1,000 reactors. The subtitle of this multivolume set is *The Synthesis and Multivolume Actions of Organic Compounds*. Includes over 20,000 literature references citing original source material.

394. Comprehensive Virology. Heinz Fraenkel-Conrat; Robert R. Wagner. Plenum, 1974-, multivolume set, currently 16 Vols. available with more planned. LC: 77-84583. ISBN: Vol.1, 0-306-35141-2. Cost: Each volume varies in price from $19.50 to $49.50.

This is a monumental treatise that may eventually have as many as 25 volumes. The volumes currently available are entitled as follows: Vol.1, *Descriptive Catalogue of Viruses*; Vol.2, *Reproduction: Small and Intermediate RNA Viruses*; Vol.3, *Reproduction: DNA Animal Viruses*; Vol.4, *Reproduction: Large RNA Viruses*; Vol.5, *Structure and Assembly: Virions, Pseudovirions, and Intraviral Nucleic Acids*; Vol.6, *Structure and Assembly of Small RNA Viruses*; Vol.7, *Reproduction: Bacterial DNA Viruses*; Vol.8, *Regulation and Genetics: Bacterial DNA Viruses*; Vol.9, *Regulation and Genetics: Genetics of Animal Viruses*; Vol.10, *Regulation and Genetics; Viral Gene Expression and Integration*; Vol.11, *Regulation and Genetics: Plant Viruses*; Vol.12, *Newly Characterized Protist and Invertebrate Viruses*; Vol.13, *Structure and Assembly: Primary, Secondary, Tertiary, Quarternary Structures*; Vol.14, *Newly Characterized Vertebrate Viruses*; Vol.15, *Virus-Host Interactions-Immunity to Viruses*; Vol.16, *Virus Host Interactions*.

395. Compressed Air and Gas Data. Charles W. Gibbs. Ingersoll-Rand Co, 2d ed., 1971, various paging. Date of First Edition: 1919 as *Compressed Air Data*. Cost: $12.00.

The primary purpose of this handbook is to provide information for those engineers and scientists who are interested in the use, transmission and compression of air and gas. It includes information on many different types of compressors such as dynamic compressors, thermal compressors, vane-type rotary compressors, reciprocating compressors, helical-lobe rotary compressors, and portable and booster compressors. In addition, this manual also provides information on air lift pumping, pneumatic tools, pressure losses in gas piping, vacuum equipment, gas coolers, and many tables of reference data on the compression of gas and air.

396. Compressed Air and Gas Handbook. John P. Rollins. Compressed Air and Gas Institute, 4th ed., 1973, approx. 700 pp. various paging. LC: 72-93043. Date of First Edition: 1947.

This handbook is an excellent source of information on compressors and the use of compressed air and gas in manufacturing and industry. Included are discussions on such compressors as positive displacement compressors, single- and double-acting reciprocating compressors, helical screw-type compressors, liquid piston-type rotary compressors, portable air compressors, and dynamic compressors. Information on pneumatic tools such as screwdrivers and nut-setters, saws and files, backfill tampers, paving breakers, rock drills, and sump pumps is also included. In addition, there is a section in the appendix on the sound measurement from pneumatic equipment.

397. Compressed Gas Handbook. John S. Kunkle; Samuel D. Wilson; Richard A. Cota (National Aeronautics and Space Administration). U.S. Government Printing Office, 1969, 560 pp. LC: 68-62085. Accession: N69-26987. Tech Report: NASA SP-3045. SuDoc: NAS1.21 no. 3045. Cost: $4.75.

This handbook, prepared by the National Aeronautics and Space Administration, on the theory and application of high-pressure compressible flow systems, provides thermodynamic data for hydrogen, oxygen, nitrogen, helium and air. In addition, such information as properties of gases, fluid flow equations, adiabatic and isentropic flows, compressible flow, nonadiabatic flow in pipes is also included in this handbook. A good mixture of theory and application makes this handbook an excellent ready-reference source.

398. Compressibility Data for Helium at $0°$ C and Pressure to 800 Atmospheres. R.E. Barieau; T.C. Briggs; B.J. Dalton. Helium Research Center, 1969, 58 pp. Accession: N69-34873. Tech Report: BM-R1-7287.

399. Compressibility Data for Helium Over the Temperature Range $-5°$ C to $80°$ C and at Pressures to 800 Atmospheres. T.C. Briggs. Bureau of Mines, 1970, 42 pp. Accession: N70-20678. Tech Report: BM-R1-7352.

Contains tabulated compressibility data for helium and mixtures containing helium.

400. Computation of Molecular Formulas for Mass Spectrometry. Joshua Lederberg. Holden-Day Inc, 1964, 78 pp. ISBN: 0-8162-4981-4. Cost: $13.50.

This is a compilation of computer-generated tables that will be useful in mass spectrometry analysis. Precise molecular weights are provided for various formulas of organic molecules. Includes tables of mass values, decimal residues of molecular weights, and satellite peaks derived from the naturally occurring isotopes.

401. Computed Tomography of the Brain; Atlas of Normal Anatomy. G. Salomon; Y.P. Huang. Springer (available from Otto Harrassowitz), 1978. ISBN: 3-540-08825-3.

402. Computer Aided Data Book of Vapor-Liquid Equilibria. M. Hirata; S. Ohe; K. Nagahama. Elsevier, 1975, 944 pp. ISBN: 0-444-99855-1. Cost: $107.25.

403. Computer Compilation of Molecular Weights and Percentage Compositions for Organic Compounds. M.J.S. Dewar; R. Jones. Pergamon Press Inc, 1969, 508 pp. LC: 69-19974. ISBN: 0-08-01207-X. Cost: $68.00.

Provides tables of organic compounds with corresponding molecular weights and percentage compositions.

404. Computer Handbook. H.D. Huskey; G.A. Korn. McGraw-Hill, 1962, 1,288 pp. LC: 60-15286. ISBN: 0-07-031477-2. Cost: $42.65.

405. Computer Program for Obtaining Thermodynamic and Transport Properties of Air and Products of Combustion of ASTM-A-1 Fuel and Air. S. Hippensteele; R.S. Colladay. NASA (available from NTIS), 1978, 56 pp. Accession: N78-20351. Tech Report: NASA-TP-1160; E-9371. Cost: $3.50 microform; $8.00 hard copy.

The program calculates temperature, enthalpy, viscosity, molecular weight, specific heat, thermal conductivity, Prandtl number, and entropy.

406. Computer Programming Handbook. Peter Stark. TAB Books, 1975. LC: 75-24688. ISBN: 0-8306-5752-5. Cost: $12.95.

407. Computer Security Handbook. Douglas Hoyt (ed) (Computer Security Research Group). Macmillan, 1973. LC: 73-6207. Cost: $25.00.

408. Computer Technician's Handbook. Brice Ward. TAB Books, 1971. LC: 74-147384. ISBN: 0-8306-1554-7. Cost: $14.95.

409. Computerized Axial Tomography; An Anatomic Atlas of Serial Sections of the Human Body. J. Gambarelli et al. Springer (available from Otto Harrassowitz), 1977, 286 pp. ISBN: 3-540-07961-0. Cost: $105.60.

410. Computing Principles and Techniques. Bruce Vickery. Heyden (also Adam Hilger Ltd.), 1979, 182 pp. LC: 80-462422. ISBN: 0-85274-505-2. Cost: $24.50. Series: *Medical Physics Handbooks*, Vol. 2.

This book is intended for the medical physicist and discusses computing principles, programing, and the role of computers in various fields such as radiotherapy and nuclear medicine.

411. Concise Atlas of Anatomy: A Practical Handbook. C. V. Mosby, available in 1977 or 1978, 72 pp. ISBN: 0-8016-3522-5. Cost: Approx. $7.00.

412. Concise Automative Handbook. John A. Miller (trans) (Wright Patterson Air Force Base). NTIS, 6th ed., 1974, 565 pp. Accession: AD-776894. Tech Report: FTD-HT-23-504-74.

413. Concise Handbook of Community Psychiatry and Community Mental Health. Leopold Bellak. Grune & Stratton, 1974. ISBN: 0-8089-0833-2. Cost: $14.50.

414. Concise Handbook of Respiratory Diseases. S. Farzan; Doris Hunsinger; Mary Phillips. Reston (Prentice-Hall), 1978, Vol.1, 304 pp. ISBN: 0-87909-180-0. Cost: $15.95.

This handbook covers a wide range of pulmonary diseases.

415. Concise Handbook on Aging. Robert N. Butler; Myrna I. Lewis. C. V. Mosby, 1978, 300 pp. ISBN: 0-8016-0922-4. Cost: $9.50 paper copy.

416. Concise Handbook on Surveys for Grid Construction. N.N. Severyanov (Foreign Technology Division, Wright-Patterson Air Force Base). NTIS, translation of Russian monograph (selected chapters) 1972, Jan. 24, 1975, 143 pp. Accession: AD/A-007 173. Tech Report: FTD-HC-23-1542-74. Date of First Edition: 1972. Cost: $6.00 paper; $2.25 microfiche.

This handbook discusses such things as geological engineering surveying, aerial surveying and mapping, and general field surveying methods and problems.

417. Concise World Atlas of Geology and Mineral Deposits. Duncan R. Derry. Mining Journal Books, London (also John Wiley and Sons, Inc), 1980, 110 pp. LC: 80-675233. ISBN: 0-470-26996-0. Cost: $61.95.

Provides a geological map of the world in 10 sheets.

418. Concrete Bridge Designer's Manual. E. Pennells. Scholium International Inc, 1978, 164 pp. ISBN: 0-7210-1083-0. Cost: $37.50.

This book deals with bridge decks, loadings, reinforced and prestressed concrete, as well as structural analysis of bridge decks.

419. Concrete Construction Handbook. Joseph J. Waddell. McGraw-Hill, 2d ed., 1974, 960 pp. LC: 73-17385. ISBN: 0-07-067654-2. Date of First Edition: 1968. Cost: $45.00.

This comprehensive reference source covers, in great detail, the field of concrete construction engineering including such areas as properties of concrete, finishing and curing, precast and prestressed concrete, repairing and mixing concrete. Each section is written by a recognized expert in his/her respective field, and the latest information is provided in this constantly changing field of technology. An excellent index is provided to help the user find such information as joints in concrete, mixers, grouting of concrete, vacuum processing of concrete, and construction of ornamental objects from concrete. For related works, please refer to such titles as *Concrete Manual; Prestresed Precast Concrete Design Handbook for Standard Products; Prestressed Concrete Designer's Handbook*, the general civil engineering handbooks, and *Handbook of Concrete Engineering*.

420. Concrete Engineering Handbook. W.S. Lalonde; M. Janes. McGraw-Hill, 1962, 1,172 pp. LC: 59-13934. ISBN: 0-07-036089-8. Cost: $48.75.

421. Concrete Estimating Handbook. Michael F. Kenny. Van Nostrand Reinhold, 1975, 170 pp. LC: 74-76075. ISBN: 0-442-12160-1. Cost: $19.95.

Includes measurement and re-use of formwork, reinforcing steel, comparison charts for welded wire mesh, and mensuration guidelines.

422. Concrete Inspection Manual. Joseph J. Waddell. International Conference of Building Officials, 1976, 332 pp. LC: 76-47961.

Covers such topics as durability, strength of concrete, fresh concrete, portland cement, aggregates, formwork, slabs, pavements, curing, and steel reinforcement. Also discusses such topics as the effects of hot and cold weather, precast concrete, prestressed concrete, waterproofing, and dampproofing.

423. Concrete Manual; A Water Resources Technical Publication. U.S. Bureau of Reclamation. U.S. Government Printing Office, 8th ed., 1975, 627 pp. SuDoc: I27.19/2:C74/974. Date of First Edition: 1936. Cost: $11.40.

This excellent handbook includes such information as the properties of concrete, among which are durability, strength, elasticity, thermal properties, creep, water tightness, and workability. Designed primarily for usage by the Bureau of Reclamation in concrete construction projects, this handbook provides such information as quality and gradation of aggregates, concrete mixes, transporting concrete, precast concrete pipe, repairing concrete, and vacuum-processed concrete. There are innumerable illustrations, tables, and figures, as well as information on various tests that have been made on concrete materials. For related works, please refer to such titles as *Concrete Construction Handbook*; *Prestressed Concrete Designer's Handbook*; *Prestressed Precast Concrete Design Handbook for Standard Products*; the general civil engineering handbooks, and *Handbook of Concrete Engineering*.

424. Concrete Masonry Handbook for Architects, Engineers, Builders. Frank A. Randall Jr; William C. Panarese. Portland Cement Association, 4th ed., 1976. LC: 75-44908.

This handbook has an extensive listing of tables on such topics as physical properties, portion specifications, moisture content, and load specifications of concrete. A bibliography is provided on page 185 and includes various references, as well as recommended practices, tests and codes. Data are provided on such subjects as surface texture of concrete masonry, fire resistance, sound absorption in walls, joints, and cracking.

425. Concrete, Masonry, and Brickwork: A Practical Handbook for the Homeowner and Small Builder. U.S. Army. Dover, 1975. LC: 75-12130. ISBN: 0-486-23203-4. Cost: $4.00.

Originally prepared by the U.S. Army under the title: *Concrete and Masonry*.

426. Consolidated Index of Selected Property Values: Physical Chemistry and Thermodynamics. Office of Critical Tables, National Academy of Sciences-National Research Council, 1962, 274 pp. Series: *NAS-NRC Publication*, 976.

This publication indexes 6 compilations that provide thermodynamic properties data. The 6 publications are as follows: *Selected Values of Chemical Thermodynamic Properties; Selected Values of Properties of Hydrocarbons and Related Compounds; Selected Values of Properties of Chemical Compounds; Contributions to the Data on Theoretical Metallurgy; Selected Values for the Thermodynamic Properties of Metals and Alloys*; and *Thermodynamic Properties of the Elements*.

427. Constants of Diatomic Molecules. Klaus-Peter Huber; Gerhard Herzberg. Van Nostrand Reinhold, 736 pp. Cost: $27.50.

Includes constants of over 900 diatomic molecules. Such information as electronic energies, vibrational constants, rotational constants, and observed transitions is covered.

428. Constitution of Binary Alloys. Max Hansen. McGraw-Hill, 2d ed., 1958, 1,305 pp. LC: 63-22429. ISBN: 0-07-026050-8. Cost: $75.00.

First Supplement published in 1965 (910 pp.) R. P. Elliott, $44.50, ISBN: 0-07-019189-1. Second supplement published in 1969 (720 pp.), F. Shunk. $44.95, ISBN: 0-07-057315-8. Over 4,000 binary systems are covered in this compilation. Phase diagrams are included for many of the alloys, in addition to solubility, crystal structure melting point, and eutectic point. See also *Handbook of Binary Phase Diagrams*.

429. Construction and Applications of Conformal Maps. E.F. Beckenbach (ed) (National Bureau of Standards). NTIS, 280 pp. Accession: PB-175819. Cost: $9.25. Series: NBS AMS18.

430. Construction Business Handbook: A Practical Guide to Accounting, Credit, Finance, Insurance, and Law for the Construction Industry. Robert F. Cushman. McGraw-Hill, 1978, 704 pp. ISBN: 0-07-014982-8. Cost: $32. 50.

Practical information on negotiating contracts, insurance coverage, financial aid, bidding, and taxes.

431. Construction Industry Handbook. R.A. Burgess et al. Cahners Books (also Construction Press, Ltd), 2d ed., 1973, 465 pp. ISBN: 0-8436-0119-1 (also 0-904406-63-6). Cost: $24.95.

Includes information on the properties of building materials.

432. Construction Industry Production Manual. Taylor F. Winslow. Craftsman Book Co, 1972, 176 pp. ISBN: 0-910460-04-3. Cost: $6.00.

This manual includes personnel-hour tables for estimating various types of construction jobs. These labor tables are useful in estimating how much time is required and how large a crew is required to accomplish a given task, such as framing a roof, hanging lighting fixtures, and moving a certain amount of soil.

433. Construction Inspection Handbook. James J. O'Brien. Van Nostrand Reinhold, 1974, 512 pp. LC: 74-1019. ISBN: 0-442-26258-2. Cost: $24.50.

This publication is concerned with quality control in construction. It covers such topics as roofing systems, foundations, pilings, and HVAC systems.

434. Construction Manual: Concrete and Formwork. T.W. Love. Craftsman Book Co, 1973, 176 pp. ISBN: 0-910460-03-5. Cost: $4.25.

Construction information is provided on the amount of concrete that is required and the amount of formwork that is required to accomplish a particular task. Information is presented on reinforcing concrete and curing concrete, as well as practical information such as how deep to dig the footing.

435. Construction Sealants and Adhesives. John Philip Cook. Wiley-Interscience, 1970. LC: 70-121905. ISBN: 0-471-16900-5. Cost: $26.95.

Essentially, this handbook provides an evaluation of different types of sealants and adhesives for construction purposes.

436. Construction Specifications Handbook. Hans W. Meier. Prentice-Hall Inc, 2d ed., loose-leaf. Cost: $49.95.

This loose-leaf service provides models for writing construction specifications for numerous topics, including such subjects as concrete formwork, metal decking, swimming pools, elevators, plumbing, fire sprinkler systems, waterproofing, dampproofing, excavation, and carpeting.

437. Consumer's Energy Handbook. Peter Norback; Craig Norback. Van Nostrand Reinhold, 1981, 272 pp. ISBN: 0-442-26066-0. Cost: $12.95.

This is basically a source book explaining how to conserve energy. Insulation and heating systems for buildings are discussed along with such topics as solar energy, batteries, and cars of the future. A listing of societies and associations concerned with energy conservation is included.

438. Contamination Control Handbook. Sandia Corp. NTIS, 1969, 314 pp. Accession: N70-13566. Tech Report: NASA-CR-61264; NASA-SP-5076.

439. Contractor's Management Handbook. J. O'Brien; R. Zilly; T. Graham. McGraw-Hill, 1971, 800 pp. LC: 76-169190. ISBN: 0-07-047565-2. Cost: $32.45.

440. Contributions on Partially Balanced Incomplete Block Designs with Two Associate Classes. W.H. Clatworthy. NTIS, 1956, 70 pp. Accession: PB-251902. Cost: $4.50. Series: NBS AMS47.

441. Contributions to the Data on Theoretical Metallurgy. U.S. Bureau of Mines. U.S. Government Printing Office, published as U.S. Bureau of Mines Bulletins between 1932-1976, multivolume set, 16 volumes.

442. Contributions to the Solution of Systems of Linear Equations and the Determination of Eigenvalues. Olga Taussky (ed) (National Bureau of Standards). National Bureau of Standards, 1954, 139 pp. Accession: PB-251901. Cost: $6.00. Series: NBS AMS39.

443. Control and Removal of Radioactive Contamination in Laboratories. National Bureau of Standards. NCRP Publications. Cost: $2.00. Series: AMS Handbook 48.

Available from NCRP Publications, PO Box 30175, Washington, DC 20014 as NCRP Report 8.

444. Control Engineers' Handbook: Servomechanisms, Regulators, and Automatic Feedback Control Systems. J. Truxal. McGraw-Hill, 1958, 1,078 pp. LC: 57-6411. ISBN: 0-07-065308-9. Cost: $49.50.

445. Control of Oil Pollution on the Sea and Inland Waters. J. Wardley-Smith. Graham and Trotman Ltd, (distributed in U.S. by International Publications Service), 1976. ISBN: 0-686-67327-1. Cost: $37.50.

This is a comprehensive handbook on oil spills and the effect of oil spills on land and water. A discussion is provided on the different methods of treating oil spills and the advantages and disadvantages of these methods.

446. Convenience and Fast Food Handbook. Marvin E. Thorner. AVI Publishing Co, 1973, 358 pp. ISBN: 0-87055-134-5. Cost: $24.50.

This guide to the fast food industry includes such information as food preparation, storage, processing, and quality control.

447. Conversion Tables For SI Metrication. William J. Semioli; Paul B. Shubert (trans). Industrial Press, 1974, 460 pp. ISBN: 0-8311-1104-6. Cost: $25.00.

This compilation features computer-generated tables containing nearly 70,000 conversion values.

448. Conversion Tables. Lefax Publishing Co, loose-leaf binding. ISBN: 0-685-14129-2. Cost: $3.00. Series: *Lefax Data Books*, No. 630.

449. Copper Wire Tables. NTIS. Cost: $4.00. Series: NBS Handbook 100.

National Bureau of Standards. Handbook 100, February 21, 1966 (supersedes Circular 31).

450. Correlation Tables for the Structural Determination of Organic Compounds. M. Pestemer. Verlag Chemie, 1975, 157 pp. ISBN: 3-527-25531-1. Cost: DM 98.

This is a compilation of spectral data from the ultraviolet and visible absorption region of approximately 2,300 organic compounds. This information will be very useful for the identification of molecules or molecular groups.

451. Corrosion and Wear Handbook for Water-Cooled Reactors. D.J. DePaul. McGraw-Hill, 1957, 293 pp. LC: 57-12896.

452. Corrosion Control Handbook. Energy Communications, Inc. Petroleum Engineer Publishing Co, 4th ed., 1975, 336 pp. LC: 74-19929.

This handbook contains information on corrosion, as it relates to offshore drilling and pipelines. Various protective coatings are discussed, and charts are included.

453. Corrosion Data Survey. NACE Publications, 4th ed., 1967. Cost: $60.00.

Corrosion information is presented for 50 materials used in the chemical and petrochemical industries. Newest data points, charts, and tables are included.

454. Corrosion Guide. Erich Rabald. Elsevier Publishing Co, 2d ed., 1968, 900 pp. LC: 67-19853. ISBN: 0-444-40465-1. Date of First Edition: 1951. Cost: $146.25.

This reference work provides a considerable amount of data on the corrosion of many and varied materials and includes information on their corrosive agents. This practical guide is for the engineer who needs information on how metals, alloys, and nonmetallic materials behave in working conditions where corrosive agents exist. Data on the resistance of materials to corrosive agents are provided. In addition, an extensive bibliography has been prepared at the beginning of this handbook in order to acquaint the user with sources of additional information on corrosion. For related works, please refer to *Corrosion-Resistant Materials Handbook; Waldman's Engineering Alloys,* and the *Metal Handbook.*

455. Corrosion Handbook. Herbert H. Uhlig. John Wiley and Sons (sponsored by the Electrochemical Society), 1948, 1,188 pp. ISBN: 0-471-89562-8. Cost: $47.50.

456. Corrosion Resistance Tables; Metals, Plastics, Non-Metallics, and Rubbers. Philip A. Schweitzer. Marcel Dekker Inc, 1976, 1,232 pp. LC: 76-46718. ISBN: 0-8247-6488-9. Cost: $109.75.

Provides data that would be helpful in making decisions regarding the selection of materials for construction. Corrosion is shown as a function of temperatue in this compilation.

457. Corrosion Resistant Materials Handbook. Ibert Mellan. Noyes Data Corp, 3d ed., 1976, 685 pp. LC: 71-128680. ISBN: 0-8155-0628-7. Date of First Edition: 1966. Cost: $39.00.

This handbook is similar to the *Corrosion Guide* in that it includes many types of the materials, such as metals, alloys, glass, wood, polymers, ceramics, rubbers, and it indicates the effects corrosive substances have on these materials. This handbook is filled with over 150 tables providing a considerable amount of data on corrosion-resistant materials. Trade names and company names and addresses are provided so that the user may obtain more information on the corrosion-resistant materials. For related works, please refer to *Metals Handbook; ASME Handbook; Metals Reference Book,* and the *CRC Handbook of Materials Science.*

458. Corrosion. L.L. Shreir. Halsted Press (Butterworths), 2d ed., 1976, 2 Vols., 1,000 pp. each Vol. ISBN: 0-408-00267-0. Cost: $199.00.

459. Cost-Benefit Analysis: A Handbook. Peter A. Sassone; William A. Schaffer. Academic Press, 1978, 182 pp. LC: 77-6612. ISBN: 0-12-619350-9. Cost: $13.50.

460. Cotton Ginners Handbook. T.J. Johnston et al. U.S. Government Printing Office, 1977, 110 pp. SuDoc: A1.76:503. Cost: $2.50.

This handbook is geared toward the large-scale ginners. It covers the cleaning of harvested cotton for marketing.

461. Craniofacial Embryology: Dental Practitioner Handbook No. 15. Geoffrey H. Sperber. Year Book Medical Publishers, 2d ed., 1976, 187 pp. ISBN: 0-8151-8114-0. Cost: $12.50 paper.

462. CRC Atlas of Scintimaging. Henry N. Wellman. CRC Press, 1977, 8 Vols. Cost: Section 1 and Section 2, $295.00; Section 3, $900.00.

This is a diagnostic tool to be used for those interested in clinical nuclear medicine. Over 500 scintimages are presented as examples of clinical cases. Section 3 includes scintimage transparencies in 11 x 14 format suitable for lightbox viewing.

463. CRC Composite Index for CRC Handbooks. CRC Press Inc, 2d ed., 1977, 1,250 pp. ISBN: 0-8493-0282-X. Cost: $199.50.

This compilation, which indexes the 49 handbooks published by the CRC Press, has over 25,000 entries.

464. CRC Handbook of Agricultural Productivity. Miloslav Rechcigl. CRC Press, 1981, Vol.1, 464 pp.; Vol.2, 416 pp. ISBN: Vol.1, 0-8493-3961-8; Vol.2, 0-8493-3963-4. Cost: Vol.1, $72.95; Vol.2, $64.95.

This is a 2-volume set entitled as follows: Vol.1, *Plant Productivity*; Vol.2, *Animal Productivity*. This handbook is concerned with factors affecting production of both crops and livestock. Soil environment, crop rotation, irrigation, and genetic damage are discussed with regard to plants. Animal productivity is presented along with such topics as disease, nutrition, climate, and pollution.

465. CRC Handbook of Analytical Toxicology. Irving Sunshine. CRC Press, Inc, 1969, 1,081 pp. LC: 69-20046. ISBN: 0-8493-3551-5. Cost: $59.95.

The *CRC Handbook of Analytical Toxicology* lists many different chemicals, drugs, pesticides, air pollutants, and water pollutants and provides analytical data that help to identify these substances. Chromatographic data are provided along with such other data as melting points, ultraviolet spectra data and infrared information. Many literature references are included along with a comprehensive index for easy retrieval of data. The reader is also referred to such handbooks as the *CRC Atlas of Spectral Data and Physical Constants for Organic Compounds* and the *CRC Handbook of Chromatography* for additional information on chemical analysis and identification.

466. CRC Handbook of Antibiotic Compounds. Janos Berdy. Chemical Rubber Co, Vol.1-4, 1980; Vol.5-7, 1981; Vol.8-10, tent 1982, 10 vols. ISBN: Vol.1, 0-8493-3451-9; Vol.7, 0-8493-3458-6. Cost: Vol.1, $64.95; Vol.2, $74.95; Vol.3, $59.95; Vol.4, part 1 $74.95 and part 2 $59.95; Vol.5, $74.95; Vol.6, $64.95; Vol.7, $49.95.

This handbook is a compilation of information on over 65,000 antibiotics. Such data as name, synonyms, producing organism, chemical type, molecular formula, weight, color, optical rotation, ultraviolet spectrum, solubility rating, stability, toxicity, and antimicrobial activity are included.

467. CRC Handbook of Bimolecular and Termolecular Gas Reactions. J.A. Kerr; S.J. Moss. CRC Press, 1981, Vol.1, 576 pp.; Vol.2, 256 pp. ISBN: Vol.1, 0-8493-0375-3; Vol.2, 0-8493-0376-1. Cost: Vol.1, $79.95; Vol.2, $49.95.

This 2-volume set contains information on quantitative kinetic data on bimolecular and termolecular gas phase reactions. This handbook also includes data on combustion reactions and gas phase ozonolysis.

468. CRC Handbook of Biochemistry and Molecular Biology. Gerald D. Fasman. CRC Press, Inc, 3d ed., 1975, 8 Vols. approx. 4,642 pp.; Index Vol., 295 pp. LC: 75-29514. ISBN: 0-87819-503-3 (set); Index volume 0-8493-0511-X. Cost: Section A, Vol.1, $59.95; Vol.2, $79.95; Vol.3, $69.95; Section B, Vol.1 $69.95; Vol.2, $79.95; Section C, Vol.1, $69.95; Section D, Vol.1 $69.95; Vol.2, $59.95; Index Volume $50.00.

This comprehensive multivolume handbook is divided into 4 major sections, each section having one or more volumes. The sections are entitled as follows: Section A *Proteins*; Section B *Nucleic Acids*; Section C *Lipids, Carbohydrates*; Section D *Physical and Chemical Data*. This handbook includes such subjects as RNA and DNA, hormones, amino acids, enzymes, sugars, and fatty acids. Physical and chemical data are provided for many substances including density, dielectric constants, solubility, spectral data, temperature coefficients, and some chromatographic data. This handbook consists of numerous tables and has been prepared by a very impressive list of experts in the fields of biochemistry and molecular biology. An index volume is available for this handbook.

469. CRC Handbook of Biochemistry in Aging. James R. Florini. Chemical Rubber Co, 1981, 272 pp. ISBN: 0-8493-3141-2. Cost: $54.95.

Information on biology of aging is presented, including tissue composition, DNA replication, chromatin, energy metabolism, and changes in hormones.

470. CRC Handbook of Biosolar Resources. Oskar R. Zaborsky; Thomas A. McClure. CRC Press, 1981, Vol.1, Part 1, 704 pp.; Vol.1, Part 2, 592 pp.; Vol.2, 608 pp. ISBN: Vol.1, Part 1, 0-8493-3471-3; Vol.1, Part 2, 0-8493-3472-1; Vol.2, 0-8493-3473-X. Cost: Vol.1, Parts 1 and 2 $79.95; Vol.2, $76.95.

This 2-volume set is entitled as follows: Vol.1, Part 1, *Light Absorption Processes and Electron Transport Systems.* (Zaborsky). Vol.1, Part 2, *General Characterisitcs of Photosynthetic Organisms.* (Zaborsky). Vol.2, *Resource Materials.* (McClure). This multivolume handbook is a reference source on renewable resources derived from biomass. Such topics are photosynthesis, microorganisms, and various biological processes are discussed.

471. CRC Handbook of Chemical Synonyms and Trade Names. William Gardner. CRC Press, 8th ed., 1978, 776 pp. Cost: $59.95.

Contains over 35,000 entries, including plastics, alloys, and pharmaceuticals.

472. CRC Handbook of Chemistry and Physics: A Ready-Reference Book of Chemical and Physical Data. Robert C. Weast. CRC Press, 63rd ed., 1982-83, 2,432 pp. LC: 13-11056. ISBN: 0-8493-0463-6. Date of First Edition: 1918. Cost: $59.95.

The *Handbook of Chemistry and Physics* is a standard 1-vol. ready-reference handbook that is updated annually, providing the latest and most accurate information for the chemical and physical sciences. One of the major features is the tabulated data on the physical constants of organic and inorganic compounds. Such information as density, specific gravity, melting point, boiling point, solubility, and index of refraction is provided. Also, this handbook includes data on such subjects as the isotopes; the planets, satellites, and the Earth; amino acids; carbohydrates; and thermal conductivity of various organic and inorganic compounds. An excellent index and table of contents is provided for easy access to this data. Also includes physical constants of the alkali metals and table for Doppler linewidths. *See also Lange's Handbook of Chemistry.*

473. CRC Handbook of Chromatography. Gunter Zweig; Joseph Sherma. CRC Press, Inc, 1972, 2 Vols., Vol.1, 784 pp.; Vol.2, 343 pp. LC: 76-163067. ISBN: Vol.1, 0-8493-0561-6; Vol.2, 0-87819-562-9. Cost: Vol.1, $59.95; Vol.2, $49.95.

This 2-volume work provides chromatographic data for approximately 12,000 chemical components. Four types of chromatographic methods are used: thin-layer, liquid, paper, and gas. The handbook is very well indexed, and literature sources are provided for the investigator who needs to refer to the original reference. Readers who are interested in identifying chemicals are also referred to the *CRC Atlas of Spectral Data and Physical Constants for Organic Compounds*, and the *CRC Handbook of Tables for Organic Compound Identification.*

474. CRC Handbook of Clinical Engineering. Barry N. Feinberg. Chemical Rubber Co, 1980, 384 pp. ISBN: 0-8493-0244-7. Cost: $59.95. Series: *CRC Series in Engineering in Medicine and Biology.*

Covers such topics as hospital electrical systems, artificial kidneys, medical instrumentation, and defibrillators.

475. CRC Handbook of Clinical Laboratory Data. Willard R. Faulkner et al. Chemical Rubber Co, 2d ed., 1968, 710 pp. LC: 68-54212. ISBN: 0-87819-722-2. Cost: $28.00.

Provides tables of normal values of substances in blood and other body fluids. Covers such topics as blood banking, hematology, microbiology, electrophoretic fractionation, and serum proteins. Includes the theory and diagnostic value of 110 clinical measurements and substances.

476. CRC Handbook of Digital Systems Design for Scientists and Engineers: Design with Analog, Digital, and LSI. Wen C. Lin. CRC Press, 1981, 272 pp. ISBN: 0-8493-0671-X. Cost: $49.95.

This handbook covers information on linear circuits, switching circuits, logic circuits, and analog-digital-analog conversion.

477. CRC Handbook of Electrophoresis. L.A. Lewis; J.J. Opplt. CRC Press, Inc, 1980, Vol.1, 336 pp. ISBN: 0-8493-0571-3. Cost: Vol.1, $56.95; Vol.2, $59.95.

Currently one section (more sections are planned, each section consisting of one or more volumes) entitled: *Electrophoresis of Lipoprotiens.*

478. CRC Handbook of Energy Utilization in Agriculture. David Pimentel. Chemical Rubber Co, 1980, 496 pp. ISBN: 0-8493-2661-3. Cost: $69.95.

This handbook discusses the energy inputs and outputs for agricultural production, livestock production, and marine fishery production. Data in this handbook are presented in the context of agriculture and forestry being major solar energy conversion systems.

479. CRC Handbook of Engineering in Medicine and Biology. David G. Fleming; Barry N. Feinberg. CRC Press Inc, 1976, Vol.1, Section 1, 432 pp. LC: 75-44222. ISBN: Vol.1, Section 1, 0-87819-285-9, Section 2, 0-8493-0242-0. Cost: Vol.1, Section 1, $57.95; Section 2, $59.95.

This is a relatively new handbook that has recently been published by CRC Press, and other volumes are planned for release in the future. It includes information that is useful for biomedical applications of engineering principles, such as properties of biomedical materials, regulation of temperature in animals, electric fields, lubrication of joints, microprocessors, and measurements. It is planned that the second volume of this set include information on sophisticated life support systems and instrumentation such, as heart-lung machines and hemodialysis devices, as well as information on computer data collection and processing. The articles in this handbook are written by an impressive group of contributors who have provided a well-documented and thoroughly indexed reference work.

480. CRC Handbook of Food Additives. Thomas E. Furia. CRC Press Inc, Vol.1, 2d ed., 1973; Vol.2, 1980, Vol.1, 998 pp.; Vol.2, 432 pp. LC: 68-21741. ISBN: Vol.1, 0-8493-0542-X; Vol.2, 0-8493-0543-8. Date of First Edition: 1968. Cost: Vol.1, $69.95; Vol.2, $59.95.

A wealth of information is provided in the second edition of the *CRC Handbook of Food Additives* for those people seeking information in connection with the food processing industry. Each section is written by an expert in his/her respective field of specialization. Such subjects as vitamins, dyes, fatty acids, flavorings, sweeteners, emulsifiers, corn syrups, and food stabilizers are included. In addition, various U.S. Food and Drug Administration regulations are discussed as well as such areas as phosphates, nitrates, and industrial enzymes. The reader's attention is also directed to the *CRC Fenaroli's Handbook of Flavor Ingredients* for further information.

481. CRC Handbook of High Resolution Infrared Laboratory Spectra of Atmospheric Interest. David G. Murcray; Aaron Goldman. CRC Press, 1981, 288 pp. ISBN: 0-8493-2950-7. Cost: $49.95.

Spectra of molecular species that are present in the atmosphere as an atmospheric pollutant or as a naturally occurring chemical are included in the handbook. All spectra were obtained with a resolution of at least .05 $^{-1}$

482. CRC Handbook of Hospital Acquired Infections. Richard P. Wenzel. Chemical Rubber Co, 1981, 656 pp. ISBN: 0-8493-0202-1. Cost: $84.95; $94.95 outside USA.

This handbook concentrates on the existing infection control problems and the state of the art control of hospital acquired infections. Information is provided on antibiotic usage and antiseptics and disinfectants in hospitals. Various types of infections are discussed including hospital acquired pneumonia, hospital acquired urinary tract infections, and hospital acquired gastrointestinal infections.

483. CRC Handbook of Hospital Safety. Paul E. Stanley. Chemical Rubber Co, 1981, 416 pp. ISBN: 0-8493-0751-1. Cost: $64.95; $74.95 outside USA.

Discusses safety assurance and loss prevention as well as hazards and corrective remedies. Safety is presented in the context of various situations like the emergency room, operating room, laboratory, radiation safety, infection control, sanitation, and fire safety.

484. CRC Handbook of Immunology in Aging. Marguerite M.B. Kay; Takashi Makinodan. Chemical Rubber Co, 1981, 336 pp. ISBN: 0-8493-3144-7. Cost: $59.95; $68.95 outside U.S..

This handbook discusses such topics as genetics, marrow stem cells, aging of the immune system, dietary restrictions, infectious diseases, and neoplasia.

485. CRC Handbook of Irrigation Technology. Herman J. Finkel. CRC Press, 1981, 368 pp. ISBN: 0-8493-3231-1. Cost: $59.95.

Irrigation systems are presented for different soils and for different climates.

486. CRC Handbook of Laboratory Animal Science. Edward C. Melby; Norman H. Altman. CRC Press, Inc, Vols.1 and 2, 1974; Vol.3, 1976, Vol.1, 451 pp.; Vol.2, 523 pp.; Vol.3, 943 pp. LC: 74-19795. ISBN: Vol.1, 0-8493-0341-9; Vol.2, 0-8493-0342-7; Vol.3, 0-87819-344-8. Cost: Vol.1, $59.95; Vol.2, $64.95; Vol.3, $69.95.

This 3-volume set provides a tremendous amount of tabulated and graphic data on experimental and laboratory animals. Articles are also presented on such subjects as legislation, regulations, and various requirements that need to be met when using animals for laboratory purposes. Such information as diseases, infections, drug dosages, and physiological data of various laboratory animals are presented. Vol.3 includes a cumulative index to all 3 volumes in order to enable the user to retrieve data more easily.

487. CRC Handbook of Laboratory Safety. Norman V. Steere. Chemical Rubber Co, 2d ed., 1971, 854 pp. LC: 67-29478. ISBN: 0-8493-0352-4. Date of First Edition: 1967. Cost: $59.95.

The second edition of this handy reference book provides a considerable amount of material on handling dangerous chemicals, protection from radiation, ultraviolet radiation, lasers, and laboratory infections that may develop from handling experimental animals. Information is also available on burns from acids and other toxic chemicals, dangers of mercury vapors, dangers from electric shock, cryogenic equipment, inhalation of vapors, and proper ventilation equipment, including hoods and air conditioning. This handbook has an excellent index and also has various tables of additional information on hazardous chemicals. For related works, please refer to *Dangerous Properties of Industrial Materials; Handbook of Ventilation for Contaminant Control*, and *Handbook of Reactive Chemical Hazards*.

488. CRC Handbook of Lasers with Selected Data on Optical Technology. Robert J. Pressley. CRC Press, 1971, 631 pp. LC: 72-163066. ISBN: 0-87819-381-2. Cost: $49.95.

This handbook covers a wide range of lasers, including neutral and ionized gas lasers, dye lasers, injection lasers, and molecular gas lasers. Each chapter is written by a well-known expert in his/her particular area of specialization, most authors having industrial affiliations. An excellent collection of tables on gas lasers listed by wavelengths is provided, in addition to such optical information as magneto-optic properties of various materials, holography, raman scattering, and organic and inorganic photochromic materials.

489. CRC Handbook of Marine Science. J. Robert Moore. CRC Press, Inc, 1974, 4 Vols. (more being planned); Vol.1, 640 pp.; Vol.2, 390 pp.; Section B, Vol.1, 216 pp.; Section B, Vol.2, 240 pp. LC: 73-88624. ISBN: Vol.1, 0-87819-389-8; Vol.2, 0-87819-390-1; Section B, Vol.2, 0-87819-391-X; Section B, Vol.1, 0-8493-0214-5. Cost: Vol.1, $59.95; Vol.2, $53.95; Section B, Vol.1, $49.95; Section B, Vol.2, $49.95.

This multivolume set brings together a massive amount of data regarding the oceans, and the fauna and flora that inhabit the marine environment. Such information as ocean currents, salinity tables, plankton, topographic data, and underwater cables can be found in this collection. In addition, the volume entitled *Compounds from Marine Organisms* will be of interest, for it lists approximately 500 organic compounds that have been obtained from marine sources. Fishing information, distribution and migration habits are also included in this thoroughly documented and well-indexed reference work. The reader is also referred to the *Handbook of Ocean and Underwater Engineering*.

490. CRC Handbook of Mass Spectra of Drugs. Irving Sunshine. Chemical Rubber Co, 1981, 464 pp. ISBN: 0-8493-3572-8. Cost: $69.95.

This is a very comprehensive compilation of mass spectrometry data. Of interest will be the molecular weight index which arranges the spectra by ascending molecular weight as well as the 8-peak index which arranges the spectra according to the most intense ion. Over 500 mass spectra curves are included.

491. CRC Handbook of Materials Science. Charles T. Lynch. Chemical Rubber Co. Press, Vol.1, 1974, Vols.2 and 3, 1975, Vol.1, 752 pp.; Vol.2, 440 pp.; Vol.3, 642 pp. LC: 73-90240. ISBN: Complete set, 0-87819-234-4; Vol.1, 0-87819-231-X; Vol.2, 0-87819-232-8; Vol.3, 0-87819-233-6; Vol.4, 0-8493-0234-X. Cost: Vol.1, $69.95; Vol.2, $49.95; Vol.3, $64.95; Vol.4, $69.95.

This 4-volume reference book includes volumes entitled: Vol.1, *General Properties*; Vol.2, *Metals, Composites and Refractory Materials*; Vol.3, *Nonmetallic Materials and Applications*; Vol.4, *Wood*. A wealth of tabulated data are provided in this handbook including information on such materials as nuclear materials, graphites, polymers, aluminum, ferrous alloys, nickel, and glass. The physical properties of these materials are tabulated along with phase diagrams and standards. Each volume is thoroughly indexed, and a section is provided for finding additional information on materials, including literature searching services and information analysis centers. For related works, please refer to

Metals Handbook; ASME Handbook; Metals Reference Book; and *Thermophysical Properties of Matter*.

492. CRC Handbook of Mathematical Sciences. William H. Beyer. Chemical Rubber Co, 5th ed., 1978, 992 pp. LC: 78-10602. ISBN: 0-8493-0655-8. Date of First Edition: 1962. Cost: $49.95.

This handbook was formerly the *CRC Handbook of Tables for Mathematics*. This 1-volume reference source is a massive collection of mathematical information including such subjects as trigonometry and celestial mechanics. Tables, equations, formulas, and notations are provided for every conceivable area of mathematics. For example, information is provided on such subjects as algebra, logarithms, hyperbolic functions, geometry, calculus, differential equations, and probability and statistics. In addition, interest tables are presented plus such information as mortality tables, calendar data, and an index that provides easy access. Also includes tables on Laplace transforms, Fourier transforms, Bessel transforms, and Poisson distributions. See also: *Engineering Mathematics Handbook, CRC Standard Mathematical Tables*, and *Handbook of Mathematical Tables and Formulas*.

493. CRC Handbook of Medical Physics. Robert G. Waggener et al. Chemical Rubber Co, 1980, 500 pp. ISBN: 0-8493-0525-X. Cost: $59.95.

Discusses such topics as ultrasound, lasers, microwaves, thermography, and effects of ionizing radiation. Also includes a glossary of medical terms and anatomy terms.

494. CRC Handbook of Microbiology, Condensed Paperback Edition. Allen I. Laskin; Hubert Lechevalier. CRC Press, condensed paperback ed., 1974, 930 pp. LC: 74-12937. ISBN: 0-87819-585-8. Cost: $19.95.

This 930-page resource is a condensed version of the 4-volume set of the first edition of the *CRC Handbook of Microbiology*, thus making it much less expensive, but containing a sufficient amount of information for general usage by most students. This edition includes most of the information contained in the unabridged edition except for that provided on microbial products, which is found in Vol.4. As in the larger edition, information is presented on such microorganisms as protozoa, fungi, algae, viruses, bacteria, etc. Please refer to the annotation for the 4-volume set for additional information.

495. CRC Handbook of Microbiology. Allen I. Laskin; Hubert Lechevalier. CRC Press, Inc, 2d ed., Vol.1, 1977; Vol.2, 1978, multivolume set (future volumes planned), Vol.1, 768 pp.; Vol.2, 888 pp., Vol.3, 1,040 pp., Vol.4, 720 pp. ISBN: Vol.1, 0-8493-7201-1; Vol.2, 0-8493-7202-X; Vol.3, 8493-7203-8; Vol.4, 0-8493-7204-6. Cost: Vol.1, $69.95; Vol.2, $79.95; Vol.3, $74.95; Vol.4, $64.95.

The 2d edition of the *CRC Handbook of Microbiology* will consist of multiple volumes with the first 4 volumes entitled as follows: Vol.1, *Bacteria*; Vol.2, *Fungi, Algae, Protozoa, and Viruses*; Vol.3, *Microbial Composition: Amino Acids, Proteins, and Nucleic Acids*; Vol.4, *Microbial Composition: Carbohydrates, Lipids, and Minerals*. This handbook covers such information as the properties of microorganisms and presents data on antibiotics, enzymes, DNA, and aliphatic and aromatic compounds.

496. CRC Handbook of Nutrition and Food; Selected Data for Experimental and Applied Human, Animal, Microbial, and Plant Nutrition. Miloslav Rechcigl Jr (trans). CRC Press, Inc, 1977-, approx. 500 pp. ISBN: 0-8493-2721-0. Cost: Approx. $55.00.

497. CRC Handbook of Nutritional Requirements in a Functional Context. Miloslav Rechcigl. Chemical Rubber Co, 1981, Vol.1, 560 pp.; Vol.2, 624 pp. ISBN: Vol.1, 0-8493-3956-1; Vol.2, 0-8493-3958-8. Cost: Vol.1, $82.95; Vol.2, $87.95. Series: Section D of the *CRC Handbook Series in Nutrition and Food.*

This 2-volume set is entitled as follows: Vol.1, *Development and Conditions of Physiological Stress.* Vol.2, *Hematopoiesis, Metabolic Function, and Resistance to Physical Stress.* This handbook relates nutritional requirements with specific processes and systems, such as the development of nervous tissue, bone tissue, muscle tissue, adipose tissue, and the reproductive system. Also covers such topics as hematopoiesis, metabolic functions, and resistance to physical and physiological stress.

498. CRC Handbook of Nutritive Value of Processed Food. Miloslav Rechcigl. Chemical Rubber Co, 1981, Vol.1, 624 pp.; Vol. 2, 496 pp. ISBN: Vol.1, 0-8493-3951-0; Vol.2, 0-8493-3953-7. Cost: Vol.1, $79.95; Vol.2, $68.95.

This 2-volume set is entitled as follows: Vol.1, *Food for Human Use*; Vol.2, *Animal Feedstuffs.* This handbook discusses the effects that processing food has on the nutritional value of the end product. Such topics as heat processing, canning, freezing, freeze-drying, fermentation, enzyme treatment, irradiation, and microwave cooking are discussed in this 2-volume set.

499. CRC Handbook of Pest Management in Agriculture. David Pimentel. Chemical Rubber Co, 1981, Vol.1, 624 pp.; Vol.2, 528 pp.; Vol.3, 640 pp. ISBN: Vol.1, 0-8493-3841-7; Vol.2, 0-8493-3842-5; Vol.3, 0-8493-3843-3. Cost: Vol.1, $79.95; Vol.2, $69.95; Vol.3, $76.95.

This 3-volume set includes information on pesticides as well as natural means of controlling pests. Pest management is discussed in connection with numerous crops, such as alfalfa, wheat, strawberries, etc. Genetic resistance for the control of plant disease is also discussed along with the use of sex attractants in insect control.

500. CRC Handbook of Physical Properties of Rocks. Robert S. Carmichael. Chemical Rubber Co, 1981, 512 pp. ISBN: 0-8493-0226-9. Cost: $74.95.

Tabulated data are provided on mineral composition, electrical properties, seismic velocities, magnetic properties, and spectroscopic properties of rocks.

501. CRC Handbook of Physiology in Aging. Edward J. Masoro. Chemical Rubber Co, 1981, 464 pp. ISBN: 0-8493-3143-9. Cost: $72.95. Series: *CRC Series in Aging.*

Provides information on the various organ systems of the body, such as the nervous system, the endocrine system, and the cardiovascular system. Also covers such topics as metabolism, temperature regulation, exercise, and gastrointestinal physiology.

502. CRC Handbook of Radiation Doses in Nuclear Medicine and Diagnostic X-ray. James G. Kereiakes; Marvin Rosenstein. Chemical Rubber Co, 1980, 264 pp. ISBN: 0-8493-3245-1. Cost: $49.95.

Data are provided in tabular form on dosimetry from radiopharmaceuticals and from diagnostic X-rays.

503. CRC Handbook of Radiation Measurement and Protection. Allen Brodsky. CRC Press, Inc, 1979-, Section A, Vol.1, 720 pp.; Section A, Vol.2, 416 pp. ISBN: Section A, Vol.1, 0-8493-3756-9; Section A, Vol.2, 0-8493-3757-7; Section B, 0-8493-3765-8; Section C, 0-8493-3775-5. Cost: Section A, Vol.1, $79.95; Section A, Vol.2, $54.95.

The 3 sections are entitled as follows: Section A, *General Scientific and Engineering Information*; Vol.1, *Physical Science and Engineering Data*; Vol.2, *Biological and Mathematical Data*; Section B, *Medical Radiological Physics and Safety*; Section C, *Health Physics Methods in Industrial, Academic, and Research Facilities.*

504. CRC Handbook of Radioactive Nuclides. Yen Wang. Chemical Rubber Co, 1969, 960 pp. LC: 75-81089. ISBN: 0-87819-521-1. Cost: $49.95.

This handbook provides an excellent collection of tabulated data in the field of radiology and nuclear medicine. Physical and nuclear properties of radioisotopes are presented along with information on dosages, injury, and radiation protection. Equipment that is used to measure, monitor, and handle radioactive nuclides is discussed. Information is also provided for the application of radio-nuclides in industrial situations and medical diagnosis such as brain scanning and cardiovascular disease. In addition, information on the disposal of radioactive wastes is included as well as mathematical tables, equations, and a thorough index. See also *Nuclear Engineering Handbook; Neutron Cross Sections,* and *Reactor Handbook.*

505. CRC Handbook of Reactive Chemical Hazards. L. Bretherick. Butterworth, London, 2d ed., 1979. ISBN: 0-408-70927-8. Cost: $130.00.

This handbook provides information on approximately 4,000 chemicals, compounds, and mixtures that can be potentially hazardous if certain interactions occur. The chemical instability data are presented in the major section where the chemicals are listed in chemical formula sequence. The chemical literature is generously cited, and original bibliographic citations are given so that the investigator can pursue the information for further study. An index is provided for easy access to this chemical hazard data. The reader is referred to the *Chemical Technology Handbook,* which was published by the American Chemical Society; the *CRC Handbook of Laboratory Safety*; and to *Dangerous Properties of Industrial Materials* by N. Irving Sax.

506. CRC Handbook of Spectrophotometric Data of Drugs. Irving Sunshine. Chemical Rubber Co, 1981, 416 pp. ISBN: 0-8493-3571-X. Cost: $64.95. Series: *CRC Series in Analytical Toxicology.*

This handbook will be useful for identifying unknown substances through such techniques as ultraviolet spectroscopy, infrared spectroscopy, flourescence and phosphorescence spectroscopy. The data on known substances are arranged by ascending wavelengths. Molecular structure and absorption properties are also listed for comparison purposes.

507. CRC Handbook of Spectroscopy. J.W. Robinson. CRC Press Inc, 1974, Vol.1, 913 pp.; Vol.2, 578 pp.; Vol.3, 560 pp. LC: 73-77524. ISBN: Vol.1, 0-87819-331-6; Vol.2, 0-87819-332-4; Vol.3, 0-8493-0333-8. Cost: Vol.1, $69.95; Vol.2, $59.95; Vol.3, $74.95.

This is a handy 3-volume reference source that provides information that will help scientists and engineers to identify various materials and compounds. This handbook provides information on such subjects as infrared, ultraviolet, and Raman spectroscopy; electron spin and nuclear magnetic resonance; emission, atomic, flame and X-ray spectroscopy; and mass spectral tables including such chemical groups as ketones, esters, alcohols, and many others. This handbook provides 125 pages of indexes to help retrieve the information on spectroscopy. For a comprehensive collection of spectra data, see the *CRC Atlas of Spectral Data and Physical Constants for Organic Compounds.*

508. CRC Handbook of Tables for Applied Engineering Science. Ray E. Bolz; George L. Tuve. Chemical Rubber Co, 2d ed., 1973, 1,150 pp. LC: 75-117044. ISBN: 0-8493-0252-8. Date of First Edition: 1970. Cost: $49.95.

This reference work is a handy one-volume source of information for critically evaluated engineering data. The handbook includes such information as properties of solids, liquids, and gases; nuclear cross-section data; combustion; heating, refrigeration, and air conditioning; fluid mechanics; heat and mass transfer data; and ultrasonics. A special effort has been made to provide information in metric SI units and conventional units, as well as to provide conversion factors. An excellent index is also included in this handbook. Engineers will find very useful the section on publishers' addresses and engineering organizations' addresses. Also refer to *Handbook of Engineering Fundamentals* and the *Engineering Manual.*

509. CRC Handbook of Tables for Organic Compound Identification. Zvi Rappoport. CRC Press (Blackwell Scientific, Oxford), 3rd ed., 1967, 560 pp. LC: 63-19660. ISBN: 0-87819-303-0 (also 0-8493-0303-6). Date of First Edition: 1960. Cost: $49.95.

This handbook provides physical constants on many different organic compounds which are tabulated and organized by chemical groups, such as alcohols, ketones, esters, alkenes, etc. Such data as boiling point, refractive index, density, and derivative properties are provided for each organic compound so that their identification can be made. Logarithmic tables, carbohydrate properties and other data are included along with an index for the organic compounds examined in the tables. For readers interested in identifying chemicals, please refer also to such handbooks as the *Atlas of Spectral Data and Physical Constants for Organic Compounds* and the *Handbook of Chromatography.*

510. CRC Handbook of Tables for Probability and Statistics. William H. Beyer. Chemical Rubber Co, 2d ed., 1968, 656 pp. LC: 66-17301. ISBN: 0-8493-0692-2. Date of First Edition: 1966. Cost: $39.95.

This handbook is one of the best available collections of tables and data on probability and statistics. Theories and methods of probability distributions are presented, along with information on such subjects as standard deviation, regression and correlation, variance tables, finite differences, nonparametric statistics, correlation coefficients, and numerous theorems and formulas. A section on mathematical tables is presented in this well-documented and well-indexed reference source. See also *Handbook of Tables for Mathematics; A Table of Series and Products* and the *Handbook of Mathematical Tables and Formulas.*

511. CRC Handbook of Tables of Commercial Thermal Processes for Low-Acid Canned Foods. C.R. Stumbo et al. Chemical Rubber Co, 1981, Vol.1, 560 pp.; Vol.2, 544 pp. ISBN: Vol.1, 0-8493-2961-2; Vol.2, 0-8493-2963-9. Cost: $69.95 per Vol.

These tables are to be used as a guide in thermal processing of foods in commercial sterilization. A considerable amount of data are provided on conduction-heating products, safe thermal processes, and convection-heating products.

512. CRC Handbook of Tables of Functions for Applied Optics. Leo Levi. CRC Press, 1974, 640 pp. LC: 73-88627. ISBN: 0-87819-371-5. Cost: $69.95.

This handbook consists of 640 pages of tables on mathematical functions that are useful to scientists and engineers working in the field of applied optical system design. Included are tables on such subjects as blackbody radiation, Fresnel integrals, mono- and poly-chromatic radiation, and line spread functions. In addition, such data as modulation transfer functions, linear interpolation functions, and photometric units are presented in this collection of optical tables. Of related interest is the *Handbook of Lasers with Select Data on Optical Technology.*

513. CRC Handbook of Terpenoids, Monoterpenoids. Sukh Dev; J.S. Yadav; Anubhav Narula. Chemical Rubber Co, 1981, Vol.1, 272 pp.; Vol.2, 528 pp. ISBN: Vol.1, 0-8493-3601-5; Vol.2, 0-8493-3602-3. Cost: Vol.1, $49.95; Vol.2, $69.95.

Information is provided on terpine and monoterpine chemistry. Included are such data as identification of terpines and monoterpines, spectral data, biosynthesis, pharmacological properties, and various uses and applications.

514. CRC Handbook of Transportation and Marketing in Agriculture. Essex E. Finney. Chemical Rubber Co, 1981, Vol.1, 464 pp.; Vol.2, 496 pp. ISBN: Vol.1, 0-8493-3851-4; Vol.2, 0-8493-3852-2. Cost: $59.95 per Vol. Series: *CRC Series in Agriculture.*

This 2-volume set covers marketing information for agricultural products such as poultry, peanuts, dairy products, cereals, grain, soybeans, cotton, tobacco. Marketing costs, transportation, wholesaling, and price spreads are also discussed.

515. CRC Handbook of Zoonoses. James H. Steele. CRC Press, Inc, 1979-, approx. 600 pp. per Vol. ISBN: Vol.1, Section 1, 0-8493-2906-X. Cost: Vol.1, Section 1, $74.95.

This handbook currently includes 3 sections entitled as follows: Section 1, *Bacterial Zoonoses;* Section 2, *Viral Zoonoses;* Section 3, *Parasitic Zoonoses.*

516. CRC Handbook Series in Clinical Laboratory Science. David Seligson (ed). CRC Press, Inc, 1977, multivolume set, approx. 500 pp. per Vol. Cost: Approx. $59.95 per Vol.

This handbook includes sections entitled as follows: *Hematology, Nuclear Medicine, Blood Banking, Microbiology, Pathology, Toxicology, Clinical Chemistry, Immunology, Virology, and Rickettsiology.*

517. CRC Handbook Series in Inorganic Electrochemistry. Louis Meites; Peter Zuman; Ananthakrishnan Narayanan. Chemical Rubber Co, Vol.1, 1980; Vol.2, 1981, Vol.1, 512 pp.; Vol.2, 544 pp. ISBN: Vol.1, 0-8493-0361-3; Vol.2, 0-8493-0362-1. Cost: $69.95 per Vol.

This is a comprehensive compilation of data in 13 tables on electrochemical behavior of ions in compounds in various supporting electrolytes, in aqueous and nonaqueous solvents, and in fused salts. In all cases, the data are backed by the key to literature citations.

CRC Handbook Series in Marine Science. *See* **CRC Handbook of Marine Science.**

CRC Handbook Series in Nutrition and Food. *See* **CRC Handbook of Nutrition and Food.**

518. CRC Handbook Series in Organic Electrochemistry. Louis Meites; Petr Zuman. CRC Press, Vols.1 and 2,

1977; Vols.3 and 4, 1979, Vol.1, 896 pp.; Vol.2, 640 pp.; Vol.3, 680 pp.; Vol.4, 496 pp. ISBN: 0-8493-7251-8 (2 vol set); Vol.3, 0-8493-7223-2; Vol.4, 0-8493-7224-0. Cost: 2-Vol. set, $135.00; Vol.3, $59.95; Vol.4, $59.95.

Vols.1 and 2 contain data from 3,800 chemical compounds. Vol.3 provides information on electrochemical behavior of 1,309 organic and organometallic compounds. Also includes *Chemical Abstracts* registry number.

CRC Handbook Series in Zoonoses. *See* **CRC Handbook of Zoonoses.**

519. CRC Manual of Nuclear Medicine Procedures. John W. Keyes Jr. Chemical Rubber Co, 3d ed., 1978, 224 pp. ISBN: 0-8493-0707-4. Cost: $29.95.

This manual discusses radioisotope-imaging equipment and the various procedures that are used for different systems of the body. For example, procedures are presented for the cardiovascular system, pulmonary system, central nervous system, gastrointestinal system, and endocrine system. Radiation safety precautions are discussed with regard to handling radionuclides and waste procedures.

CRC Mass Spectrometry of Pesticides and Pollutants. See: *Mass Spectrometry of Pesticides and Pollutants.*

520. CRC Standard Mathematical Tables. William H. Beyer; Samuel M. Selby. Chemical Rubber Co, Press, 26th ed., 1981, 618 pp. LC: 30-4052. ISBN: 0-8493-0626-4. Date of First Edition: 1931. Cost: $24.95.

This is a handy, inexpensive reference source that has been through 26 different editions enabling the publisher to provide the latest in the field of mathematics. Although this handbook is not quite as comprehensive as *CRC Handbook of Tables for Mathematics*, it does provide such information as 5-place trigonometric function tables, logarithmic tables, calculus with integrals, Laplace transforms, and interest tables. A thoroughly prepared index, conversion factors, and metric information make this handbook very useful. See also *Handbook of Mathematical Tables and Formulas, Engineering Mathematics Handbook*, and the *CRC Handbook of Tables for Probability and Statistics*.

521. Creation of a Ceramics Handbook. TLSP: Final Report. W.J. Craft. North Carolina Agricultural and Technical State University, Greensboro, NC (available from NTIS), 1976, 290 pp. Accession: N76-20274. Tech Report: NASA-CR-146575. Cost: $9.25 hard copy.

This is a result of an extensive literature search on a number of data bases. The results of these searches are arranged by material properties, including mechanical, electrical, electromagnetic, and fracture. Such ceramic materials as alumina, magnesium oxide, silicon nitride, and silicon carbide are listed.

522. Creep-Rupture Data for the Refractory Metals to High Temperatures. J.B. Conway; P.N. Flagella.. Gordon and Breach, Science Publishers Inc, 1971, 798 pp. ISBN: 0-677-02660-9. Cost: $161.75.

523. Critical Analysis of the Heat Capacity Data of the Literature and Evaluation of Thermodynamic Properties of Copper, Silver, and Gold from 0 to 300 K. G.T. Furukawa; M.L. Reilly; W.G. Saba. National Bureau of Standards, 1968, 54 pp. Accession: N70-71850. Tech Report: NSRDS- NBS-18.

524. Critical Assessment of Thermochemical Data for Transition Metal-Silicon Systems. (Thermodynamic Properties of Transition Metal/Silicon Binary Alloys.) T.G. Chart. National Physical Lab, Teddington, England (available from NTIS), 1972, 69 pp. Accession: N73-12970. Tech Report: NPL-CHEM-18. Cost: $5.50.

Data on the thermodynamic properties of metal transition/silicon systems are provided with information on values for enthalpies of formation and partial enthalpies of solution. Gibbs' free energy information is also included.

525. Critical Evaluation of Equilibrium Constants in Solution. Part A: Stability Constants of Metal Complexes. Critical Survey of Stability Constants of EDTA Complexes. G. Anderegg. Pergamon Press, 1977, 42 pp. ISBN: 0-08-022009-6. Cost: $7.00. Series: *IUPAC Chemical Data Series*, No. 14.

526. Critical Evaluation of Equilibrium Constants Involving 8-Hydroxyquinoline and Its Metal Chelates. J. Stary. Pergamon Press Inc, 1979, 40 pp. LC: 78-41196. ISBN: 0-08-023929-3. Cost: $12.00. Series: *IUPAC Chemical Data Series*, Vol. 24.

At head of title: International Union of Pure and Applied Chemistry, Analytical Chemistry Division, Commission on Equilibrium Data; *Critical Evaluation of Equilibrium Constants in Solution*; Part A, *Stability Constants of Metal Complexes*. Provides a critical evaluation of the disassociation of oxine, the stability of oxine, distribution of oxine, and the liquid-liquid distribution of metal oxinates.

527. Critical Evaluation of Rate Data for Homogeneous Gas-Phase Reactions of Interest in High-Temperature Systems. D.L. Baulch; D.D. Drysdale; A.C. Lloyd. Available from D.L. Baulch, Leeds University, England, 1968-69, Part 1, 30 pp.; Part 2, 52 pp.; Part 3, 72 pp. Accession: Part 1, N69-31364; Part 2, N69-22468; Part 3, N69-39305. Tech Report: Rept-1; Rept-2; Rept-3.

This compilation includes 3 reports dealing with the high temperature reaction rate of hydrogen and oxygen, and carbon monoxide.

528. Critical Evaluation of some Equilibrium Constants Involving Alkylammonium Extractants. A.S. Kertes. Pergamon Press Inc, 1977. ISBN: 0-08-021591-2. Cost: $6.00. Series: *IUPAC Chemical Data Series*, Vol. 2.

At head of title: *Critical Evaluation of Equilibrium Constants in Solution*-Part B. Presents data on the critical evaluation of formation constants of alkylammonium salts and the aggregation constants of alkylammonium salts. Published as a report to IUPAC Commission Vol. 6-IUPAC Commission on Equilibrium Data.

529. Critical Evaluation of some Equilibrium Constants Involving Organophosphorus Extractants. Y. Marcus. Pergamon Press Inc, 1974, 100 pp. ISBN: 0-08-02084-X. Cost: $14.50. Series: *IUPAC Chemical Data Series*, Vol. 1.

Published as a report to IUPAC Commission Vol. 6, IUPAC Commission on Equilibrium Data.

530. Critical Micelle Concentrations of Aqueous Surfactant Systems. P. Mukerjee; K.J. Mysels (National Bureau of Standards). NTIS, 1971. Accession: COM 71-50203. Cost: $8.00. Series: NSRDS-NBS-36.

531. Critical Review of Equilibrium Data for Proton-and-Metal Complexes of 1,10-Phenanthroline, 2,2'-Bipyridyl and Related Compounds. W.A.E. McBryde. Pergamon Press Inc, 1978, 85 pp. LC: 77-30483. ISBN: 0-08-022344-3. Cost: $11.00. Series: *IUPAC Chemical Data Series*, Vol. 17.

At head of title: International Union of Pure and Applied Chemistry, Analytical Chemistry Division; Commission on Equilibrium Data; *Critical Evaluation of Equilibrium Constants in Solution*; Part A, *Stability Constants of Metal Complexes*. Acidity constants for 1,10-phenanthroline, 2,2'-bipyridinium ions and their derivatives are presented. The data include values in nonaqueous or mixed solids. Enthalpies and entropies of formation for the proton- and metal-ion complexes are also presented.

532. Critical Review of Ultraviolet Photo-Absorption Cross Sections for Molecules of Astrophysical and Aeronomic Interest. R.D. Hudson (National Bureau of Standards). U.S. Government Printing Office, 1971. SuDoc: C13.48:38. Cost: $1.50. Series: NSRDS-NBS-38.

533. Critical Solution Temperatures. Alfred W. Francis. American Chemical Society, 1961, 246 pp. LC: 61-11857. ISBN: 0-8412-0032-7. Cost: $22.00. Series: *Advances in Chemistry Series*, No. 31.

This is a collection of information on solvents including over 6,000 systems. The information is useful in determining whether 2 liquids mix. Seventy percent of the information is on hydrocarbons and the remaining portion is on nonhydrocarbon solvents.

534. Critical Stability Constants. Robert M. Smith; Arthur E. Martell. Plenum, Vol.1, 1974; Vol.2, 1975; Vol.3, 1977; Vol.4, 1976, Vol.1, 469 pp; Vol.2, 415 pp; Vol.3, 495 pp; Vol.4, 247 pp. LC: 74-10610. ISBN: Vol.1, 0-306-35211-7; Vol.2, 0-306-35212-5; Vol.3, 0-306-35213-3; Vol.4, 0-306-35214-1. Cost: Vol.1, $35.00; Vol.2, $45.00; Vol.3, $49.50; Vol.4, $35.00.

The 4 volumes are entitled as follows: Vol.1, *Amino Acids*; Vol.2, *Amines*; Vol.3, *Other Organic Ligands*; Vol.4, *Inorganic Complexes*. This 4-volume set contains stability constants of ligands, metal complex equilibria data and information on metal ions. An earlier edition was entitled *Stability Constants of Metal-ion Complexes*, The Chemical Society, 1964. NOTE: Pergamon Press, under the auspices of the Analytical Chemistry Division of the IUPAC has published the second supplement to the 2d ed. of *Stability Constants of Metal-ion Complexes*. Part A, *Inorganic Ligands*, 1980; Part B, *Organic Ligands*, 1979, each volume $150.00

535. Critical Survey of Stability Constants and Related Thermodynamic Data of Fluoride Complexes in Aqueous Solution. A.M. Bond; G.T. Hefter. Pergamon Press Inc, 1980, 80 pp. LC: 80-40413. ISBN: 0-08-022377-X. Cost: $25.00. Series: *IUAPC Chemical Data Series*, Vol. 27.

Enthalpy and entropy data are presented along with stability constants data of fluoride complexes and aqueous solutions.

536. Critical Surveys of Data Sources: Electrical and Magnetic Properties of Metals. M.J. Carr, et al. National Bureau of Standards (available from NTIS), 1976, 92 pp. LC: 76-18879. Accession: N77-19247; PB-258557. Tech Report: NBS-SP-396-4. Cost: $3.50 microform; $9.50 hard copy.

This is actually a directory of sources of numerical data on the electrical and magnetic properties of various metals.

537. Critical Surveys of Data Sources: Mechanical Properties of Metals. R.B. Gavert; R.L. Moore; J.H. Westbrook. General Electric Co (available from NTIS), 1974, 88 pp. LC: 74-8090. Accession: N75-21450. Tech Report: NBS-SP-396-1; Com-74-51060/3. Cost: $2.25 microform.

Provides information as to where data can be found on the mechanical properties of metals and alloys. There is actually no property data included but rather a listing of sources where the data can be obtained.

538. Criticality Handbook. R.D. Carter et al (Atlantic-Richfield Hanford Co). NTIS, 1968-71, 3 Vols. Tech Report: ARH-600.

539. Critically Evaluated Transition Probabilities for Ba I and II. B.M. Miles; W.L. Wiese (National Bureau of Standards). NTIS, 1969. Accession: AD-681351. Cost: $4.00. Series: NBS Tech. Note 474.

540. Cross Point System ESK 400 E. P. Binder. Heyden and Sons, 1976, 390 pp. ISBN: 0-85501-244-4. Cost: $50.30.

This handbook provides information on Private Automatic Branch Exchanges (PABX). This will be of interest to the communication and telephone industry.

541. CRSI Design Handbook, Vol. 2 R.C. Reese. Concrete Reinforcing Steel Institute, 1965, various paging.

542. CRSI Handbook Ultimate Strength Design. Concrete Reinforcing Steel Institute. Concrete Reinforcing Steel Intitute, 2d ed., 1980, various paging.

543. CRSI Handbook, Based upon the 1977 ACI Building Code. Concrete Reinforcing Steel Institute. Concrete Reinforcing Steel Institute, 4th ed., 1970, 800 pp.

544. CRSI Handbook. CRSI Engineering Practice Committee. Concrete Reinforcing Steel Institute, 4th ed., 1980, irregular pagination. Date of First Edition: 1952. Cost: $29.00.

The Concrete Reinforcing Steel Institute has presented an excellent guide on the structural properties of concrete. Information is presented on load capacity tables for flat slabs, flat plates, and joist construction. Such topics as pilings, cantilevered retaining walls, and column design are also discussed.

545. Cryogenic Materials Data Handbook, Revised. Fred R. Schwartzberg et al (Martin Marietta Corp). NTIS, 1970, rev. of report dated August 1964, AD-609562, 2 Vols. Accession: Vol.1, AD-713619; Vol.2, AD-713620. Tech Report: AFML-TDR-64-280.

546. Cryogenics Handbook. Beverly Law. Ann Arbor Science Publishers, Inc, 1980, 304 pp. ISBN: 0-86103-021-4. Cost: $39.00.

This handbook contains a listing of manufacturers, suppliers, and research organizations involved in low temperature scientific activity. Also includes several useful tables and conversion factors.

547. Crystal Data, Classification of Substances by Space Groups and Their Identification From Cell Dimensions. J.D.H. Donnay; Werner Nowacki. Geological Society of

America, 1954, 719 pp. (out of print). Series: *G.S.A. Memoir*, 60.

548. Crystal Data; Determinative Tables. J.D.H. Donnay; Helen M. Ondik. U.S. Department of Commerce, National Bureau of Standards and the Joint Committee on Powder Diffraction Standards, 3d ed., 1972 (Vol.1, 1972; Vol.2, 1973), 4 Vols. LC: 77-187758. Cost: Vol.1, $30.00; Vol.2, $50.00; Vol.3, $65.00; Vol.4, 65.00.

This set of tables includes 4 volumes entitled as follows: Vol.1, *Organic Compounds* (thru 1966); Vol.2, *Inorganic Compounds* (thru 1966); Vol.3, *Organics, Organometallics* (1967-74); Vol.4, *Inorganic Metals, Minerals* (1967-69). Includes over 47,000 entries listing such data as axial ratios, cell dimensions, and crystal structure. Available from: Joint Committee on Powder Diffraction Standards, 1601 Park Lane, Swarthmore, PA 19081.

549. Crystal Structure Transformations in Binary Halides. C.N.R. Rao; M. Natarajan (National Bureau of Standards). NTIS, 1972. Accession: COM 72-50849. Cost: $4.50. Series: NSRDS-NBS-41.

550. Crystal Structure Transformations in Inorganic Nitrites, Nitrates, and Carbonates. C.N.R. Rao; B. Prakash; M. Natarajan (National Bureau of Standards). U.S. Government Printing Office, 1975. SuDoc: C13.48:53. Cost: $1.15. Series: NSRDS-NBS-53.

551. Crystal Structure Transformations in Inorganic Sulfates, Phosphates, Perchlorates, and Chromates. B. Prakash; C.N.R. Rao (National Bureau of Standards). U.S. Government Printing Office, 1975. SuDoc: C13.48:56. Cost: $0.85. Series: NSRDS-NBS-56.

552. Crystal Structures. Ralph W. Wyckoff. Wiley-Interscience, 2d ed., 1963-71, 6 Vols. (Vol.6 has 2 parts). LC: 63-22897. ISBN: Vol.1, 0-470-96860-5. Cost: Vol.1, $30.75; Vol.2, $40.75; Vol.3, $51.75; Vol.4, $40.75; Vol.5, $47.50; Vol.6, Part 1, $40.25; Vol.6, Part 2, $54.50.

553. Cumulative Index to the Mössbauer Effect Data Indexes. John G. Stevens; Virginia Stevens; William Gettys. Plenum, 1980, 358 pp. ISBN: 0-306-65150-5. Cost: $75.00.

This is an index to a 9-volume set entitled *Mössbauer Effect Data Indexes*. The index provides 3 approaches including isotopes, subject, and author.

554. Curium Data Sheets. E.E. Ketchen; S.J. Rimshaw. NTIS, 1969, 40 pp. Accession: N69-25947. Tech Report: ORNL-4357.

These data sheets include curium data tabulations, including composition, critical mass, thermophysical and chemical properties, biological tolerances, and shielding information. An update to the *Curium Data Sheets* was published in 1973, edited by T. A. Butler, E. E. Ketchen and J. R. Distefano, 46 pp, accession number N76-72023, and technical report number ORNL-4910.

555. Current Drug Handbook 1976-78. Mary W. Falconer et al (comps). Saunders, 1976, 279 pp. LC: 58-6390. ISBN: 0-7216-3567-9. Cost: $6.50.

556. Current Therapy 1979. Howard Conn. W.B. Saunders Co, 1979, 1,000 pp. Cost: $27.00.

Presents 300 medical conditions with corresponding therapeutic guidelines.

557. Cutting for Construction: A Handbook of Methods and Applications of Hard Cutting and Breaking On Site. D. Lazenby; P. Phillips. Halsted Press, 1978, 116 pp. LC: 78-040610. ISBN: 0-470-26437-3. Cost: $22.95.

Describes tools and techniques that are used in cutting and demolition jobs during the construction process.

558. Cutting's Handbook of Pharmacology: The Actions and Uses of Drugs. T.Z. Craigie. Appleton-Century-Crofts, 6th ed., 1979, 697 pp. Cost: $14.95.

The emphasis of this book is on the pharmacology and the chemical structure of drugs.

559. Cyclic Designs. J.A. John; F.W. Wolock; H.A. David (National Bureau of Standards). U.S. Government Printing Office, 1972, 79 pp. SuDoc: C13.32:62. Cost: $1.20. Series: NBS AMS62.

560. Damage Tolerant Design Handbook. A Compilation of Fracture and Crack-Growth Data for High-Strength Alloys. Ohio Metals and Ceramics Information Center (available from NTIS), 1972, 3 Vols. (basic volume and supplements 1 and 2). Accession: AD-753774; Supplement 1, AD-772810; Supplment 2, AD-A010392. Cost: $75.00 microfiche.

Several high-strength alloys, selected for their potential aircraft applications, and their fracture mechanics data, have been presented in this handbook. The following alloys are included: nickel-base alloy (718), titanium alloys, aluminum alloys, alloy steels, and stainless steels.

561. Dana's Manual of Mineralogy. Cornelius S. Hurlbut Jr; Cornelius Klein (original author was James Dwight Dana). John Wiley and Sons, 19th ed., 1977, 532 pp. LC: 77-1131. ISBN: 0-471-42226-6. Cost: $19.95.

562. Dangerous Properties of Industrial Materials. N. Irving Sax. Van Nostrand Reinhold, 5th ed., 1979, 1,118 pp. LC: 78-20812. ISBN: 0-442-27373-8. Date of First Edition: 1951, entitled: *Handbook of Dangerous Materials*. Cost: $57.50.

This monumental work has gone through 5 different editions and now includes information on approximately 16,000 industrial materials that are potentially hazardous. There are many sections in this handbook written by recognized authorities on such subjects as radiation problems, cancer, noise, air pollutants, fire dangers, microwaves, and food additives. The major tabulation in this handbook follows the narrative sections and provides information on the chemicals such as various fire, explosion, toxic and disaster ratings; physical constants, antidotal information, and formulas. In addition, such information as regulations, various related legislation, shipping, handling, and labeling of hazardous materials is included, as well as many bibliographic references to original source material. Also refer to such related titles as the *Handbook of Reactive Chemical Hazards*, the *Chemical Technology Handbook* and the *CRC Handbook of Laboratory Safety*. Updating service now available as *Dangerous Properties of Industrial Materials Report* at $96.00/yr.

563. Darkroom Handbook: A Complete Guide to the Best Design, Construction, and Equipment. Dennis Curtin; Jow DeMaio. Van Nostrand Reinhold, 1979, 192 pp. ISBN: 0-930764-08-0. Cost: $17.95.

This book deals primarily with designing and equipping a professional darkroom. Includes information on the design plans for plumbing systems, tables, and sinks.

564. Data Acquisition and Conversion Handbook. Datel-Intersil, Inc, 242 pp. Cost: $3.95 (outside U.S., add $2.75 for shipping).

This handbook contains over 315 illustrations and 40 tables on data conversion circuits and systems. Also presents information on A/D and D/A converters and high speed operational amplifiers.

565. Data and Formulae for Engineering Students. J.C. Anderson et al. Pergamon Pres, 2d ed., 1969, 54 pp. ISBN: 0-08-013989-2. Cost: $7.25.

566. Data Book for Civil Engineers. Elwyn E. Seelye. John Wiley and Sons, 3d ed., 1957-60, 3 Vols. LC: Vol.1, 58-5932; Vol.2, 57-5932. ISBN: Vol.1, 0-471-77286-0; Vol.2, 0-471-77319-0; Vol.3, 0-471-77352-2. Cost: Vol. 1, $57.95; Vol. 2, $51.95; Vol. 3, $29.95.

Vol.1 is entitled *Design*; Vol.2, *Specifications and Cost*; and Vol.3, *Field Practice*.

567. Data Book for Electronic Technicians and Engineers. John D. Lenk. Prentice-Hall, 1968, 185 pp. LC: 68-15840. ISBN: 0-13-197160-3. Cost: $14.95.

568. Data Book for Pipe Fitters and Pipe Welders. Edward Harold Williamson. Bailey Bros. and Swinfen, 1977. LC: 77-361925.

Provides set-out tables for fitting and welding pipes.

569. Data Book on Hydrocarbons: Application to Process Engineering. J.B. Maxwell. R.C. Krieger Publishing Co Inc, rep. of 1950 ed., 1975, 268 pp. ISBN: 0-88275-257-X. Cost: $13.50.

570. Data Collection for Nonelectronic Reliability Handbook. William Yurkowsky. NTIS, 1968, Vol.1, 145 pp.; Vol.3, Sec.1, 549 pp.; Vol.4, Sec.2, 76 pp.; Vol.5, Sec.3, 147 pp. Accession: Vol.1, AD-841-106. Tech Report: Vol.1, FR-68-16-84. Cost: $3.50 per Vol., microform.

This handbook presents information on 600 different types of nonelectric parts. Included is such information as failure rate data, stress level data, and statistical distributions.

571. Data for Biochemical Research. R.M. Dawson. Oxford University Press, 2d ed., 1969. ISBN: 0-19-855338-2. Cost: $19.95.

572. Data for Plasmas in Local Thermodynamic Equilibrium. Hans-Werner Drawin; Paul Felenbok. Gauthier-Villars, Paris, 1965, 503 pp. LC: 65-56119.

573. Data Handbook for Clay Materials and Other Non-Metallic Minerals. H. Van Olphen; J.J. Fripiat. Pergamon Press, 1979, 346 pp. LC: 78-41214. ISBN: 0-08-022850-X. Cost: $76.00.

This is a compilation of physical, chemical, and mineralogical data on clay and other nonmetallic materials. Data resulting from chemical analysis, X-ray fluorescence, X-ray diffraction, cation exchange capacity, surface area, electron microscopy, and thermal analysis information are provided. In addition, infrared spectroscopy, Raman spectroscopy, and electron spin resonance

data are included. The nonmetallic materials are gibbsite, calcite, magnesite, and gypsum.

574. Data Handling for Science and Technology. An Overview and Sourcebook. Stephen A. Rossmassler; David G. Watson. Elsevier-North Holland Publishing Co, 1980, 184 pp. ISBN: 0-444-86012-6. Cost: DM 55.00.

This handbook was sponsored by Codata and UNESCO.

575. Data of Geochemistry. U.S. Geological Survey, 6th ed., 1962. Series: *U.S. Geological Survey. Professional Paper*, 440.

576. Daylight Illumination-Color-Contrast Tables for Full-Form Objects. M.R. Nagel; H. Quenzel; W. Kweta; R. Wendling. Academic Press Inc, 1978, 494 pp. ISBN: 0-12-513750-8. Cost: $43.00.

Provides tables that present the computation of luminance distribution, computation of irradiance, and illuminance.

577. DC Motors, Speed Controls, Servo Systems; An Engineering Handbook. Electro-craft Corp. U.S.A. Pergamon Press, 3d ed., 1977, various paging. LC: 76-56647. ISBN: 0-08-021715-X. Cost: $33.00.

578. DC-DC Converters. General Electric Co. NTIS, 1974, 50 pp. Accession: N75-21513. Tech Report: NASA-CR-120728. Cost: $3.75 hard copy.

Test data and various tables are presented on the electrical and mechanical characteristics of DC-DC converters.

579. De Laval Engineering Handbook. Hans Gartmann. McGraw-Hill, 3d ed., 1970, 512 pp. (various paging). LC: 73-107288. ISBN: 07-022908-2. Date of First Edition: 1947. Cost: $24.50.

This handbook, originally prepared by De Laval Turbine, Inc., provides a considerable amount of engineering information on such subjects as fluid mechanics, turbines, transmissions, condensers, gears, pumps, and filtering. The 3rd edition of the *De Laval Engineering Handbook* is a handy source of information for quick look-up engineering data needed to solve practical design problems. Most of the data are in tabular or graphic form, and an excellent index is provided. For related works, refer to such titles as the *Gas Turbine Engineering Handbook*, the *CRC Handbook of Tables for Applied Engineering Science*, the *Engineering Manual* and the *Engineers' Manual*.

580. Defense Nuclear Agency Reaction Rate Handbook. Marlyn H. Bortner; Theodore Baurer. NTIS, 2d ed., 1972, 1,033 pp. Accession: AD-763 699; various revisions No. 1, AD-764 303; No. 6, AD-A024 362; No. 7, AD-A062 474. Cost: $1.45 microform; $21.00 hard copy.

This handbook presents information regarding the atmosphere during major disturbances such as nuclear bursts or solar flare impingements. The chemical and physical mechanisms of these disturbances are discussed in terms of fluid mechanics, energy balance, chemical kinetics, equilibrium data, photochemical processes, electron collision processes, and charged-particle recombination processes.

581. Design Charts for Reinforced Concrete Sections. R. Walther; B. Houriet. Presses Polytechniques Romandes, 1977. Cost: SFr. 80.00 per Vol.

This is a 2-volume set. Vol.1 is entitled *Design Charts for Reinforced Concrete Sections: Solid Sections*. Vol.2 is entitled *Design*

Charts for Reinforced Concrete Sections: Hollow Sections, copyright date 1980. Vol.1 consists of 201 charts for solid circular and rectangular reinforced concrete sections subjected to different types of bending, such as biaxial bending, combined normal force, and pure bending. Vol.2 contains concrete sections, and 52 charts for circular hollow reinforced concrete sections.

582. Design Graphs for Concrete Shell Roofs. C.B. Wilby. Applied Science Publishers Ltd., 1980, 148 pp. ISBN: 0-85334-899-5. Cost: $37.50.

Graphs and tables are presented on the design of cylindrical reinforced concrete shell roofs. Such topics as symmetrical shells, north-light shell roofs and boundary conditions are discussed.

583. Design Handbook for Phase Change Thermal Control and Energy Storage Devices. W.R. Humphries; E.T. Griggs. National Aeronautics and Space Administration, 1977, 255 pp. Accession: N78-15434. Tech Report: NASA-TR-1074. Cost: $3.50 microform; $20.00 hard copy.

Phase change thermal control information is presented using straight metal fins as the filler and paraffin as the phase change material. Performance in zero-G and one-G fields is examined.

584. Design Handbook of Wastewater Systems; Domestic, Industrial, Commercial. Brian L. Goodman. Technomic Publishing Co., Inc, 1971, 160 pp. LC: 77-161057. ISBN: 0-87762-065-2. Cost: $20.00.

This manual covers the full range of wastewater treatment from filter design, biological treatment, pretreatment to chlorination and solids disposal.

585. Design Handbook: In Accordance with the Strength Design Method of ACI 318-77. American Concrete Institute, 2d ed., Vol.1, 1981; Vol.2, 1978, Vol.1, 544 pp.; Vol.2, 228 pp. LC: 78-50814. ISBN: Vol.1, 0-685-85091-9; Vol.2, 0-685-85093-5. Cost: Vol.1, $59.95; Vol.2, $43.00. Series: *ACI Special Publication*, (81) (Vol. 1); *ACI Special Publication*, 17 A (78) (Vol. 2)..

Vol.1 is entitled *Beams, Slabs, Brackets, Footings and Pile Caps*. Vol.2 is entitled *Columns*. Tables and graphs are presented to assist in the calculation of axial load, uniaxial bending, biaxial bending, load-moment strength, and other topics dealing with reinforced concrete. Many design examples are presented along with design aids and commentary, which include equations and derivations.

586. Design Manual—Roller and Silent Chain Drives. Roller and Silent Chain Section, American Chain Association, 1974, 99 pp. Date of First Edition: 1955. Cost: $10.00.

587. Design of Free-Air Ionization Chambers. National Bureau of Standards. NTIS. Accession: COM 73-10872. Cost: $3.50. Series: NBS Handbook 64.

588. Design Tables for Beams on Elastic Foundations and Related Structural Problems. K.T. Sundara; Raja Iyengar; S. Anantha Ramu. International Ideas, Inc, 1979. ISBN: 0-85334-841-3. Cost: $38.90.

Tables are presented on influence coefficients for determining the distribution of pressure, bending moment, and shear force in a beam on an elastic foundation.

589. Design Tables for Timber Roof Structures. Halasz et al, 1972, 410 pp. Cost: DM 69.

Includes 45 illustrations, 28 tables. Text is in German, English, and French.

590. Designer's Handbook of Pressure Sensing Devices. Jerry L. Lyons. Van Nostrand Reinhold, 1980, 304 pp. LC: 79-24468. ISBN: 0-442-24964-0. Cost: $27.50.

This is basically a guide to the selection, specification, and usage of pressure-sensing devices, such as pressure switches, pressure transducers, diaphragms, and liquid level switches.

Desk-Top Data Bank Series. *See* Foams; Films, Sheets and Laminates; Chemical Resistance Data Books; Specifications for Adhesives; Elastomers; Plastics for Electronics; Specifications for Plastics; Commercial Names and Sources; Tool Steels; Stainless Steels.

591. Determination of Configurations by Chemical Methods. Kagan. Heyden, 1977. ISBN: 0-9940010-2-9. Cost: $39.00. Series: *Stereochemistry, Fundamentals and Methods Series*, Vol. 3.

592. Determination of Configurations by Dipole Moments, CD or ORD. Kagan. Heyden, 1977. ISBN: 0-9940010-1-0. Cost: $57.00. Series: *Stereochemistry, Fundamentals and Methods Series*, Vol. 2.

593. Determination of Configurations by Spectrometric Methods. Kagan. Heyden, 1977. ISBN: 0-9940010-0-2. Cost: $69.00. Series: *Stereochemistry, Fundamentals and Methods Series*, Vol. 1.

594. Determination of Thermodynamic Properties of AeroZINE-50. Final Report. J.P. Copeland; J.A. Simmons. NTIS, 1968, 125 pp. Accession: N69-16220. Tech Report: NASA-CR-92463.

Title is followed by, *Thermodynamic Properties of AeroZINE at Pressures .001 to 60 Psia and Temperatures from Minus 100 to 250 Degrees*. This report also has an appendix D that was issued in 1968, 58 pp., Accession No.: N69-16889, Technical Report No.: NASA-CR-92464. These reports cover such thermodynamic properties as pressure gradients, temperature gradients, enthalpy, entropy, and temperature effects.

595. Development of MIL-HDBK-5 Design Allowable Properties and Fatigue-Crack Propagation Data for Several Aerospace Materials—Stainless Steels, Cobalt Alloys, and Aluminum Alloys. P.E. Ruff; S.H. Smith. NTIS, 1977, 208 pp. Accession: N78-30174; AD-A054430. Tech Report: AFML-TR-77-162. Cost: $3.50 microform; $17.00 hard copy.

This report provides considerable information on aerospace materials including mechanical properties, metal fatigue, crack propagation, modulus of elasticity, and various stress-strain relationships.

596. Diabetes Today—A Handbook for the Clinical Team. J. Ireland et al. HM + M Publishers Ltd, 1979, 256 pp. ISBN: 0-85602-053-2. Cost: $16.50.

Provides information on diabetic control and information regarding the adjustment of the patient's blood glucose.

597. Diagnostic Handbook of Speech Pathology. Williams and Wilkins, 1979, 387 pp. LC: 78-15594. Cost: $17.95.

This is more of a graduate-level textbook than a handbook, but it does provide information as a modest reference on diagnostic procedures in speech pathology.

Dictionary of Organic Compounds. *See* **Heilbron's Dictionary of Organic Compounds.**

598. Diffusion and Defect Data. Trans Tech Publications. Trans Tech Publications, 1967-, multivolume set. Cost: Vols.1-20, $600.00; Vols.21-26, $344.00.

This multivolume set, currently in 26 volumes, covers both metals and nonmetals. It provides extended abstracts and a collection of properties and data on innumerable materials. Such topics as diffusional mass transport, irradiation effects, point defects, ionic conduction, line, and planar defects are discussed. Numerous materials such as semiconductors, ceramics, organic crystals, ionic substances, halides, and inert-gas solids are presented. In addition, diffusion and migration in solids and liquids, color centers, ion implantation, and stacking faults are discussed.

599. Digital Computer User's Handbook. M. Klerer; C.A. Korn. McGraw-Hill, 1967, 750 pp. LC: 66-19285. ISBN: 0-07-035043-4. Cost: $39.75.

600. Digital Detailed Data. Mark R. Klein (Reliability Analysis Center). NTIS, 1976, 234 pp. Accession: ADA-033937. Tech Report: RAC-MDR-4. Cost: $50.00.

601. Digital Filter Design Handbook. F.J. Taylor. Marcell Dekker, Inc, in press (1982). ISBN: 08247-1357-5. Series: *Electrical Engineering and Electronics Series*, Vol. 12.

This book is currently in press and promises to be an excellent source of information on the design of digital filters.

602. Digital Integrated Circuit D.A.T.A. Book. Derivation and Tabulation Associates, Inc. DATA, Inc, a continuation issued semiannually, pages vary per issue. Cost: $44.00 per year by subscription.

603. Directory of Electronic Circuits. Matthew Mandl. Prentice-Hall, rev. enlarged 2d ed., 1977, 352 pp. ISBN: 0-13-214924-9. Cost: $17.95.

Contains schematics on more than 200 basic electronic circuits such as signal processors, digital circuits and control system circuits.

604. Disability and Rehabilitation Handbook. Robert M. Goldenson. McGraw-Hill, 1978, 736 pp. LC: 78-1441. ISBN: 0-07-023658-5. Cost: $24.50.

605. Discontinued Integrated Circuit D.A.T.A. Book. Derivation and Tabulation Associates, Inc. DATA, Inc, a continuation issued semiannually, pages vary per issue. Cost: $44.00 per year by subscription.

606. Discontinued Thyristor D.A.T.A. Book. Derivation and Tabulation Associates, Inc. DATA, Inc, a continuation issued annually, pages vary per issue. Cost: $12.75 per year by subscription.

607. Discontinued Transistor D.A.T.A. Book. Derivation and Tabulation Associates, Inc. DATA, a continuation issued annually, pages vary per issue. Cost: $18.50 per year by subscription.

608. Dissociation Constants of Inorganic Acids and Bases in Aqueous Solution. D.D. Perrin. Pergamon Press Inc, 1969, 110 pp. ISBN: 0-08-020825-8. Cost: $10.50. Series: *Chemical Data Series*, Vol. 19.

609. Dissociation in Heavy Particle Collisions. G.W. McClure; J.M. Peek. Wiley-Interscience, 1972. Cost: $13.95.

610. Distillation Equilibrium Data. Ju-chin Chu. Van Nostrand Reinhold, 1950, 312 pp. Cost: $37.50.

611. Distribution and Transportation Handbook. Harry J. Bruce. Cahners Books, 1971, 393 pp. LC: 76-132669. ISBN: 0-8436-1400-5. Cost: $15.00.

612. Doctor's Lawyer: A Legal Handbook for Doctors. Marc J. Lane. C.C. Thomas, 1974. ISBN: 0-398-02988-1. Cost: $12.75.

613. DOE Facilities Solar Design Handbook. Public Document Distribution Center, 1978, 169 pp. (loose-leaf). SuDoc: E 1.32:0006/1. Cost: $3.75.

Includes information on solar heating and cooling, system components, and solar energy economics.

614. Dow's Fire and Explosion Index Hazard Classification Guide. American Institute of Chemical Engineers, 5th ed., 1980, 57 pp. LC: 80-29237. ISBN: 0-8169-0194-5. Date of First Edition: 1964. Cost: $20.00.

Provides such information as procedures for risk analysis calculation, determining hazard factors, determining fire and explosion index, loss control credit factors, and property damage calculations.

615. Drainage Handbook for Surface Mining. West Virginia Department of Natural Resources, 1975, 75 pp. plus appendices.

Covers such topics as sediment control, acid water, land stabilization, waste disposal, building sediment dams, and designing spillways. In addition, sediment ponds, earth embankments, gabion sediment dams, crib sediment dams, excavated sediment channels, and stone check dams are discussed. Numerous figures, tables, and charts are provided on rock riprap flumes, pipe flow dimensions, spillway hydraulics, and parabolic waterway design.

616. Drift Mobilities and Conduction Band Energies of Excess Electrons in Dielectric Liquids. A.O. Allen (National Bureau of Standards). U.S. Government Printing Office, 1976. SuDoc: C13.46:58. Cost: $0.70. Series: NSRDS-NBS-58.

617. Drilling and Drilling Fluids. George V. Chilingarian; Paul Vorabutr. Elsevier-North Holland Publishing Co, 1980, 776 pp. ISBN: 0-444-41667-9. Cost: $136.50. Series: *Developments in Petroleum Science*, 11.

This comprehensive volume covers such topics as rotary drilling, clays, rheology, emulsion drilling fluids, oil-base drilling fluids, polymer drilling fluids, friction reducers, and corrosion.

618. Drilling Data Handbook. French Publishing Association, Inc, 1978, 448 pp. ISBN: 2-7108-033408. Cost: $46.80.

Information is provided in metric and Anglo-American units.

619. Drug Dosage in Laboratory Animals; A Handbook. C.D. Barnes; L.G. Eltherington. University of California Press, 2d rev. and enlarged ed., 1973, 341 pp. LC: 74-155402. ISBN: 0-520-02273-4. Cost: $13.75.

620. Drug Identification Guide. Van Nostrand Reinhold, 4th ed., 1980, 116 pp. ISBN: 0-442-84075-6. Cost: $4.95 soft copy.

A useful guide that will help to identify over 1,000 drugs, capsules, and tablets. These drugs are shown in actual size and in full color.

621. Drug Information Services Handbook. G. Edward Collins; Herman L. Lazarus. Publishing Sciences Group, 1975. LC: 74-82441. ISBN: 0-88416-024-6. Cost: $10.00.

622. Drug Interactions Index, 1978-1979. Eric W. Martin. J. B. Lippincott, Co, 1979, 312 pp. LC: 78-15654.. Cost: $10.95.

Provides information on drug interactions.

623. Drugs in Current Use and New Drugs 1977. Walter Modell. Springer Publishing Co, 22nd ed, 1977, 175 pp. ISBN: 0-8261-0155-0. Cost: $6.50 paper copy.

624. Drugs in Dentistry: Dental Practitioner Handbook No. 9. L.W. Kay. Year Book Medical Publishers, 2d ed, 1975, 332 pp. ISBN: 0-8151-4998-0. Cost: $17.50.

625. Dry Kiln Handbook. Jack L. Bachrich. Miller Freeman Publishing, Inc, 1980, 400 pp. ISBN: 0-87930-087-6. Cost: $50.00.

626. Dynamical Properties of Solids. G.K. Horton; A.A. Maradudin. Elsevier-North Holland Publishing Co, 1974-80, Vol.1, 662 pp. ISBN: Vol.1, 0-7204-0278-6. Cost: Vol.1, $109.75.

This 4-volume set covers numerous properties of solids, and is a vast collection of materials and properties including crystalline solids, as well as other types of solids. Vol.1 is entitled *Crystalline Solids, Fundamentals*, 1974. Vol.2 is entitled *Crystalline Solids, Applications.*, 1975, 536 pp., ISBN 0-7204-0284-0, $102.50. Vol.3, 1980, 318 pp., ISBN 0-444-85314-6, $58.50. Vol.4, 1980, 450 pp., ISBN 0-444-85315-4, $78.00.

627. E/MJ Operating Handbook of Mineral Surface Mining and Exploration. Richard Hoppe. McGraw-Hill, 1978, 446 pp. LC: 78-4501. ISBN: 0-07-019518-8. Cost: $19.50. Series: *E/MJ Library of Operating Handbooks*, Vol. 2.

This handbook is actually a collection of short articles on numerous topics ranging from surface mining to blasting. Numerous minerals are discussed in this compilation including bauxite, uranium, iron ore, and copper.

628. E/MJ Operating Handbook of Mineral Underground Mining. Robert Sisselman. McGraw-Hill, 1979, 458 pp. ISBN: 0-07-019521-8. Cost: $19.50. Series: *E/MJ Library of Operating Handbooks*, Vol. 3.

This publication is a collection of articles on underground mining, blasting, use of explosives, materials handling, and transportation of ores. In addition, this handbook covers safety of underground mining, dust control, ventilation, wet suppression systems, and fire fighting equipment.

629. E/MJ Second Operating Handbook of Mineral Processing. Lane White (Mining Information Services). McGraw-Hill, 1980, 509 pp. LC: 80-18292. ISBN: 0-07-606684-3. Cost: $19.50.

This handbook includes many flow charts and tables as well as information on hydrometallurgical and pyrometallurgical processing. An excellent index is provided. Compiled from articles in E/MJ for 1977-79.

630. Earth I: The Upper Atmosphere, Ionosphere and Magnetosphere. Charlotte W. Gordon et al. Gordon and Breach, 1977, 416 pp. ISBN: 0-677-16100-X. Series: *Handbook of Astronomy, Astrophysics and Geophysics*, Vol. 1.

This is basically a handbook on astronomy and astrophysics. Covers such topics as the Van Allen belt, polar-cap absorption, auroral particle precipitation, and hydrogen in the upper atmosphere.

631. Earth Manual; A Water Resources Technical Publication, A Guide to the Use of Soils as Foundations and as Construction Materials for Hydraulic Structures. U.S. Department of the Interior, Bureau of Reclamation. U.S. Government Printing Office, 2d ed., 1974, 810 pp.

632. Earth Science Manual. Edward E. Lyon et al. William C. Brown Co, Publishers, 4th ed., 1976, 224 pp. ISBN: 0-697-05079-3. Cost: $7.95.

633. Economic Analysis Handbook. Public Documents Distriubtion Center, 1975, rep. 1978, 147 pp. SuDoc: D 209.14: EC 7. Cost: $3.25.

This is a handbook on how to prepare a cost/benefit analysis. This publication was developed for use by the U.S. Navy, however, it has wide applications in the civilian and private sector.

634. Effect of Hazardous Material Spills on Biological Treatment Processes. Andrew P. Pajak et al. NTIS, 1977, 204 pp. Accession: PB-376 724. Tech Report: EPA/600/2-77/236. Cost: $3.50 microform; $17.00 hard copy.

Over 250 chemical substances are listed in alphabetical order with the corresponding effects on biological treatment processes such as the activated sludge treatment process.

635. Efficient Electricity Use: A Reference Book on Energy Management for Engineers, Architects, Planners and Managers. Craig B. Smith. Pergamon Press Inc, 2d ed., 1978, 778 pp. LC: 75-44373. ISBN: 0-08-023227-2. Date of First Edition: 1976. Cost: $44.00.

This book is on energy conservation for the building industry. Such topics as ventilation, air conditioning, heat recovery, electromechanical energy, and energy management are discussed. Contains over 500 literature references, conversion factors, and a considerable amount of engineering data on energy conservation.

636. Eight-Place Tables of Trigonometric Functions for Every Second of Arc. Jean Peters. Chelsea Publishing Co, 1939 (renewed 1965), 954 pp. LC: 63-23072. ISBN: 0-

8284-0185-3. Date of First Edition: 1939. Cost: $33.00 for thumb-index edition.

As stated in the title, this mathematical handbook consists of *Eight-Place Tables of Trigonometric Functions for Every Second of Arc*. Such trigonometric functions as sines, cosines, tangents, and cotangents are covered. The appendix includes 21 place tables for sines and cosines. The thumb-index edition is easy to use, and the trigonometric functions can be located quickly. Also refer to such related works as the *CRC Standard Mathematical Tables*, the *CRC Handbook of Tables for Mathematics*, and various publication series of the U.S. National Bureau of Standards such as *Applied Mathematics Series* numbers AMS 5, AMS 32, AMS 40, AMS 43, Columbia University Press Series number CUP5, and Mathematical Tables Series numbers MT 3, MT 4, MT 5, MT 13, and MT 16.

637. Elastic Constants of Crystals. Hillard B. Huntington. Academic Press, Inc, 1958 (rep. 1964), 139 pp. ISBN: 0-12-608456-4. Cost: $12.00 (reprinted).

The focus of this information is primarily on the elastic constants of inorganic single crystals. Elastic constants are provided for various materials including alkali halides, aluminum, indium, diamonds, and some piezoelectric materials. Variation of elastic constants due to changes in temperature and pressure are also discussed.

638. Elastically Supported Beams and Plates, Cylindrical Shells. K.H. Wölfer. International Publishing Service, 4th ed., 1978, 536 pp. of tables and 180 pp. of text. ISBN: 3-7625-0778-3. Cost: $56.00.

This is a collection of tables of coefficients for bending moment, shear force, and bearing pressure for elastically supported beams, plates, and cylindrical shells. The tables and diagrams are based on the modulus of subgrade reaction method.

639. Elastomers. Society of Plastics Engineers. International Plastics Selector, Inc, 2d ed., 1979, over 990 pp. Cost: $41.65 members; $50.95 nonmembers. Series: *Desk Top Data Bank*.

Property data are provided on over 3,400 elastomers. Natural rubber specifications, and aging test data are also listed.

640. Electric Conductivity of Ferroelectrics. V.M. Gurevich. NTIS, Russian translation for the National Bureau of Standards, 1969, trans. 1971. Accession: TT-70-50180. Cost: $10.50.

641. Electric Motor Handbook. E.H. Werninck. McGraw-Hill, 1978, 648 pp. LC: 77-30474. ISBN: 0-07-084488-7. Cost: $34.50.

Provides information on AC and DC motors, brushless motors, and linear actuators.

642. Electric, Electronic Handbook. Department of Defense. Superintendent of Documents, 1978, 256 pp. SuDoc: D1.6/2:EL 2. Cost: $4.75.

This is a Department of Defense handbook that would be useful in estimating person-hour requirements for electrical installations and electronics projects.

643. Electrical and Electronics Products World Safety Mutual Speedy Finder. Media International Promotions Inc, 400 pp. Cost: $100.00. Series: *World Standards Mutual Speedy Finder*, Vol. 5.

Compares tables of 6 countries and the International Electrotechnical Commission standards. Countries included are U.S., West Germany, Canada, Australia, United Kingdom, and Japan. Numerous safety standards are included for such equipment as television receivers, radio receivers, audio receivers, household appliances, tape recorders, and amplifiers.

644. Electrical and Electronics World Standards Mutual Speedy Finder. Media International Promotions, Inc, 1,150 pp. Cost: $95.00. Series: *World Standards Mutual Speedy Finder*, Vol. 2.

Indexes over 18,000 standards on electrical and electronics subjects. Covers 5 different countries including the U.S., West Germany, France, United Kingdom, and Japan. It has 3 indexes—standards number index, key word index, and products-parts name index.

645. Electrical Design, Safety and Energy Conservation. Albert Thumann. Fairmont Press Inc, 1977. Cost: $24.95.

This book has a wealth of information on electrical design for industrial plants and includes such things as lighting systems, power distribution, and conduits. This book will be helpful in complying with OSHA Standards for electrical installations.

646. Electrical Engineer's Handbook. Arthur Liebers. Key Books, 1968, 319 pp. LC: 68-2240.

647. Electrical Engineer's Reference Book. M.G. Say. Butterworths (also Transatlantic), 13th ed., 1973, various paging. ISBN: 0-408-70289-3. Date of First Edition: 1945. Cost: $52.00.

The 13th edition of the *Electrical Engineer's Reference Book* is an excellent source of information on such electrical subjects as generators, electrical networks, conversion of various types of energy plants into electrical power, electrical transmission, electrical materials, transformers, electromagnetism. This handbook includes information on various applications of electrical power, plus it provides the electrical calculations and data in SI units (and occasionally in conventional units), and it provides an index for retrieval of the electrical data. For related works, please refer to the *Standard Handbook for Electrical Engineers*, the *Electronics Engineers' Handbook*, the *Basic Electronic Instrument Handbook*, the *Handbook of Electronic Packaging*, the *Handbook of Materials and Processes for Electronics*, the *Handbook of Modern Electronic Data*, *Industrial Electornics Handbook*, and the *Handbook of Semiconductor Electronics*.

648. Electrical Engineering Handbook. Siemens Atkiengelsellschaft. Heyden and Sons, 1976, 749 pp. ISBN: 0-85502-231-5. Cost: $33.00.

This is an English translation of a German handbook. The emphasis is on electrical power, but the book does cover other subjects, such as automation and control engineering. Also covers synchronous machines, electric motors, circuit breakers, and electrochemistry.

649. Electrical Engineers' Handbook. H. Pender et al. John Wiley and Sons, 4th ed., 1949, 2 Vols. LC: 49-11664. ISBN: Vol.1, 0-471-67881-3; Vol.2, 0-471-67848-1. Date of First Edition: 1914. Cost: Vol.1, $31.95; Vol.2, $35.50.

650. Electrical Estimating Handbook. Irving M. Cohen. Van Nostrand Reinhold, 1975, 121 pp. LC: 74-78570. ISBN: 0-442-12152-0. Cost: $19.95.

This handbook, primarily designed for the construction industry, provides information on estimating electrical needs, explains electrical specifications and includes tables that will assist in more accurate electrical systems.

651. Electrical Installations Handbook, Part I and II. G.G. Seip. Heyden and Sons Inc, 2-Vols., 1,316 pp. ISBN: 0-85501-260-9. Cost: $115.50.

This handbook covers such topics as power supply and distribution systems, switching systems, electricity meters, cables and wires, and illumination systems. Various installations are discussed such as electrical installation for hospitals, hotels, theaters, and airports.

652. Electrical Materials Handbook. Allegheny Ludlum Steel Corp. Allegheny Ludlum Steel Corp, 1961.

653. Electrical Properties of Printed Wiring Boards. C.W. Jennings. NTIS, 1976, 61 pp. Accession: N77-17381. Tech Report: SAND-75-0663. Cost: $3.50 microform; $8.00 hard copy.

Test results are tabulated for printed circuit boards. Such properties as voltage holdoff, current carrying capacity, and insulation resistance are listed.

654. Electrical Properties of Solids: Surface Preparation and Methods of Measurement. T.F. Connolly. Plenum Publishing Corp, 1972, 96 pp. LC: 74-133469. ISBN: 0-306-68324-5. Cost: $14.50. Series: *Solid State Physics Literature Guides*, Vol. 4.

This book is organized by properties and materials and concentrates primarily on semiconductors. Also includes some metals, alloys, insulators, and dielectric materials. Such topics as resistivity and conductivity, dielectrics, surface preparation, and the Hall effect are discussed.

655. Electrical Specifications for Building Construction. John E. Traister. Reston Publishing Co, Inc (division of Prentice-Hall, Inc), 1978, 216 pp. LC: 77-21708. ISBN: 0-87909-214-9. Cost: $18.95.

This book provides specifications for the electrification of buildings. For example, such topics as power generation, heating, cooling, controls and instrumentation, lighting, power transmission, and communications are discussed.

656. Electrical Wiring Handbook. Edward L. Safford. TAB Books, 1980, 432 pp. LC: 80-14369. ISBN: 0-8306-9932-5. Cost: $16.95.

This handbook provides information on circuits and electrical systems layout, electrical lighting, and electrical renovation. Wiring of various electrical systems is discussed, including security systems.

657. Electro-Optics Handbook. Glenn R. Elion; Herbert A. Elion. Marcel Dekker, 1979, 367 pp. LC: 79-20585. ISBN: 0-8247-6879-5. Cost: $39.75. Series: *Electro-Optics Series*, Vol. 3.

This handbook provides information on underwater optical transmission, lasers, image and camera tubes, and various optical formulas and optical materials. In addition, information is presented on computer modeling of electro-optic components. Electro-optic displays and arrays are also treated in the handbook, and numerous bibliographic references are provided.

658. Electrodiagnosis: A Handbook for Neurologists. Mario P. Smorto; J.V. Basmajian. Harper and Row, 1977. ISBN: 0-06-142410-2. Cost: $16.95.

659. Electrolytic Conductance and Conductances of the Halogen Acids in Water. W.J. Hamer; H.J. Dewane (National Bureau of Standards). NTIS, 1970. Accession: PB-192183. Series: NSRDS-NBS-33.

660. Electromagnetic Interference and Compatibility. Don White Consultants, Inc. Don White Consultants, Inc, 6 Vols., Vol. 1, 580 pp.; Vol. II, 450 pp.; Vol. III, 800 pp.; Vol. IV, 400 pp.; Vol. V, 700 pp.; Vol. VI, 1,035 pp. Cost: Vol. 1, $45.00; Vol. II, $68.00; Vol. III, $85.00; Vol. IV, $62.00; Vol. V, $72.00; Vol. VI, $45.00 ($339.00 per set). Series: *Handbook Series on EMI/EMC*.

661. Electromagnetic-Pulse Handbook for Electric Power Systems. Edward F. Vance (Stanford Research Inst). NTIS, final report, Feb. 4, 1975, 340 pp. Accession: AD-A009228. Tech Report: DNA-3466F. Cost: $2.25 microfiche; $10.00 paper copy.

This handbook covers electromagnetic-pulse (EMP) for the power user as well as the effects of EMP on electric power systems. Included are data and formulas for evaluating coupling of electromagnetic-pulse to power transmission and distribution lines, coupling through the service entrance, transient coupling through transformers, and lightning-arrester firing characteristics. Also treated are testing, grounding, and EMP protective measures.

662. Electron Fractography Handbook. A. Phillips et al. Metals and Ceramics Information Center, 1,212 pp. Accession: MCIC-HB-08. Cost: $90.00.

This handbook includes over 15,000 figures having over 2,000 fractographs. The alloys selected for this handbook are primarily structural metals of interest to the aerospace industry. For example, aluminum, steel, titanium, and nickel-based alloys are tested under a wide range of known environmental conditions. A bibliography and references are included for further reading.

663. Electron Impact Excitation of Atoms. B.L. Moiseiwitsch; S.J. Smith (National Bureau of Standards). NTIS, 1968. Cost: $5.50. Series: NSRDS-NBS-25.

664. Electron Microscopic Atlas in Ophthalmology. E. Yamada; S. Shikano. Georg Thieme, 1973, 375 pp. ISBN: 3-13-498801-1. Cost: DM 178.

Contains 234 illustrations.

665. Electron Microscopy: A Handbook for Biologists. Edgar Howard Mercer. Blackwell Scientific (also available from Lippincott in paper edition), 3d ed., 1972, 145 pp. LC: 72-305487. ISBN: 0-632-08330-1. Cost: $4.50 paper copy.

Microtechnique is discussed along with methods of preparation of materials to be viewed by an electron microscope. For example, staining methods and embedding media receive coverage in this handbook.

666. Electronic Absorption and Internal and External Vibrational Data of Atomic and Molecular Ions Doped in Alkali Halide Crystals. J.C. Jain; A.V.R. Warrier; S.K. Agarwal (National Bureau of Standards). U.S. Government Printing Office, 1974. SuDoc: C13.48:52. Cost: $1.15. Series: NSRDS-NBS-52.

667. Electronic Circuit Design Handbook. TAB Books, 4th ed., 1971. LC: 65-24823. ISBN: 0-8306-1101-0. Cost: $17.95.

668. Electronic Circuits Manual. John Markus. McGraw-Hill, 1971, 988 pp. LC: 70-152007. ISBN: 0-07-040444-5. Cost: $41.50.

669. Electronic Components Handbook. Keith Henney; Craig Walsh. McGraw-Hill, 1957-59, 3 Vols.

670. Electronic Components Handbook. Thomas H. Jones. Reston (division of Prentice-Hall), 1978, 391 pp. LC: 77-22341. ISBN: 0-87909-222-X. Cost: $24.95.

This book contains numerous illustrations and excellent photographs of different electronic components. Such components as relays, switches, resistors, and transformers are discussed.

671. Electronic Components: A Survey of the Most Used Component Types, Their Characteristic Data, Applications and Failure Rates. J. Biegel. NTIS, 1976, 125 pp. Accession: N77-12293. Tech Report: ECR-62. Cost: $3.50 microform; $11.00 hard copy.

This book is written in Danish with an English summary. Data are presented on electronic components to provide circuit designers with the necessary information regarding characteristic applications and failure rates. Advantages and disadvantages of certain types of components are indicated.

672. Electronic Conversions, Symbols and Formulas. Rufus P. Turner. TAB Books, 1975, 224 pp. LC: 75-31464. ISBN: 0-8306-5750-9 (hardbound); 0-8306-4750-3 (paperbound). Cost: $8.95.

673. Electronic Data Handbook. General Dynamics/Astronautics. NTIS, 1961, 63 pp. Accession: AD-832 958. Tech Report: GDA-AE61-D476. Cost: $3.50 microform; $8.00 hard copy.

This handbook is primarily a guide for the electronics technician. Includes information on soldering, various formulas, diagrams, standards, and safety procedures. Also consists of diagrams with color codes.

674. Electronic Data Reference Manual. Matthew Mandl. Reston Publishing Co, (division of Prentice-Hall, Inc), 1979, 304 pp. ISBN: 0-8359-1641-3. Cost: $18.95.

This manual consists of several chapters dealing with such topics as digital data circuitry, power supplies, oscillators, transistors, linear circuits, transmission lines, antennas, telemetry, and pulse code modulation. Various tables, equations, diagrams and symbols are included for the user's convenience. In addition, electronic color codes are also included.

675. Electronic Databook: A Guide for Designers. Rudolf F. Graf. Van Nostrand Reinhold, 2d ed., 1974. ISBN: 0-442-22796-5. Date of First Edition: 1971. Cost: $11.95.

676. Electronic Designer's Handbook: A Practical Guide to Transistor Circuit Design. T.K. Hemingway. TAB Books, 3d ed., 1979. ISBN: 0-8306-1038-3. Cost: $9.95.

677. Electronic Drafting and Design. Nicholas M. Raskhodoff. Prentice-Hall, 3d ed., 1977. ISBN: 0-13-250613-0. Cost: $18.95.

678. Electronic Engineers Master. Herman Publishing, Inc, annual. LC: 58-9813. ISBN: 0-89047-046-4. Cost: $50.00.

This publication is a listing of electronic equipment that is sold directly to the OEM (Original Equipment Manufacturers) market.

Electronic Engineers' Reference Book. *See* **Handbook of Electronic Engineering.**

679. Electronic Filter Design Handbook. A.B. Williams. McGraw-Hill, 1980, 544 pp. LC: 80-11998. ISBN: 0-07-070430-9. Cost: $32.50.

This handbook provides examples and schematics as well as tables of numerical values for designing filters for simple networks as well as complex configurations. Information is presented for the design of active and passive electronic filters and especially elliptical function filters.

680. Electronic Imaging Techniques; A Handbook of Conventional and Computer-Controlled Animation, Optical, and Editing Processes. Eli L. Levitan. Van Nostrand Reinhold, 1977, 195 pp. LC: 76-24376. ISBN: 0-442-24771-0. Cost: $17.95.

681. Electronic Measurements and Instrumentation. Bernard M. Oliver; John M. Cage. McGraw-Hill, 1971. ISBN: 0-07-047650-0. Cost: $44.50.

See also, *Handbook of Electronic Instruments and Measurement Techniques.*

682. Electronic Properties of Composite Materials. M.A. Leeds. IFI/Plenum Data Corp, 1972, 104 pp. Accession: A73-17872. Cost: $15.00. Series: *Handbook of Electronic Materials,* Vol. 9.

Numerous tables on the electronic properties of composite materials are presented along with such data as arc resistance, arc tracking, resistance, dielectric constants, dissipation factor, electrical resistivity, and electrical conductivity. In addition, such properties as thermal conductivity and the linear thermal expansion coefficient are presented. Fiber-reinforced polymers, laminates, eutectics, ceramics, and metal composites are included in this handbook.

683. Electronic Properties of Materials: A Guide to the Literature. H. Thayne Johnson et al. IFI/Plenum Press, Vol.1, 1965; Vol.2, 1967; Vol.3, 1971, Vol.1, 1,681 pp.; Vol.2, 1,800 pp.; Vol.3, 1,895 pp. LC: 65-12176. ISBN: Vol.1, 0-306-68221-4; Vol.2, 0-306-68222-2; Vol.3, 0-306-68223. Cost: $195.00 per Vol.; $495.00 per set.

684. Electronic Properties Research Literature Retrieval Guide: 1972-1976. J.F. Chaney; T.M. Putnam. Plenum Publishing Corp, 1979, 4 vols. LC: 79-16802. ISBN: 0-306-68010-6. Cost: $330.00.

This is a comprehensive compilation that was prepared by the Center for Information and Numerical Data Analysis (CINDAS) at Purdue University. This 4-volume comprehensive work provides numerous bibliographic references for locating electronic properties of various types of materials. This guide covers all types of materials, ranging from composites, polymers, rocks to metals, ceramics, and soils. Numerous types of electronic properties can be found through this guide, such as dielectric constant, absorption coefficient, electric hysteresis, electrical resistivity, magnetoelectric properties, Hall coefficient, luminescence properties, photoelectronic properties, refractive index, thermal electronic properties, and piezoelectric properties.

685. Electronic Scanning Radar Systems (ESRS) Design Handbook. Peter J. Kahrilas. Artech House, Inc, 1976, 372 pp. LC: 75-43051. ISBN: 0-89006-023-1. Cost: $30.00. Series: *Artech Radar Library.*

686. Electronic Technician-Dealer Magazine Editors, TV Troubleshooter's Handbook. *Electronic Technician/Dealer,* eds (available from TAB Books), 3d ed., 1975, 448 pp. ISBN: 0-8306-5821-1. Cost: $8.95.

687. Electronic Test and Measurement Handbook. John J. Schultz. TAB Books, 1969, 224 pp. LC: 79-94453. ISBN: 0-8306-9506-0. Cost: $7.95.

688. Electronic, Magnetic, and Thermal Properties of Solid Materials. K. Shröder. Marcel Dekker Inc, 1978, 600 pp. ISBN: 0-8247-6487-0. Cost: $55.00. Series: *Electrical Engineering and Electronics Series,* Vol. 5.

689. Electronics Data Handbook. Martin Clifford. TAB Books, 2d ed., 1972, 255 pp. LC: 72-82250. ISBN: 0-8306-2118-X. Cost: $8.95.

690. Electronics Designers' Handbook. L.J. Giacoletto. McGraw-Hill, 2d ed., 1977, 2,344 pp. LC: 77-793. ISBN: 0-07-023149-4. Date of First Edition: 1957. Cost: $56.50.

Has over 1,700 illustrations including information on solid state devices, amplifiers, navigational systems, integrated circuits, DC circuits, Servo systems, vacuum tubes, and television receivers. First edition edited by Robert W. Landee, et al.

691. Electronics Engineers' Handbook. Donald G. Fink; Alexander A. McKenzie. McGraw-Hill, 1975, approx. 2,150 pp. (various paging). LC: 74-32456. ISBN: 0-07-020980-4. Date of First Edition: 1975. Cost: $57.50.

The emphasis of this handbook is on electronics, or the microcosm of the electrical world, including information on such subjects as circuits, amplifiers, semiconductors, components, transducers, oscillators, etc. A detailed table of contents heads each section, and many tables are provided as well as numerous bibliographic references throughout the handbook. In addition, this reference source includes such subjects as antennas, radar, television, acoustics, and recording systems. For related works, please refer to the *Electrical Engineer's Reference Book,* the *Basic Electronic Instrument Handbook,* the *Handbook of Electronic Packaging,* the *Handbook of Materials and Processes for Electronics,* the *Handbook of Modern Electronic Data; Industrial Electronics Handbook,* the *Handbook of Semiconductor Electronics,* and the *Standard Handbook for Electrical Engineers.* Also refer to the *Handbook for Electronics Engineering Technicians; Electronic Conversions, Symbols and Formulas,* and *Electronics Handbook.*

692. Electronics Handbook. Clyde Herrick. Goodyear Publishing Co., Inc, 1975, 225 pp. LC: 74-15621. ISBN: 0-87620-266-0. Cost: $20.50.

The *Electronics Handbook* contains a multitude of practical information on such subjects as circuits, oscilloscopes, amplifiers, resistors, transistors and other semiconductors. The information is presented in graphs, charts, tables, figures, and an index is provided in order to help retrieve the proper information from the text. A glossary is also provided to assist the user in understanding the electronic terminology. For related works, please refer to

the *Electronics Engineers' Handbook,* the *Electrical Engineer's Reference Book,* the *Basic Electronic Instrument Handbook,* the *Handbook of Electronic Packaging,* the *Handbook of Materials and Processes for Electronics,* the *Handbook of Modern Electronic Data; Industrial Electronics Handbook,* the *Handbook of Semiconductor Electronics,* and the *Standard Handbook for Electronics Engineering Technicians* and *Electronic Conversions, Symbols and Formulas.*

693. Electronics Reference Databook. Norman H. Crowhurst. TAB Books, 1969, 227 pp. LC: 69-14553. ISBN: 0-8306-9488-9. Cost: $7.95.

694. Electroplaters' Process Control Handbook. D. Gardner Foulke. R.E. Krieger Publishing Co, Inc, 2d ed., 1975, 429 pp. LC: 74-13010. ISBN: 0-88275-213-8. Cost: $22.50.

This handbook provides information on standardized procedures and test methods for the control of electrodeposition processes. Various methods of analysis are discussed, including spectroscopy, colorimetric analysis, spectrophotometric analysis, potentiometric titrations, X-ray diffraction analysis, and polarographic techniques. The adhesion of electrodeposits is presented as well as various corrosion tests, surface finishing techniques and plating specifications. Solution analysis is presented and many different solutions are discussed, including pickling solutions, anodizing solutions, phosphating solutions, and various plating solutions, such as cobalt solutions, bronze-plating solutions, silver-plating solutions, and zinc-plating solutions.

695. Electroplating Engineering Handbook. A. Kenneth Graham. Van Nostrand Reinhold, 3d ed., 1971, 845 pp. LC: 76-129204. ISBN: 0-442-22791-4. Date of First Edition: 1954. Cost: $39.50.

The third edition of the *Electroplating Engineering Handbook* contains a great deal of information on the latest advances in electroplating engineering. The book consists of many articles written by recognized experts in their respective fields. This handbook contains such information as surface preparation cleaning of metals, waste water treatment, and safety requirements in the processing of metals. It also includes a lot of practical information on the design and layout of electroplating facilities, as well as a number of engineering tables and an index to help retrieve the information. For related works, please refer to the *Metals Handbook; ASME Handbook; Woldman's Engineering Alloys; Handbook of Aluminum; Metals Reference Book,* and the *Aerospace Structrual Metals Handbook.*

696. Elevated Temperature Properties of Selected Superalloys. R.J. Favor; D.P. Moon; R.C. Simon. American Society for Testing and Materials, (Issued under the auspices of the ASTM-ASME-MPC Joint Committee on Effect of Temperature on the Properties of Metals and the Defense Metals Information Center), 1968, 358 pp. Accession: A69-19913. Cost: $8.80 members; $11.00 nonmembers. Series: DS 7-S1.

Mechanical properties are provided for several high-strength alloys. Chemical composition, creep rupture strength, and heat treatment information is provided over a range of temperatures.

697. Elevations from Zenith Distances (Machine Computation with 6-Place Natural Tangent Tables) 0°-45°. A.C. Poling. National Geodetic Survey, 1947, 30 pp. Series: *U.S. Coast and Geodetic Survey Publication,* G-56.

698. Emergency Care Handbook. Burton A. Waisbren. Drake Publishers, 1975. ISBN: 0-87749-848-2. Cost: $8.95.

699. Empirical Correlations of Properties of Graphites (for Reactor Applications). C. Meyers. NTIS, 1968, 36 pp. Accession: N69-17116. Tech Report: GAMD-8738.

This collection of tables presents information on the properties of graphites for usage in nuclear reactors. Various chemical properties are listed.

700. Encyclopedia of Chemical Trademarks and Synonyms. H. Bennett. Chemical Publishing Co, Vol.1, 1981; Vol.2, 1981; Vol.3, 1982, approx. 400 pp. per Vol. ISBN: Vol.1, 0-8206-0286-8. Cost: Vol.1, $65.00.

In addition to trademarks and synonyms, this 3-volume set also includes chemical compositions, abbreviated names and acronyms. All types of chemicals and chemical compounds are included in this comprehensive work, such as cosmetics, plastics, drugs, foods, adhesives, detergents, alloys, and textiles.

701. Encyclopedia of Common Natural Ingredients Used in Food, Drugs, and Cosmetics. A.Y. Leung. Wiley-Interscience, 1980, 409 pp. LC: 79-25998. ISBN: 0-471-04954-9. Cost: $47.00.

Provides information on more than 300 natural ingredients that are used in processing foods and drugs or in cosmetics. A considerable amount of information is presented on various botanicals.

702. Encyclopedia of Electronic Circuits. Leo G. Sands; Donald R. Mackenroth. Prentice-Hall, 1975. ISBN: 0-13-275404-5. Cost: $15.95.

This is a vast collection of different types of circuits including signal generator circuits, audio controlled circuits, logic circuits, biasing circuits, power supply circuits, amplifier circuits, coupling circuits, and frequency conversion circuits. Schematic diagrams are presented for trouble shooting or building these electronic circuits.

703. Encyclopedia/Handbook of Materials, Parts and Finishes. H.R. Clauser. Technomic Publishing Co, 1976, 575 pp. LC: 75-43010. ISBN: 0-87762-189-6. Cost: $35.00.

This handbook is a collection of approximately 300 different articles describing materials, parts and finishes that are used by engineering designers. It provides information on many materials including metals, wood, textiles, ceramics, plastics, composite materials, and different coatings. Engineering properties, processing characteristics, and production information is provided for each material or coating. For related works, please refer to such tites as the *CRC Handbook of Materials Science; Materials Handbook; Materials Handling Handbook; Handbook of Engineering Materials; Dangerous Properties of Industrial Materials; Material Properties Handbook; Handbook of Materials and Processes for Electronics*, and the *Handbook of Engineering Materials*.

704. Encyclopedic Handbook of Medical Psychology. S. Krauss. Butterworths, 1976, 600 pp. ISBN: 0-407-00044-5. Cost: $24.95.

705. Energy Cogeneration Handbook: Criteria for Central Plant Design. George Polimeros. Industrial Press, 1981, 247 pp. LC: 80-27186. ISBN: 0-8311-1130-5. Cost: $39.50.

This handbook emphasizes energy conversion, utilization, and conservation in the design of efficient central energy production plants. Such topics as pumping systems, chiller arrangement, power cycles, heat balancing, distribution systems, chiller arrangement, power cycles, heat balancing, distribution systems, construction of central plants, and water expansion of chilled water are presented. This book also presents information on organic Rankine cycles and information on burning of waste as a power source. Extensive references are provided in this handbook.

706. Energy Conservation Idea Handbook. Academy for Educational Development. American Council on Education, 1981, 171 pp. ISBN: 0-8268-1458-1. Cost: $15.00.

Describes 500 ideas as to how institutions can reduce energy consumption and provide means of energy conservation.

707. Energy Conservation Self-Evaluation Manual. Fairmont Press (available from Van Nostrand Reinhold), 1980. ISBN: 0-915-58622-3. Cost: $16.95.

This handbook helps to provide suggestions that will assist in the evaluation of energy consumption in any facility, and provide guidance to plant managers to help reduce energy consumption.

708. Energy Conversion Systems Reference Handbook. W.R. Menetrey et al (Wright-Patterson Air Force Base). NTIS, 1960, 11 Vols. Accession: AD-257357. Tech Report: WADD-60-699.

709. Energy Handbook. Robert L. Loftness. Van Nostrand Reinhold, 1978, 741 pp. LC: 77-18190. ISBN: 0-442-24836-9. Cost: $52.50.

This handbook contains sections which cover nuclear power, solar, geothermal and fossil fuels. Note: price includes indexes.

710. Energy Interrelationships: A Handbook of Tables and Conversion Factors for Combining and Comparing International Energy Data. Nathaniel B. Guyol. National Energy Information Center (available from NTIS), 1977, 64 pp. Accession: PB-269 034. Tech Report: PEA/B-77/166. Cost: $3.50 microform; $8.00 hard copy.

This handbook represents an effort to compare international energy data on such topics as coal, peat, lignite, natural gas, coke, petroleum products, gasoline, fuel oil, jet engine fuels, mine gases, and solid fuels. Calorific value and thermal efficiency are listed for these fuels. A bibliography has been prepared listing sources from which the data have been derived.

711. Energy Levels of Light Nuclei, A=18-20. F. Ajzenberg-Selove. Elsevier, 1972. ISBN: 0-7204-0255-7. Cost: $17.50.

This is a reprint from *Nuclear Physics*, Vol.190A, No.1.

712. Energy Management Handbook for Petroleum Refineries, Gas Processing and Petrochemical Plants. Gulf Publishing Co, 1979, 280 pp. ISBN: 0-87201-262-X. Cost: $16.95.

713. Energy Management Handbook. W.C. Turner. Wiley-Interscience, 1981, 784 pp. LC: 81-10351. ISBN: 0-471-08252-X. Cost: $52.95.

714. Energy Managers' Handbook. Gordon A. Payne. IPC Science and Technology Press, Ltd, 2d ed., 1980, 164 pp. ISBN: 0-86103-033-8. Cost: $19.50.

This compilation covers combustion plant operation, heat distribution, and ways to reduce energy costs.

715. Energy Reference Handbook. N.C. McNerney; Thomas F.P. Sullivan. Government Institutes, Inc, 2d ed., 1978, 352 pp. ISBN: 0-86587-081-0. Cost: $18.50.

This book is essentially a dictionary of terms in the field of energy including such areas as ocean energy, geothermal shale oil, gas, coal, and nuclear energy. In addition to being a dictionary of energy terms, this handbook also includes information that is useful in making energy calculations. For example, conversion tables, metric tables, various abbreviations and information on constants are included. For related works, please refer to such titles as *Keystone Coal Industry Manual; Gas Engineers' Handbook; Handbook of Natural Gas Engineering; Hydrogen: Its Technology and Implications; Reactor Handbook; Engineering Tables for Energy Operations; Petroleum Production Handbook; Nuclear Engineering Handbook*, and *Hydroelectric Handbook*.

716. Energy Source Book. Center for Compliance Information. Aspen Systems Corp, 1978, 736 pp. Cost: $49.50.

Covers energy reserves and estimates of future production. Provides information on demand, supply and outlook on such energy sources as oil shale, tar sands, wind energy, ocean energy, biomass conversion, solar energy, hydrogen energy, geothermal energy, coal, and fuel cells.

717. Energy Technology Handbook. Douglas M. Considine. McGraw-Hill, 1977, 1,857 pp (various paging). LC: 76-17653. ISBN: 0-07-012430-2. Cost: $58.50.

Provides information on coal conversion, natural gas, petroleum, hydropower nuclear power, solar energy, and geothermal energy sources. Also provides information on reducing the environmental impact of energy conversion processes.

718. Energy Users Databook. H.B. Locke. Crane, Russak, and Co, 1981, 150 pp. ISBN: 0-8448-1402-4. Cost: $25.00.

This databook contains information on fuels, combustion, heat transfer, steam, furnaces, boilers, and various energy instruments. Many formulas, figures, tables, and graphs are used to present this data on energy management.

719. Energy-Saving Guidebook. George E. Springer; Gene E. Smith. Technomic Publishing Co, Inc, 1975, 104 pp. LC: 74-81585. ISBN: 0-87762-147-0. Cost: $3.95.

Contains charts showing fuel savings by using different types of insulation in structures. Also provides information on fuel savings and driving vehicles under different conditions.

720. Engine Specification Manual: American Passenger Cars. F and J Publishing Corp, 8th ed., 1976, various paging. ISBN: 0-89311-004-3. Cost: $9.95.

721. Engineer's Companion: A Concise Handbook of Engineering Fundamentals. Mott Souders. John Wiley and Sons, 1966, 426 pp. LC: 65-26851. ISBN: 0-471-81395-8. Cost: $24.50.

This handbook provides ready reference on heat transfer, fluid mechanics, electricity, and physics.

722. Engineer's Guide to High-Temperature Materials. Francis Jacob Clauss. Addison-Wesley, 1969, 401 pp. LC: 79-81299. ISBN: 0-201-01055-0. Cost: $15.95.

723. Engineer's Handbook. Lefax Publishing Co, 2 Vols., loose-leaf binding. Cost: $19.00 per set. Series: *Lefax Technical Manuals*, No. 787.

724. Engineer's Handbook. Arthur Liebers. Key, 1968, 319 pp. LC: 68-2240.

This hanbook includes many formulas and methods that would be useful for students of engineering.

725. Engineer's Manual. Public Documents Distribution Center, 1963 (rep. 1977), 194 pp. SuDoc: D209.14: En 3. S/N: 008-050-00136-4.

This manual was originally designed for the Navy, but it appears to have wider applications. It specifically addresses various engineering performance standards and discusses the work sampling technique.

726. Engineer's Manual. Ralph G. Hudson. John Wiley and Sons, Inc, 2d ed., 1939, 340 pp. ISBN: 0-471-41844-7. Date of First Edition: 1917, renewed 1944. Cost: $16.95.

For related works of a general engineering nature, please refer to the following titles: *Engineering Manual; Handbook of Engineering Fundamentals; CRC Handbook of Tables for Applied Engineering Science; Engineering Tables and Data*; and *De Laval Engineering Handbook*.

727. Engineer's Pocket Book of Tables, Formulae, and Memoranda. George Newnes, Ltd, 5th ed., 1964, 614 pp.

728. Engineered Performance Standards: Public Works Maintenance. U.S. Department of the Navy. Superintendent of Documents, Government Printing Office, 1962-75 (rep. 1977), 8-Vol. set. SuDoc: D 209.14. Cost: $25.55 (set).

This multivolume set is a collection of publications that are used in estimating the Navy's maintenance operations at its shore establishments. Such topics as wharf building, pipe fitting, plumbing, rigging, carpentry, engineering, and machine shop data are presented.

729. Engineering and Industrial Graphics Handbook. George Rowbotham. McGraw-Hill, 1982, 608 pp. LC: 81-12399. ISBN: 0-07-054080-2. Cost: $49.50.

Presents information on industrial and engineering graphics that will be useful in improving communications. Such topics as dimensioning, layout, electronic drafting, and the economics of drafting are discussed.

730. Engineering and Technical Handbook. Donald G. McNeese; Albert L. Hoag. Prentice-Hall, 1957, 376 pp. LC: 57-6690. ISBN: 0-13-277434-8. Cost: $19.95.

731. Engineering Compendium on Radiation Shielding. R.G. Jaeger et al. Springer-Verlag (Sponsored by the International Atomic Energy Agency), 1968-, 3 Vols. LC: 68-19816. ISBN: Vol.1, 0-387-04080-3; Vol.2, 3-540-05075-2; Vol.3, 0-387-05076-0. Cost: Vol.1, $152.50; Vol.2, $166.20; Vol.3, $152.50.

This compendium consists of 3 volumes entitled as follows: Vol.1, *Shielding Fundamentals and Methods*; Vol.2, *Shielding Materials and Design*; Vol.3, *Shield Design and Engineering*.

732. Engineering Data Book. Natural Gas Processors Supplies Association. Gas Processors Suppliers Asociation, 9th ed., 1972.

733. Engineering Data for Aluminum Structures. Aluminum Association, 3d ed., 1975, 104 pp. Date of First Edition: 1969. Series: *Aluminum Construction Manual*, Section 3.

Information is presented on various properties of aluminum, including thermal conductivity, electrical conductivity, thermal expansion, and tensile properties. Information is also presented on various structural shapes of aluminum, threaded fasteners, commercial roofing and siding, casting, and welded construction.

734. Engineering Data for Product Design. Douglas C. Greenwood. Conquest Publications, 1961, 448 pp. ISBN: 0-07-02436-1. Cost: $29.50.

Covers a lot of practical information on product design. Includes information on metals, nonmetallic materials, beams, torque, bearings, shafts, gears, vibration, mechanical control, hydraulics, and pneumatics, springs.

735. Engineering Data on New Aerospace Structural Materials. Vol. 1. O.L. Deel; H. Mindlin (Battelle Columbus Labs). NTIS, 1972, Vol.1, 133 pp.; Vol.2, 184 pp. Accession: Vol.1, N73-22517, AD-755407; Vol.2, N73-22518, AD-755408. Tech Report: Vol.1, AFML-TR-72-196-vol-1; Vol.2, AFML-TR-72-196-vol-2.

This work covers such metals as aluminum alloys, nickel alloys, stainless steels, and titanium alloys. The following properties have been listed: tension, compression, shear, bend, impact, fracture toughness, fatigue, creep, stress-rupture, stress corrosion.

736. Engineering Design Handbook. Hydraulic Fluids. Army Materiel Command. NTIS, 1971, 288 pp. Accession: AD-884519. Tech Report: AMCP-706-123. SuDoc: D101.22/3:706-123.

737. Engineering Design Handbook: Aerodynamics. Army Materiel Command. NTIS. Accession: AD-830377. SuDoc: D101.22/3: No.. Cost: $6.00.

738. Engineering Design Handbook: Ammunition, Section 1, Artillery Ammunition—General, with Table of Contents, Glossary, and Index for Series. Army Materiel Command. NTIS. Accession: AD-830290. Tech Report: AMCP-706-244. SuDoc: D101.22/3:706-244. Cost: $6.00.

739. Engineering Design Handbook: Ammunition, Section 2, Design for Terminal Effects. Army Materiel Command. NTIS. Accession: AD-389 304. Tech Report: AMCP-706-245(c). SuDoc: D101.22/3:706-245(c). Cost: $12.50.

740. Engineering Design Handbook: Ammunition, Section 4, Design for Projection. Army Materiel Command. NTIS, 1964, 190 pp. Accession: AD-830296. Tech Report: AMCP-706-247. SuDoc: D101.22/3:706-247. Cost: $9.00.

741. Engineering Design Handbook: Ammunition, Section 5, Inspection Aspects of Artillery Ammunition Design. Army Materiel Command. NTIS. Accession: AD-830284. Tech Report: AMCP-706-248. SuDoc: D101.22/3:706-248. Cost: $4.50.

742. Engineering Design Handbook: Ammunition, Section 6, Manufacture of Metallic Components of Artillery Ammunition. Army Materiel Command. NTIS, 1964, 50 pp. Accession: AD-830266. Tech Report: AMCP 706-249. SuDoc: D101.22/3:706-249. Cost: $5.25.

743. Engineering Design Handbook: Armor and Its Applications (u). Army Materiel Command. NTIS. Tech Report: AMCP-706-170(s). SuDoc: D101.22/3:706-170(s).

744. Engineering Design Handbook: Automatic Weapons. Army Materiel Command. NTIS, 1970, (various paging). Accession: AD-868578. Tech Report: AMCP-706-260. SuDoc: D101.22/3:706-260. Cost: $12.00.

745. Engineering Design Handbook: Automotive Series. Automotive Bodies and Hulls. Army Materiel Command. NTIS, 1970, 406 pp. Accession: AD-873103. Tech Report: AMC-PAM-706-357. SuDoc: D101.22/3:706-357. Cost: $13.25.

746. Engineering Design Handbook: Automotive Series. Automotive Suspensions. Army Materiel Command. NTIS, 1967, 456 pp. Accession: AD-817023. Tech Report: AMC-PAM-706-356. SuDoc: D101.22/3:706-356.

747. Engineering Design Handbook: Automotive Series. The Automotive Assembly. Army Materiel Command. NTIS, 1965, 511 pp. Accession: AD-830268. Tech Report: AMC-PAM-706-355. SuDoc: D101.22/3:706-355.

748. Engineering Design Handbook: Ballistic Missile Series; Aerodynamics. Army Materiel Command. NTIS, 1965, 99 pp. Accession: AD-830 377. Tech Report: AMC-PAM-706-283. Cost: $3.50 microfiche; $9.50 hard copy.

Design information is presented with regard to ballistic missile flight. It specifically includes information on high speed, high temperature, and high altitude effects.

749. Engineering Design Handbook: Bottom Carriages. Army Materiel Command. NTIS. Accession: AD-830396. Tech Report: AMCP-706-344. SuDoc: D101.22/3:706-344. Cost: $4.50.

750. Engineering Design Handbook: Carriages and Mounts Series. Equilibrators. Army Materiel Command. NTIS, supersedes Report No. AMCP-706-345, AD-830293, Dec. 16, 1970, 70 pp. Accession: AD/A-003 347. Tech Report: AMCP-706-345. SuDoc: D101.22/3:706-345. Date of First Edition: April 1960. Cost: $2.25 microfiche; $4.50 paper copy.

This handbook covers equilibrators on artillery equipment. It deals with unbalance of tipping parts or with azimuth equilibrators which compensate for the effect of tilt on the mounts. Various comparisons of equilibrators are also included.

751. Engineering Design Handbook: Carriages and Mounts—General. Army Materiel Command. NTIS. Accession: AD-830276. Tech Report: AMCP-706-340. SuDoc: D101.22/3:706-340. Cost: $4.50.

752. Engineering Design Handbook: Compensating Elements. Army Materiel Command. NTIS. Accession: AD-830275. Tech Report: AMCP-706-331. SuDoc: D101.22/3:706-331. Cost: $6.50.

753. Engineering Design Handbook: Computer Aided Design of Mechanical Systems, Part One. Army Materiel Command. NTIS, 1973, 346 pp. Accession: AD-767826. Tech Report: AMCP 706-192. SuDoc: D101.22/3:706-192. Cost: $12.00.

754. Engineering Design Handbook: Computer Aided Design of Mechanical Systems, Part Two. Army Materiel Command. NTIS. Tech Report: AMCP-706-193. SuDoc: D101.22/3:706-193.

755. Engineering Design Handbook: Cradles. Army Materiel Command. NTIS. Accession: AD-830277. Tech Report: AMCP-706-341. SuDoc: D101.22/3:706-341. Cost: $5.25.

756. Engineering Design Handbook: Criteria for Environmental Control of Mobile Systems. Army Materiel Command. NTIS, 1971, 179 pp. Accession: AD-889588. Tech Report: AMCP-706-120. SuDoc: D101.22/3:706-120. Cost: $9.00.

757. Engineering Design Handbook: Design for Air Transport and Airdrop of Materiel. Army Materiel Command. NTIS, 1967, 349 pp. Accession: AD-830262. Tech Report: AMC-PAM-706-130. SuDoc: D101.22/3:706-130. Cost: $12.00.

758. Engineering Design Handbook: Design for Control of Projectile Flight Characteristics. Army Materiel Command. NTIS, 1966, (various paging). Accession: AD-801509. Tech Report: AMCP-706-242 (replaces AMCP-706-246). SuDoc: D101.22/3:706-242. Cost: $7.25.

759. Engineering Design Handbook: Design Guidance for Producibility. Army Materiel Command. NTIS. Tech Report: AMCP-706-100. SuDoc: D101.22/3:706-100.

760. Engineering Design Handbook: Design of Aerodynamically Stabilized Free Rockets. Army Materiel Command. NTIS. Accession: AD-840582. Tech Report: AMCP-706-280. SuDoc: D101.22/3:706-280. Cost: $9.25.

761. Engineering Design Handbook: Development Guide for Reliability. Part Six: Mathematical Appendix and Glossary. Army Materiel Command. NTIS, 1976, 101 pp. Tech Report: AMCP-706-200. Cost: $3.50 microform; $11.00 hard copy.

Covers such topics as binomial distribution, Poisson distribution, probability distributions, exponential distributions, beta distribution, gamma distribution, and Bayesian statistics.

762. Engineering Design Handbook: Editorial Style Guide: A Publisher's Guide for Editors, Writers, and Compositors of Scientific and Technical Reports. Army Materiel Command. NTIS, 1975, 192 pp. Accession: AD-A021627. Tech Report: NWC-Ad Pub-157. SuDoc: D101.22/3:157. Cost: $12.50.

763. Engineering Design Handbook: Electrical Wire and Cable. Army Materiel Command. NTIS, 1969, 297 pp. Accession: AD-865109. Tech Report: AMC-PAM-706-125. SuDoc: D101.22/3:706-125.

764. Engineering Design Handbook: Elements of Aircraft and Missile Propulsion. Army Materiel Command. NTIS. Accession: AD-861082. SuDoc: D101.22/3:. Cost: $21.25.

765. Engineering Design Handbook: Elements of Armament Engineering, Part One, Sources of Energy. Army Materiel Command. NTIS, 1964, (various paging). Accession: AD-830272. Tech Report: AMCP-706-106. SuDoc: D101.22/3:706-106. Cost: $6.50.

766. Engineering Design Handbook: Elements of Armament Engineering, Part Three, Weapon Systems and Components. Army Materiel Command. NTIS. Accession: AD-830288. Tech Report: AMCP-706-108. SuDoc: D101.22/3:706-108. Cost: $9.25.

767. Engineering Design Handbook: Elements of Armament Engineering, Part Two, Ballitics. Army Materiel Command. NTIS, 1963, (various paging). Accession: AD-830287. Tech Report: AMCP-706-107. SuDoc: D101.22/3:706-107. Cost: $9.50.

768. Engineering Design Handbook: Elements of Terminal Ballistics, Part One, Kill Mechanisms and Vulnerability. Army Materiel Command. NTIS. Accession: AD-389219. Tech Report: AMCP-706-160(c). SuDoc: D101.22/3:706-160(c). Cost: $15.00.

769. Engineering Design Handbook: Elements of Terminal Ballistics, Part Three. Application to Missile and Space Targets (u). Army Materiel Command. NTIS. Tech Report: AMCP-706-162 (SRD). SuDoc: D101.22/3:706-162.

770. Engineering Design Handbook: Elements of Terminal Ballistics, Part Two, Collection and Analysis of Data Concerning Targets. Army Materiel Command. NTIS. Accession: AD-289318. Tech Report: AMCP-706-161(c). SuDoc: D101.22/3:706-161(c). Cost: $7.00.

771. Engineering Design Handbook: Elevating Mechanisms. Army Materiel Command. NTIS. Accession: AD-830301. Tech Report: AMCP-706-346. SuDoc: D101.22/3:706-346. Cost: $4.50.

772. Engineering Design Handbook: Environmental Series, Part 1, Basic Environmental Concepts. Army Materiel Command. NTIS, July 1974, (supersedes AMC Pamphlet 706-115 October, 1969), 225 pp. Accession: AD-784999. Tech Report: AMCP-706-115. SuDoc: D101.22/3:706-115/2.

773. Engineering Design Handbook: Environmental Series, Part Three, Induced Environmental Factors. Army Materiel Command. NTIS, 1976, 401 pp. Accession: AD-A023512. Tech Report: AMCP-706-117. SuDoc: D101.22/3:706-117. Cost: $17.50.

774. Engineering Design Handbook: Environmental Series. Part Five. Glossary of Environmental Terms. Army Materiel Command. NTIS, July 31, 1975, 204 pp, part 5 of 5. Accession: AD-A015180. Tech Report: AMCP-706-119. SuDoc: D101.22/3:706-119. Cost: $12.50 microfiche; $12.50 paper copy.

775. Engineering Design Handbook: Environmental Series. Part Four. Life Cycle Environments. Army Materiel Command. NTIS, March 31, 1975, 407 pp, part 4 of 5. Accession: AD-A015179. Tech Report: AMCP-706-118. SuDoc: D101.22/3:706-118. Cost: $17.50 paper; $17.50 microfiche.

776. Engineering Design Handbook: Environmental Series. Part Two. Natural Environmental Factors. Army Materiel Command. NTIS, Part 2, April, 1975 (see also Part 1, AD-784999), 962 pp. Accession: AD-A012648. Tech Report: AMCP-706-116. SuDoc: D101.22/3:706-116. Cost: $25.00 microfiche; $25.00 paper copy.

This handbook discusses environmental factors that are associated with natural phenomena, such as humidity, pressure, wind, fog and whiteout, solid precipitation, terrain, microbiological organisms, temperature, ozone, solar radiation, macrobiological organisms, salt fog, rain, salt and salt water.

777. Engineering Design Handbook: Experimental Statistics, Section 1, Basic Concepts and Analysis of Measurement Data. Army Materiel Command. NTIS, 1969, various paging. Accession: AD-865421. Tech Report: AMCP-760-110. SuDoc: D101.22/3:706-110. Cost: $9.00.

778. Engineering Design Handbook: Experimental Statistics, Section 2, Analysis of Enumerative and Classificatory Data. Army Materiel Command. NTIS, 1969, supersedes report dated April, 1965, AD-830-294, 74 pp. Accession: AD-865422. Tech Report: AMC-PAM-706-111. SuDoc: D101.22/3:706-111. Cost: $5.25.

779. Engineering Design Handbook: Experimental Statistics, Section 3, Planning and Analysis of Comparative Experiments. Army Materiel Command. NTIS, 1969, supersedes report dated 30 April 1965, AMC-PAM-706-112. Accession: AD-865423. Tech Report: AMC-PAM-706-112. SuDoc: D101.22/3:706-112. Cost: $6.00.

780. Engineering Design Handbook: Experimental Statistics, Section 4, Special Topics. Army Materiel Command. NTIS, 1966, supersedes report dated 30 March 1966, AD-830397, AMC-PAM-705-114, 109 pp. Accession: AD-865424. Tech Report: AMC-PAM-706-113. SuDoc: D101.22/3:706-113. Cost: $6.50.

781. Engineering Design Handbook: Experimental Statistics, Section 5, Tables. Army Materiel Command. NTIS, 1969, supersedes report dated 31 July 1963, AD-830295, 97 pp. Accession: AD-865425. Tech Report: MC-PAM-706-114. SuDoc: D101.22/3:706-114. Cost: $6.00.

782. Engineering Design Handbook: Explosions in Air, Part Two (U). Army Materiel Command. NTIS. Tech Report: AMCP-706-182 (SRD). SuDoc: D101.22/3:706-182.

Part 2 of this handbook covers classified aspects of air blast technology, shock tubes, and response of structures to blast loading.

783. Engineering Design Handbook: Explosions in Air. Part One. Army Materiel Command. NTIS, July 15, 1974, 335 pp, part 1 of 2. Accession: AD/A-003817. Tech Report: AMCP-706-181. SuDoc: D101.22/3:706-181. Cost: $2.25 microfiche; $10.00 paper copy.

Part 1 of this handbook covers air blast theory, blast scaling, air blast transducers, photography of air blast waves, and data reduction techniques.

784. Engineering Design Handbook: Explosives Series. Explosive Trains. Army Materiel Command. NTIS, 1974, 241 pp. Accession: AD-777482, supersedes report AD-462254. Tech Report: AMCP-706-179. SuDoc: D101.22/3:706-179.

785. Engineering Design Handbook: Explosives Series. Properties of Explosives of Military Interest. Army Materiel Command. NTIS, 1971, supersedes report dated 22 March 1967, including change 1, AD-814964, 404 p. Accession: AD-764340. Tech Report: AMCP-706-177. SuDoc: D101.22/3:706-177.

786. Engineering Design Handbook: Fabric Design. Army Materiel Command. NTIS. Tech Report: 706-300. SuDoc: D101.22/3:706-300.

787. Engineering Design Handbook: Fire Control Computing Systems. Army Materiel Command. NTIS. Accession: AD-879465. Tech Report: AMCP-706-329. SuDoc: D101.22/3:706-329. Cost: $15.00.

788. Engineering Design Handbook: Fire Control Systems—General. Army Materiel Command. NTIS. Accession: AD-830809. Tech Report: AMCP-706-327. SuDoc: D101.22/3:706-327. Cost: $15.25.

789. Engineering Design Handbook: Fundamentals of Ballistic Impact Dynamics, Part One. Army Materiel Command. NTIS. Tech Report: AMCP-706-158. SuDoc: D101.22/3:706-158.

790. Engineering Design Handbook: Fundamentals of Ballistic Impact Dynamics, Part Two (U). Army Materiel Command. NTIS. Tech Report: AMCP-706-159(S). SuDoc: D101.22/3:706-159(S).

791. Engineering Design Handbook: Fuzes, Proximity, Electrical, Part Five. Army Materiel Command. NTIS. Accession: AD-389296. Tech Report: AMCP-706-215(C). SuDoc: D101.22/3:706-215(C). Cost: $6.25.

792. Engineering Design Handbook: Fuzes, Proximity, Electrical, Part Four (U). Army Materiel Command. NTIS. Tech Report: AMCP-706-214(S). SuDoc: D101.22/3:706-214(S).

793. Engineering Design Handbook: Fuzes, Proximity, Electrical, Part One. Army Materiel Command. NTIS. Accession: AD-389295. Tech Report: AMCP-706-211(C). SuDoc: D101.22/3:706-211(C). Cost: $4.75.

794. Engineering Design Handbook: Fuzes, Proximity, Electrical, Part Three (U). Army Materiel Command. NTIS. Tech Report: AMCP-706-213(S). SuDoc: D101.22/3:706-213(S).

795. Engineering Design Handbook: Fuzes, Proximity, Electrical, Part Two (U). Army Materiel Command. NTIS. Tech Report: AMCP-706-212(S). SuDoc: D101.22/3:706-212(S).

796. Engineering Design Handbook: Fuzes. Army Materiel Command. NTIS, 1969, various paging. Accession: AD-889245. Tech Report: AMCP-706-210. SuDoc: D101.22/3:706-210. Cost: $9.25.

797. Engineering Design Handbook: Grenades (U). Army Materiel Command. NTIS. Tech Report: AMCP-706-240(C). SuDoc: D101.22/3:706-240(C).

798. Engineering Design Handbook: Guns—General. Army Materiel Command. NTIS, 1964, various paging. Accession: AD-830303. Tech Report: AMCP-706-250. SuDoc: D101.22/3:706-250. Cost: $6.00.

799. Engineering Design Handbook: Hardening Weapon Systems Against RF Energy. Army Materiel Command. NTIS. Accession: AD-894810. Tech Report: AMCP-706-235. SuDoc: D101.22/3:706-235. Cost: $10.75.

800. Engineering Design Handbook: Helicopter Engineering. Part One. Preliminary Design. Army Materiel Command. NTIS, August 30, 1974, 775 pp. Accession: AD/A-002 007. Tech Report: AMCP-706-201. SuDoc: D101.22/3:706-201. Cost: $2.25 microfiche; $18.75 paper copy.

Army helicopter design requirements are covered in this handbook for all missions under visual flight rule operation. The basic aerial vehicle is discussed but such things as sensors, cargo-handling equipment, weapons, power plants, generators, and batteries are excluded from the handbook.

801. Engineering Design Handbook: Helicopter Engineering. Part Three. Qualification Assurance. Army Materiel Command. NTIS, 1972, 334 pp. Accession: AD-901657. Tech Report: AMCP-706-203. SuDoc: D101.22/3:706-203.

802. Engineering Design Handbook: Helicopter Performance Testing. Army Materiel Command. NTIS (or U.S. Government Printing Office), 1974, 263 pp. Accession: AD-785000. Tech Report: AMCP-706-204. SuDoc: D101.22/3:706-204.

803. Engineering Design Handbook: Infrared Military System, Part Two (U). Army Materiel Command. NTIS. Tech Report: AMCP-706-128(S). SuDoc: D101.22/3:706-128(S).

804. Engineering Design Handbook: Infrared Military Systems, Part One. Army Materiel Command. NTIS, 1973, 569 pp. Accession: AD-763495, supersedes AD-885227. Tech Report: AMCP-706-127. SuDoc: D101.22/3:706-127. Cost: $16.25.

805. Engineering Design Handbook: Interior Ballistics of Guns. Army Materiel Command. NTIS, 1965, various paging. Accession: AD-462060. Tech Report: AMCP-706-150. SuDoc: D101.22/3:706-150. Cost: $8.00.

806. Engineering Design Handbook: Joining of Advanced Composites. Andrew Devine (Army Materiel Command). NTIS, 1979, 147 pp. Accession: AD-A072 362. Tech Report: DARCOM-P-706-316. Cost: $16.50 microfiche; $16.50 hard copy.

Covers such topics as joining techniques for advanced composites, fiber-reinforced composites, bonded joints, adhesive bonding, stresses related to these joints, and stress analysis.

807. Engineering Design Handbook: Liquid-Filled Projectile Design. Army Materiel Command. NTIS, 1969, various paging. Accession: AD-853719. Tech Report: AMCP-706-165. SuDoc: D101.22/3:706-165. Cost: $7.50.

808. Engineering Design Handbook: Maintainability Engineering Theory and Practice (METAP). Army Materiel Command. NTIS, 1976, 459 pp. Accession: AD-A026006. Tech Report: AMCP-706-133. SuDoc: D101.22/3:706-133. Cost: $17.50.

809. Engineering Design Handbook: Maintainability Guide for Design. Army Materiel Command. NTIS, 1972, 448 pp. Accession: AD-754202. Tech Report: AMCP-706-134. SuDoc: D101.22/3:706-134.

810. Engineering Design Handbook: Maintenance Engineering Techniques. Army Materiel Command. U.S. Government Printing Office, June, 1975, 554 pp. Tech Report: AMCP-706-132. SuDoc: D101.22/3:706-132.

811. Engineering Design Handbook: Metric Conversion Guide. Army Materiel Command. NTIS, 1976, 146 pp. Accession: AD-A029902. Tech Report: AMCP-706-470. SuDoc: D101.22/3:706-470. Cost: $9.00.

812. Engineering Design Handbook: Military Pyrotechnics Series, Part Four, Design of Ammunition for Pyrotechnic Effects. Army Materiel Command. NTIS, March 15, 1974, 290 pp, 1 of 5 Vols. Accession: AD-A000 821, supersedes AD-780673. Tech Report: AMCP-706-188. SuDoc: D101.22/3:706-188. Cost: $2.25 microfiche; $9.25 paper copy.

This volume, 1 of 5 in this series, covers the area of pyrotechnic ammunitions concentrating on the engineering aspects of the terminal effects. It also considers producibility, reliability, maintainability, human factors, safety, and cost, as well as design with respect to performance.

813. Engineering Design Handbook: Military Pyrotechnics, Part Five, Bibliography. Army Materiel Command. NTIS. Accession: AD-803864. Tech Report: AMCP-706-189. SuDoc: D101.22/3:706-189. Cost: $6.00.

814. Engineering Design Handbook: Military Pyrotechnics, Part One, Theory and Application. Army Materiel Command. NTIS, 1967, various paging. Accession: AD-817071. Tech Report: AMCP-706-185. SuDoc: D101.22/3:706-185. Cost: $10.75.

815. Engineering Design Handbook: Military Pyrotechnics, Part Three, Properties of Materials Used in Pyrotechnic Compositions. Army Materiel Command. NTIS. Accession: AD-830394. Tech Report: AMCP-706-187. SuDoc: D101.22/3:706-187. Cost: $12.50.

816. Engineering Design Handbook: Military Pyrotechnics, Part Two, Safety, Procedures and Glossary. Army Materiel Command. NTIS, 1963, 58 pp. Accession: AD-830371. Tech Report: AMCP-706-186. SuDoc: D101.22/3:706-186. Cost: $5.25.

817. Engineering Design Handbook: Military Vehicle Electrical Systems. Army Materiel Command. NTIS, 1974, 593 pp. Accession: AD-783697. Tech Report: AMCP-706-360. SuDoc: D101.22/3:706-360.

818. Engineering Design Handbook: Military Vehicle Power Plant Cooling. Army Materiel Command. NTIS, June 1975, 641 pp. Accession: AD-A013769. Tech Report: AMCP-706-361. SuDoc: D101.22/3:706-361. Cost: $20.00 microfiche; $20.00 paper copy.

The primary subject of this handbook is vehicle cooling systems for the Armed Forces. It will be helpful in the design of new military vehicles, in that it provides information with regard to cooling systems and their interrelationships with the total vehicle system. Such areas as heat transfer, cooling fans, temperature control, and fluid flow are covered in this handbook.

819. Engineering Design Handbook: Muzzle Devices. Army Materiel Command. NTIS, 1968, various paging. Accession: AD-838748. Tech Report: AMCP-706-251. SuDoc: D101.22/3:706-251. Cost: $7.25.

820. Engineering Design Handbook: Packaging and Pack Engineering. Army Materiel Command. NTIS, 1972, 380 pp (various paging). Accession: AD-901533. Tech Report: AMCP-706-121, supersedes AMC-PAM-706-121 (1964). SuDoc: D101.22/3:706-121. Cost: $13.00.

821. Engineering Design Handbook: Principles of Explosive Behavior. Army Materiel Command. NTIS. Accession: AD-900260. Tech Report: AMCP-706-180. SuDoc: D101.22/3:706-180. Cost: $13.00.

822. Engineering Design Handbook: Propellant Actuated Devices. Army Materiel Command. NTIS, 1975, 198 pp. Accession: AD-A016716, supersedes AD-449769. Tech Report: AMCP-706-270. SuDoc: D101.22/3:706-270. Cost: $12.50 microfiche; $12.50 paper copy.

823. Engineering Design Handbook: Recoil Systems. Army Materiel Command. NTIS. Accession: AD-830281. Tech Report: AMCP-706-342. SuDoc: D101.22/3:706-342. Cost: $6.00.

824. Engineering Design Handbook: Reliable Military Electronics. Army Materiel Command. NTIS, 1976, 435 pp. Accession: AD-A025665. Tech Report: AMCP-706-124. SuDoc: D101.22/3:706-124. Cost: $15.00.

825. Engineering Design Handbook: Rotational Molding of Plastic Powders. Army Materiel Command. NTIS, 1975, 68 pp. Accession: AD-A013178. Tech Report: AMCP-706-312. SuDoc: D101.22/3:706-312. Cost: $15.00.

826. Engineering Design Handbook: SABOT Technology Engineering. Army Materiel Command. NTIS. Accession: AD-903789. Tech Report: AMCP-706-445. SuDoc: D101.22/3:706-445. Cost: $11.00.

827. Engineering Design Handbook: Servomechanisms, Section Four, Power Elements and System Design. Army Materiel Command. NTIS, 1965, rev. ed. in prep. Accession: AD-830283. Tech Report: AMCP-706-139. SuDoc: D101.22/3:706-139.

828. Engineering Design Handbook: Servomechanisms, Section One, Theory. Army Materiel Command. NTIS, 1965, rev. ed. in prep. Tech Report: AMCP-706-136. SuDoc: D101.22/3:706-136.

829. Engineering Design Handbook: Servomechanisms, Section Three, Amplification. Army Materiel Command. NTIS, 1965, rev. ed. in prep. Accession: AD-830274. Tech Report: AMCP-706-138. SuDoc: D101.22/3:706-138.

830. Engineering Design Handbook: Servomechanisms, Section Two, Measurement and Signal Converters. Army Materiel Command. NTIS, 1965, rev. ed. in prep. Accession: AD-830263. Tech Report: AMCP-706-137. SuDoc: D101.22/3:706-137.

831. Engineering Design Handbook: Short Fiber Plastic Base Composites. Army Materiel Command. NTIS, 1975, 113 pp. Accession: AD-A015181. Tech Report: AMCP-706-313. SuDoc: D101.22/3:706-313. Cost: $12.50.

832. Engineering Design Handbook: Solid Propellants, Part One. Army Materiel Command. NTIS, 1964, 121 pp. Accession: AD-830265. Tech Report: AMCP-706-175. SuDoc: D101.22/3:706-175. Cost: $7.25.

833. Engineering Design Handbook: Spectral Characteristics of Muzzle Flash. Army Materiel Command. NTIS. Accession: AD-818532. Tech Report: AMCP-706-255. SuDoc: D101.22/3:706-255. Cost: $6.50.

834. Engineering Design Handbook: Structures. Army Materiel Command. NTIS. Accession: AD-830267. Tech Report: AMCP-706-286. SuDoc: D101.22/3:706-286. Cost: $6.00.

835. Engineering Design Handbook: System Analysis and Cost-Effectiveness. Army Materiel Command. NTIS, 1971, 410 pp. Accession: AD-884151. Tech Report: AMCP-706-191. SuDoc: D101.22/3:706-191. Cost: $13.25.

836. Engineering Design Handbook: Tables of the Cumulative Binomial Probabilities. Army Materiel Command. NTIS, 1971, 577 pp. Accession: AD-903967. Tech Report: AMCP-706-109. SuDoc: D101.22/3:706-109. Cost: $19.00.

Covers the binomial distribution and the incomplete beta function ratio.

837. Engineering Design Handbook: Terminal Ballistics. Army Materiel Command. NTIS, 1976, 232 pp. Accession: AD-A021833. Tech Report: NWC-TP-5780. SuDoc: D101.22/3:5780. Cost: $12.50.

838. Engineering Design Handbook: The Monte Carlo Method of Evaluating Integrals. Army Materiel Command. NTIS, 1975, 210 pp. Accession: AD-A005891. Tech Report: NWC-TP-5714. SuDoc: D101.22/3:5714. Cost: $7.25.

839. Engineering Design Handbook: Timing Systems and Components. Army Materiel Command. NTIS, 1975, 491 pp. Accession: AD-A020020. Tech Report: AMCP-706-205. SuDoc: D101.22/3:706-205. Cost: $17.50.

840. Engineering Design Handbook: Top Carriages. Army Materiel Command. NTIS. Accession: AD-830393. Tech Report: AMCP-706-343. SuDoc: D101.22/3:706-343. Cost: $4.50.

841. Engineering Design Handbook: Trajectories, Different Effects, and Data for Projectiles. Army Materiel Command. NTIS. Accession: AD-830264. Tech Report: AMCP-706-140. SuDoc: D101.22/3:706-140. Cost: $5.25.

842. Engineering Design Handbook: Traversing Mechanisms. Army Materiel Command. NTIS. Accession: AD-830291. Tech Report: AMCP-706-347. SuDoc: D101.22/3:706-347. Cost: $4.50.

843. Engineering Design Handbook: Value Engineering. Army Materiel Command. NTIS. Accession: AD-894478. Tech Report: AMCP-706-104. SuDoc: D101.22/3:706-104. Cost: $6.00.

844. Engineering Design Handbook: Warheads-General. Army Materiel Command. NTIS. Accession: AD-501329. Tech Report: AMCP-706-290(C). SuDoc: D101.22/3:706-290(C). Cost: $12.50.

845. Engineering Design Handbook: Weapon System Effectiveness (U). Army Materiel Command. NTIS. Tech Report: 706-281(SRD). SuDoc: D101.22/3:706-281 (SRD).

846. Engineering Design Handbook: Wheeled Amphibians. Army Materiel Command. NTIS, 1971, 425 pp. Accession: AD-881357. Tech Report: AMCP-706-350. SuDoc: D101.22/3:706-350. Cost: $13.25.

847. Engineering Encyclopedia. Franklin D. Jones; Paul B. Schubert. Industrial Press, 3d ed., 1963, 1431 pp. LC: 63-10415. ISBN: 0-8311-1017-1. Cost: $28.00.

This is an excellent single-volume work that provides information on such topics as air gauges, rotary files, corrosion fatigue, cold extrusion, ceramic cutting, cermet, surface finishing, and heat transfer.

848. Engineering Formulas and Tables. Lefax Publishing Co, loose-leaf binding. Cost: $9.50. Series: *Lefax Technical Manuals*, No. 783.

849. Engineering Formulas for Cement Manufacturers. Kurt E. Peray. Chemical Publishing Co, Inc, 1979, 325 pp. ISBN: 0-8206-0245-0. Cost: $32.50.

This represents a substantial collection of formulas and data with regard to cement manufacturing. Includes information on such topics as chemistry of cement, heat balance, alkali balance, and kilns. Also includes formulas for such topics as fan engineering, fluid flow, heat transfer, and dust collection.

850. Engineering Formulas. Kurt Gieck. MaGraw-Hill, 3d ed., 1979, 433 pp. LC: 79-19687. ISBN: 0-07-000000-0. Date of First Edition: 1971. Cost: $11.95.

Provides information which will assist the user in formulating engineering calculations. Includes information on hydraulics, heat transfer, Fourier series, and permutations.

851. Engineering Handbook of Space Technology. A.A. Alatyrtsev. NTIS, translated from the Russian by Foreign Technology Division, Wright Patterson Air Force Base, 1971, 713 pp. Accession: AD-727912. Tech Report: FTD-HC-23-456-70.

852. Engineering Handbook. Drop Spillways, Section 11 Soil Conservation Service. NTIS, 1968, 173 pp. Accession: PB-243645. Cost: $2.25 microfiche; $6.25 paper copy.

853. Engineering Handbook. Hydraulics, Section Five. Soil Conservation Service. NTIS, 1956, 244 pp. Accession: PB-243644. Cost: $2.25 microfiche; $7.50 paper copy.

854. Engineering Handbook. Structural Design, Section Six. Soil Conservation Service. NTIS, 1964, 227 pp. Accession: PB-243890. Cost: $2.25 microfiche; $7.50 paper copy.

855. Engineering Handbook: Chute Spillways, Section 14. Soil Conservation Service. NTIS, 1955, 248 pp. Accession: PB-243646. Cost: $2.25 microfiche; $7.50 paper copy.

856. Engineering Manual: A Practical Reference of Data and Methods in Architectural, Chemical, Civil, Electrical, Mechanical, and Nuclear Engineering. Robert H. Perry. McGraw-Hill, 3d ed., 1976, 800 pp. (various paging). LC: 76-12514. ISBN: 0-07-049476-2. Date of First Edition: 1959. Cost: $24.50.

The *Engineering Manual* is a very interesting collection of general engineering information in the fields that are mentioned in the subtitle. This handbook includes many tables and data in such fields as thermodynamics, heat transfer, fluid dynamics, etc. A very good collection of mathematical tables is included for the convenience of the engineer. This handbook has an excellent index to retrieve the information as well as a very detailed table of contents at the beginning of each chapter. In addition to the other basic engineering principles, this handbook includes up-to-date information on environmental problems and energy conversion techniques. For other general engineering handbooks, please refer to the following titles: *The Engineer's Manual; De Laval Engineering Handbook; Handbook of Engineering Fundamentals; CRC Handbook of Tables for Applied Engineering Science;* and *Engineering Tables and Data.*

857. Engineering Materials Handbook. C.L. Mantell. McGraw-Hill, 1958, 1,960 pp. LC: 56-11720. ISBN: 0-07-040025-3. Cost: $41.50.

858. Engineering Mathematics Handbook. Jan J. Tuma. McGraw-Hill, 2d ed., 1979, 394 pp. LC: 77-17786. ISBN: 0-07-065429-8. Cost: $21.95.

This one-volume reference source provides an excellent collection of mathematics tables, formulas, and equations for the practicing engineer or scientist. This handbook includes such subjects as differential equations, calculus, Laplace transforms, and many mathematical functions. It is well-indexed and provides and excellent bibliography and list of references. For related works, please refer to the following titles: *Handbook of Mathematical Tables and Formulas, A Table of Series and Products; CRC Handbook of Tables for Mathematics, Eight-Place Tables of Trigonometric Functions for Every Second of Arc; CRC Standard Mathematical Tables; CRC Handbook of Tables for Probability and Statistics,* and various publications of the U.S. National Bureau of Standards including such series as the Applied Mathematics Series, The Mathematical Table Series, and the Columbia University Press Series.

859. Engineering Properties of Ceramics. Data Book to Guide Materials Selection for Structural Applications. James F. Lynch et al (Battelle Memorial Institute). NTIS, 1966, 660 pp. Accession: AD-803765. Tech Report: AFML-TR-66-52.

860. Engineering Properties of Selected Ceramic Materials. J.F. Lynch et al. American Ceramic Society, Inc, 1966, loose-leaf service with updates (various paging). Tech Report: AFML-TR-66-52.

Supersedes the following title: *Refractrory Ceramics for Aerospace*, 1964, American Ceramics Society.

861. Engineering Property Data on Selected Ceramics, Vol. One, Nitrides. Metals and Ceramics Information Center. NTIS, 1976, 111 pp. Accession: AD-A023773. Tech Report: MCIC-HB-07-Vol. 1. Cost: $20.00.

Includes 232 references, 44 figures, and 26 tables. Replaces Section 5.3, Nitrides, of the 1966 Databook.

862. Engineering Sciences Data. Engineering Sciences Data Unit. ESDU Marketing Ltd, 1940, Currently 120 Vols., loose-leaf service. Cost: $522.00 per Vol.

This multi-volume set is an excellent source of information on evaluated data. This comprehensive work includes a vast amount of evaluated data on the stress and strength of structures. In addition, information is provided on heat transfer, tribology, fluid mechanics, acoustic fatigue, aerospace structures, and various vapour pressures and thermodynamic properties of compounds. A 373 page index is included with this service and is published annually. This key word index provides access to the complete range of 790 data items (evaluated reports) and also provides a table of physical properties of over 1,600 elements and pure compounds.

863. Engineering Sciences Data: Mechanical Engineering Series: Fluid Mechanics, Internal Flow. Engineering Sciences Data Unit (London), 5 Vols., 1972-76.

864. Engineering Steel Chains - Applications Handbook. Engineering Steel Chain Division, American Chain Association. American Chain Association, 1973, 102 pp. Cost: $10.00.

865. Engineering Tables and Data. A.M Howatson; P.G. Lund; J.D. Todd. Chapman and Hall (also available from Halsted Press), 1972, 168 pp. ISBN: 412-11550-6. Cost: $7.00.

This handbook, a collection of engineering tables, data, formulas, and equations is very useful to the practicing engineer and to students of engineering. It includes information on such subjects as the properties of solids, liquids, and gases; statics and dynamics; thermodynamics and heat transfer; fluid mechanics; engineering stress and strain; and electricity. A representative collection of mathematical tables is included as well as a list of bibliographic references for further investigation. For related works in the area of general engineering tables, please refer to the following titles: *Handbook of Engineering Fundamentals; Engineering Manual; CRC Handbook of Tables for Applied Engineering Science; De Laval Engineering Handbook*, and *The Engineers' Manual*.

866. Engineering Tables for Energy Operators. J.E. (Hank) Kastrop. Energy Communications, Inc, 1973, 312 pp. LC: 73-91128.

This handbook represents a collection of formulas, tables, equations, properties data, charts, calculations, and curves that primarily reflect the oil and gas operations such as pipeline transmission, processing, drilling, and production. This handbook includes information on such subjects as displacement tables for pumps, dimensions of different sizes of tubes, hydraulic capacities for different sizes of pipes, welding information for pipelines and natural gas compressibility factors. Such subjects as presure drop in oil and gas lines, hydrostatic pressure, steam generators, corrosion inhibitors, and various properties are also discussed. In addition, information is provided for drilling operations, conversion factors for the metric system, interest tables, and ways of calculating future gas and oil reserves. For related works in the field of energy technology, please refer to the following titles: *Petroleum Production Handbook; Nuclear Engineering Handbook; Hydroelectric Handbook; Energy Reference Handbook; Keystone Coal Industry Manual; Gas Engineer's Handbook; Handbook of Natural Gas Engineering; Hydrogen: Its Technology and Implications*.

867. Engineers' Metric Manual and Buyers' Guide. D.S. Lock. Pergamon Press, 1975, 350 pp. LC: 72-97365. ISBN: 0-08-018220-8. Cost: $200.00.

This hanbook providing metric information that has been used in the British Engineering Industry, will have wide applications everywhere. Dimensions for fasteners, bearings, fluid sealing, coupling systems, and machine tools are provided.

868. Engineers' Relay Handbook. National Association of Relay Manufacturers. Hayden Book Co, 2d ed., 1969, 335 pp. LC: 66-23643.

869. Enthalpy-Entropy Diagram for Steam. Basil Blackwell. Basil Blackwell (distributed by International Scholarly Book Services), 1977. ISBN: 0-631-97470-9. Cost: $2.00.

870. Environment Regulation Handbook: Air Pollution, Land Use, Mobile Sources, NEPA, Noise, Pesticides, Radioactive Materials, Solid Wastes, Water Pollution. Environment Information Center, Inc. Environment Information Center, Inc, 1973-, 2,500 pp. plus; loose-leaf service with monthly updates.

This loose-leaf service covers laws and regulations dealing with the environment. For related works, please refer to the the *Environmental Law Handbook* and *Environmental Law Reporter*.

871. Environment Sanitation Handbook. NASA. NTIS, 1973, 48 pp. Accession: N73-32020. Tech Report: NASA-TM-X-69522.

872. Environmental Assessment and Impact Statement Handbook. Paul N. Cheremisinoff; Angelo C. Morresi. Ann Arbor Science Publishers, 1977, 438 pp. LC: 76-050989. ISBN: 0-250-40158-4. Cost: $37.50.

Covers the requirements associated with the National Environmental Policy Act of 1965. Guidelines, groundrules procedures and format are discussed with regard to the preparation of environmental impact statements.

873. Environmental Biology. Philip L. Altman; Dorothy S. Dittmer. Federation of American Societies for Experimental Biology, 1966, 694 pp. LC: 66-25792. ISBN: 0-913822-03-5. Cost: $20.00.

Includes 190 tables and covers such topics as temperature, atmospheric pressures, pollution, radiant energy, and biorhythms

874. Environmental Engineers' Handbook. Bezla G. Liptazk. Chilton Book Co, 1974, Vol.1, 2,018 pp.; Vol.2, 1,340 pp.; Vol.3, 1,130 pp. LC: 72-14241. ISBN: Vol.1, 0-8019-5692-7; Vol.2, 0-8019-5693-5; Vol.3, 0-8019-5694-3. Cost: Vol.1, $39.00; Vol.2, $35.00; Vol.3, $35.00.

875. Environmental Health Monitoring Manual. United States Steel Corp, 1973, various paging.

This is a reference guide that was designed primarily for and by the United States Steel Corp. to train plant personnel to avoid toxic hazards and harmful physical conditions. It emphasizes the maintenance and testing of instrumentation, and discusses such topics as sampling of particulates, sampling of gases and vapors, calibration of air sampling instruments, measurement of noise, and heat stress. Illumination and its measurement are also discussed.

876. Environmental Impact Analysis Handbook. John G. Rau; David C. Wooten. McGraw-Hill, 704 pp. ISBN: 0-07-051217-5. Cost: $36.50.

This is basically a guide for preparing environmental impact statements. It is designed primarily to help planners assess noise, air, and water quality, and the energy impact of all projects.

877. Environmental Impact Data Book. Jack Gordon et al. Ann Arbor Science Publishers, 1979, 866 pp. LC: 77-92596. ISBN: 0-250-40212-2. Cost: $49.95.

This is a guide for preparing environmental impact statements. It provides information in table format to help quantify certain types of concepts. Discusses such topics as air quality, water resources, noise, ecosystems, toxic substances, energy, transportation, and social economic impact.

878. Environmental Impact Handbook. Robert W. Burchell; David Listokin. Center for Urban Policy Research, 1975. ISBN: 0-88285-024-5. Cost: $8.95.

879. Environmental Law Handbook. J. Gordon Arbuckle et al. Government Institutes, Inc, 6th ed., 1979, 368 pp. Date of First Edition: 1973. Cost: $38.50.

This book is revised annually and includes an analysis of regulations and laws relating to such environmental problems as noise, pesticides, air pollution, water pollution, and solid waste disposal.

880. Environmental Phosphorus Handbook. Edward J. Griffith et al. Wiley-Interscience, 1973, 718 pp. LC: 72-11574. ISBN: 0-471-32779-4. Cost: $47.50.

This handbook covers such topics as the geochemistry of phosphorus, phosphate deposits, and the analysis of phosphorus and its compounds.

881. Environmental Science Handbook for Architects and Builders. S.V. Szokolay. Halsted Press, 1980, 532 pp. LC: 79-25004. ISBN: 0-470-26904-9. Cost: $76.95.

Covers such topics as sound, space, light, and heat with regard to building design. A bibliography and index are included.

882. Environmental Sources and Emissions Handbook-1975. Marshall Sittig. Noyes Data Corp, 1975, 521 pp. LC: 75-800. ISBN: 0-8155-0568-X. Cost: $36.00.

This handbook containing 382 tables, covers various processes that produce pollutants.

883. Enzyme Handbook. Thomas E. Barman. Springer-Verlag, 1969, 2 Vols. LC: 69-19293. ISBN: 0-387-04423-X. Cost: $49.20.

NOTE: Supplement 1 to the *Enzyme Handbook* was published in 1974, ISBN: 0-387-06761-2, $25.00.

884. EPA Manual of Chemical Methods for Pesticides and Devices. Association of Official Analytical Chemists. Association of Official Analytical Chemists, 1979, 1,056 pp. Cost: $40.00.

More than 350 individual infrared spectra of pesticide compounds are included in this manual. In addition, there is a substantial bibliography on pesticide formulations, as well as a section on thin layer chromatographic analysis.

885. EPA/NIH Mass Spectral Data Base. S.R. Heller; G.W.A. Milne. Bureau of Standards (available from Superintendent of Documents), 1978. LC: 78-606178.

886. EPIC Energy Conservation Program Guide for Industry and Commerce. National Bureau of Standards. U.S. Government Printing Office. SuDoc: C13.11:115. Cost: $2.90. Series: NBS Handbook 115.

887. EPIC Supplement 1. National Bureau of Standards. U.S. Government Printing Office. SuDoc: C13.11:115/1. Cost: $2.25. Series: NBS Handbook 115.

888. Epilepsy Handbook. Louis D. Boshes; Frederic A. Gibbs. C.C. Thomas, 2d ed., 1972. ISBN: 0-398-02194-5. Cost: $11.75.

889. Equation of State for Use with Sodium. Part 3: Tables of Thermodynamic Properties. M.A.H.G. Alderson. NTIS, 1975, 182 pp. Accession: N77-76794. Tech Report: SRD-R-50-PT-3.

This is a collection of tables on the equation of state for use with sodium. Includes information on critical temperatures, enthalpy, and entropy.

890. Equilibrium and Thermodynamic Properties of Mixed Plasmas. M. Capitelli; E. Ficocelli; E. Molinari. Available from NTIS, 1968-73, 3 vols., Vol.1, 173 pp.; Vol.2, 139 pp.; Vol.3, 195 pp. Accession: Vol.1, N76-72366; Vol.2, N71-23083; Vol.3, N73-25764.

The titles of Vols.2 and 3 are as follows: Vol.2, *Argon-Oxygen Plasmas at .01-10 Atmospheres, Between 2000 K and 35000 K*; Vol.3, *Argon-Hydrogen Plasmas at .01-100 Atmospheres, Between 2000 K and 35000 K*. This 3-volume set covers mixed plasmas, equilibrium compositions, and thermodynamic properties for nitrogen mixtures with helium, argon, and xenon. In addition, equilibrium composition, and thermodynamic properties of mixed plasmas are also presented for argon-oxygen plasmas and argon-hydrogen plasmas.

891. Equilibrium Constants of Liquid-Liquid Distribution Reactions. Y. Marcus. Pergamon, 1974-78, 4 vols. ISBN: Vol.1, 0-08-020828-2; Vol.2, 0-08-020829-0; Vol.3, 0-08-022032-0; Vol.4, 0-08-022343-5. Cost: Vol.1, $24.00; Vol.2, $14.50; Vol.3, $7.20; Vol.4, $37.00.

This compilation, sponsored by the International Union of Pure and Applied Chemistry, consists of 3 parts entitled as follows:

Part 1 *Organophosphorus Extractants*; Part 2 *Alkylammonium Salt Extractants*; Part 3 *Compound Forming Extractants, Solvating Solvents and Inert Solvents*; Part 4 *Chelating Extractants*. Each volume is in the IUPAC Chemical Data Series: Part 1 is Vol. 3, Part 2 is Vol. 4, Part 3 is Vol. 15, and Part 4 is Vol. 18.

892. Equipment Design Handbook for Refineries and Chemical Plants. Frank L. Evans Jr. Gulf Publishing Co, 2d ed., Vol.1, 1979; Vol.2, 1980. ISBN: Vol.1, 0-87201-254-9; Vol.2, 0-87201-255-7. Cost: Vol.1, $24.95; Vol.2, $32.95.

This 2-volume set covers such subjects as compressors, ejectors, pumps, refrigeration, and nonrotating equipment as it relates to oil refining and chemical processing.

893. Equivalent Valves Reference Manual. Oildom Publishing Co. of Texas (also available from Gulf Publishing), 17th ed., 1974, 167 pp. Cost: $30.00.

This is a listing of various types of valves produced by different manufacturers. It includes data on their differences and similarities. The valves are listed by type, material composition, and the pressure ratings of each of the valves. For related works, please refer to such titles as *ISA Handbook of Control Vales, Handbook of Valves*, and *ISA Control Valve Compendium*.

894. Essential Formulae for Electrical Engineers. Noel M. Morris. Halsted Press, 1974, 26 pp. LC: 74-8677. ISBN: 0-470-61565-6. Cost: $3.95.

895. Eurohealth Handbook: 1978 Edition. Robert S. First (ed). Hanover Publications, 1978, 786 pp. Cost: $300.00.

Covers health care facilities in 18 European countries including hospitals and medical services and government health agencies.

896. Evaluation and Compilation of Thermodynamic Properties of High Temperature Chemical Species. M.W. Chase. NTIS, 1978, 12 pp. Accession: N78-32205; AD-A055660. Tech Report: AFOSR-78-1060TR. Cost: $3.50 microform; $5.00 hard copy.

This set of tables represents an update of the critically evaluated *JANAF Thermochemical Tables*. Entropy, enthalpy, and other thermodynamic properties are presented at high temperatures. Note: see *JANAF Thermochemical Tables*.

897. Evaluations of the Elevated Temperature Tensile and Creep-rupture Properties of C-Mo, Mn-Mo and Mn-Mo-Ni Steels. G.V. Smith. American Soceity for Testing and Materials, 1971, 94 pp. LC: 70-174977. ISBN: 0-8031-0113-9. Cost: $6.25. Series: *ASTM Data Series Publication*, DS 47.

This set of tables covers the various properties of several alloys of steel. Carbon-molybdenum, manganese-molybdenum steels, and manganese-molybdenum-nickel steels are included in the tables. Such properties as yield strength, tensile strength, elongation, reduction of area, creep, and rupture properties are presented.

898. Examination of Farm Milk Tanks. National Bureau of Standards. NTIS, May 22, 1964. Accession: COM 72-10619. Cost: $4.00. Series: NBS Handbook 98.

899. Examination of Liquefied Petroleum Gas Liquid-Measuring Devices. National Bureau of Standards. NTIS, April 23, 1965. Cost: $3.50. Series: NBS Handbook 99.

900. Examination of Liquefied Petroleum Gas Vapor-Measuring Devices. National Bureau of Standards. U.S. Government Printing Office. SuDoc: C13.11:117. Cost: $0.75. Series: NBS Handbook 117.

901. Examination of Water for Pollution Control: A Handbook for Management and Analysis. M.J. Suess. Pergamon Press, 1980, 3-Vol. set, approx. 1,700 pp. ISBN: 0-08-025255-9-h. Cost: $325.00.

The 3 volumes are entitled as follows: Vol.1, *Sampling Data Analysis and Laboratory Equipment*; Vol.2, *Physical, Chemical and Radiological Examination*; Vol.3, *Biological, Bacteriological and Virological Examination*. This 3-volume work was prepared under the direction of the World Health Organization and includes such topics as sampling data analysis, radiological examination, physical and chemical examination of water samples, and bacteriological and virological examination of water samples, as well as the identification of metallic ions and nonmetallic constituents in water samples.

902. Examination of Weighing Equipment. National Bureau of Standards. NTIS, March 1, 1965. Accession: COM 73-10635. Cost: $9.25. Series: NBS Handbook 94 (supersedes Handbook 37).

903. Examination Procedure Outlines for Commercial W&M Devices. National Bureau of Standards. NTIS. Accession: COM 73-50836. Cost: $5.00. Series: NBS Handbook 112.

904. Excavation and Grading Handbook. Nick Capachi. Craftsman Books, 1978, 320 pp. LC: 78-3850. ISBN: 0-910460-54-X. Cost: $12.25.

This handbook covers such information as excavation around utility lines, building narrow embankments, removal of old road services, widening roads, setting grades for cuts, building curbs and pipelines, paving highways, and laying sewers. Such topics as trenching, shoring, drain pipe, and highway excavation are also treated.

905. Excavation Handbook. Horace K. Church. McGraw-Hill, 1980, 976 pp. LC: 80-19630. ISBN: 0-07-010840-4. Cost: $49.50.

Covers such topics as fragmentation of rock, hauling rock, dumping and compacting excavation, construction of haul roads, machinery used in excavation, and different types of rock formations, ores, and minerals.

906. Executive's Handbook to Minicomputers. Robert Allen Bonelli. Petrocelli (division of McGraw-Hill), 1979, 320 pp. LC: 78-25584. ISBN: 0-07-091045-6; 0-89433-090-X (Petrocelli). Cost: $12.00.

This is a guide to minicomputer systems and a discussion of such topics as central processors, main memory, various terminals, and other peripheral devices. The handbook also presents applications of software and discusses programing languages.

907. Experimental Statistics. National Bureau of Standards. U.S. Government Printing Office, Aug. 1, 1963. SuDoc: C13.11:91. Cost: $8.45. Series: NBS Handbook 91.

908. Experiments in the Computation of Conformal Maps. J. Todd (ed) (National Bureau of Standards).

NTIS, 1955, 61 pp. Accession: PB-175819. Cost: $4.50. Series: NBS AMS42.

NOTE: Related to NBS AMS18.

909. Facilities and Plant Engineering Handbook. Bernard T. Lewis; J.P. Marron. MGraw-Hill, 1973, 1,024 pp (various paging). LC: 73-15565. ISBN: 0-07-037560-7. Cost: $49.50.

Information is provided in this handbook on such subjects as plant security, elevator maintenance, painting, preventive maintenance, inventory control, roofing, utilities, air conditioning, and safety measures. An index and bibliography are included.

910. Faith, Keyes, and Clark's Industrial Chemicals. William Lawrence Faith et al. Wiley-Interscience, 4th ed., 1975, 904 pp. LC: 75-17951. ISBN: 0-471-54964-9. Cost: $65.00.

The first 3 editions of this handbook were published under the title *Industrial Chemicals.* This handbook lists many different chemicals that are used in industry and provides information about these chemicals, such as processes, the economics associated with the chemical, flow diagrams, etc. The handbook follows a dictionary format and lists the chemicals in alphabetical sequence. An index is provided, in addition to a listing of companies that make these chemicals. Metric units have been used in presenting the chemical and physical properties of the chemicals, and a section of conversion tables has been provided.

911. Farinograph Handbook. American Association of Cereal Chemists. 80 pp. ISBN: 0-913250-01-5. Cost: $15.00.

This handbook deals with the physical design, theory, and proper utilization of the Farinograph. It discusses the measurement of the dough-forming properties of different wheats under various controlled conditions of temperature.

912. Farm Builder's Handbook. R.J. Lytle et al. McGraw Hill, 3d ed., 1978, 288 pp. LC: 78-73354. ISBN: 0-07-039276-5. Date of First Edition: 1969. Cost: $24.95.

913. Farm Chemical Handbook (title varies). Meister. Date of First Edition: 1908.

914. Fat-soluble Vitamins. Hector F. DeLuca. Plenum Publishing Corp, 1978, 287 pp. LC: 78-2009. ISBN: 0-306-33582-4. Cost: $27.50. Series: *Handbook of Lipid Research,* Vol. 2.

This collection of information provides data on vitamin A, vitamin D, vitamin E, and vitamin K with regard to fat-soluble vitamins. An extensive reference list is included, plus an index.

915. Fatigue Design Handbook: A Guide for Product Design and Development Engineers. James A. Graham et al. Society of Automotive Engineers, 1968, 132 pp. LC: 67-29346.

This handbook includes a considerable number of figures and tables on fatigue design. Such information as stress/strain properties are included, in addition to material strength evaluation, fatigue properties and identifying fatigue failures, as well as loads on such components as gears, wheels, chains, and axles. Fatigue data are presented on such materials as steels, aluminum and iron, as well as information on stress formulas for springs. This book is intended for those people who have a prior knowledge of fatigue problems.

916. Fatty Acids and Glycerides. Arnis Kukis. Plenum Publishing Corp, 1978, 469 pp. LC: 77-25277. ISBN: 0-306-33581-6. Cost: $35.00. Series: *Handbook of Lipid Research,* Vol. 1.

Provides data on lipid chemistry for the fatty acids and glycerides. A detailed account is made of the analytical behavior and composition of natural and synthetic fatty acids and acylglycerols. Composition of selected dietary fats, oils, margarine, and butter is included.

917. Fenaroli's Handbook of Flavor Ingredients. Thomas E. Furia; Nicolo Bellenca. CRC Press, Inc, 2d ed., 1975, 2 Vols., Vol.1, 560 pp.; Vol.2, 944 pp. LC: 72-152143. ISBN: Vol.1, 0-87819-534-3; Vol.2, 0-87819-532-7. Date of First Edition: 1971. Cost: Vol.1, $19.75; Vol.2, $19.95 (price reduced from earlier eds.).

The second edition of the 2-volume set of the *CRC Fenaroli's Handbook of Flavor Ingredients* provides information on approximately 1,200 natural and synthetic flavors. Physical and chemical characteristics are presented for the natural flavors and such properties as melting point, refractive index, solubility, etc. are presented for the synthetic flavors. In addition to the tabulated data, this handbook includes many articles on flavorings written by recognized experts, plus literature reference sources for further study, and a well prepared index. The reader's attention is also directed to the *CRC Handbook of Food Additives* for further information on flavors.

918. Ferroelectrics Literature Index. T.F. Connolly; Donald T. Hawkins. Plenum Publishing Corp, 1974, 705 pp. LC: 74-4363. ISBN: 0-306-68326-1. Series: *Solid State Physics Literature Guides,* Vol. 6.

This is a bibliographic guide to the literature on ferroelectrics and includes over 3,681 references from the literature mainly between mid-1969 through December 1972. It is organized on a materials basis and includes such materials as barium titanate, lead titanate, strontium titanate, lead zirconate, lithium niobate, tungsten trioxide, selenites and selenates, potassium ferrocyanide, rochelle salt, triglycine sulfate, hydrogen halides, and lead germanate. A permuted title index serves as an access point to these ferroelectric materials.

919. Ferromagnetic Materials: A Handbook on the Properties of Magnetically Ordered Substances. E.P. Wohlfarth. Elsevier-North Holland Publishing Co, 1980, Vol.1, 589 pp.; Vol.2, 559 pp. LC: 79-9308. Cost: Dfl. 178.50 (Dutch guilders) per volume.

This 2-volume set is eventually projected to be 4 volumes. This is a guide to the literature on ferromagnetic materials, and provides an excellent access point to the available research articles.

920. Ferromagnetic-Core Design and Application Handbook. M.F. DeMaw. Prentice-Hall, Inc, 1981, 256 pp. LC: 80-16136. ISBN: 0-13-314088-1. Cost: $19.95.

This book deals with electronic circuits and magnetic-core devices. Such topics as narrow-band and broad-band transformers, inductors, ferrite loop antennas, ferrite beads, and RF chokes are discussed. Permanent-magnet materials are also presented, in addition to magnetic core materials, from low frequencies through UHF frequencies.

921. FET Applications Handbook. Jerry Eimbinder. TAB Books, 2d ed., 1970, 350 pp. LC: 75-114708. ISBN: 0-8306-0240-2. Cost: $14.95.

922. Fiber, Fabrics and Flames: Instructor's Handbook with References. U.S. Department of Health, Education, and Welfare. U.S. Government Printing Office, 1970.

923. Fiberglass Reinforced Plastics Deskbook. Nicholas P. Cheremisinoff; Paul N. Cheremisinoff. Ann Arbor Science Publishers, 1978, 300 pp. LC: 0-250-40245-9. Cost: $37.50.

Covers the topic of fiberglass-reinforced plastics, their manufacture and application. The chemistry of fiberglass-reinforced plastics and resins is also presented, along with its economic assessment and projected usage. Design methods and scale-up models are also presented for the convenience of the user.

924. Fibers, Films, Plastics and Rubbers: A Handbook of Common Polymers. Butterworth, 1971.

Replaced by *Handbook of Common Polymers*. For related works, please refer to *Polymer Handbook* and *SPI Plastic Engineering Handbook*.

925. Field Crop Diseases Handbook. Robert F. Nyvall. AVI Publishing Co, 1979, 465 pp. ISBN: 0-87055-336-4. Cost: $47.50.

This is a guide to the identification and control of over 800 crop diseases. Photographs of over 190 diseases are included. This book covers diseases that are common among such plants as alfalfa, barley, buckwheat, cotton, corn, flax, oats, peanuts, rice, rye, sorghum, soybeans, sugar beets, sunflowers, tobacco, and wheat. An index and glossary are included for the user's convenience.

926. Field Engineer's Manual. Robert O. Parmley. McGraw-Hill, 1981, 612 pp. LC: 80-28187. ISBN: 0-07-048513-5. Cost: $24.50.

This manual consists of information dealing with the building and construction industry. It includes mathematical formulas, conversion tables, metric system units, and discusses such topics as heating and ventilation systems, welding, surveying, drainage, hydraulics, sewage collection, water supply, fire protection, water storage, and energy usage.

927. Fiesers' Reagents for Organic Synthesis. Mary Fieser; Louis Fieser. John Wiley and Sons, Inc, 1967-81, 9 Vols. ISBN: Vol.1, 0-471-25875-X. ISSN: 0271-616X. Series: *Reagents for Organic Synthesis Series*.

This is a continuing series currently in 9 volumes, and contains information on reagents for organic synthesis. Each volume includes information on the reagents, an index on the reagents according to types, an author index, and a subject index.

928. Films, Sheets, and Laminates. International Plastics Selector, 1979, 775 pp. Cost: $49.95. Series: *Desk-Top Data Bank*.

This handbook covers over 1,500 films, sheets, and laminates, and provides such information as properties, manufacturers, generic type, and commercial name. It has several cross indexes for easy access.

929. Fire and Flame Retardant Polymers. Albert Yehaskel. Noyes Data Corp, 1979, 482 pp. LC: 78-70742. ISBN: 0-8155-0733-X. Cost: $45.00. Series: *Chemical Technology Review*, 122.

About 250 processes are described with regard to the usage of polymers to retard fire. For example, fabricated plastics, adhesives, wood, paints, paper, and elastomers are described as being fire resistant when processed with particular types of polymers. Such polymers as polyurethanes, polyesters, phenolics, polypropylene, epoxies, silicones, polystyrenes, polyamides, acrylics, polycarbonates, and PVC are used to render certain types of materials fire resistant.

930. Fire Investigation Handbook. Francis L. Brannigan, et al. National Bureau of Standards, 1980, 197 pp. LC: 80-600095. Cost: $8.00. Series: *NBS HandbooK* No. 134.

Covers such topics as investigation of causes of fire, ignition sources, and chemistry and physics of fire. Also includes a list of sources for further information.

931. Fire Property Data—Cellular Plastics. Herbert G. Nadeau and staff. Technomic Publishing, 1981, 167 pp. Cost: $25.00.

Data are provided for foamed plastics used for insulation. Such plastics as polyurethanes, isocyanurates, polystyrenes, and phenolic are discussed. Several types of data are provided, including infrared spectra, thermal gravimetric analysis, heat release rate, autoignition, mass loss rate, peak mass burning rate, and stoichiometric data.

932. Fire Protection Guide on Hazardous Materials. Amy E. Dean; Keith Tower. National Fire Protection Assoc, 7th ed., 1975, various paging. LC: 78-59832. ISBN: 0-87765-130-2. Cost: $12.50. Series: *NFPA* No. SPP-I C.

933. Fire Protection Handbook. Gordon P. McKinnon; Keith Tower. National Fire Protection Assoc, 15th ed., 1981, approx. 1,400 pp. (various paging). LC: 62-12655. ISBN: 0-87765-199-X. Cost: $53.95.

Covers all types of fire hazards including properties of dangerous chemicals, gases and flammable liquids. Also discusses fire extinguishers, alarms, and water supply. Contains over 1,000 illustrations.

934. Fire Resistant Textiles Handbook. W.A. Reeves; G.L. Drake; R.M. Perkins. Technomic Publishing Co, 1974, 276 pp. LC: 73-82116. ISBN: 0-87762-088-1. Cost: $35.00.

935. Fire Retardant Coated Fabrics Formulations Handbook. Vijay Mohan Bhatnagar. Technomic Publishing Co, 1974. Series: *Progress in Fire Retardancy*, Vol.4, 1974, (out of print).

This handbook deals with the fire retardant chemical formulations for the protection of textiles and fabrics.

936. Fire Retardant Formulations Handbook. Vijay Mohan Bhatnagar. Technomic Publishing Co, 1972, 245 pp. LC: 72-80324. ISBN: 0-87762-090-3. Cost: $20.00. Series: *Progress in Fire Retardancy*, Vol.1.

937. Fire Retardant Polyurethanes: Formulations Handbook. Vijay Mohan Bhatnagar. Technomic Publishing Co, 1977, 300 pp. LC: 72-91704. ISBN: 0-87762-217-5. Cost: $9.95 (price reduced from earlier ed.). Series: *Progress in Fire Retardancy*, Vol.8.

Provides a year-by-year account of patents and articles on fire retardant urethanes. Includes 95 formulations, through 1975.

938. Firedamp Drainage: A Handbook for the Coal Mining Industries. Commission of the European Communities; Coal Directorate. Verlag Gluckauf, 1980, 416 pp. Cost: DM 138.

This handbook includes over 200 illustrations describing the problem of methane drainage in coal mining. This book is available in 3 different languages: English, German, and French.

939. First Class Radiotelephone License Handbook. Edward M. Noll. Howard W. Sams and Co, Inc, 4th ed., 1974, 416 pp. LC: 74-15459. ISBN: 0-672-21144-0. Cost: $8.50 paper.

This is basically a study guide for the radiotelephone license examination required by the Federal Communications Commission.

940. First Spectrum of Hafnium (Hf I). William F. Meggers; Charlotte E. Moore (National Bureau of Standards). U.S. Government Printing Office, 1976. SuDoc: C13.44:153. Cost: $1.35. Series: NBS Monograph 153.

941. Fish Farming Handbook. E. Evan Brown; John B. Gratzek. AVI Publishing Co, Inc, 1980, 392 pp. ISBN: 0-87055-341-0. Cost: $24.50.

This handbook presents all types of information on aquaculture, including fish diseases, fish nutrition, building of raceways, tanks, controlling pH, and fertilization of different varieties of fish. Covers such fish as catfish, trout, eel, salmon, and goldfish.

942. Fisheries Handbook of Engineering Requirements and Biological Criteria. Milo C. Bell. Fisheries-Engineering Research Program, U.S. Army Corps of Engineers, 1973.

Includes information on the biological needs of fish and the engineering design requirements for the construction and maintenance of fisheries.

943. Flammability Handbook for Plastics. Carlos J. Hilado. Technomic Publishing Co, 2d ed., 1974, 201 pp. LC: 74-82519. ISBN: 0-87762-139-X. Cost: $29.00.

Includes heat release, ignition, fire gases, thermal conductivity, and combustion information. Data on fire retardants and limiting oxygen index of various materials are presented.

944. Flammability Test Methods Handbook. Carlos J. Hilado. Technomic Publishing Co, 1973, 313 pp. LC: 73-90088. ISBN: 0-87762-127-6. Cost: $25.00.

Covers tests for flammability of gases, liquids, films, fibers, automotive interiors, and building construction materials.

945. Flash Point Index of Trade Name Liquids. National Fire Protection Association, 9th ed., 1978, 470 pp. LC: 78-54003. ISBN: 0-87765-127-2. Date of First Edition: 1954. Cost: $5.50.

Covers over 8,800 trade name liquids with their corresponding flash points, principle use, manufacturer, and reference source.

946. Flat and Corrugated Diaphragm Design Handbook. Mario Di Giovanni. Marcel Dekker, 1982, 424 pp. ISBN: 0-8247-1281-1. Cost: $55.00. Series: *Mechanical Engineering Series*, Vol. 11.

This handbook is concerned with the calculation of performance characteristics of flat and corrugated diaphragms. Covers such topics as materials used to make diaphragms, thermal effects on diaphragm materials, lateral pressures on diaphragms, calculations on deflection, transverse loading of diaphragms, and fre-

quency response. Stress analysis, overload protection, semiconductor diaphragms, curve fitting, and the design of corrugated bellows information is also presented.

947. Flight Profile Performance Handbook: Vol. 1-9. Army TRADOC Systems Ananlysts Activity, White Sands Missle Range, New Mexico (available from NTIS), 1978. Accession: Vol.1, AD-A060 549. Cost: $3.50 per Vol. microform; $9.50 hard copy.

Covers such topics as fluid flow, aviation fuels, helicopters, aerodynamics, velocity, various types of aircraft, and the effects of weather, and barometric pressure on flight.

948. Flight Training Handbook. Federal Aviation Administration. Government Printing Office, 1980, 325 pp. ISBN: 0-89100-165-4. Cost: $7.50.

This handbook provides information and guidance to pilots on the performance, procedures, and maneuvers which have specific applications in pilot operations.

949. Flowmeter Computation Handbook. American Society of Mechanical Engineers, 1961, 169 pp. Cost: $12.50; ASME members $10.00.

Provides information on the application and theory of fluid metering.

950. Fluid Power Handbook and Directory. *Hydraulics and Pneumatics*, eds (available from Industrial Publishing Co), 12th ed., 1978-79, various paging. LC: 58-30725.

This compilation, more of a buyer's guide and product directory than a handbook, contains well tabulated, difficult to obtain information. The section on design data, of particular interest to the engineer, includes information about the various components of fluid power devices such as compressors, hydraulic pumps, motors, and valves. Information is also provided on different types of circuits that are used in fluid power systems, as well as the different areas of fluid power applications, such as the aerospace industry, computers, and metal working machines. An index is provided, along with a listing of trade names and manufacturers.

951. Fluid Properties Handbook. R. Gershman; J.T. Osugi; A.L. Sherman. NTIS, 1971. Accession: B71-10078. Tech Report: M-FS-21169.

This revised handbook presents quantitative data in the form of graphs and charts pertaining to thermodynamic properties of specific cryogenic fluids and several metals. Covers such topics as cryogenic fluids, fluid dynamics, liquid phases, thermodynamic properties, and vapor phases. Covers such fluids as helium, hydrogen, various metals, methylhydrazine, nitrogen, nitrogen tetroxide, oxygen, and water.

952. Foams. International Plastics Selector, Inc, 2d ed., 1978, 825 pp. Cost: $49.95. Series: *Desk-Top Data Bank*.

Information is provided in this tabulation on data available on thermoplastic and thermosetting foams. Such topics as liquid foaming systems, fabricated shapes, industry applications, additives, and aging test data are provided. In addition, data on structural foams, foam composites, and expandable beads are presented.

953. Food Additive Tables. M. Fondu. Elsevier, updated ed., 1980, 162 pp. ISBN: 0-444-41937-3. Cost: $98.00.

954. Food Beverage Service Handbook: A Complete Guide to Hot and Cold Soft Drinks. M.E. Thorner; R.J.

Herzberg. AVI Publishing Co, 1970. LC: 79-125088. ISBN: 0-87055-087-X. Cost: $21.00.

955. Food Chemicals Codex. National Academy Press, 3d ed., 1981, 735 pp. ISBN: 0-309-03090-0. Cost: $45.00.

956. Food Law Handbook. H.W. Schultz. AVI Publishing Co, Inc, 1981, 675 pp. ISBN: 0-87055-372-0. Cost: $79.50.

This publication provides information on federal laws which pertain to food. Brief annotations of each law are listed. Such topics as food additives, color additives, alcohol, pasteurized milk, poultry products, packaging and labeling, fungicides, insecticides, grain standards, and food standards are discussed.

957. For Good Measure: A Complete Compendium of International Weights and Measures. William D. Johnstone. Holt, Rinehart and Winston, 1975. LC: 74-18447. ISBN: 0-03-013946-5. Cost: $12.95.

958. Foreman's Handbook. Carl Heyel. McGraw-Hill, 4th ed., 1967, 591 pp. LC: 67-13907. ISBN: 07-028627-2. Date of First Edition: 1943. Cost: $25.50.

The emphasis of this handbook is on practical information that would be helpful to a foreman in management situations. Information is provided on such subjects as cost control, personnel management, scheduling, applications of electronic data processing, and safety considerations. Carl Heyel has written a section on *zero defects* that would be of interest to any foreman.

959. Forestry Handbook. Reginald D. Forbes et al. Ronald Press Co, 1955, 1,201 pp. (various paging). LC: 55-6815. ISBN: 0-8260-3155-2. Cost: $31.50.

This handbook, prepared under the general direction of the Society of American Foresters, provides a considerable amount of practical information in the area of forestry management and its related subjects. This handbook includes useful information on such subjects, as surveying and aerial photography, forest fire information, various types of soils and geologic configurations, and building and maintaining roads in forests. The physical and chemical properties of wood are included as well as such subjects as water sheds and wood technology. This handbook contains many bibliographies on specialized subjects as well as an index, formulas, tables, and conversion factors that will make it a useful addition to a technical library. For related works, please refer to the general civil engineering handbooks including such titles as *Civil Engineering Handbook; Civil Engineer's Reference Book; Standard Handbook for Civil Engineers,* and the *Handbook of Environmental Civil Engineering.*

960. Forging Design Handbook. S.A. Sheridan. American Society for Metals, 1972, 318 pp. LC: 72-83400. Accession: AD-743700.

This handbook is designed primarily for the aerospace design community and it provides information on high-strength parts. Covers material selection and design problems involving corners and fillets, webs, cavities, and flash and trim.

961. Forging Equipment, Materials, and Practices. Taylan Altan et al (Metals and Ceramics Information Center). NTIS, 1973, 501 pp. Accession: AD-771 344. Tech Report: MCIC-HB-03. Cost: $25.00 microform or hard copy.

Forging information is provided for various different metals. Describes in-process quality control and selection of forging equipment. Such topics as forge presses, heat treatment, hydraulic presses, plastic deformation, and refractory metals are discussed.

962. Forging Industry Handbook. Jon E. Jenson. Forging Industry Association, 1966, 518 pp.

963. Formation of C-C Bonds. Jean Mathieu; Jean Weil-Raynal; D.H.R. Barton. Heyden and Sons Inc, 1973-79, 3 Vols. (Vol.1, 519 pp.; Vol.2, 655 pp.). LC: 73-36113. ISBN: Vol.1, 0-88416-050-5; Vol.3, 0-88416-098-X. Cost: Vol.1, $60.00; Vol.2, $85.00; Vol.3, $60.00.

The volumes are entitled as follows: Vol.1, *Introduction of One Functional Carbon Atom;* Vol.2, *Introduction of a Carbon Chain or an Aromatic Ring;* Vol.3, *Introduction of an a-Functional Carbon Chain and Special Condensation Reactions.* This 3-volume series provides information on the formation of carbon coupling reactions. Over 3,000 formula schemes are presented, as well as over 500 tables which provide information on these reactions. Such reactions as hydroxymethyllation, formylation, carboxymethylation, cyanation, vinylation, alkynylation, alkylidenation, and thioacylation are described in this book.

964. Formula Manual. Norman H. Stark. Stark Research Corp, 1973, 82 pp.

The Formula Manual is a rather interesting collection of chemical formulas on everyday household products such as shampoos, detergents, floor waxes, and furniture polishes. In addition, products such as automotive chemicals including windshield cleaners, antifreeze, and radiator corrosion inhibitors are included. These formulas can easily be followed by the handy do-it-yourselfer.

965. Formulaire du Beton Arme, Tome 2. M. Courtland; P. Lebelle. Editions Eyrolles, 2d ed., 1976, 735 pp. Date of First Edition: 1953.

This French volume presents numerous tables and calculations on reinforced concrete.

966. Formulary of Cosmetic Preparations. Michael Ash; Irene Ash. Chemical Publishing Co, Inc, 1978, 486 pp. ISBN: 0-8206-0218-3. Cost: $25.00.

This formulary provides a collection of cosmetic formulas and describes methods for their preparation. Includes shampoos, deodorants, makeup, and lotions.

967. Formulary of Detergents and Other Cleaning Agents. Michael Ash; Irene Ash. Chemical Publishing Co, 1980, 356 pp. ISBN: 0-8206-0247-7. Cost: $25.00.

This book includes formulas for cleaning agents that are listed both by use and by chemical constituents. Various cleansers, solvent cleaners, industrial floor cleaners, dry-cleaning agents, and household cleaners are listed. Instructions are provided for the preparation of each cleaning agent.

968. Formulary of Paints and Other Coatings. Michael Ash; Irene Ash. Chemical Publishing Co, Inc, 1978, Vol.1, 434 pp. ISBN: 0-8206-0248-5. Cost: $25.00.

This book is a collection of formulas for paints. Information is provided on the chemical constituents of the paints and the different characteristics and usage of the paints. Vol.2 will soon be available.

969. Formulas for Natural Frequency and Mode Shape. Robert D. Blevins. Van Nostrand Reinhold, 1979, 492 pp. LC: 79-556. ISBN: 0-442-20710-7. Cost: $32.50.

This is a reference book providing formulas and principles of vibration of structural fluid systems. Such topics as traverse vibration of flexure beams, longitudinal vibration, torsional vibration of beams and shafts, traverse vibration of shear beams, structural vibrations in a fluid, and various properties of solids, liquids, and gases are presented in this publication. This book provides over 80 tables and 600 frames of formulas with graphic illustrations. Over 100 plate geometries are presented with their natural frequencies.

970. Formulas for Stress and Strain. Raymond J. Roark; Warren C. Young. McGraw-Hill, 5th ed., 1976, 624 pp. LC: 75-26612. ISBN: 0-07-053031-9. Cost: $32.50.

Provides loadings and supports on elements with various boundary conditions.

971. Foseco Foundryman's Handbook. Foseco International, Ltd. Pergamon Press, 8th ed., 1976. ISBN: 0-08-018020-5. Cost: $8.25.

972. Foundation Engineering Handbook. Hans F. Winterkorn; H.Y. Fang. Van Nostrand Reinhold, 1975, 752 pp. LC: 74-1066. ISBN: 0-442-29564-2. Cost: $42.50.

Covers properties of soils, floating foundations, foundation vibrations, and earthquake effects.

973. Foundation Engineering. G. Leonards. McGraw-Hill, 1962, 1,146 pp. LC: 61-9112. ISBN: 0-07-037198-9. Cost: $48.50.

974. Foundryman's Handbook. Foundry Services Ltd. Pergamon, 8th ed., 1975, 390 pp. ISBN: 0-08-018020-5. Cost: $13.75.

Covers such topics as light casting alloys, iron castings, die casting, and nonferrous casting alloys. Information is also presented in metric measurement and SI units.

975. Four-Six Semiconducting Compounds—Data Tables. M. Neuberger. NTIS, 1969, 111 pp. Accession: N70-24608; AD-699260. Tech Report: EPIC-S-12.

This volume lists information on the properties of conducting compounds. Such topics as germanium compounds, crystal structure, tellurium compounds, lead compounds, electrical properties, mechanical properties, magnetic properties, and optical properties are presented.

976. Fourier Expansions: A Collection of Formulas. Fritz Oberhettinger. Academic Press, 1973, 72 pp. ISBN: 0-12-523640-9. Cost: $21.50.

977. Fourier Transforms of Distributions and Their Inversions: A Collection of Tables. Fritz Oberhettinger. Academic Press, 1973, 180 pp. ISBN: 0-12-523650-6. Cost: $21.00.

978. Fowler's Electrical Engineer's Pocket Book. William Henry Fowler. Scientific Publishing Co, 1966.

979. Fowler's Mechanical Engineer's Pocket Book. William Henry Fowler. Scientific Publishing Co, 1966.

980. Fowler's Mechanics' and Machinists' Pocket Book: A Synopsis of Practical Rules for Fitters, Turners, Millwrights, Erectors, Pattern Makers, Foundrymen, Draughtsmen, Apprentices, Students, etc. William Henry Fowler. Scientific Publishing Co, 1966.

981. Fractional Factorial Designs for Experiments with Factors at Two and Three Levels. W.S. Connor; Shirley Young (National Bureau of Standards). NTIS, 1961, 65 pp. Accession: AD-700470. Cost: $4.50. Series: NBS AMS58.

982. Fractional Factorial Experiment Designs for Factors at Three Levels. W.S. Connor; M. Zelen (National Bureau of Standards). NTIS, 1959, 35 pp. Accession: COM 73-11111. Cost: $4.00. Series: NBS AMS54.

NOTE: NBS AMS48 and AMS58 are related.

983. Fractional Factorial Experimental Designs for Factors at Two Levels. National Bureau of Standards. NTIS, 1957, 85 pp. Accession: PB-176119. Cost: $5.00. Series: AMS48.

NOTE: Sequels are AMS54 and AMS58.

984. French's Index of Differential Diagnosis. Francis Dudley Hart. Year Book Medical Publishers, 10th ed., 1973. Cost: $49.95.

985. Fuel Economy Handbook. W. Short. Graham and Trotman, 2d ed., 1980, 317 pp. ISBN: 0-86010-130-4. Cost: £12.50.

986. Fuel Oil Manual. Paul F. Schmidt. Industrial Press, 3d ed., 1969, 263 pp. LC: 69-10507. ISBN: 0-8311-3014-8. Cost: $19.00.

987. Fuel Properties. Naval Air Systems Command. NTIS, 1967, 202 pp. Accession: N76-73750.

This compilation represents a collection of data and properties on aircraft fuels. Such fuels as JP-4 jet fuel, JP-5 jet fuel, and JP-6 jet fuel are included, along with their chemical properties.

988. Fulmer Materials Optimizer. General Electric Co, 8-Vol. set, loose-leaf service, 3 updates per year. Cost: $990.00 plus an annual $260.00 for the updating service.

This 8-volume loose-leaf service represents an effort to provide an information system for selection of engineering materials. Properties data, forming methods, and costs are presented for metals, plastics and ceramics.

989. Functional Histology: A Text and Colour Atlas. P. Wheater; G. Burkitt; V. Daniels; P. Deakin. Churchill Livingstone, Inc, 1979, 288 pp. ISBN: 0-443-01657-7. Cost: $22.00.

This atlas includes over 387 full-color illustrations relating histologic structure to function. Such topics as skeletal tissues, the immune system, the respiratory system, the urinary system, the circulatory system, nervous tissue, the endocrine glands, and blood are presented in this atlas. These illustrations are either light micrographs, electron micrographs, or light drawings.

990. Fundamental Formulas of Physics. Donald H. Menzel. Dover, 2d ed., 1960, 2 Vols. ISBN: Vol.1, 0-486-60595-7; Vol.2, 0-486-60596-5. Cost: Vol.1, $5.00; Vol.2, $5.00.

991. Fundamental Measures and Constants for Science and Technology. Frederick D. Rossini. CRC Press, 1974, 132 pp. LC: 74-11696. ISBN: 0-8493-5079-4 (formerly 0-87819-051-1). Cost: $34.95.

992. Fundamentals of Analytical Flame Spectroscopy. Cornelis T. Alkemande; Roland Herrmann. Halsted Press (division of John Wiley and Sons), 1979, 442 pp. LC: 79-4376. ISBN: 0-470-26710-0. Cost: $82.95.

This publication includes an atlas of 49 spectrograms, a table of 7,500 wavelengths of spectral lines, and bands observable in flames. Covers emission spectroscopy, atomic absorption spectroscopy, and atomic fluorescence spectroscopy.

993. Fundamentals of Radiation Dosimetry. J.R. Greening. Heyden and Sons, 1981, 172 pp. ISBN: 0-9960020-5-7. Cost: $27.00. Series: *Medical Physics Handbook*, 6.

Provides information on radiation therapy, radiation protection, radiation physics, and radiodiagnosis.

994. Further Contributions to the Solution of Simultaneous Linear Equations and the Determination of Eigenvalues. National Bureau of Standards. NTIS, 1958, 81 pp. Accession: PB-251903. Cost: $5.00. Series: NBS AMS49.

995. Gamma Rays of the Radionuclides: Tables for Applied Gamma Ray Spectrometry. Gerhard Erdtmann; Werner Soyka. Verlag Chemie, 1979, 862 pp. ISBN: 0-89573-022-7. Cost: $141.20. Series: *Topical Presentations in Nuclear Chemistry*, Vol. 6.

These tables include 2,055 radionuclides and 48,000 gamma-ray lines. This is a compilation of all gamma- and X-rays of all known radionuclides.

996. Gamma-ray Angular Correlations. R.D. Gill. Academic Press, 1975, 238 pp. ISBN: 0-12-283850-5. Cost: $39.50.

997. Gas Chromatographic Data Compilation. American Society for Testing and Materials. American Society for Testing and Materials, 1967. Cost: $40.00. Series: ASTM-AMD-25A; ASTM Publication Code No. 10-025010-39.

The *First Supplement* has been published as ASTM-AMD-25A-S1, 1971, 726 pp., Publication Code No. 10-025011-39, $40.00. This supplement includes approximately 20,000 items and the original publication contained chromatographic data on 23,000 items. *Gas Chromatographic Data on Magnetic Tape-AMD-26* contains all the data on AMD-25A and AMD-25A-S1.

998. Gas Engineers Handbook: Fuel Gas Engineering Practices. C. George Segeler. Industrial Press, Inc, 1965, 1,550 pp (various paging). LC: 65-17328. ISBN: 0-8311-3011-3. Cost: $63.00.

The *Gas Engineers Handbook*, a very useful compilation of data on natural gas, including information on such subjects as liquified natural gas, transmission and distribution of natural gas, a discussion of the many different uses of natural gas, and the various chemicals that can be produced from processing natural gas. There is an abundance of tables and charts listing the many properties of gases, air, and water. There is also a collection of mathematical and conversion tables and many and varied tables listing the physical constants of substances related to natural gas. In addition, information is provided on such subjects as corrosion of pipelines, gas odorization, the production of gas, and the removal of such elements as sulfur dioxide and nitrogen from natural gas. This handbook is very well indexed and many bibliographic references are provided for further investigation. For related titles on energy, please refer to such handbooks as the *Handbook of Natural Gas Engineering; Hydrogen: Its Technology and Implications; Reactor Handbook; Engineering Tables for Energy Operators; Petroleum Production Handbook; Nuclear En-*

gineering Handbook; Hydroelectric Handbook; Energy Reference Handbook; and the *Keystone Coal Industry Manual.*

999. Gas Phase Reaction Kinetics of Neutral Oxygen Species. H.S. Johnston (National Bureau of Standards). Office of Standard Reference Data, 1968. Series: NSRDS-NBS-20.

1000. Gas Tables: Thermodynamic Properties of Air Products of Combustion and Component Gases, Compressible Flow Functions. Joseph H. Keenan; Joseph Kaye. 2d ed., 1980, 217 pp. LC: 79-15098. ISBN: 0-471-02207-1. Cost: $27.95.

This is a thorough revision of the 1948 *Tables* with improved accuracy and uniformity. Tables for all of the important individual combustion-product species are presented, along with various percentage-theoretical air mixtures.

1001. Gas Turbine Cycle Calculations: Thermodynamic Data Tables for Air and Combustion Products for Three Systems of Units. M.S. Chappell; E.P. Cockshutt. National Research Council of Canada (available NTIS), 1974, 175 pp. Accession: N75-14758. Tech Report: NRC-14300 LR-579. Cost: $6.25.

Procedures are presented for calculating the thermodynamic properties of air and combustion products in open cycle combustion turbines. The set of tables consists of specific heat information, enthalpy, and entropy data of dry air.

1002. Gas Turbine Engineering Handbook. John W. Sawyer. Gas Turbine Publications, Inc, 1966, 552 pp. LC: 66-30711.

This handbook includes a considerable amount of information on the design and manufacture of gas turbines and the various applications of gas turbines, such as in the aviation industry, pipeline industry, and the nuclear and railroad industries. Information is provided about fuels for gas turbines, heat exchangers, and noise control of gas turbine systems. Various codes and specifications are also included in addition to a subject index and an advertiser's index.

1003. Gasohol Handbook. V. Daniel Hunt. Industrial Press, 1981, 576 pp. ISBN: 0-8311-1137-2. Cost: $29.50.

This book discusses such topics as production processes, specific plant design, cost data, environmental issues, and available feed stocks. The economics of ethanol is provided for the full range of systems from small-, intermediate-, to large-scale systems.

1004. Gear Handbook: The Design, Manufacture and Application of Gears. D.W. Dudley. McGraw-Hill, 1962, 937 pp. LC: 61-7304. ISBN: 0-07-017902-6. Cost: $52.50.

1005. Geigy Scientific Tables. CIBA Pharmaceutical Co, 8th ed., 1982, multi-vol. set. Cost: Vol.1, $19.95.

Vol.2 is entitled *Statistics and Mathematics*, and Vol.3 is entitled *Physical Chemistry, Hematology, and Biometrics.* Subsequent volumes are planned that will deal with metabolism, endocrinology, circulation, and respiration. Vol. 1 deals with units of measurement, body fluids, organic composition of the body, and nutrition. Such topics as chemical composition of body fluids, viscosity, pH, specific gravity, and freezing points of approximately 20 body fluids are presented. These properties are provided for such fluids as tears, bile, synovial fluid, and gastric juices. Information on electrolytes, renal function and basal metabolism are also presented.

1006. General Class Amateur License Handbook. (title varies) H.S. Pyle. Sams, 2d ed., 1968, 143 pp. LC: 68-20563. ISBN: 0-672-20639-0. Cost: $3.95 paper copy.

1007. General Engineering Handbook. C.E. O'Rourke. McGraw-Hill, 2d ed., 1940, 1,120 pp. LC: 40-13535.

1008. General Handbook for Radiation Monitoring. Los Alamos Scientific Laboratory; Atomic Energy Commission. U.S. Government Printing Office, 3d ed., 1958, 180 pp. Tech Report: LA-1835.

1009. General Safety Standards for Installations Using Non-Medical X-Ray and Sealed Gamma Ray Sources, Energies Up to 10 MeV. National Bureau of Standards. U.S. Government Printing Office. SuDoc: C13.11:114. Cost: $.90. Series: NBS Handbook 114 (supercedes H93).

1010. Geochemical Tables. H.J. Rosler; H. Lange. Elsevier, 1972, 468 pp. LC: 79-132143. ISBN: 0-444-40894-0. Cost: $63.50.

1011. Geodetic and Grid Angles—State Coordinate Systems. L.G. Simmons. U.S. Department of Commerce, Coast and Geodetic Survey, 1968, 5 pp. Series: *ESSA Technical Report* C and GS 36.

1012. Geographic Atlas of World Weeds. L. Holm. John Wiley and Sons, 1979, 500 pp. LC: 78-24280. ISBN: 0-471-04393-1. Cost: $30.00.

1013. Geological Time Table. F.W. Van Eysinga. Elsevier, 3d rev. ed., 1976. ISBN: 0-444-41362-6. Cost: $4.15.

1014. George C. Marshall Space Flight Center High Reynolds Number Wind Tunnel Technical Handbook. H.S. Gwin; NASA. NTIS, 1973, 48 pp. Accession: N74-19890. Tech Report: NASA-TM-X-64831.

1015. Gesundheitsschädliche Arbeitsstoffe Toxikologisch-Arbeitsmedizinische Berundung Von MAK-Werten. D. Henschler. Verlag Chemie, 1st issue, 1972, loose-leaf service. Cost: DM 58.

The translated title is *Harmful Chemicals—The Toxicological and Medical Establishment of MAK-Values.* This German compilation of tables consists of information on the tested and accepted MAK (maximum concentration at place of work) values, and is published on an annual basis in loose-leaf format. This work is concerned with the measurement and testing of maximum exposure levels of harmful and toxic chemicals.

1016. Giant Handbook of Electronic Circuits. Raymond A. Collins. TAB Books, 1980, 882 pp. LC: 80-21319. ISBN: 0-8306-9673-3. Cost: $24.95.

This handbook is a collection of electronic circuits for all types of applications such as ham radios, CBs, television, AM/FM, SSB, and computers.

1017. Glass Engineering Handbook. E. Shand. McGraw-Hill, 2d ed., 1958, 471 pp. LC: 58-10006. ISBN: 0-07-056395. Cost: $45.00.

1018. Gmelin's Handbuch Der Anorganischen Chemie. L. Gmelin. Verlag, Chemie, Vol.1, 1924; Supplements, 1937, For prices, please consult publisher.

This is a very comprehensive multivolume treatise on inorganic chemistry. The information presented is critically evaluated, and literature citations are provided to the original references where the research was reported. The publisher has announced that by the end of 1981 all volumes will be in English or partially in English.

1019. Gould Battery Handbook. Gustav A. Mueller ed. Gould, Inc, 1973, 359 pp. LC: 73-76463.

1020. Grafter's Handbook. Robert John Garner. Oxford University Press, 4th ed., 1979. ISBN: 0-19-519024-6. Cost: $16.95.

1021. Graphs and Hypergraphs. C. Berge. Elsevier, 2d rev. ed., 1976, 528 pp. LC: 72-88288. ISBN: 0-7204-0479. Cost: $49.00. Series: *Mathematical Library*, No. 6.

1022. Graphs and Questionnaires. C.F. Picard. Elsevier, 1980, 432 pp. ISBN: 0-444-85239-5. Cost: $39.00. Series: *Mathematical Studies Series*, Vol. 32.

1023. Graphs and Tables for Use in Radiology. F. Wachsman; G. Drexler. Springer-Verlag, 2d ed., 1976, 240 pp. ISBN: 0-387-07809-6. Cost: $19.70.

1024. Graphs of the Velocity of Flowing Water in Rivers, Canals and Pipes. Cimpa. Wilhelm Ernst and Sohn, 1976, 37 pp. Cost: DM 18. Series: *Bauingenieur-Praxis*, Vol. 123.

This text is available in English, German, and Spanish. It includes 20 tables on the velocity of flowing water in rivers, canals, and pipes.

1025. Graphs, Data Structures, Algorithms. M. Nagel; H.J. Schneider. Hanser (available from Otto Harrassowitz, 6200 Wiesbaden), 1979, 340 pp. Cost: DM 48.

1026. Gray's Anatomy of the Human Body. Henry Gray; C.M. Goss (ed). Lea and Febiger, 29th ed., rev., 1973, 1,466 pp. LC: 73-170735. ISBN: 0-8121-0377-7. Cost: $28.50.

1027. Gray's Anatomy. Roger Warwick; Peter L. Wiliams. W.B. Saunders, 35th British ed., 1973, 1,471 pp. ISBN: 0-7216-9127-7. Cost: $38.50.

1028. Green's Functions and Transfer Functions Handbook. A.G. Putkovskiy, (translated by L.W. Longdon). Ellis Horwood, LTD (available from Halsted Press, division of Wiley), 1982, 236 pp. LC: 82-3073. ISBN: 0-85312-447-7. Cost: $64.95.

This book was originally published in 1979 in Moscow and covers such topics as differential equations, Green's functions, Laplace transformations, Eigen values, and transfer functions.

1029. Grounds Maintenance Handbook. Herbert S. Conover. McGraw-Hill, 3d ed., 1977, 631 pp. LC: 76-18901. ISBN: 0-07-012412-4. Cost: $24.75.

1030. Groundwater Monitoring Handbook. General Electric Co, 1980, 440 pp. LC: 80-82885. ISBN: 0-931-69014-5. Cost: $150.00.

This book discusses groundwater pollution and the mobility of polluted groundwater, sources and causes of pollution, and mechanisms of pollution. It provides information on sanitary landfills, hazardous waste disposal, oxidation ponds, and other types of hydrologic data.

1031. Group Constants for Nuclear Reactor Calculations. I.I. Bondarenko. Consultants Bureau (Plenum), translation 1964, 151 pp. LC: 64-23252.

1032. Guide to Classification in Geology. J.W. Murray. Halsted Press (division of John Wiley and Sons), 1981, 120 pp. LC: 80-41094. ISBN: 0-470-27090-X. Cost: $34.95.

1033. Guide to Mathematical Tables. Aleksandr Vasil'evich Lebedev; R.M.A. Fedorova. Pergamon Press, 1960, also includes supplements.

Guide to Mathematical Tables: Supplement No.1 by N.M. Burunova, 1960 (Pergamon Mathematical Tables Series: Vol. 6). ISBN: 0-08-009244-6, $16.75.

1034. Guide to Plastics: Property and Specfication Charts. McGraw-Hill, 1980, 316 pp.

Includes 500 plasticizers, 600 stabilizers, with details on properties and a comparative analysis of 47 films and 19 foam types. List such properties as tensile strength, compressive strength, flexural yield strength, impact strength, compressive modulus, and thermoconductivity.

1035. Guide to Tables of the Normal Probability Integral. National Bureau of Standards. NTIS, 1952, 16 pp. Accession: PB-178392. Cost: $3.50. Series: NBS AMS21.

1036. Guide to World Screw Threads. P.A. Sidders. Industrial Press, 1969, 318 pp. LC: 71-185990. ISBN: 0-8311-1092-9. Cost: $19.50.

This is a reference guide providing comprehensive coverage of all different types of screw threads that have been standardized throughout the world. Includes Unified and American thread series, American pipe threads, British threads of Whitworth, and non-Whitworth forms, and ISO metric threads.

1037. Guidebook of Electronic Circuits. John Markus. McGraw-Hill, 1974, 1,067 pp. LC: 74-9616. ISBN: 0-07-040445-3. Cost: $41.50.

This guidebook is a remarkable collection of different types of electronic circuits and includes over 3,600 circuits for all types of electronic devices. For example, circuits are provided for such devices as citizens band circuits, burglar alarm circuits, fire alarm circuits, intercom circuits, timer circuits, and telephone circuits. A name index, as well as a subject index, is provided for the convenience of the user.

1038. Guidebook: Toxic Substances Control Act. George S. Dominquez. CRC Press, 1977, 350 pp. ISBN: 0-8493-5321-1. Cost: $59.95.

1039. Gunite: A Handbook for Engineers. T.F. Ryan. Cement and Concrete Association, 1973, 63 pp. ISBN: 0-7210-0820-8. Cost: £1.50.

Covers such topics as refractory gunite, architectural applications, and equipment currently used to perform high quality work.

1040. Gypsum Construction Handbook: With Product and Construction Standards. United States Gypsum Co, 1978, 472 pp. Cost: $3.50.

1041. Handbook and Atlas of Gastro-Intestinal Exfoliative Cytology. Joao C. Prolla; Joseph B. Kirsner. University of Chicago Press, 1973. LC: 72-80621. ISBN: 0-226-68451-2. Cost: $19.50.

1042. Handbook and Charting Manual for Student Nurses. Alice L. Price. C.V. Mosby, 5th ed., 1971, 220 pp. ISBN: 0-8016-4047-4. Cost: $7.95.

1043. Handbook and Standard for Manufacturing Safer Consumer Products. Superintendent of Documents. Government Printing Office, 1977, 82 pp. SuDoc: Y3.C76/3:8 M31/977. Cost: $2.30.

This publication discusses the implementation of the Consumer Safety Commission's System Standard.

1044. Handbook for Aircraft Mechanics. P.S. Shevelko et al. Voenizdat, Moscow (available from NTIS), 3d rev. and enlarged ed., 1974, 592 pp. Accession: A75-23366.

This handbook, in Russian, presents information on aeronautical engineering and techniques of aircraft design. Flight mechanics for both fixed and rotary winged aircraft are discussed, along with such topics as aircraft communication, navigation, and control. In addition, reference data are supplied on the properties of metals, alloys, oils, greases, fuels, and lubricants.

1045. Handbook for Analytical Quality Control in Water and Wastewater Laboratories. U.S. Environmental Protection Agency, Analytical Quality Control Laboratory, 1972, loose-leaf.

1046. Handbook for Automatic Computation, Vol. 1, Part A and Part B: Translation of Algol 60. H. Rutishauser; A.A. Grau et al. Springer-Verlag, Vol.1, Part A and B, 1967, Vol.1, Part A, 323 pp; Vol.1, Part B, 397 pp. LC: Vol.1, Part A, 67-13537; Vol.1, Part B, 70-383792. ISBN: Vol.1, Part A, 0-387-03826-4; Vol.1, Part B, 0-387-03828-0. Cost: Vol.1, Part A $35.40; Vol.1, Part B $52.60. Series: *Grundlehren der Mathematischen Wissenschaften*, Part A 135; Part B 137.

Vol.2, *Linear Algebra*. J. H. Wilkinson and C. Rernsch. LC 75-154804 (Grundlehren der Mathematischen Wissenschaften, Vol. 186), 1971. ISBN 0-387-05414-6. $40.50.

1047. Handbook for Chemical Technicians. Howard J. Strauss; Milton Kaufman. McGraw-Hill, 1977, 512 pp. LC: 76-10459. ISBN: 0-07-062164-0. Cost: $23.50.

1048. Handbook for Comprehensive Pediatric Nursing. Martha U. Barnard et al. McGraw-Hill, 1980, 336 pp. ISBN: 0-07-003740-X. Cost: $9.95.

Covers such topics as health maintenance, fluid balance, common therapeutic and diagnostic procedures, and different types of pediatric emergencies. Various charts and reference tables are contained in this handbook.

1049. Handbook for Computing Elementary Functions. Lazar Aronovich Liusternik; G.J. Tee trans. Pergamon Press, 1965, 251 pp. LC: 64-22370.

1050. Handbook for CRPL Ionospheric Predictions Based on Numerical Methods of Mapping. National Bureau of Standards. NTIS, Dec. 21, 1962. Accession: PB-188-654. Cost: $4.50. Series: NBS Handbook 90.

1051. Handbook for Dairymen. Anthony Colletti. Iowa State University Press, 1963. ISBN: 0-8138-0740-9. Cost: $6.00.

1052. Handbook for Dental Identification. Lester L. Luntz; Phyllys Luntz. Lippincott, 1973, 194 pp. LC: 73-4304. ISBN: 0-397-50315-6. Cost: $14.00.

1053. Handbook for Differential Diagnosis of Infectious Disease. Jonas Shulman; David Schlossberg. Prentice-Hall, 304 pp. ISBN: 0-8385-3616-6(X). Cost: $13.50 paper.

The causes and treatment of such diseases as rash, fever, monoarthritis, and jaundice are discussed. Numerous charts are also included in this handbook, listing the common causes of infectious disease.

1054. Handbook for Differential Diagnosis of Neurologic Signs and Symptoms. Kenneth M. Heilman et al. Appleton-Century-Crofts, 1977. ISBN: 0-8385-3617-4. Cost: $7.50 paper copy.

1055. Handbook for Electronic Circuit Design. Campbell Loudoun. Reston (div of Prentice-Hall), 1978, 276 pp. LC: 77-23981. ISBN: 0-87909-334-X. Cost: $18.95.

Covers oscillator circuit design, AF and RF power amplifiers, nonsinusoidal oscillator circuit design, wave filter design and amplifier bias stabilization.

1056. Handbook for Electronic Engineers and Technicians. Harry Elliott Thomas. Prentice-Hall, 1965, 427 pp. LC: 65-22188.

1057. Handbook for Electronics Engineering Technicians. Milton Kaufman; Arthur H. Seidman. McGraw-Hill, 1976, 515 pp (various paging). LC: 75-40457. ISBN: 0-07-033401-3. Cost: $25.50.

The *Handbook for Electronics Engineering Technicians* includes many tables, charts, and equations on various electronic subjects such as diodes, circuits, amplifiers, batteries, resistors and many other semiconductors. In addition, such subjects as temperature in semiconductors; various measuring and testing devices; silicon controlled rectifiers; light emitting diodes; vacuum tubes and transformers receive coverage in this well-indexed handbook. For related works, please refer to the *Electronics Engineers' Handbook*; the *Electrical Engineer's Reference Book*; the *Basic Eelectronics Instrument Handbook*; the *Electronics Handbook*; the *Handbook of Electronic Packaging*; the *Handbook of Materials and Processes for Electronics*; the *Handbook of Modern Electronic Data; Industrial Electronics Handbook*; the *Handbook of Semiconductor Electronics*; and the *Standard Handbook for Electrical Engineers*. Also refer to *Electronic Conversions, Symbols and Formulas*.

1058. Handbook for Emergency Medical Personnel. Donald R. Snyder; Paul R. Palmer. McGraw-Hill, 1978, 224 pp. ISBN: 0-07-059518-6. Cost: $7.95.

1059. Handbook for Engineers. K.P. Yakovlev et al; J.J. Cornish trans; G.O. Harding trans. Pergamon Press, 1965-67, 2 vols. LC: 63-10014 (GB). ISBN: Vol.1, 0-08-010129-1; Vol.2, 0-08-011183-1. Cost: Vol.1, $8.50; Vol.2, $11.50.

The *Handbook for Engineers* consists of 2 volumes entitled as follows: Vol.1 *Mathematics and Physics*; Vol.2 *Mechanics, Strength of Materials, and the Theory of Mechanisms and Machines.*

1060. Handbook for Environmental Impact Analysis. R.K. Jain; L.V. Urban; G.S. Stacey (Army Construction Enginering Research Laboratory). NTIS, Final report, Sept. 1974, 215 pp. Accession: AD/A-006 241. Tech Report: CERL-TR-E-59. Cost: $2.25 microfiche; $7.75 paper copy.

This handbook consists of a step-by-step guide to prepare environmental impact statements and assessments based upon the requirements of the National Environmental Policy Act of 1969 and the guidelines recommended by the President's Council on Environmental Quality. The handbook is designed specifically for army personnel, but its format and procedural explanations would be useful in many other applications concerning environmental impact statements.

1061. Handbook for Environmental Planning: The Social Consequences of Environmental Change. James McEvoy III (ed). Wiley-Interscience, 1977, 336 pp. LC: 76-57239. ISBN: 0-471-58389-8. Cost: $28.00.

Covers environmental impact statements.

1062. Handbook for Evaluating Ecological Effects of Pollution at DARCOM Installations. Carlos F.A. Pinkham; David A. Gauthier. NTIS, 1979, 7 Vols. Accession: Vol.3, AD-A089 879; Vol.4, AD-A089 880; Vol.5, AD-A089 881; Vol.6, AD-A089 882; Vol.7, AD-A089 922. Cost: $3.50 per Vol. microfiche; Vols.3 and 7, $6.50 hard copy; Vols.4 and 5, $9.50 hard copy; Vol.6, $12.50 hard copy.

This 7-volume set discusses such topics as sampling techniques, aquatic pollution surveys, terrestrial pollution surveys, and unexpected declines in animal population. This handbook outlines the different responsibilities for environmental scientists and also provides directions for paraecologists. Such information as library literature searches, conducting preliminary investigations, and handling data are presented.

1063. Handbook for Infectious Disease Management. Cornellis A. Kolff; R.C. Sanchez. Addison-Wesley Publishing Co Inc, 1979, 280 pp. Cost: $11.95.

Discusses prevention, diagnosis, management, and control of infectious diseases.

1064. Handbook for Information Systems and Services. Pauline Atherton. UNIPUB (division of R R Bowker), 1978, 259 pp. ISBN: 92-3-101457-9. Cost: $21.75.

Discusses such topics as handling of scientific and technical data, equipment used in information systems, professional training, and facilities available.

1065. Handbook for Linear Regression. Mary Sue Younger. Duxbury Press (division of Wadsworth Inc), 1979, 569 pp. LC: 78-24267. Cost: $20.95.

1066. Handbook for Manufacturing Entrepreneurs. Robert S. Morrison. Western Reserve Press Inc, 1973. LC: 73-91212. Cost: $17.50.

1067. Handbook for Pesticide Disposal by Common Chemical Methods. C.C. Shih; C.F. Dal Porto. U.S. Environmental Protection Agency (available from NTIS), 1975, 103 pp. Accession: PB-252864. Series: *Environmental Protection Publication*, SW-112c.

1068. Handbook for Planet Observers. Gunter Dietmar Roth. Van Nostrand Reinhold, 1970, 205 pp. LC: 73-126882; 74-497607. ISBN: 571-08345-5.

1069. Handbook for Pre-Stressed Concrete Bridges. S. Sarkar. Structural Engineering Research Centre, 1969, 239 pp. LC: 70-905772. Cost: $30.00.

Bibliography on page 240

1070. Handbook for Professional Divers. Ronald M. Titcombe. Lippincott, 1973, 292 pp. ISBN: 0-397-58128-9. Cost: $19.00.

1071. Handbook for Prospectors: A Revision of Handbook for Prospectors and Operators of Small Mines. M.W. Von Bernewitz; R.M. Pearl. McGraw-Hill, 5th ed., 1973, 576 pp. LC: 72-11749. ISBN: 0-07-049025. Cost: $21.95.

Handbook for Prospectors and Operators of Small Mines. *See* Handbook for Prospectors.

1072. Handbook for Radio Engineering Managers. J.F. Ross. Butterworths Publishing Inc, 1980, 947 pp. ISBN: 0-408-00424-X. Cost: $94.95.

This is a vast collection of information on radio installations ranging from safety practice, fires, and corrosion, to the economics of operating a radio facility. Such topics as radiation hazards, X-rays, ultraviolet radiation, lightning protection, and transformer problems are discussed. Specifications for equipment are also listed, such as transmitting antennas, broad-band radio communication relay equipment, broadcast transmitters, and dual channel television.

1073. Handbook for Radiologic Technologists and Special Procedures Nurses in Radiology. N.W. Powell. C.C. Thomas, 1974. ISBN: 0-398-03066-9. Cost: $8.75.

1074. Handbook for Remedial Speech. H.J. Heltman. Expression, 1973. Cost: $4.00.

1075. Handbook for Research and General Practice. D.S. Eimerl; A.J. Laidlaw. Longman, 2d ed., 1969. ISBN: 0-443-00606-7. Cost: $7.75.

1076. Handbook for Sampling and Sample Preservation of Water and Wastewater. J.H. Moser; K.R. Huibregise. NTIS, 1976, 278 pp. Accession: N77-76530; PB-259946. Tech Report: EPA-600/4-76-049.

Discusses such topics as pollution monitoring, flow measurement, preserving samples, and measuring chemical and physical properties of samples.

1077. Handbook for the Engineering Structural Analysis of Solid Propellants. Chemical Propulsion Information Agency; J.Edmund Fitzgerald; William R. Hufferd. NTIS, 1971, 870 pp. Accession: AD 887478. Tech Report: CPIA-Pub-214. Cost: $31.00.

1078. Handbook for the Identification of Insects of Medical Importance. John Smart. The British Museum of Natural History, 4th ed., 1965, 303 pp. LC: 66-70347.

1079. Handbook for the Laryngectomee. Robert L. Keith. Interstate Printers & Publishers, Inc, 1974. LC: 73-93688. Cost: $1.95.

1080. Handbook for the Operation and Maintenance of Air Pollution Control Equipment. Frank L. Cross; Howard E. Hesketh. Technomic Publishing Co, 1975, 285 pp. LC: 74-33843. ISBN: 0-87762-160-8. Cost: $35.00.

Essentially, this handbook is a collection of articles on air pollution equipment written to enable supervisors and managers to reduce costs. Information is provided on such subjects as electrostatic precipitators, absorption and adsorption equipment, incinerators, and different types of filters and collectors. Emphasis is placed on the installation of air pollution control equipment as well as on ways to reduce costs in the operation and maintenance of this equipment. A bibliography appears on pages 282-285. For related works, please refer to such titles as *Pollution Engineering Practice Handbook; Handbook of Environmental Civil Engineering; CRC Handbook of Environmental Control; Air Pollution*, and *The Environmental Engineers' Handbook*.

1081. Handbook for the Orthopaedic Assistant. Franz Richard Schneider. Mosby, 2d ed., 1976. LC: 76-4850. ISBN: 0-8016-4351-1. Cost: $11.95.

1082. Handbook for the Practice of Pediatric Psychology. J.M. Tuma. John Wiley and Sons, 1982, 400 pp. LC: 81-11567. ISBN: 0-471-06284-7. Cost: $25.00. Series: *Wiley Series on Personality Processes.*

1083. Handbook for the Rolling Resistance of Pneumatic Tires. S.K. Clark; R.N. Dodge. Institute of Science and Technology, Industrial Development Div, Univ of Michigan, 1979, 78 pp. ISBN: 0-686-72867-X. Cost: $12.00.

This handbook emphasizes a method of measuring tire rolling resistance. Information is provided for different types of tires on vehicle weights and tire pressures. Energy conservation is a major consideration in this book. Over 30 figures and 9 tables are included.

1084. Handbook for the Technologist of Nuclear Medicine. Julia S. Hudak. C.C. Thomas, 1971, 157 pp. LC: 75-161165. ISBN: 0-398-02178-3. Cost: $8.75.

1085. Handbook for the Young Diabetic. A.E. Fischer; D.L. Horstmann. Stratton Intercontinental Medical Book Co, 4th ed., 1972. LC: 64-22057. ISBN: 0-913258-01-6. Cost: $5.95.

1086. Handbook for Theoretical Computation of Line Intensities in Atomic Spectra. I.B. Levinson; A.A. Nikitin; Z. Lerman trans. Davey (Israel Program for Scientific Translation), 1965, 242 pp. LC: 65-700.

1087. Handbook for Thin Film Technology. L. Maissel; R. Glang. McGraw-Hill, 1970, 800 pp. LC: 73-79497. ISBN: 0-07-039742-2. Cost: $38.50.

1088. Handbook for Total Quality Assurance. Edward M. Stiles. National Foremen's Inst, 1964, 144 pp. LC: 64-22218.

1089. Handbook for Transistors. John D. Lenk. Prentice-Hall, 1975, 296 pp. LC: 75-2294. ISBN: 0-13-382259-1. Cost: $18.95.

This handbook is directed primarily to the upper-level electronics technician or engineer who would like a better understanding of transistor technoloy. This book discusses many different types of transistors including field effect transistors, uni-junction transistors, and bipolar transistor. Testing data are presented for different types of transistors as well as data on temperatures and frequencies as they relate to transistors. Information is provided for transistors in circuits, and design considerations are presented. This handbook includes an index, and explanations are included to help the user understand different equations presented.

1090. Handbook for Ultimate Strength Design of Reinforced Concrete Members. V.K. Ghanekar et al. Structural Engineering Research Centre, Roorkee, U.P., 2d ed., 1970, 147 pp. LC: 75-8077.

1091. Handbook for Underwater Explosive Excavation in Rock and Coral. J.S. Brower; Lawrence W. Hallanger. Gulf Publishing Co, Date not set. ISBN: 0-87201-881-4. Cost: $19.95.

1092. Handbook for Welding Design. C. Rowland Harman; Institute of Welding, Ltd. Pitman, 2d ed.,1967, Vol.1, 249 pp.

1093. Handbook in Diagnostic Teaching: A Learning Disabilities Approach. Philip H. Mann; Patricia Suiter. Allyn & Bacon, Inc, 1974. ISBN: 0-205-04416-6. Cost: $25.95.

1094. Handbook of Abnormal Psychology. H.J. Eysenck. Knapp Press, (div of Knapp Communications), 2d ed., 1973. LC: 72-97452. ISBN: 0-912736-13-5. Cost: $42.00.

1095. Handbook of Accident Prevention. National Safety Council Staff. National Safety Council, 4th ed., 1970. ISBN: 0-87912-035-5. Cost: $3.15.

1096. Handbook of Acoustical Enclosures and Barriers. Richard K. Miller; Wayne V. Montone. Fairmont Press, Inc (division of Van Nostrand Reinhold). ISBN: 0-915-58606-1. Cost: $29.95.

Engineering design information is presented in the form of equations and tables to enable the designer to predict noise attenuation of different types of enclosures.

1097. Handbook of Active Filters. David E. Johnson; J.R. Johnson; Harry P. Moore. Prentice-Hall, 1980, 244 pp. LC: 79-10373. ISBN: 0-13-372409-3. Cost: $22.95.

This handbook covers such filters as low-pass filters, high-pass filters, band-pass filters, and band-reject filters. In addition, Butterworth filters and Chebyshev filters are discussed, along with such topics as infinite-gain multiple-feedback filters. A detailed treatment of integrated circuit operational amplifiers is also presented.

1098. Handbook of Adhesive Bonding. C.V. Cagle. McGraw-Hill, 1973, 800 pp. LC: 73-2875. ISBN: 0-07-009588-4. Cost: $55.00.

Provides information on adhesive selection. Covers properties, chemical composition characteristics and applications. Includes information on bonding- to aluminum, titanium, lead, steel, and ceramics.

1099. Handbook of Adhesives. Irving Skeist. Van' Nostrand Reinhold Co, 2nd ed by Litton Educational Publishing, Inc, 2nd ed, 1977, 948 pp. LC: 76-18057. ISBN: 0-442-27634-6. Date of First Edition: 1962. Cost: $42.50.

Covers bonding, formulations, metalizing of plastics and critical surface tension of polymers.

1100. Handbook of Advanced Composites. George Lubin. Van Nostrand Reinhold, 2d ed., 1981, 750 pp. ISBN: 0-442-24897-0. Cost: $54.50.

This handbook covers new information on modern processes and materials used in the preparation of advanced composites. Topics such as fibers, parting agents, processing, environmental resistance, and design are discussed. Various tables are presented with data on the use of plastics and plastics technology in the composites industry.

1101. Handbook of Advanced Solid-State Trouble-Shooting. Miles Ritter-Sanders Jr. Reston Publishing Co, 1977, 255 pp. LC: 77-23190. ISBN: 0-87909-321-8. Cost: $17.95.

Miles Ritter-Sanders is a pseudonym for Robert Gordon Middleton.

1102. Handbook of Advanced Wastewater Treatment. Russell L. Culp; George Mack Wesner; Gordon L. Culp. Van Nostrand Reinhold, 2d ed., 1978, 632 pp. LC: 77-24483. ISBN: 0-442-21784-6. Cost: $35.50.

Covers the use and regeneration of powdered activated carbon, selective ion exchange, disinfection, breakpoint chlorination demineralization, and chemical sludge handling.

1103. Handbook of Aerosol Technology. Paul Amsdon Sanders. Van Nostrand Reinhold, 1979, 526 pp. LC: 78-18428. ISBN: 0-442-27348-7. Cost: $29.95.

The first edition of this book was published in 1970 under the title *Principles of Aerosol Technology*. The new edition covers such topics as alternative fluorocarbons and their properties, fluorocarbon toxicology, hydrocarbons, compressed gases, flammability, aerosol valves, and various containers and filling equipment. In addition, such topics as vapor pressure, spray characteristics, and solvency are discussed as well as emulsions, foams, and suspensions.

1104. Handbook of Aging. James Birren. Van Nostrand Reinhold, 1977, 3-Vol. set. ISBN: 0-442-20790-5. Cost: $96.00.

1105. Handbook of Aging. Elliott D. Smith. Barnes and Noble, 1973. ISBN: 0-06-463332-2. Cost: $1.95.

1106. Handbook of Agricultural Charts. U.S. Department of Agriculture. U.S. Government Printing Office, 1963.

Handbook of Agricultural Enginering. *See* **Nogyo Doboku Handbook.**

1107. Handbook of Agricultural Pests: Aphids, Thrips, Mites, Snails, and Slugs. Tokuwo Kono. State of California, Dept of Food and Agriculture, 1977, 205 pp.

Handbook of Agricultural Productivity. *See* **CRC Handbook of Agricultural Productivity.**

1108. Handbook of Air Conditioning System Design. Carrier Air Conditioning Co. McGraw-Hill, 1966, 786 pp.

(various paging). LC: 65-17650. ISBN: 07-010090-X. Date of First Edition: 1960. Cost: $45.00.

Carrier Air Conditioning Company has assembled a considerable amount of information for the design engineer about air conditioning systems. Included is design information on air distribution systems, refrigerants, fans, and refrigeration machines. Calculations for load estimating is simplified by the provision of data on solar heat gain, ventilation, and psychrometrics. An index is provided.

1109. Handbook of Air Conditioning, Heating and Ventilating. Eugene Stamper; Richard L. Koral. Industrial Press, 3d ed., 1979, 1,420 pp. LC: 78-71559. ISBN: 0-8311-1124-0. Date of First Edition: 1959. Cost: $59.00.

Covers automatic control applications, unit ventilators, plastic pipes, noise reduction, heat exchangers, and heat pumps.

1110. Handbook of Air Pollution Analysis. Roger Perry; Robert J. Young. Chapman and Hall, Ltd, 1977, 506 pp. LC: 77-12646. ISBN: 0-412-12660-5; 0-470-993616-2 (Halsted). Cost: $69.95.

This handbook discusses different types of pollutants and methods of sampling and analyzing these pollutants. Such methods as gravimetric techniques, light reflectance methods, aerosol particle counters, density gradient separation, and infrared techniques are discussed. Such metal analysis methods as radioactivation methods, emission spectrography, polarography, atomic absorption spectrometry, and anodic stripping voltametry are presented. Such pollutants as volatile lead compounds, mercury compounds, nitrogen compounds, sulfur compounds, hydrocarbons, carbon monoxide, halogen-containing compounds, fluorides and chlorine are discussed. References are provided at the end of each chapter for further reading.

1111. Handbook of Air Pollution. J.P. Sheeley. U.S. Government Printing Office, 1968, various paging. Accession: PB-190247. Series: *Public Health Service Publication*, No. 999-AP-46.

1112. Handbook of Air Quality in the United States. Wilfrid Bach; Anders Daniels. Oriental Publishing Co, 1975, 235 pp. LC: 75-8390. Cost: $14.95.

This handbook presents federal air quality control standards and regulations and relates them to an analysis of the ambient air quality data. An inventory of emissions is provided along with a national distribution of air pollution levels. Air pollution levels are provided for such pollutants as hydrocarbons, sulfur dioxide, suspended particulates, photochemical oxidants, etc. An index is included along with references and a list of some of the most recent publications on air pollution.

1113. Handbook of Aircraft Materials. V.G. Aleksandrov. Izdatelstvo Transport, Moscow (available from NTIS), 1972, 328 pp. Accession: A72-40459.

This Russian handbook is a compilation of physicochemical and mechanical properties of different types of steels, nonferrous alloys, cermets, plastics, lubricants, nickel alloys, and refractory alloys.

1114. Handbook of Algebraic and Trigonometric Functions. Allan Herbert Lytel. Sams, 1964, 160 pp. LC: 64-25044.

1115. Handbook of Alkaloids and Alkaloid-Containing Plants. Robert F. Raffauf. John Wiley and Sons, 1970,

1,261 pp. LC: 73-113713. ISBN: 0-471-70478-4. Cost: $95.00.

1116. Handbook of Aluminum. Alcan Aluminum Corp, 3d ed., 1970, 251 pp. LC: 76-11401. Date of First Edition: 1957. Cost: $5.00.

Many tables are in this handbook including some of the mechanical and physical properties of aluminum as well as conversion factors and chemical composition limits. A considerable amount of data are provided on aluminum including fabrication forming, various aluminum alloys, the machining of aluminum, finishing, and jointing. There is an index, several appendices and data on corrosion resistance.

1117. Handbook of Analysis of Organic Solvents. V. Sedivec; J. Flek. Halsted Press, trans., 1976, 455 pp. LC: 75-44239. ISBN: 0-470-15010-6. Cost: $55.95.

1118. Handbook of Analysis of Synthetic Polymers and Plastics. J. Urbanski et al; G. Gordon Cameron (trans). Halsted Press, 1976, 400 pp. LC: 76-5880. ISBN: 0-470-15081-5. Cost: $91.95.

1119. Handbook of Analytical Chemistry. L. Meites. McGraw-Hill, 1963, 1,788 pp. (various paging). LC: 61-15915. ISBN: 0-07-041336-3. Cost: $75.00.

1120. Handbook of Analytical Control of Iron and Steel Production. T.S. Harrison. Halsted Press (division of John Wiley and Sons), 1979, 602 pp. LC: 78-41222. ISBN: 0-470-26538-8. Cost: $114.95.

Covers such topics as refractory materials, coke-oven biproducts, lubricants and oils as they relate to iron and steel production, pollution problems, and water supply. In addition, such subjects as metallurgical analysis, steel-making slags, iron ore sinter, and sampling methods are discussed.

1121. Handbook of Analytical Derivatization Reactions. Daniel R. Knapp. John Wiley and Sons, 1979, 882 pp. LC: 78-12944. ISBN: 0-471-03469-X. Cost: $51.50.

This handbook includes information for preparing derivatives of compounds for chromatographic and mass spectrometric analysis. This book is restricted to methods involving covalent derivatives of organic compounds. Such compounds as amino compounds, carboxylic acids, fatty acids, peptides, aldehydes, ketones, steroids, prostaglandins, carbohydrates, and nucleotides are presented in this handbook. A comprehensive indexing system has been included to make it easy for the user to gain access to the intellectual content of this compilation.

1122. Handbook of Analytical Design for Wear. C.W. MacGregor et al. IFI/Plenum Press, 1964, 106 pp. LC: 64-8816. ISBN: 0-306-65107-6. Cost: $35.00.

1123. Handbook of Anion Determination. W. John Williams. Butterworths Publishing, Inc, 1979, 630 pp. LC: 78-40553. ISBN: 0-408-71306-2. Cost: $130.00.

This handbook covers the methods for separations and selective determinations for anions. Includes all different types of anions, such as halogen anions, phosphorus oxyanions, and sulfur anions. References to the periodical literature are included.

Handbook of Antibiotic Compounds. See *CRC Handbook of Antibiotic Compounds.*

1124. Handbook of Applicable Mathematics. Walter Ledermann. John Wiley and Sons, Vols.1 and 2, 1980; Vol. 3, 1981; Vols.4-6 in preparation, Vol.1, 544 pp.; Vol.2, 444 pp.; Vol.3, 592 pp. LC: 79-42724. ISBN: Vol.1, 0-471-27704-5; Vol.2, 0-471-27821-1; Vol.3, 0-471-27947-1. Cost: $91.50 per Vol. or $77.50 subscription price.

This 6-volume set is entitled: Vol.1, *Algebra.*, Walter Ledermann and Steven Vajda; Vol.2, *Probability.*, Emlyn Lloyd; Vol.3, *Numerical Methods.*, Robert Churchhouse; Vol.4, *Analysis.*, Walter Ledermann and Steven Vajda; Vol.5, *Geometry and Combinatorics.*, Walter Ledermann and Steven Vajda; Vol.6, *Statistics., Emlyn Lloyd. This handbook will eventually consist of 6 volumes covering such topics as algebra, probability, numerical methods, mathematical analysis, geometry, combinatorics, and statistics. A comprehensive bibliography is provided for in-depth study. This work is specifically designed for the nonmathematician.*

1125. Handbook of Applied Hydraulics. C.V. Davis; K.E. Sorenson. McGraw-Hill, 3d ed., 1969, 1,216 pp. (various paging). LC: 67-25809. ISBN: 0-07-015538-0. Date of First Edition: 1942. Cost: $56.50.

This handbook contains a considerble amount of information on hydraulic engineering including such areas as the design of dams, spillways, and canals. Information on flood control and water supply and distribution is provided in addition to irrigation systems, sewage, and water treatment. An index is provided along with references cited throughout the text.

1126. Handbook of Applied Hydrology: A Compendium of Water-Resources Technology. Ven Te Chow (W.T. Chow). McGraw-Hill, 1964, 1,468 pp. (various paging). LC: 63-13931. ISBN: 0-07-010774-2. Cost: $57.50.

Essentially, this handbook has to do with water resources technology and, as a result, it crosses many fields of knowledge and brings together many related subjects such as oceanography, agronomy, and climatology. This handbook has gathered together a considerable amount of information on such subjects as the sources of water, sedimentation, the hydrology of various different types of land and areas of population, flood control, and reservoirs. Such engineering information as fluid mechanics and soil mechanics is provided in this well-indexed handbook. For related works, please refer to the *Handbook on the Principles of Hydrology.*

1127. Handbook of Applied Instrumentation. D.M. Considine; S.D. Ross. McGraw-Hill, 1964, 1,156 pp. (various paging). LC: 62-21926. ISBN: 0-07-012426-4. Cost: $42.85.

This handbook is an excellent compendium of different devices used to measure physical characteristics, mechanical characteristics, and most any other characteristic exhibited by any type of material. For example, such subjects as temperature, acoustics, radiation, and moisture content are discussed along with the devices that are used to measure these characteristics. In addition, instrumentation and measurement are discussed from the various points of view of different industries such as the food industry, the pulp and paper industry and the steel production industry. Emphasis is placed on how-to information. For related works, please refer to the following title: *Processed Instruments and Controls Handbook.*

1128. Handbook of Applied Mathematics. Martin Ernest Jansson et al; Edward E. Grazda et al (eds). Van Nostrand Reinhold, 4th ed., 1966, 1,119 pp. LC: 66-9325. Cost: $32.50.

1129. Handbook of Applied Mathematics: Selected Results and Methods. Carl Eric Pearson. Van Nostrand Reinhold, 1975, 1,266 pp. ISBN: 0-442-26493-3. Cost: $39.50.

1130. Handbook of Applied Psychology. Douglas H. Fryer; Edwin R. Henry. Johnson Reprint Co, (subs of Harcourt, Brace and Jovanovich), 1969. ISBN: 0-384-17070-6. Cost: $42.00.

1131. Handbook of Architectural Detaiis for Commercial Buildings. Joseph DeChiara. McGraw-Hill, 1980, 512 pp. ISBN: 0-07-016215-8. Cost: $29.50.

This handbook includes over 500 illustrations of architectural details taken from actual working drawings.

1132. Handbook of Asbestos Textiles. Asbestos Textile Institue, 3d ed., 1967. ISBN: 0-87245-025-2. Cost: $3.00.

Also available from Textile Book Service.

1133. Handbook of Astronautical Engineering. Heinz Hermann Koelle. McGraw-Hill, 1961, various paging. LC: 61-7305.

This handbook includes information on the design of space vehicles, space flight, propulsion systems, and any other relevant data that would be helpful to the astronautical engineer. Navigation, space vehicles, communication systems, and different types of propulsion systems such as solid propellants, liquid propellants, and even solar propulsion are discussed. This handbook includes an index, and many bibliographic references are provided for those who would like additional information on the subject.

1134. Handbook of Astronautics. N. Ya. Kondrat'ev; V.A. Odinstov; R. Hardin (trans). Israel Program for Scientific Translations, 1966. ISBN: 0-914326-08-2. Cost: $15.00.

This Russian translation contains a considerable amount of information on space flight and navigation in space.

1135. Handbook of Astronautics. S.W. Smith (British Interplanatary Society). Dufour Editions, Inc, 2d ed., 1969, 128 pp. LC: 67-75819 (GB). ISBN: 0-8023-1151-2. Cost: $6.25.

Also available from the University of London Press.

1136. Handbook of Astronomy, Astrophysics, and Geophysics. C. W. Gordon; V. Canuto. Gordon and Breach, Vol.1, 1978 (others in preparation), Vol.1, 420 pp. ISBN: Vol.1, 0-677-16100-X. Cost: $81.50.

This handbook began in 1978 with Vol.1 and several additional volumes are planned for the future. This handbook series covers such topics as the ionosphere, magnetosphere, cosmic physics, the planetary system, and stellar evolution.

1137. Handbook of Atomic Data. S. Frago et al. Elsevier, 1976, 2d impression 1979, 552 pp. LC: 76-16162. ISBN: 0-444-41461-4. Cost: $105.00.

1138. Handbook of Audio Circuit Design. Derek Cameron. Reston (division of Prentice-Hall), 1978, 255 pp. LC: 77-16819. ISBN: 0-87909-362-5. Cost: $18.95.

Covers such topics as audio amplifier design, speaker circuits, telephone circuits, and power amplifiers. In addition, there is a color-coded listing for resistors and capacitors in the appendix.

1139. Handbook of Auditory and Vestibular Research Methods. C.A. Smith; J.A. Vernon. C.C. Thomas, 1975. ISBN: 0-398-03231-9. Cost: $43.75.

1140. Handbook of Auditory Perceptual Training. Cora L. Reagan. C. C. Thomas, 1973. ISBN: 0-398-02885-0. Cost: $7.95.

1141. Handbook of Auger Electron Spectroscopy: A Reference Book of Standard Data for Identification and Interpretation of Auger Electron Spectroscopy Data. Lawrence E. Davis et al. Physical Electronics Industries, 2d ed., 1976, 252 pp. LC: 76-364786. Date of First Edition: 1972.

1142. Handbook of Automated Analysis: Continuous Flow Techniques. William A. Coakley. Marcel Dekker, Inc, 1981, 160 pp. ISBN: 0-8247-1392-3. Cost: $19.75.

This handbook covers various continuous flow techniques as a method of analysis. More specifically, the handbook evaluates the Auto Analyzer as a device used for continuous flow analysis. Analysis of drugs, trace metals, pesticides, hydrocarbons, and pollutants are discussed. In addition, such topics as solvent extraction, quality control, and different types of samples suitable for analysis are presented.

1143. Handbook of Automated Electronic Clinical Analysis. Harry E. Thomas. Prentice-Hall, 1979, 372 pp. ISBN: 0-8359-2735-0. Cost: $40.50.

1144. Handbook of Automation, Computation, and Control. Eugene M. Grabbe et al. John Wiley and Sons, 1958-61, 3 Vols.. LC: 58-10800. ISBN: Vol.1, 0-471-32010-2; Vol.2, 0-471-32043-9; Vol.3, 0-471-32076-5. Cost: $31.50 each.

The *Handbook of Automation, Computation, and Control* consists of 3 volumes entitled as follows: Vol.1, *Control Fundamentals*; Vol.2, *Computers and Data Processing*; Vol.3, *Systems and Components*. This 3 volume-set is an excellent collection of information on the design and use of computers. Information is provided on such subjects as systems engineering, the different types of industrial applications, and the basic fundamentals of computation that provide the necessary background for a thorough understanding of the subject.

1145. Handbook of Bacteriological Technique. Francis Joseph Baker; M.R. Breach. Appleton, (also Butterworth and Co), 2d ed., 1967, 482 pp. LC: 68-1340. ISBN: 0-407-72801-5. Cost: $15.50.

Handbook of Bacteriology. *See* **Medical Microbiology.**

1146. Handbook of Basic Circuits: TV-FM-AM Matthew Mandl. Macmillan Publishing Co, 1956, 365 pp.

1147. Handbook of Basic Electronic Troubleshooting. John Lenk. Prentice-Hall, 1977, 272 pp. LC: 76-7538. ISBN: 0-13-372482-4. Cost: $17.95.

Discusses methods for analyzing trouble systems, localizing faulty circuits, and isolating defective components. Has over 160 illustrations of wiring diagrams.

1148. Handbook of Basic Microtechnique. Peter Gray. McGraw-Hill, 3d ed., 1964, 302 pp. LC: 63-22430. ISBN: 0-07-024206-2. Cost: $19.00.

1149. Handbook of Basic Pharmacokinetics. W.A. Ritschel. Drug Intelligence Publications, 2d ed., 1980, 470 pp. LC: 79-90428. ISBN: 0-914768-34-4. Cost: $19.50.

1150. Handbook of Basic Transistor Circuits and Measurements. Richard D. Thornton et al. John Wiley and Sons, 1966, 156 pp. LC: 66-21061. ISBN: 0-471-86502-8. Cost: $7.95. Series: *Semiconductor Electronic Education Committee*, Vol. 7.

1151. Handbook of Behavioral Neurobiology. Plenum, 1978, 5 Vols. LC: 78-17238. ISBN: Vol.1, 0-306-35191-9; Vol.2, 0-306-35192-7; Vol.3, 0-306-40218-1; Vol.4, 0-306-40585-7; Vol.5, 0-306-40613-6. Cost: Vol.1, $39.50; Vol.2, $35.00; Vol.3, $39.50; Vol.4, $45.00; Vol.5, $45.00.

The 5 volumes in this set are entitled: Vol.1, *Sensory Integration.*, R. Bruce Masterson; Vol.2, *Neuropsychology.*, edited by Michael S. Gazzaniga; Vol.3, *Social Behavior and Communications.*, edited by Peter Marler and J.G. Vandenberg; Vol.4, *Biological Rhythms.*, edited by Jurgen Aschoff; Vol.5, *Motor Coordination.*, edited by Arnold L. Towe and Erich S. Luschei. This is a comprehensive series in behavioral neurobiology, and covers such topics as neuropsychology, neurophysiology, brain damage, motor coordination, and communication.

Handbook of Bimolecular and Termolecular Gas Reactions. *See* **CRC Handbook of Bimolecular and Termolecular Gas Reactions.**

1152. Handbook of Binary Diagrams. W.G. Moffatt. General Electric, 1977, loose-leaf service, updated twice annually, 3 binders, $175.00 plus $68.00 per year for the updating service.. ISBN: 0-931690-005.

This new handbook is a collection of over 800 binary phase diagrams which provide information on heattreatment and the constitution of various alloys. In addition to the semiannual updating service, a special feature of this handbook is the cross index, which helps to interrelate information on over 2,000 binary systems. This handbook is a must for any library that provides metallurgical information. For related works, please refer to the *Metals Handbook*. See also *Constitution of Binary Alloys*, and *Index to Binary Phase Collections*. These diagrams supplement Hansen, Elliot, and Shunk.

1153. Handbook of Binary Metallic Systems: Structure and Qualities. N.V. Ageev. Israel Program for Scientific Translations (available from NTIS), 1966-67, 2 Vols.. Tech Report: Vol.1, TT 66-51149; Vol.2, TT 66-51150.

1154. Handbook of Biochemistry and Biophysics. Henry Clarence Damm; P.K. Besch. World Publications, 1966, 736 pp. LC: 65-25773.

Handbook of Biochemistry in Aging. See *CRC Handbook of Chemistry in Aging.*

Handbook of Biological Data. *See* **Biology Data Book.**

1155. Handbook of Biological Illustration. Frances W. Zweifel. University of Chicago Press, 1961. LC: 61-19734. ISBN: 0-226-99699-9. Cost: $3.95.

1156. Handbook of Biological Psychiatry. Herman M. Van Praag. Marcel Dekker, Inc, 1979, 6 Vols. ISBN: Vol.1, 0-8247-6835-3; Vol.2, 0-8247-6892-2; Vol.3, 0-8247-6965-1; Vol.4, 0-8247-6966-X; Vol.5, 0-8247-6967-8;

Vol.6, 0-8247-6968-6. Cost: Vol.1, $45.00; Vol.2, $46.50; Vol.3, $29.50; Vol.4, $110.00; Vol.5, $28.95; Vol.6, $67.00. Series: *Experimental and Clinical Psychiatry Series*.

This 6-volume set consists of the following titles: Part I *Disciplines Relevant to Biological Psychiatry*. Part II *Brain Mechanisms and Abnormal Behavior—Psychophysiology*. Part III *Brain Mechanisms and Abnormal Behavior—Genetics and Neuroendocrinology*. Part IV *Brain Mechanisms and Abnormal Behavior—Chemistry*. Part V *Drug Treatment in Psychiatry*. Part VI *Practical Application of Psychotropic Drugs and Other Biological Treatments*. This handbook series discusses such topics as electrolyte metabolism, alcohol tolerance, Parkinson's disease, Huntington's chorea, schizophrenia, sleep disorders, and pharmacokinetics.

1157. Handbook of Biological Wastewater Treatment; Evaluation, Performance, and Cost. Henry H. Benjes Jr. Garland STPM Press, 1980, 192 pp. LC: 79-12394. ISBN: 0-8240-7089-5. Cost: $19.95.

Covers several different wastewater treatment methods including suspended-growth and attached-growth systems. Also includes the Activated Biofilter system of wastewater treatment.

1158. Handbook of Biomedical Instrumentation and Measurement. Harry Elliott Thomas. Reston Publishing Co, 1974, 550 pp. LC: 74-13804. ISBN: 0-87909-323-4. Cost: $29.95.

This handbook includes information on the electronic equipment used for medical purposes with primary emphasis on cardiac support equipment, but also including equipment used in such areas as blood analysis, pulmonary support apparatus, X-ray equipment, and muscle instrumentation. Some background information in anatomy and physiology is provided at the beginning of the handbook so that the user can better understand the electronic equipment and its applications to medicine. An index is provided along with an extensive appendix listing various data associated with biomedical instrumentation.

1159. Handbook of Biomedical Plastics. Henry Lee; Kris Neville. Pasadena Technology Press, 1971, various paging. LC: 74-29647.

Handbook of Biosolar Resources. *See* **CRC Handbook of Biosolar Resources.**

1160. Handbook of Blunt Body Aerodynamics. John A. Darling (Naval Ordnance Laboratory, White Oak, Md). NTIS, 1973, 154 pp. Accession: AD-776586. Tech Report: NOLTR-73-225.

1161. Handbook of Bovine Obstetrics. V. Sloss; J.H. Dufty. Williams and Wilkins, 1980, 208 pp. LC: 79-25031. Cost: $29.95.

1162. Handbook of Brittle Material Design Technology. W.H. Dukes. NTIS, 1971, 161 pp. Accession: N71-20027. Tech Report: AGARD-AG-152-71.

This handbook provides structural data on brittle nonmetallic materials for use in designing re-entry vehicles. Such topics as elastic deformation, structural failure, mechanical properties and refractory materials are discussed.

1163. Handbook of Brittle Material Design Technology. W. Dukes. NATO, Advisory Group for Aerospace Research Development, 1971, 155 pp. LC: 70-594284. Series: *AGARDograph*, 152.

1164. Handbook of BS 5337: 1976 (The Structural Use of Concrete for Retaining Aqueous Liquids). R.D. Anchor; A.W. Hill; B.P. Hughes. Viewpoint Publications, 1979, 60 pp. ISBN: 0-7210-1078-4. Cost: $20.00.

This handbook is a guide to British Standard 5337, 1976. It includes such information as reinforced concrete, prestressed concrete, detailing joints, prestressing tendons, and inspection and testing of structures. Distributed by Scholium International.

1165. Handbook of Building Security Planning and Design. Peter S Hopf. McGraw-Hill, 1979, 576 pp. LC: 78-21636. ISBN: 0-07-03016-9. Cost: $34.50.

This handbook includes numerous diagrams, charts, checklists, and a security product directory of products and vendors of security equipment. Such topics as electronic security systems, lighting, surveillance cameras, vaults, fire alarms, and locks are covered in this handbook. Unique security problems associated with such institutions as libraries, schools, computer installations, warehouses, private residences, and industrial plants are also discussed.

1166. Handbook of Business Formulas and Controls. Spencer A. Tucker. McGraw-Hill, 1979, 384 pp. LC: 78-26737. ISBN: 0-07-065421-2. Cost: $24.95.

This handbook provides useful information to facilitate decision-making with the aid of formulas and ratios. Covers objective methods for making decisions regarding pricing, productive capacity, profitability analysis, modelling of profit structures, and marketing distribution. Also provides information for the development of ratios for controlling manufacturing, sales, and finances.

1167. Handbook of Cancer Immunology. Harold Waters. Garland Press, 1978, 9 volumes, Vol.1, 352 pp; Vol.2, 284 pp.; Vol.3, 442 pp.; Vol.4, 343 pp.; Vol.5, 486 pp.; Vol.6, 320 pp.; Vol.7, 384 pp.; Vol.8, 496 pp., Vol.9, 272 pp. LC: Vol.1, 76-52693; Vol.2, 77-18080; Vol.3, 77-18079; Vol.4, 77-18079; Vol.5, 77-25702; Vol.6, 80-773; Vol.7, 80-774; Vol.8, 80-775; Vol.9, 80-776. ISBN: Vol.1, 0-8240-9864-1; Vol.2, 0-8240-7001-0; Vol.3, 0-8240-7002-X; Vol.4, 0-8240-7003-8; Vol.5, 0-8240-7004-6; Vol.6, 0-8240-7110-7; Vol.7, 0-8240-7111-5; Vol.8, 0-8240-7112-3; Vol.9, 0-8240-7113-1. Cost: Vol.1, $50.00; Vol.2, $42.50; Vol.3, $65.00; Vol.4, $50.00; Vol.5, $70.00, Vol.6, $48.00; Vol.7, $57.50; Vol.8, $70.00, Vol.9, $40.00.

This 9-vol. set is entitled: Vol.1, *Basic Cancer—Related Immunology*; Vol.2, *Cellular Escape from Immune Destruction*; Vol.3, *Immune Status in Cancer Treatment and Prognosis—Part A*; Vol.4, *Immune Status in Cancer Treatment—Part B*; Vol.5, *Immunotherapy*; Vol.6, *Lymphoid Cell Subpopulations*; Vol.7, *Immune Function and Dysfunction in Relation to Cancer*; Vol.8, *Tumor Antigens*; Vol.9, *Humoral Immunity in Relation to Cancer*. Such topics as lymphocyte production, bone marrow, antibodies, and antigens are discussed in this comprehensive series. Immune response to tumors, Hodgkins disease, and leukemia are among the topics discussed.

1168. Handbook of Cane Sugar Engineering. E. Hugot. American Elsevier, trans. 2d ed., 1972, 1,079 pp. LC: 79-135484. ISBN: 0-444-40896-7. Cost: $219.50.

1169. Handbook of Canine Electrocardiography. Gary R. Bolton. Saunders, 1975. LC: 74-17749. ISBN: 0-7216-1838-3. Cost: $22.50.

1170. Handbook of Cardiology for Nurses. Walter Modell et al. Springer Publishers, 5th ed., 1966. LC: 66-17768. ISBN: 0-8261-0025-2. Cost: $5.95.

1171. Handbook of Cast Iron Pipe for Water, Gas, Sewerage and Industrial Service. Cast Iron Pipe Research Association, 2d ed., 1952, 444 pp. Date of First Edition: 1927. Cost: $5.00.

This useful handbook presents such topics as the flow of water in cast iron pipe, laying cast iron pipe, flanged pipe fittings for water, and various standards and specifications. Please also refer to the following title, *Handbook: Ductile Iron Pipe, Cast Iron Pipe.*

1172. Handbook of Catalyst Manufacture. M. Sittig. Noyes Data Corp, 1978, 474 pp. LC: 77-15217. ISBN: 0-8155-0686-4. Cost: $48.00. Series: *Chemical Technology Review*, no. 98.

This is basically a collection of 415 patents on catalyst manufacturing. Such topics as alkylations, catalytic oxidations, steam reforming and methanation, dehydrogenations, and cracking processes are presented.

1173. Handbook of Cavitation Erosion. A. Thiruvengadam; Hydronautics, Inc. NTIS, 1974, 320 pp. Accession: AD-787073. Tech Report: TR-7301-1.

1174. Handbook of Cell and Organ Culture. Donald Joseph Merchant et al. Burgess, 1964, 263 pp. LC: 64-22573. ISBN: 0-8087-1313-2. Cost: $12.95.

1175. Handbook of Cellular Chemistry. Annabelle Cohen. Mosby (distributed in U.K. by Kimpton), 1975, 158 pp. LC: 74-14780. ISBN: 0-8016-1007-9. Cost: $10.50.

1176. Handbook of Chemical Data. F.W. Atack. Reinhold Publishing Co, 1957-, annual. LC: 58-21.

Supersedes in part the *Chemists' Yearbook*, being a separate publication for tables of constants.

1177. Handbook of Chemical Lasers. R.W. Gross; J.F. Bott. Wiley-Interscience, 1976, 744 pp. LC: 76-6865. ISBN: 0-471-32804-9. Cost: $57.50.

Covers the kinetics of hydrogen—halide chemical lasers, continuous; wave hydrogen—halide lasers, gas dynamics of supersonic mixing lasers; metal—atom oxidation lasers. Includes bibliographical information and index.

1178. Handbook of Chemical Microscopy. Emile Monnin Chamot; Clyde Walter Mason. Wiley-Interscience, Vol.1, 3d ed, Vol.2, 2d ed, 1958, Vol.1, 502 pp.; Vol.2, 439 pp. LC: 58-12706. ISBN: Vol.1, 0-471-14355-3; Vol.2, 0-471-04122-X. Date of First Edition: 1978. Cost: Vol.1, $35.00; Vol.2, $33.95.

The *Handbook of Chemical Microscopy* consists of 2 volumes entitled as follows: Vol.1, *Principles and Use of Microscopes and Accessories*; Vol.2, *Chemical Methods and Inorganic Qualitative Analysis.*

1179. Handbook of Chemistry and Physics. M.V. George et al. Van Nostrand Reinhold, 2d ed., 1970, 367 pp. LC: 76-125954. ISBN: 0-442-068020-4. Cost: £3.

Includes bibliographies.

Handbook of Chemistry. *See* **Lange's Handbook of Chemistry.**

1180. Handbook of Child Nursing Care. M.J. Wallace. John Wiley and Sons, 1971. LC: 75-134041. ISBN: 0-471-91850-4. Cost: $5.50.

1181. Handbook of Chlorination: For Potable Water, Waste-Water, Cooling Water, Industrial Processes, and Swimming Pools. George Clifford White. Van Nostrand Reinhold Co, 1972, 752 pp. LC: 70-178857. ISBN: 0-442-29398-4. Cost: $39.50.

The editor of this handbook has gathered together a wealth of information on chlorine and its use as a disinfectant in the treatment of water. The handbook includes information on the properties of chlorine and its various uses for particular categories of water such as swimming pools, waste water, and cooling water. In addition, chlorination equipment is discussed. This handbook has an index and various references for further investigation. For related works, please refer to the *Handbook of Swimming Pool Construction, Maintenance, and Sanitation*, the *Water Treatment Handbook*, and *Water Quality and Treatment.*

1182. Handbook of Circuit Analysis, Language and Techniques. Randall W. Jensen; Lawrence P. McNamee. Prentice-Hall, 1976, 809 pp. LC: 75-23235. ISBN: 0-13-372649-5. Cost: $37.95.

This handbook has a considerable amount of information about electronic circuit design of data processing equipment. It also contains a discussion of programing language for computers. Includes bibliographic references.

1183. Handbook of Clinical Audiology. Jack Katz. Williams and Wilkins, 2d ed., 1978, 834 pp. ISBN: 0-683-04550-4. Cost: $32.00.

1184. Handbook of Clinical Behavior Therapy. S.M. Turner; K.S. Calhoun; H.E. Adams. John Wiley and Sons, 1981, 776 pp. LC: 80-16841. ISBN: 0-471-04178-5. ISSN: 0195-4008. Cost: $35.00. Series: *Wiley Series on Personality Processes.*

1185. Handbook of Clinical Dermatoglyphs. M.S. Elbualy; Joan D. Schindeler. University of Miami Press, 1971. LC: 71-143458. ISBN: 0-87024-191-5. Cost: $7.95.

1186. Handbook of Clinical Dietetics. American Dietetic Association. Yale University Press, 1981, 139 pp. LC: 80-11317. Cost: $20.00.

Information is presented on diets, nutrition practice for better health, and disease prevention.

1187. Handbook of Clinical Drug Data. James E. Knoben et al. Drug Intelligence Publications, 3d ed., 1973, 305 pp. LC: 73-85166. ISBN: 0-914768-11-5. Cost: $4.75.

1188. Handbook of Clinical Endodontics. Richard Bence; Franklin S. Weine (ed). Mosby, 1976. LC: 76-4582. ISBN: 0-8016-0586-5. Cost: $12.95.

Handbook of Clinical Engineering. *See* **CRC Handbook of Clinical Engineering.**

1189. Handbook of Clinical Impedance Audiometry. James Jerger. American Electromedics Corp, 1975, 235 pp. LC: 75-18681.

This handbook includes bibliographies and index.

1190. Handbook of Clinical Neurology. P.J. Vinken; G.W. Bruyn. North-Holland Publishing Co (also Elsevier), 1969- , multivolume set, 37 volumes in 1982. LC: 68-8297. ISBN: Vol.1, 0-444-10298-1 (2d reprint Vol.1, 0-7204-7201-6). Cost: Consult publisher for prices of each volume.

1191. Handbook of Clinical Neuropsychology. S.B. Filskov; T.J. Boll. Wiley-Interscience, 1980, 768 pp. LC: 80-15392. ISBN: 0-471-04802-X. ISSN: 0195-4008. Cost: $30.00. Series: *Wiley Series on Personality Processes.*

Covers such topics as epilepsy, alcoholism, aging, and cerebral laterality.

1192. Handbook of Clinical Pathology. Joseph A. Sisson. Lippincott, 1976, 524 pp. LC: 75-21273. ISBN: 0-397-50346-6. Cost: $18.00.

1193. Handbook of Clinical Psychobiology and Pathology. Sanford I. Cohen; Robert N. Ross. Hemisphere Publishing Corp, 1982. ISBN: Vol.1, 0-89116-173-2; Vol.2, 0-89116-174-0; 2-volume set 0-89116-172-4. Cost: Vol.1, $25.00; Vol.2, $35.00; 2-Vol. set $55.00. Series: *The Series in Clinical and Community Psychology.*

Covers the physiological and chemical mechanisms involved in brain functions as it relates to psychological disturbances.

1194. Handbook of Clinical Ultrasound. M.de Vlieger; J.H. Holmes. Wiley Medical, 1978, 970 pp. LC: 78-14458. ISBN: 0-471-02744-8. Cost: $122.00.

1195. Handbook of Clinical Veterinary Pharmacology. Dan Upson. Bonner Springs. ISBN: 0-93-507815-0. Cost: $17.50.

1196. Handbook of Coaxial Microwave Measurements. David A. Gray. General Radio Co, 1968, 163 pp.

1197. Handbook of Colorimetry. Arthur C. Hardy. MIT Press, 1936, 87 pp. ISBN: 0-262-08001-X. Cost: $40.00.

Covers trichromatic coefficients, spectrophotometry, and tristimulus values.

1198. Handbook of Commercial Scientific Instruments. Claude Veillon; Wesley Wendlandt. Marcel Dekker, Vol.1, 1972; Vol.2, 1974, Vol.1, 174 pp. LC: 72-87851. ISBN: Vol.1, 0-8247-1700-7; Vol.2, 0-8247-6060-3. Cost: Vol.1, $15.50; Vol.2, $16.25.

Vol.1, *Atomic Absorption*, Vellion; Vol.2, *Thermoanalytical Techniques*, Wendlandt.

1199. Handbook of Commercial Sound Installations. Leo G. Sands. Audel, 1965. LC: 65-19026. ISBN: 0-672-23126-3. Cost: $5.95.

1200. Handbook of Common Methods in Limnology. Owen T. Lind. Mosby, 1974. LC: 74-8422. ISBN: 0-8016-3017-7. Cost: $11.95.

1201. Handbook of Common Polymers. W.T. Roff; J.R. Scott. CRC Press, 1971.

Replaces *Fibres, Plastics, and Rubbers: A Handbook of Common Polymers.* Provides information on the properties of polymers and their applications. For related works, please refer to the *Polymer Handbook*, and *SPI Plastics Engineering Handbook.*

1202. Handbook of Communications. Ithiel D. Pool; Wilbur Schramm. Rand Corp, 1973. ISBN: 0-528-68714-X. Cost: $40.00.

1203. Handbook of Community Health. Murray Grant. Lea and Febiger, 2d ed., 290 pp. LC: 74-3029. ISBN: 0-8121-0495-1. Cost: $7.50 paper copy.

1204. Handbook of Community Medicine. A.M. Nelson et al. Year Book Medical Publishers, 1975, 364 pp. ISBN: 0-8151-6346-0. Cost: $29.95.

1205. Handbook of Community Mental Health. Stuart E. Golann; Carl Eisdorfer. Prentice-Hall, 1972. ISBN: 0-13-377242-X. Cost: $29.95.

1206. Handbook of Comparative World Steel Standards. International Technical Information Institute (distributed by ASTM), 450 pp. Cost: $125.00 plus $14.00 handling ASTM PCN (13-1100080-01).

Compares the steel standards of 6 countries, including the U.S., United Kingdom, West Germany, Japan, France, and the U.S.S.R. Includes chemical composition, tensile strength, and yield point.

1207. Handbook of Components for Electronics. Charles A. Harper. McGraw-Hill, 1977, 1,088 pp. LC: 76-26117. ISBN: 0-07-026682-4. Cost: $42.50.

1208. Handbook of Components in Solvent Extractions. Alfred W. Francis. Gordon and Breach, Science Publishers, 1972, 544 pp. LC: 72-78013. ISBN: 0-677-03080-0. Cost: $104.00.

1209. Handbook of Composite Construction Engineering. Gajanan M. Sabnis. Van Nostrand Reinhold, 1979, 380 pp. LC: 78-18354. Cost: $27.50.

This handbook describes the different combinations of structural steel, concrete, and wood that meet specific stress and construction situations. Such topics as load capacity, light gauge steel framework, and design techniques for in-fill walls and frames are presented.

1210. Handbook of Compositions at Thermodynamic Equilibrium. Charles R. Noddings; G.M. Mullet. John Wiley (available from Krieger), 1965. LC: 65-14730. ISBN: 0-470-64163-0. Cost: $27.00.

Consists of approximately 587 tables.

1211. Handbook of Compressed Gases. Gene R. Hawes et al (Compressed Gas Association). Van Nostrand Reinhold, 2d ed., 1981, 521 pp. LC: 80-19969. ISBN: 0-442-25419-9. Cost: $44.50.

Contains information on handling, shipping and storing compressed gases in compliance with safety standards. Covers properties, mixtures and manufacture of compressed gases.

1212. Handbook of Computer Maintenance and Trouble-shooting. Byron W. Maguire. Reston Publishing Co, Inc, 1973, 366 pp. LC: 73-8932. ISBN: 0-87909-324-2. Cost: $18.00.

1213. Handbook of Computer Management. Ronald B. Yearsley; G.M.R. Graham. Halsted Press, 1973, 328 pp. LC: 72-9028. ISBN: 0-470-97720-5. Cost: $19.75.

1214. Handbook of Computer-Aided Composition. A.H. Phillips. Marcel Dekker Inc, 1980, 456 pp. ISBN: 0-8247-6963-5. Cost: $55.00. Series: *Books in Library and Information Science Series*, Vol. 31.

1215. Handbook of Concrete Culvert Pipe Hydraulics. Portland Cement Association. Portland Cement Association, 1964, 267 pp. LC: 64-22189. ISBN: 0-89312-073-1. Cost: $10.00.

1216. Handbook of Concrete Design. Lefax Publishing Co, loose-leaf binding. Cost: $8.00. Series: *Lefax Technical Manuals*, No. 790.

1217. Handbook of Concrete Engineering. Mark Fintel. Van Nostrand Reinhold, 1974, 802 pp. LC: 74-4045. ISBN: 0-442-22393-5. Cost: $47.50.

The main emphasis of this handbook is on the design of structures using reinforced concrete. Such structures as shelves, dams, pipes, high-rise buildings, and footings are discussed. In addition, materials properties are presented including fire resistance, behavior in marine environments, and mechanical behavior. Many tables and illustrations are provided along with an extensive index. For related works, please refer to the *ACI Manual of Concrete Practice; Concrete Construction Handbook; Concrete Manual.*

1218. Handbook of Congenital Malformations. Alan Rubin. Saunders, 1967, 398 pp. LC: 66-18503. ISBN: 0-7216-7780-0. Cost: $14.00.

1219. Handbook of Conical Antennas and Scatterers. Robert M. Bevensee. Gordon and Breach, Science Publishers, 1973, 173 pp. LC: 71-172793. ISBN: 0-677-00480-X. Cost: $52.25.

Includes computation of Legender and spherical Bessel functions. Covers coaxial horn antennas and formulas for time response of an antenna or scatterer.

1220. Handbook of Coniferae and Ginkgoaceae. William Dallimore; A.B. Jackson; S.G. Harrison (rev, 4th ed). St. Martins (also available from E. Arnold and Co), 4th ed, 1967, 729 pp. LC: 67-11838.

1221. Handbook of Construction Equipment Maintenance. Lindley R. Higgins. McGraw-Hill, 1979, 1,088 pp. LC: 79-12393. ISBN: 0-07-028764-3. Cost: $42.50.

This book contains information, diagrams, and tables on preventive maintenance of construction equipment. Such topics as hydraulic cranes, asphalt equipment, motor graders, and crushing equipment are discussed. Such maintenance technologies as welding, lubrication, and resurfacing are treated, as well as the maintenance of such equipment as concrete pumps, compactors, and bulldozers.

1222. Handbook of Construction Management and Organization. Joseph P. Frein. Van Nostrand Reinhold, 2d ed., 1980, 704 pp. LC: 79-9540. ISBN: 0-442-22475-3. Cost: $42.50.

Covers site investigation to contract completion.

1223. Handbook of Construction Management. Laurence E. Reiner. Prentice-Hall, 1972, 339 pp. LC: 77-38240. ISBN: 0-13-377267-5. Cost: $17.95.

1224. Handbook of Construction Operations Forms and Formats. Robert Carlsen; James McHugh. Prentice-Hall, Inc, 1978, 343 pp. ISBN: 0-13-377218-7. Cost: $49.95.

This loose-leaf service provides models of forms, reports, records, and worksheets that will be useful in the construction business. Forms for such topics as site evaluation, material receipt, engineering, procurement, and subcontracting are described in this work.

1225. Handbook of Construction Resources and Support Services. J.A. MacDonald. John Wiley and Sons, Inc, 1981, 593 pp. LC: 79-90746. ISBN: 0-471-09354-8. Cost: $49.50 paper.

This compilation lists over 1,500 organizations that can provide information on the construction industry. Such organizations as research facilities, testing facilities, state, regional, and federal departments, trade associations, and private consulting firms are listed with a brief annotation as to their expertise, describing services, publications, and library holdings. Such topics as seismology, hydrology, weather, fire controls, soil conservation, water resources, energy, environmental sciences, and site planning are representative of subjects that have corresponding organizations that could supply pertinent information. There is a subject index as well as a geographic index for the convenience of the user.

1226. Handbook of Continuous Casting. E. Herrmann; D. Hoffmann. Otto Harrassowitz, 1980, 742 pp. ISBN: 3-87017-134-0. Cost: DM 1,250 cloth.

This handbook covers the latest techniques for continuous casting of steel and nonferrous metals. It includes descriptions of over 3,700 patents related to continuous casting. Such topics as different types of molds, open-ended tubular molds, hollow castings in open-ended molds, filaments, electroslag melting and casting, electromagnetically stirring the molten metal, cooling, and shaping are presented. Such metals as iron and steel, aluminum, magnesium, copper, uranium, and titanium are discussed in this handbook.

1227. Handbook of Controls and Instrumentation. John D. Lenk. Prentice-Hall, Inc, 1980, 336 pp. LC: 79-25280. ISBN: 0-13-377069-9. Cost: $19.95.

This handbook covers a wide range of controls and instrumentation that measure or monitor such conditions as acceleration, altitude, displacement, torque, force, fluid flow, or any type of motion. Includes over 280 illustrations, charts, and schematics.

1228. Handbook of Conveyor and Elevator Belting. Goodyear Tire and Rubber Co, 1975, 200 pp.

This handbook covers all aspects of conveyor belt construction, capacity of conveyor belts, and selection of conveyor belts. Numerous elastomer materials are discussed, including natural rubber, ethylene propylene, chloroprene, urethane, fluoroelastomers, butyl, and chlorobutyl. In addition, such topics as power requirements, belt tension, pulleys, motors, and conveyor loading are discussed. Also covers grain elevators, grain conveyors, package conveying, installation, splicing, and safety in connection with conveyor belts.

1229. Handbook of Corrosion Experiments. National Association of Corrosion Engineers. Cost: $2.50.

This is a collection of experiments describing reactions during the corrosion of certain types of materials.

1230. Handbook of Corrosion Resistant Piping. Philip A. Schweitzer. Industrial Press, 1969, 358 pp. LC: 70-92785. ISBN: 0-8311-3016-4. Cost: $35.00.

Provides a rapid means of evaluating and comparing all materials in designing piping systems. Covers plastic piping systems, metallic piping systems, lined piping systems, and pipe system economics.

1231. Handbook of Corrosion Testing and Evaluation. W.H. Ailor Jr. John Wiley and Sons, 1972, 888 pp. LC: 74-162423. ISBN: 0-471-00985-7. Cost: $75.50. Series: *Corrosion Monograph Series.*

Includes information on corrosion testing such as galvanic corrosion test methods, electrochemical methods, testing *in vivo*, stress corrosion cracking, testing in high temperatures, and testing in hot brine loops. In addition, such information as testing nuclear materials in aqueous environments, liquid-metal test procedures, cavitation damage, erosion, seawater tests, and testing in fresh water are also discussed. The usage of radioactive tracers for inhibitor evaluation and numerous methods of evaluating corrosion effects are presented.

1232. Handbook of Critical Care Medicine. Max Harry Weil; R.J. Henning. Fischer Medical, 1979, 205 pp. Cost: $19.95.

1233. Handbook of Critical Care. James L. Berk et al. Little, Brown & Co, 1976. ISBN: 0-316-09170-7. Cost: $12.50.

1234. Handbook of Cryosurgery. Richard J. Ablin. Marcel Dekker, 1980, 424 pp. ISBN: 0-8247-6981-3. Cost: $59.75. Series: *Science and Practice of Surgery Series*, Vol. 1.

Covers such topics as the usage of cryosurgery in oral disease, neurosurgery, gynecologic malignant tumors, and benign and malignant cutaneous lesions. The usage of cryosurgery in veterinary medicine is also discussed. In addition, such topics as urologic tumors, anorectal cryosurgery, and otolaryngologic disease receive coverage in this handbook.

1235. Handbook of Crystal and Mineral Collection. William B. Sanborn. Gembooks, 1966, 81 pp. LC: 65-19980. ISBN: 0-910652-05-8. Cost: $2.00.

1236. Handbook of Culvert and Drainage Practice. ARMCO Drainage and Metal Products Inc, 1950, 474 pp. Date of First Edition: 1930. Cost: $3.00.

This is a very thorough coverage of surface drainage and subsurface drainage. Numerous topics are treated in this handbook including field installation, tunneling methods, retaining walls, and special surface drainage problems, such as airports, bridges, and roadbeds. Subsurface drainage in connection with highway roadbeds, railway subdrainage, municipal subdrainage, and agricultural subdrainage is also discussed. The hydraulics of culverts, sewers, and subdrains are presented along with different types of drainage structures, such as flexible pipe, concrete pipe-arches, and corrugated metal.

1237. Handbook of Dam Engineering. Alfred R. Golzez. Van Nostrand Reinhold, 1977, 793 pp. LC: 77-8687. ISBN: 0-442-22752-3. Cost: $57.50.

1238. Handbook of Data Processing Administration, Operations and Procedures. S.R. Mixon. American Management Assoc, 1976. LC: 75-38914. ISBN: 0-8144-5400-3. Cost: $24.95.

1239. Handbook of Data Processing for Libraries. Robert M. Hayes; Joseph Becker. Wiley-Interscience, 2d ed., 1974, 712 pp. LC: 74-9690. ISBN: 0-471-36483-5. Cost: $31.50.

1240. Handbook of Data Processing Management. Martin L. Rubin. Brandon/Systems Press (also Petrocelli Books, and also Auerbach Pubs), 1970-, 6 Vols. LC: 73-103412. ISBN: Vol.1, 0-88405-043-2. Cost: $175.00 per set.

This handbook consists of 6 volumes entitled as follows: Vol.1, *Introduction to the System Life Cycle*; Vol.2, *System Life Cycle Standards*; Vol.3, *System Life Cycle Standards: Forms, Method*; Vol.4, *Advanced Technology: Input and Output*; Vol.5, *Advanced Technology: Systems Concept*; Vol.6 *Data Processing Adminstration.*

1241. Handbook of Decomposition Methods in Analytical Chemistry. R. Bock; Iain L. Marr. John Wiley and Sons, 1979, 444 pp. LC: 78-70559. ISBN: 0-470-26501-9. Cost: $94.95.

This publication is a revised and expanded translation of the following German title: *Augschlussmethoden Anorganischen und Organischen Chemie*, 1972. Includes information on decomposition, oxidizing procedures, and desolution without chemical reaction. Includes many bibliographic references to literature, as well as diagrams of instruments and procedures.

1242. Handbook of Degree-Day Data for the United States. American Petroleum Institute. American Petroleum Institute, 1958, 2 vols. LC: 64-16401.

This handbook consists of 2 volumes entitled as follows: Vol.1, *Statistical Summaries*; Vol.2, *Basic Degree-Day Data.*

1243. Handbook of Dental Malpractice. L. Brent Wood. C.C. Thomas. ISBN: 0-398-02118-X. Cost: $5.50.

1244. Handbook of Derivatives for Chromatography. Karl Blau; Graham S. King. Heyden and Son, 1977. ISBN: 0-85501-206-4. Cost: $65.00.

Covers such topics as esterification, acylation, alkylation, and additional data on the preparation of derivatives for chromatography. Also covers chelation and nonchelation techniques, fluorescent derivatives, and derivatives by ketone-base condensation.

1245. Handbook of Diabetes Mellitus. Michael Brownlee. Garland Press, 1980, Vol.1, 450 pp.; Vol.2, 250 pp.; Vol.3, 340 pp.; Vol.4, 325 pp.; Vol.5, 420 pp. LC: Vols. 1-5, 79-26703. ISBN: Vol.1, 0-8240-7030-5; Vol.2, 0-8240-7214-6; Vol.3, 0-8240-7215-4; Vol.4, 0-8240-7223-5; Vol.5, 0-8240-7224-3. Cost: Vol.1, $65.00; Vol.2, $37.50; Vol.3, $50.00; Vol.4, $45.00; Vol.5, $60.00.

This 5-volume set is entitled: Vol.1, *Diabetes Mellitus Etiology/Hormone Physiology*; Vol.2, *Islet Cell Function/Insulin Action*; Vol.3, *Intermediary Metabolism and Its Regulation*; Vol.4, *Biochemical Pathology*; Vol.5, *Current and Future Therapies.*

Covers such topics as amino acid and protein metabolism, both in normal and diabetic humans, reaction of insulin on enzymes, dietary modification for diabetics, and lipids and lipoproteins in diabetes mellitus. This work is a comprehensive survey of both experimental and clinical aspects of diabetes mellitus. Includes numerous tables and an extensive list of references.

1246. Handbook of Diagnostic Cytology. Helena Everard Hughes; T.C. Dodds. Williams and Wilkins (also Longman), 1968, 266 pp. LC: 74-3084. ISBN: 0-443-00560-5. Cost: $19.50.

1247. Handbook of Differential Diagnosis. ROCOM. ROCOM, 1974-75, 3-Vol. set. ISBN: Vol.1, 0-89119-001-5. Cost: $26.50 per Vol.

The *Handbook of Differential Diagnosis* consists of 3 volumes entitled as follows: Vol.1, *The Chest*; Vol.2, *The Abdomen* (Part 1 and Part 2).; and Vol.3, *The Pelvic Region* (Part 1 and Part 2). Concerned primarily with the actual appearance, on microscopic examination, of cytological preparations. Contents include respiratory tract, gastrointestinal tract, body fluids, and effects of radiation.

1248. Handbook of Differential Thermal Analysis. Wiliam Joseph Smothers; Yao Chiang. Chemical Publishing Co, 1966, 633 pp. LC: 66-2814.

1249. Handbook of Digital Electronics. John D. Lenk. Prentice-Hall, Inc, 1981, 400 pp. ISBN: 0-13-377184-9. Cost: $21.95.

Includes information on digital logic, digital circuits, microcomputers, integrated circuits, electronic testing, and troubleshooting. Over 260 illustrations are provided, including diagrams, tables, and charts.

1250. Handbook of Digital IC Applications. David L. Heiserman. Prentice-Hall, Inc, 1980, 429 pp. ISBN: 0-13-372698-3. Cost: $22.95.

Covers such topics as multiplexers, demultiplexers, shift registers, digital logic, and repair and maintenance of integrated circuits.

Handbook of Digital Systems Design for Scientists and Engineers: Design with Analog, Digital, and LSI. *See* **CRC Handbook of Digital Systems Design for Scientists and Engineers: Design with Analog, Digital, and LSI.**

1251. Handbook of Dimensional Measurement. Francis T. Farago. Industrial Press, 1968, 416 pp. ISBN: 0-8311-1025-2. Cost: $36.50.

Covers fixed gages, electronic gages, microscopes, and optical projectors.

1252. Handbook of Diseases of Laboratory Animals. J. Malcolm Hime; Philip N. O'Donoghue. International Ideas, 1977. ISBN: 0-433-14723-7. Cost: $37.50.

1253. Handbook of Diseases of the Skin. Herbert Owen Mackey; John P. Mackey (rev). Macmillan (London), 9th ed, 1968, 424 pp. LC: 73-360086.

1254. Handbook of Drafting Techniques. John A Nelson. Van Nostrand Reinhold, 1981, 368 pp. ISBN: 0-422-28662-7 paper; 0-422-28661-9 cloth. Cost: $14.95 paper; $22.95 cloth.

Covers all types of drafting techniques from lettering to geometric construction and isometric drawing. Techniques on how to understand blueprints are described in this handy guide to drafting.

1255. Handbook of Drug and Chemical Stimulation of the Brain: Behavioral, Pharmacological and Physiological Aspects. R.D. Myers. Van Nostrand Reinhold, 1974, 760 pp. LC: 74-10564. ISBN: 0-442-25622-1. Cost: $38.50.

1256. Handbook of Drug Interactions. Gerald Swidler. Wiley-Interscience, 1971. ISBN: 0-471-83975-2. Cost: $20.50.

1257. Handbook of Drug Interactions. Edward A. Hartshorn. Drug Intelligence Publications, 3d ed., 1976. LC: 76-27057. ISBN: 0-914768-23-9. Cost: $5.85.

1258. Handbook of Drug Therapy. R.R. Miller; D.J. Greenblatt. Elsevier, 1979, 1,126 pp. ISBN: 0-444-00329-0. Cost: $35.00.

1259. Handbook of Drugs and Chemicals Used in the Treatment of Fish Disease. Nelson Herwig. Thomas Publishing Co, 1979, 272 pp. LC: 78-11764. ISBN: 039-803-852-X. Cost: $19.25.

1260. Handbook of Ear, Nose, and Throat Emergencies. M. Haskell Newman et al (eds). Medical Examination Publishing Co, 1973. ISBN: 0-87488-639-2. Cost: $8.00.

1261. Handbook of Electrical Noise: Measurement and Technology. Charles A. Vergers. TAB Books, 1979, 279 pp. LC: 79-14546. ISBN: 0-8306-9807-8, hard copy; 0-8306-1132-0, paper. Cost: $10.95 hard copy; $6.95 paper copy.

This is a practical approach to electrical noise and discusses specfic types of noise, noise measurement, and computation of electrical noise.

1262. Handbook of Electrical Systems Design Practices. George Nash. Prentice-Hall, Inc, 1980, 357 pp. ISBN: 0-13-633875-5. Cost: $22.95.

Covers such topics as lighting fixtures, power equipment, signalling systems, communication systems, wiring, and switch gear. Wiring of telephones, transformers, busway details and grounding are discussed. This handbook covers in detail the installation of electric utility systems.

1263. Handbook of Electroencephalography and Clincial Neurophysiology. Antoine Remond et al. Elsevier, Amsterdam (Issued by the International Federation of Societies for EEG and Clinical Neurophysiology), 1971-, multivolume set. LC: 73-80317. ISBN: 0-444-40996-3.

Index for Vols. 1-16, 110 pp., 1978, ISBN: 0-444-41678-1, $24.50.

1264. Handbook of Electromagnetic Compatibility. Donald R.J. White. Van Nostrand, 1981, 640 pp. ISBN: 0-442-29404-2. Cost: $38.50.

Covers test instrumentation and procedures with regard to electromagnetic compatibility. Military standards and specifications are included.

1265. Handbook of Electron Beam Welding. Robert Bakish; S.S. White. John Wiley and Sons, 1964, 269 pp. LC: 64-7538.

1266. Handbook of Electron Tube and Vacuum Techniques. Fred Rosebury. Addison-Wesley, 1965, 597 pp. LC: 65-7331.

1267. Handbook of Electronic Charts and Nomographs. Allan H. Lytel. Howard W. Sams, 1961. LC: 61-15779.

1268. Handbook of Electronic Charts, Graphs, and Tables. John D. Lenk. Pentice-Hall, 1970, 224 pp. ISBN: 0-13-377275-6. Cost: $17.95.

1269. Handbook of Electronic Circuit Design Analysis. Harry Thomas. Reston, 1972, 502 pp. LC: 72-85268. ISBN: 0-87909-328-5. Cost: $18.95.

1270. Handbook of Electronic Circuit Designs. John D. Lenk. Prentice-Hall, 1976, 307 pp. LC: 75-29384. ISBN: 0-13-377309-4. Cost: $16.95.

1271. Handbook of Electronic Circuits and Systems. Matthew Mandl. Reston Publishing Co, 1980, 416 pp. ISBN: 0-8359-2738-5. Cost: $21.95.

Presents such topics as detection systems, public entertainment systems, digital circuits, and power amplifiers, oscillators, and modulation systems. In addition, such topics as angular velocity, Kirchoff's law, digital circuits, and integrated circuits are covered. A handy resistor and capacitor color code section is also provided.

1272. Handbook of Electronic Circuits. R. Feinberg (ed). Barnes and Noble (division of Harper and Row), 1966, 195 pp. LC: 66-70352.

1273. Handbook of Electronic Circuits: Design, Operation, Applications. Graham J. Scoles. Halsted Press, 1975, 370 pp. LC: 74-13558. ISBN: 0-470-76715-4. Cost: $44.95.

1274. Handbook of Electronic Communications. Gary Miller. Prentice-Hall, Inc, 1979, 352 pp. LC: 78-11347. ISBN: 0-13-377374-4. Cost: $19.95.

1275. Handbook of Electronic Components and Circuits. John D. Lenk. Prentice-Hall, 1974, 216 pp. LC: 73-11038. ISBN: 0-13-377283-7. Cost: $17.95.

1276. Handbook of Electronic Control Circuits. John Markus. McGraw-Hill, 1959, 347 pp.

1277. Handbook of Electronic Design and Analysis Procedures Using Programmable Calculators. Bruce K. Murdock. Van Nostrand Reinhold, 1979, 544 pp. LC: 79-15122. ISBN: 0-442-26137-3. Cost: $29.50.

This book is designed primarily for the HP 67/97 and the TI 59 family of calculators. Techniques are provided for active filter design, network analysis, high frequency amplifier design, and engineering mathematics. In addition, these programs cover such topics as curve-fitting, elliptic integrals, Butterworth and Chebyshev filter design, and paramagnetic core inductor design.

1278. Handbook of Electronic Display Devices. Harry Thomas. Prentice-Hall, 1976. ISBN: 0-13-372631-2. Cost: $19.95.

1279. Handbook of Electronic Engineering. Leslie Ernest; Charles Hughes; F.W. Holland. Chemical Rubber Co, 3d ed., 1967, 1,532 pp. LC: 68-4144.

First to 3d edition published in London, 1958-1967, with the title: *Electronic Engineers' Reference Book.*

1280. Handbook of Electronic Formulas, Symbols and Definitions. John R. Brand. Van Nostrand Reinhold, 1979, 359 pp. LC: 78-26242. ISBN: 0-442-20999-1. Cost: $15.95.

Such topics as operational amplifiers, transistors, and passive circuits are presented in this handbook along with various formulas, symbols, and definitions. Information is provided on conductance, inductance, capacitance, band width, angular velocity, phase angles, and wave lengths. This is an excellent ready reference that will be very useful in quickly locating formulas with regard to electronics.

1281. Handbook of Electronic Instrumentation, Testing, and Troubleshooting. Vester Robinson. Reston, 1974, 358 pp. LC: 74-9657. ISBN: 0-87909-327-7. Cost: $15.95.

1282. Handbook of Electronic Instruments and Measurement Techniques. Harry Elliot Thomas; C.A. Clarke. Prentice-Hall, 1967, 398 pp. LC: 67-10754.

1283. Handbook of Electronic Materials. Electronic Properties Information Center. IFI/Plenum, 1971-1972, 9 Vols. ISBN: Vol.1, 0-306-67101-8; Vol.2, 0-306-67102-6; Vol.3, 0-306-67103-4; Vol.4, 0-306-67104-2; Vol.5, 0-306-67105-0; Vol.6, 0-306-67106-9; Vol.7, 0-306-67107-7; Vol.8, 0-306-67108-5; Vol.9, 0-306-6. Cost: $37.50 per Vol.

This handbook consists of 9 volumes with the following titles: Vol.1, *Optical Materials Properties*; Vol.2, *Semiconducting Compounds*; Vol.3, *Silicon Nitride for Microelectronic Applications*, Part I: *Preparation and Properties*; Vol.4, *Niobium Alloys and Compounds*; Vol.5, *Group IV Semiconducting Materials*; Vol.6, *Silicon Nitride for Microelectric Applications*, Part II: *Applications and Devices*; Vol.7, *Ternary Semiconducting Compounds—Data Tables*; Vol.8, *Linear Electro-optic Modulator Materials*; Vol.9, *Electronic Properties of Composite Materials.*

1284. Handbook of Electronic Meters: Theory and Application. John D. Lenk. Prentice-Hall, 2d ed., 1981, 240 pp. ISBN: 0-13-377333-7. Cost: $18.95.

Covers measurement, service, and troubleshooting procedures. Includes connection diagrams.

1285. Handbook of Electronic Packaging. Charles A. Harper. McGraw-Hill Book Co, 1969, 1,000 pp (various pagings). LC: 68-11235. ISBN: 0-07-026671-9. Cost: $37.50.

1286. Handbook of Electronic Safety Procedures. Edward A. Lacy. Prentice-Hall, 1977, 269 pp. LC: 75-44230. ISBN: 0-13-377341-8. Cost: $17.95.

This handbook includes information on safety measures and precautions to be taken when working with electronic devices. Various types of hazards are discussed including electric shock, underground hazards, explosive chemicals, and lightning. In addition, information is provided on safety precautions to be taken when working with lasers, equipment emitting radiation, and static electricty. A bibliography and index are provided.

1287. Handbook of Electronic Systems Design. Charles A. Harper. McGraw-Hill, 1980, 849 pp. LC: 79-19103. ISBN: 0-07-026683-2. Cost: $39.50.

Coverage in this handbook is devoted to such topics as communications, radar, navigation, and digital electronics. Computer systems, measurement and testing systems, and communication networks are also discussed.

1288. Handbook of Electronic Systems Design. Frank Weller. Reston (division of Prentice-Hall), 1978, 288 pp. LC: 77-10998. ISBN: 0-87909-322-6. Cost: $20.95.

This is an elementary handbook for technicians. It is primarily concerned with such topics as radio networks, telemetry, microprocessors, TV antennas, TV systems, and radar.

1289. Handbook of Electronic Tables and Formulas. Stanley Meacham; Donald Herrington. Sams, 5th ed., 1979. LC: 78-71889. ISBN: 0-672-21532-2. Cost: $9.95.

1290. Handbook of Electronic Tables. Martin Clifford. TAB Books, 2d ed., 1972, 224 pp. LC: 72-87458. ISBN: 0-8306-2125-3. Cost: $7.95.

1291. Handbook of Electronic Test Equipment. John D. Lenk. Prentice-Hall, 1971, 460 pp. LC: 78-135753. ISBN: 0-13-377366-3. Cost: $19.95.

1292. Handbook of Electronic Testing, Measurement and Troubleshooting. Matthew Mandel. Reston, 1975. ISBN: 0-87909-330-7. Cost: $16.95.

1293. Handbook of Electronics Calculations for Engineers and Technicians. Milton Kaufman; Arthur H. Seidman. McGraw-Hill, 1979. LC: 79-12387. ISBN: 0-07-033392-0. Cost: $19.50.

This handbook emphasizes practical worked-out problems on such electronic topics as audio amplifiers, semiconductor devices, oscillators, power supplies, feedback, digital logic, batteries, computer-aided design, microprocessors, microwaves, communication systems, and transmission lines. In addition, calculations are provided for such topics as antennas, thick-film technology, filters, and analog-digital conversion. An index is provided for these electronic calculations, and numerous schematic diagrams have been included to make this handbook easy to use.

1294. Handbook of Electronics Manufacturing Engineering. Bernard S. Matisoff. Van Nostrand Reinhold, 1978, 394 pp. ISBN: 0-442-25146-7. Cost: $34.50.

The emphasis of this handbook is on productivity in electronic manufacturing. Such manufacturing processes as wiring, soldering, harnesses, fabrication, safety, and production of plastic coated electronic modules are discussed.

1295. Handbook of Electronics Packaging Design and Engineering. Bernard S. Matisoff. Van Nostrand Reinhold, 1981, 400 pp. ISBN: 0-442-20171-0. Cost: $32.50.

Such topics as electronics packaging for thermal control, environmental protection, and reduction of radio and electromagnetic interference are discussed. In addition, such topics as fabrication processes, fasteners, subassemblies, miniaturization, wiring and cabling, printed circuit design, and materials are presented.

Handbook of Electrophoresis. *See* **CRC Handbook of Electrophoresis.**

1296. Handbook of Electroplating. W. Canning and Co. Span, E and FN, Ltd, 22nd rev. ed., 1978. ISBN: 0-419-10950-1.

1297. Handbook of Elemental Abundances in Meteorites: Reviews in Cosmochemistry and Allied Subjects. Brian Mason. Gordon and Breach, 1971, 555 pp. LC: 71-148927. ISBN: 0-677-14950-6 hard copy; 0-677-14955-7 paper copy. Cost: $117.25 hard copy.

1298. Handbook of Elementary Physics. N. Koshkin; M. Shirkevich. Gordon and Breach, 1965, 226 pp. ISBN: 0-677-20190-7. Cost: $41.00 microform edition only.

1299. Handbook of Elliptic Integrals for Engineers and Scientists. P.F. Byrd; M.D. Friedman. Springer-Verlag, 2d ed., 1971, 358 pp. LC: 72-146515. ISBN: 0-387-05318-2. Cost: $37.90.

Includes a collection of 3,000 formulae to facilitate the evaluation of integrals.

1300. Handbook of Emergency Care and Rescue. Lawrence W. Erven. Glencoe, rev. ed., 1976. ISBN: 0-02-472630-3. Cost: $14.95.

1301. Handbook of Emergency Toxicology: A Guide for the Identification, Diagnosis, and Treatment of Poisoning. Sidney Kaye. C.C. Thomas, 3d ed., 1973, 514 pp. LC: 71-91851. ISBN: 0-398-00987-2. Cost: $22.00.

1302. Handbook of Endocrine Function Tests in Adults and Children. Robert N. Alsever; Ronald W. Gotlin. Year Book Medical Publishers, 1975. ISBN: 0-8151-0125-2. Cost: $7.95.

1303. Handbook of Endocrinology: Diagnosis and Management of Endocrine and Metabolic Disorders. Richard S. Dillon. Lea and Febiger (also Kimpton), 1973, 556 pp. LC: 73-1948. ISBN: 0-8121-0419-6. Cost: $23.50 (£10.55).

1304. Handbook of Energy Audits. Albert Thumann; Richard Fessler. Fairmont Press, (division of Van Nostrand Reinhold), 1979, 440 pp. ISBN: 0-915586-18-5. Cost: $32.00.

This is a useful guide for conducting energy audits and systematically conserving energy. Covers numerous case studies and discusses such topics as energy auditing of electrical systems, heating systems, ventilating systems, and air conditioning systems. Different types of buildings and institutions are discussed, including hospitals, government buildings, and schools.

1305. Handbook of Energy Conservation for Mechanical Systems in Buildings. Robert W. Roose. Van Nostrand Reinhold, 1978, 592 pp. LC: 77-11027. ISBN: 0-442-27012-7. Cost: $39.95.

Includes planning, design, installation, utilization, maintenance, and modernization of heating, ventilating and air conditioning systems.

1306. Handbook of Energy Policy for Local Governments. Edward H. Allen. Lexington Books (division of D.C. Heath and Co), 1975, 236 pp. LC: 74-25074. ISBN: 0-669-97386-6. Cost: $18.95.

1307. Handbook of Energy Technology. V. Daniel Hunt. Van Nostrand Reinhold, 1981, 992 pp. ISBN: 0-442-22555-5. Cost: $49.50.

Over 600 photographs, drawings, tables, and charts are included in this handbook to help illustrate different types of energy technology. Such topics as coal, solar energy, geothermal energy, magnetic fusion, nuclear fusion, natural gas, and petroleum are explored in great detail.

Handbook of Energy Technology. *See* **Energy Technology Handbook.**

Handbook of Energy Utilization in Agriculture. *See* **CRC Handbook of Energy Utilization in Agriculture.**

1308. Handbook of Engineering Fundamentals. O.W. Eshbach; M. Souders. John Wiley and Sons, 3d ed., 1975, 1,568 pp. LC: 74-7467. ISBN: 0-471-24553-4. Date of First Edition: 1936. Cost: $39.50.

This handbook is an excellent source of information for general engineering equations, tables, theories, and formulas. It includes information on such subjects as thermodynamics, acoustics, heat transfer, properties of materials, mechanics, and electronics. This handbook provides an excellent index, a very detailed table of contents at the beginning of each section, and a collection of mathematical tables that would be useful for the practicing engineer. For related works on general engineering concepts, please refer to the following titles: *Engineering Tables and Data; The Engineers' Manual; The Engineering Manual; De Laval's Engineering Handbook,* and *CRC's Handbook of Tables for Applied Engineering Science.*

1309. Handbook of Engineering Graphics. H. Dale Walraven. McKnight and McKnight, 1965, 252 pp. LC: 65-19550.

Handbook of Engineering in Medicine and Biology. *See* **CRC Handbook of Engineering in Medicine and Biology.**

1310. Handbook of Engineering Materials. Douglas F. Miner; John B. Seastone. John Wiley and Sons, 1955, various paging. LC: 55-9366.

This handbook contains a considerable amount of information and data on engineering materials. Included is information on such subjects as nonmetal materials, metal materials, and materials for construction purposes. Quantitative information such as physical, mechanical properties, and chemical properties are included to help the engineer make a decision about what type of material to use for a particular design. This handbook includes a collection of mathematical tables, a thoroughly prepared index, and a detailed table of contents at the beginning of each section. For related works, please refer to *Materials Handling Handbook; Materials Handbook; Material Properties Handbook; Encyclopedia Handbook of Materials, Parts, and Finishes; CRC Handbook of Materials Science; Dangerous Properties of Industrial Materials;* and the *Handbook of Materials and Processes for Electronics.* Also refer to *Metals Handbook Building Construction Handbook; ASME Handbook* and *Woldman's Engineering Alloys.*

1311. Handbook of Engineering Mechanics. W. Flügge. McGraw-Hill, 1962, 1,632 pp. LC: 61-13165. ISBN: 0-07-021392-5. Cost: $75.00.

This handbook gathers together many of the diversified areas of engineering mechanics, including such topics as elasticity, statics, dynamics, vibrations, fluid flow, thermodynamics, boundary layers, turbulence, and cavitation. A major section of mathematics is provided for background information. Many formulas and tables are included to elucidate some of the more difficult concepts of mechanics. Many references are provided for further investigation, and the handbook is well-indexed. For related works, please refer to the *Handbook of Fluid Dynamics.*

1312. Handbook of Environment Management. Carlos J. Hilado. Technomic Publishing Co, 1972, 2 Vols.. LC: 74-174658. ISBN: Vol.1, 0-87762-067-9; Vol.2, 0-87762-068-7.

The *Handbook of Environment Management* consists of 2 volumes entitled as follows: Vol.1, *Fundamentals;* Vol.2, *Practice Predicting, Evaluating, and Controlling Effects of Pollution.*

1313. Handbook of Environmental Civil Engineering. Robert G. Zilly. Van Nostrand Reinhold, 1975, 1,029 pp. LC: 74-26993. ISBN: 0-442-29578-2. Cost: $42.50.

This volume is essentially a civil engineering handbook that seriously takes into condsideration the environmental consequences of civil engineering operations. Basic civil engineering subjects are covered including such subjects as surveying, photogrammetry, soil mechanics, engineering structures, and transportation engineering. In addition, such areas as water engineering, solid waste disposal, sewage treatment, and engineering management receive coverage. Many bibliographic references are provided for the civil engineering literature, and a convenient index has been prepared for easy retrieval of information. For related works, please refer to such titles as *Building Construction Handbook; Civil Engineering Handbook; Handbook of Heavy Construction; Transportation and Traffic Engineering Handbook; Standard Handbook for Civil Engineers; Highway Engineering Handbook; Civil Engineers' Reference Book; Environmental Engineers' Handbook; Handbook of Solid Waste Disposal; Pollution Engineering Practice Handbook; Betz Handbook of Industrial Water Conditioning,* and the *CRC Handbook of Environmental Control.*

1314. Handbook of Environmental Control. Richard G. Bond; Conrad P. Straub. CRC Press, Inc, Vol.1, 1973; Vol.2, 1973; Vol.3, 1973; Vol.4, 1974; Vol.5, 1975; Series Index, 1978, Vol.1, 576 pp.; Vol.2, 580 pp.; Vol.3, 835 pp.; Vol.4, 928 pp.; Vol.5, 440 pp. LC: 72-92118; 75-88823. ISBN: Entire set, 0-87819-270-0 Vol.1, 0-87819-271-9; Vol.2, 0-87819-272-7; Vol.3, 0-87819-273-5; Vol.4, 0-87819-274-3; Vol.5, 0-87819-275-1; Series Index, 0-8493-0279-X. Cost: Vol.1, $56.95; Vol.2, $56.95; Vol.3, $69.95; Vol.4, $69.95; Vol.5, $49.95; Series Index, $34.95.

This handbook is divided into five volumes as follows: Vol.1, *Air Pollution;* Vol.2, *Solid Waste;* Vol.3, *Water Supply and Treatment;* Vol.4, *Waste Water Treatment and Disposal;* Vol.5, *Hospital and Health Care Facilities.* This handbook covers the whole gamut of environmental problems and provides tabular data on such subjects as industrial sources of air pollutants, properties of air pollutants, water disinfection, reverse osmosis, landfills, composting, marine disposal of solid wastes, activated carbon treatment of sewage, septic tanks, and various waste products from industrial processes. In addition, such areas as radiation protection, toxic chemicals, and sterilization are examined, and conversion factors plus a good index are provided in each volume for more efficient usage. For related works, refer to such titles as *Handbook of Environmental Civil Engineering; Air Pollution*, and the *Handbook of Solid Waste Disposal.*

1315. Handbook of Environmental Data and Ecological Parameters. S.E. Jorgensen. Pergamon Press, 1979, 1,162 pp. LC: 78-41207. ISBN: 0-08-023436-4. Cost: $225.00. Series: *Environmental Sciences and Applications*, Vol. 6.

This comprehensive work covers such environmental data as the chemical composition of biological materials, thermodynamic quantities of biological materials, uptake rate for plants and animals, decomposition and photolysis rates for chemicals, and the rate coefficient for chemical and biochemical reactions in soil and water. Various equations are included for biological processes and also for chemical-physical processes in the environment. Exchange rates between sediment and water for phosphorus, nitrogen, and heavy metals and other compounds are also included. Stoichiometric ratios, temperature coefficients, and turnover time of processes in the ecosphere are also discussed. Bibliographic references are provided for further investigation and an excellent index is also included.

1316. Handbook of Environmental Data on Organic Chemicals. Karel Verschueren. Van Nostrand Reinhold, 1977, 659 pp. LC: 77-9401. ISBN: 0-442-29091-8. Cost: $39.95.

A listing of organic compounds providing such information as properties, air and water pollution factors, and biological effects of these chemicals on man, animals and the environment. Many *see references* are provided for alternative chemical names. An extensive bibliography containing 347 items is included at the end of the handbook.

1317. Handbook of Environmental Engineering. Lawrence K. Wang; Norman C. Pereira. Humana Press, Vol.1, 1979; Vol.2, 1980, 2 Vols., Vol.1, 512 pp. LC: Vol.1, 78-78033; Vol.2, 79-91087. ISBN: Vol.1, 0-89603-001-6; Vol.2, 0-89603-008-3. Cost: $59.50 per Vol.

There are currently 2 volumes in this set entitled as follows: Vol.1, *Air and Noise Pollution Control*; Vol.2, *Solid Waste Processing and Resource Recovery*. Such topics as air pollution, wet scrubbing, atmospheric pollution, ventilation, noise control, air conditioning, cyclones, and fabric filtration are discussed. This handbook also covers such topics as electrostatic precipitators, solid wastes, and various recycling processes.

1318. Handbook of Environmental Engineering. E.C. Theiss (Project Director). Wright-Patterson Air Force Base, 1962. LC: 69-69522. Tech Report: TR-61-363.

1319. Handbook of Environmental Health and Safety: Principles and Practices. Herman Koren. Pergamon, 1980, 2 Vols. LC: 79-28057. ISBN: 0-08-023900-5. Cost: $95.00.

This 2-volume set covers a wide variety of topics, including food protection, rodent control, solid waste management, swimming areas, air quality, and institutional environments. Four additional volumes are planned for the future to cover such topics as biological agents, chemical agents, physical agents, and resources. Although not many tables or illustrations are present in this work, it does include numerous bibliographic references to more specialized sources.

1320. Handbook of Environmental Impact Analysis: A New Dimension in Decision Making. Ravinder K. Jain; Lloyd V. Urban; Gary S. Stacey. Van Nostrand Reinhold, 1977, 320 pp. ISBN: 0-442-28807-7. Cost: $22.50.

1321. Handbook of Environmental Isotope Geochemistry. V.1 The Terrestrial Environment. P. Fritz; J.C. Fontes. Elsevier, 1980, 545 pp. LC: 79-21332. ISBN: 0-444-41780-X. Cost: $90.25.

This is the first volume in a series of 5 volumes dealing with environmental isotope geochemistry. Such topics as hydrology, aqueous geochemistry, biogeochemistry, and ground water are covered in this handbook.

1322. Handbook of Environmental Management. Carlos J. Hilado. Technomic Publications, Vol.1, 1972; Vol.2, 1973, Vol.1, 115 pp; Vol.2, 174 pp. LC: Vol.1, 74-174658; Vol.2, 74-174658. ISBN: Vol.1, 0-87762-067-9; Vol.2, 0-87762-094-6. Cost: $25.00 per Vol.

This handbook contains 2 volumes entitled as follows: Vol.1, *Fundamentals;* Vol.2, *Food for Mankind.* Such topics as pollution, population, water resources, nutrition, land use, fertilizers, irrigation, and pesticides are discussed in this handbook.

1323. Handbook of Environmental Monitoring. Frank L. Cross Jr. Technomic, 1974, 242 pp. LC: 73-86133. ISBN: 0-87762-104-7. Cost: $25.00.

Covers methods used in collection, measurement, and evaluation of environmental data.

1324. Handbook of Environmental Planning: The Social Consequences of Environmental Change. J. McEvoy III; T. Dietz. John Wiley and Sons, 1977, 323 pp. LC: 76-57239. ISBN: 0-471-58389-8. Cost: $23.50.

1325. Handbook of Enzymatic Methods of Analysis. George Guilbault. Marcel Dekker, 1977, 752 pp. ISBN: 0-8247-6425-0. Cost: $54.50.

1326. Handbook of Enzyme Biotechnology. Alan Wiseman. Halsted Press, 1975, 275 pp. LC: 75-2466. ISBN: 0-470-95617-8. Cost: $45.00.

1327. Handbook of Enzyme Electrophoresis in Human Genetics. Harry Harris. North-Holland, 1976, 306 pp. LC: 76-8848. ISBN: 0-7204-0610-2. Cost: $83.00.

1328. Handbook of Epoxy Resins. Henry Lee; Kris Neville. McGraw-Hill, 1967, 960 pp. (various paging). LC: 65-26165. ISBN: 0-07-036997-6. Cost: $55.00.

This handbook covers such topics as the chemistry of epoxy resins, the formulation of epoxy resins, curing mechanisms, and curing agents. Many applications of epoxy resins are discussed including adhesives, lamanites, chemically resistant sealants, and such areas as floor coatings, materials that resist corrosion, and various types of solvents. An index and references to the literature are provided along with many definitions of the terminology. For related works, please refer to the *Polymer Handbook.*

1329. Handbook of Essential Formulae and Data on Heat Transfer for Engineers. H.Y. Wong. Longman, 1977, 236 pp. LC: 77-5681. ISBN: 0-582-46050-6. Cost: $12.50.

Covers such properties as thermal radiation, heat transfer by convection and conduction. Covers boiling, condensation, and heat exchanges. Includes numerous tables.

1330. Handbook of Ethological Methods. Philip N. Lehner. Garland STPM Press, 1980, 419 pp. LC: 77-90468. ISBN: 0-8240-7024-0. Cost: $24.50.

This is basically a compilation of information on animal behavior including various methods of research and data collection. Statistical methods and testing are discussed, along with experimental manipulation of data, presentation and interpretation of experimental results.

1331. Handbook of Expanded Dental Auxiliary Practice. Francis Castano; Betsey Alden. Lippincott, 1973, 225 pp. ISBN: 0-397-54127-9. Cost: $11.75.

1332. Handbook of Experimental Immunology. Donald Mackay Weir. Lippincott, 3d ed., 1978, 3-Vol set. Cost: $45.25 per Vol.

Handbook of Experimental Pharmacology. New Series. *See* Handbuch der Experimentellen Pharmakologie. New Series.

1333. Handbook of Extraction Data for Metal Ions. A.M. Rozen. Pergamon Press, 1979. ISBN: 0-08-023598-0. Cost: $250.00.

1334. Handbook of Fabric Filter Technology. Charles E. Billings; John Wilder (GCA Corp). NTIS, 1970, 4 Vols. Accession: PB-200648. Tech Report: APTD-0690.

1335. Handbook of Fastening and Joining of Metal Parts. Vallory H. Laughner; Augustus D. Hargan. McGraw-Hill, 1956, 622 pp. Cost: $15.00.

This handbook covers such topics as welding, brazing, soldering, screw threads, nuts and bolts, collars, and couplings. Gas welding, arc welding, and resistance welding in ferrous and nonferrous alloys are discussed. In addition, resins, adhesives, and joint design are considered in this handbook.

1336. Handbook of Fatigue Testing. American Society for Testing and Materials, 1974, 212 pp. LC: 74-83946. Cost: $17.25. Series: *ASTM Special Technical Publication*, No. 566.

1337. Handbook of Feedstuffs: Production, Formulation, Medication. Rudolph Seiden; W.H. Pfander. Springer, 1957. LC: 57-9182. ISBN: 0-8261-0231-X. Cost: $14.95.

1338. Handbook of Fiber Optics: Theory and Applications. Helmut F. Wolf. Garland Publishing, Inc, 1979, 560 pp. LC: 78-31977. ISBN: 0-8240-7054-2. Cost: $52.50.

Covers such topics as optical detectors, optical connectors, endoscopy, and optical waveguides. Fiber optic communications is discussed as it relates to both digital and analog networks.

1339. Handbook of Fiberglass and Advanced Plastics Composites. George Lubin. Krieger (Van Nostrand Reinhold), reprint of 1969 ed., 1975, 894 pp. LC: 75-1316. ISBN: 0-88275-286-3. Date of First Edition: 1969. Cost: $45.00.

This handbook contains information on reinforced plastics, fibrous composites, and glass fibers. Sponsored by the Society of Plastics Engineers. Includes bibliographies and index. Includes filaments and high temperature resins.

1340. Handbook of Fillers and Reinforcements for Plastics. Harry S. Katz; John V. Milewski. Van Nostrand Reinhold, 1978, 652 pp. LC: 77-22335. ISBN: 0-442-25372-0-0-9. Cost: $47.50.

Topics covered include fire retardent filters, metallic filters, mineral filters, and magnetic filters. Also covers flake, reinforcements, and continuous filament reinforcements.

1341. Handbook of Filter Synthesis. Anatol I. Zverev. Wiley-Interscience, 1965, 576 pp. LC: 67-17352. ISBN: 0-471-98680-1. Cost: $48.95.

1342. Handbook of First Complex Prime Numbers. Ervand Kogbetliantz; Alice Krikorian. Gordon, 1972, 2 Vols. LC: 78-142082. ISBN: 0-677-02920-9. Cost: $133.00.

1343. Handbook of Fixture Design. F.W. Wilson (ed) (American Society of Tool and Manufacturing Engineers). McGraw-Hill, 1962, 479 pp. LC: 61-14046. ISBN: 0-07-001527-9. Cost: $32.50.

Includes information on manufacturing operations such as design data on fixtures for milling, drilling, boring, turning, and sawing, shaping, planing, and grinding.

1344. Handbook of Flame Spectroscopy. M.L. Parsons et al. Plenum Press, 1975, 478 pp. LC: 75-17865. ISBN: 0-306-30856-X. Cost: $42.50.

1345. Handbook of Flammability Regulations. Carlos J. Hilado. Technomic Publishing Co, 1975, 388 pp. LC: 75-7498. ISBN: 0-87762-159-4. Cost: $9.95 (priced reduced from earlier edition).

This handbook provides information on the various flammability codes that have been prepared by such organizations as the National Fire Protection Association, The International Conference of Building Officials, and The American Insurance Association. In addition, information is presented on the regulations of the various states in the United States as well as federal regulations that have been issued by such bodies as the Consumer Product Safety Commission. Even though this book does not list all the flammability regulations that exist, it certainly does present the most important regulations and codes, and will help the user of this handbook understand the complicated and overlapping network of regulations in the United States with regard to flammability and fire safety. For related works, please refer to the *Fire Protection Handbook,* the *Flammability Handbook for Plastics; Flammability Test Methods Handbook,* and the *National Fire Codes.*

1346. Handbook of Fluid Dynamics. Victor L. Streeter. McGraw-Hill, 1961, 1,228 pp. (various paging). LC: 60-13774. ISBN: 0-07-062178-0. Cost: $45.00.

This handbook gathers together the traditional fields of fluid mechanics including such topics as laminar flow, compressible flow, cavitation, and open-channel flow. Information on flow measurements is discussed along with a section on calculations of fluid flow by the use of a computer. References are generally provided at the end of each section so that the reader can obtain more information on the subject. Flow calculations and information are also provided in an article about magnetohydrodynamics. For related works, please refer to the *Handbook of Engineering Mechanics.*

1347. Handbook of Fluorescence Spectra of Aromatic Molecules. Isadore B. Berlman. Academic Press, 2d ed., 1971, 473 pp. LC: 78-154388. ISBN: 0-12-092656-3. Cost: $55.00.

1348. Handbook of Foamed Plastics. Rene J. Bender. Lake Publishing Co, 1965, 339 pp. LC: 65-13127.

1349. Handbook of Forensic Pathology. Abdullah Fatteh. Lippincott, 1973. LC: 73-14529. ISBN: 0-397-50328-8. Cost: $22.00.

1350. Handbook of Forms, Charts and Tables for the Construction Industry. John Beckmann. Prentice-Hall, Inc, 1979, 240 pp. ISBN: 0-13-378059-7. Cost: $32.95.

Over 100 forms, charts, and tables are presented to assist the user in the process of planning any type of building construction, from highways to parking structures.

1351. Handbook of Formulas for Stress and Strain. William Griffel. Ungar Publishing Co, 1966, 418 pp. LC: 66-17539. ISBN: 0-8044-4332-7. Cost: $28.00.

1352. Handbook of Formulas for the Analysis of Complex Frames and Arches. G.S. Glushkov et al. Davey (Israel Program for Scientific Translations), 1967, 351 pp. LC: 67-9267.

This work has been translated from the Russian.

1353. Handbook of Fractures. Edgar L. Ralston. Mosby, 1967. LC: 67-26299. ISBN: 0-8016-4074-1. Cost: $15.75.

1354. Handbook of Fresh Water Fishery Biology. Kenneth Dixon Carlander. Iowa State University Press, 3d facsimile ed., 1969, 1976, 2 Vols. LC: 69-18736. ISBN: Vol.1, 0-8138-2335-8; Vol.2, 0-8138-0670-4. Cost: Vol.1, $18.75; Vol.2, $19.50.

1355. Handbook of Fuel Cell Technology. Carl Berger. Prentice-Hall, 1968, 607 pp. ISBN: 0-13-378133-X. Cost: $22.95.

1356. Handbook of Fundamental Nursing Techniques. Mildred L. Montag; Alice R. Rines. John Wiley and Sons, 1976, 111 pp. LC: 76-5246. ISBN: 0-471-01475-3. Cost: $8.50.

1357. Handbook of Gas Laser Experiments. Grover Lee Rogers. Iliffe Science and Technology Publishers, Ltd, 1970, 67 pp. LC: 71-533905. ISBN: 0-592-05081-5.

1358. Handbook of Gasifiers and Gas Treatment Systems. Dravo Corp (available from NTIS), 1976, 167 pp.

1359. Handbook of Gemstone Identification. Richard T. Liddicoat Jr. Gemological Institute of America, 10th ed., 1975, 426 pp. LC: 66-6043. ISBN: 0-87311-006-4. Cost: $14.75.

1360. Handbook of General Experimental Psychology. Carl A. Murchison. Russell and Russell, 1969. LC: 70-77679. ISBN: 0-8462-1352-4. Cost: $32.50.

1361. Handbook of General Surgical Emergencies. Edward H. Sharp, 1977, 174 pp. LC: 76-23958. Cost: $13.30 (Australian).

1362. Handbook of Generalized Gas Dynamics. Robert P. Benedict; William G. Steltz. IFI/Plenum Press, 1966, 243 pp. LC: 65-25128. ISBN: 0-306-65118-1. Cost: $52.50.

1363. Handbook of Genetics. Robert C. King. Plenum Press, 1974-76, 5 Vols. LC: 76-64694. ISBN: Vol.1, 0-306-37611-3; Vol.2, 0-306-37612-1; Vol.3, 0-306-37613-X; Vol.4, 0-306-37614-8; Vol.5, 0-306-37615-6. Cost: Vol.1, $49.50; Vol.2, $49.50; Vols.3 and 4, $49.50; Vol.5, $49.50.

This handbook consists of 5 volumes with the following titles: Vol.1, *Bacteria, Bacteriophages, and Fungi*; Vol.2, *Plants, Plant Viruses, and Protists*; Vol.3, *Invertebrates of Genetic Interest*; Vol.4, *Vertebrates of Genetic Interest*; Vol.5, *Molecular Genetics*. Bibliographies and indexes are included.

1364. Handbook of Geochemistry. K.H. Wedepohl et al. Springer-Verlag, 1969-, loose-leaf binding, multiple volumes and parts. LC: 78-85402. ISBN: Vols.1 and 2, Part 1, 0-387-04512-0; Vol.2, Part 2, 0-387-04840-5; Vol.2, Part 3, 0-387-05125-2; Vol.2, Part 4, 0-387-06879-1; Vol.2, Part 5, 0-387-09022-3. Cost: By Subscription; Vol.2, Part 4, $174.60; Vol.2, Part 5, $398.50.

1365. Handbook of Geophysics and Space Environments. Shea L. Valley (U.S. Air Force, Cambridge Research Laboratory). McGraw-Hill (available from NTIS as AD-A056800), 1966, 692 pp. LC: 65-28130.

1366. Handbook of Geriatric Nutrition. Jeng M. Hsu; Robert L. Davis. Noyes Data Corp, 1981, 372 pp. ISBN: 0-8155-0880-8. Cost: $39.00.

The main emphasis of this handbook is on nutrition and diet as they affect the aging process. Such topics as protein nutrition, cholesterol metabolism, arteriosclerosis, and the effects of various vitamins on the aging process are discussed. In addition, such topics as the current status of zinc, selenium, chromium, and copper on the aging process is presented, along with fluoride metabolism, and the need for magnesium, phosphorus, and calcium.

1367. Handbook of Glass Manufacture: A Book of Reference for the Plant Executive, Technologist, and Engineer. Fay VaNisle Tooley. Books for Industry, rev. ed., 1974, 2 Vols.. LC: 74-77520. Cost: $125.00.

This 2-volume set is a comprehensive collection of data on glass manufacture and consists of many tables, diagrams, and charts that help to present the most authoritative data in all areas relating to glass manufacture. Such topics as electric melting of glass, the raw materials used in glass manufacture, refractories, and various types of instrumentation are discussed in this handbook. In addition, the physical properties of glass are included as well as sections on glass blowing, annealing, tempering, and chemical analysis of glass. An index is provided for easy retrieval of the data. For related titles, please refer to *Glass Engineering Handbook*.

1368. Handbook of Graphic Presentation. C.F. Schmid; S.E. Schmid. McGraw-Hill, 2d ed., 1979, 280 pp. LC: 78-13689. ISBN: 0-471-04724-4. Cost: $11.95.

Describes how graphs are used for various specialities. Includes information on rectilinear coordinate charts, bar charts, column charts, semilogarithmic charts, ratio charts, frequency graphs, statistical maps, and pictorial charts. Also discusses computer-aided graphics.

1369. Handbook of Gynecologic Emergencies. Cynthia W. Cooke. Medical Examination Publishing Co, 1975. ISBN: 0-87488-640-6. Cost: $8.00.

1370. Handbook of Hardness Data. A.A. Ivan'ko. International Scholarly Book Service (also available from NTIS), 1971, 70 pp. ISBN: 0-7065-1166-2. Accession: TT-50177. Cost: $10.50.

1371. Handbook of Hazardous Waste Management. Amir A. Metry. Technomic Publishing Co, 1980, 446 pp. LC: 79-92717. Cost: $45.00.

This handbook covers the full range of hazardous waste management, including such topics as chemical landfills, the incineration of hazardous waste, transporting hazardous waste, and the physical treatment of hazardous waste. For example, filtration, evaporation, freeze-crystallization, distillation, precipitation, flocculation, sedimentation, reverse osmosis, and neutralization are discussed. In addition, chemical and biological treatment of hazardous waste is presented, including such methods as photolysis, hydrolysis, chemical oxidation, liquid-solvent extraction, activated sludge process, aerated lagoon process, waste stabilization ponds, and anaerobic treatment process are discussed. Over 180 tables and 103 field charts, graphs, and diagrams are used to present this data on hazardous waste management.

1372. Handbook of Hearing Aid Measurement 1980. U.S. Government Publication, Superintendent of Documents, U.S. Government Printing Office, 1980, 212 pp. SuDoc: VA 1.22:11-69. Cost: $6.00.

This handbook evaluates numerous hearing aids that have been submitted to the Veteran's Administration for testing. Such testing data as harmonic distortion, equivalent input noise level, frequency response, sound pressure level, one-kilohertz gain, and battery drain are provided.

1373. Handbook of Heart Disease, Blood Pressure and Strokes. C. Anthony D'Alonzo. Macmillan, 1962. Cost: $0.95 paper.

1374. Handbook of Heat Transfer Media. P.L. Geiringer. Reinhold, 1962, 256 pp. LC: 62-18017. ISBN: 0-88275-498-X. Cost: $18.50.

1375. Handbook of Heat Transfer. Warren M. Rohsenow; James P. Hartnett. McGraw-Hill, 1973, 1,200 pp. (various paging). LC: 72-11529. ISBN: 0-07-053576-0. Cost: $54.50.

This handbook contains the necessary information that is required for a full understanding of the various concepts in the field of heat transfer. It covers such topics as convection, 2-phase flow, conduction, radiation, and condensation as they relate to heat transfer problems. Many solutions to heat transfer problems are presented and discussed. One of the highlights of this handbook is its collection of thermophysical properties. This work is well-indexed and includes many references to the literature. For related works, please refer to the *Heat Transfer and Fluid Flow Data Book.*

1376. Handbook of Heating, Ventilating and Air Conditioning. John Porges. Newnes Butterworths (also Transatlantic), 7th ed., 1977, various paging. ISBN: 0-408-00233-6. Date of First Edition: 1942. Cost: $36.00.

This handbook has been designed primarily for the practicing engineer who is likely to consult the data tables and charts on a daily basis. In fact, the purpose of the handbook is stated very clearly in the subtitle: *Useful Tables and Data Arranged in a Manner Convenient for Ready Reference.* The 7th edition includes data both in the imperial units and SI units. It provides information on such subjects as air and steam properties, hy-draulics, radiators, tubes and their dimensions, heat loss, and heat transfer. Various BSI standards are included in connection with heating and ventilating systems. An index has been provided, and a comprehensive bibliography has been prepared for further investigation. For related works, please refer to the following titles: *ASHRAE Handbook and Product Directory; ASHRAE Handbook of Fundamentals; IHVE Guide; Pump Handbook; ASPE 1971 Data Book; Mechanical Engineer's Reference Book; Kent's Mechanical Engineers' Handbook*, and the *Standard Handbook for Mechanical Engineers.*

1377. Handbook of Heavy Construction. John A. Havers; Frank W. Stubbs Jr. McGraw-Hill, 2d ed., 1971, 1,200 pp. (various paging). LC: 77-107297. ISBN: 0-07-027278-6. Date of First Edition: 1959. Cost: $58.50.

This handbook contains information on the different types of equipment that are used in heavy construction work such as bulldozers, scrapers, cranes, and loaders. It also contains information on different types of heavy construction such as laying pipelines and building cofferdams. Tunneling, the diversion of rivers at construction sites, and steel and timber construction are also discussed in this well-indexed one-volume reference source.

1378. Handbook of HeI Photoelectron Spectra of Fundamental Organic Compounds. K. Kimura; S. Katsumata; Y. Achiba; T. Yamazaki; S. Iwata. Halsted Press (division of John Wiley and Sons, Inc), 1981, 250 pp. LC: 81-6449. ISBN: 0-470-27200-7. Cost: $39.95.

Contains approximately 200 HeI photoelectron spectra of fundamental organic compounds.

1379. Handbook of Hematological and Blood Transfusion Technique. John William Delaney; George Garratty. Appleton (also Butterworths), 2d ed., 1969, 422 pp. LC: 70-8647. ISBN: 0-407-72851-1. Cost: $13.65.

1380. Handbook of Hematology. Harold Downey. Hafner, (rep. of 1938 ed.), 1965, 4-Vol set. ISBN: 0-02-843980-5. Cost: $165.00.

1381. Handbook of Hemophilia. K.M. Brinkhous; H.C. Hemker. Excerpta Medica, (American Elsevier), 1975, 2 Vols., 927 pp. LC: 74-21851. ISBN: 90-219-2098-0. Cost: $228.25.

Handbook of High Resolution Infrared Laboratory Spectra of Atmospheric Interest. *See* **CRC Handbook of High Resolution Infrared Laboratory Spectra of Atmospheric Interest.**

1382. Handbook of High Resolution Multinuclear NMR. C. Brevard; P. Granger. John Wiley and Sons, 1981, 240 pp. LC: 81-8603. ISBN: 0-471-06323-1. Cost: $22.50.

This handbook provides information for conducting NMR experiments on most of the elements in the periodic table. Covers such topics as relaxing time, NOE effects, and acquisition parameters.

1383. Handbook of High Vacuum Engineering. H.A. Steinherz. Van Nostrand Reinhold (available from Krieger), 1963, 368 pp. ISBN: 0-442-15513-1. Cost: $17.50.

1384. Handbook of Highway Engineering. Robert F. Baker et al. Van Nostrand Reinhold, 1975, 904 pp. LC: 74-23226. ISBN: 0-442-20520-1. Cost: $47.50.

This handbook gathers together a great deal of information on highway engineering and construction including such topics as drainage, the design of intersections, the location of the highways, and the construction of bridges and tunnels. In addition it covers such topics as soil characteristics in foundations, subgrades, the construction and design of pavements, and the removal of snow and ice from highway surfaces. This handbook is well-indexed, includes many references to publications and liberally cites publications issued by the Transportation Research Board, and the American Association of State Highway and Transportation Officials and other publications of interest to those who would like to make a further study of the subjects. For related works, please refer to *Highway Engineering Handbook; Transportation and Traffic Engineering Handbook,* and *Highway Capacity Manual.*

1385. Handbook of Histology. Karl Amos Stiles. McGraw-Hill, 5th ed., 1968, 251 pp. LC: 68-13101. ISBN: 0-07-061426-1. Cost: $8.95.

1386. Handbook of Histopathological and Histochemical Techniques. Charles Frederick Albert Culling. Butterworth and Co, 3d ed., 1974, 712 pp. LC: 74-189627. ISBN: 0-407-72901-1. Cost: £8.50 ($29.75).

1387. Handbook of Home Remodeling and Improvement. LeRoy O. Anderson. Van Nostrand Reinhold, 1978, 320 pp. ISBN: 0-442-20343-8. Cost: $14.95.

Includes information on material selection, thermal insulation, sound insulation, ventilation, and painting. Tabular data are presented on properties and uses of different types of woods, proper sizes of beams, and spans for joists and rafters, as well as insulating values for certain types of materials.

Handbook of Hospital Acquired Infections. *See* **CRC Handbook of Hospital Acquired Infections.**

1388. Handbook of Hospital Psychiatry: A Practical Guide to Therapy. Louis Linn. International Universities Press. LC: 55-8237. ISBN: 0-8236-2300-9. Cost: $22.50.

Handbook of Hospital Safety. *See* **CRC Handbook of Hospital Safety.**

1389. Handbook of Housing Systems for Designers and Developers. Laurence Stephen Cutler; Sherrie Stephen Cutler. Van Nostrand Reinhold, 1974. ISBN: 0-442-21820-6. Cost: $19.95.

1390. Handbook of Human Embryology. Richard Wheeler Haines; Ahmed Mohiuddin. Williams and Wilkins (Longman), 5th ed., 1973, 260 pp. ISBN: 0-443-00957-0. Cost: $6.00.

1391. Handbook of Human Nutritional Requirements. Unipub. Unipub, 1974.

1392. Handbook of Hydraulics, for the Solution of Hydraulic Engineering Problems. Ernest F. Brater; Horace W. King. McGraw-Hill, 6th ed., 1976, 584 pp. LC: 76-6486. ISBN: 0-07-007243-4. Date of First Edition: 1918. Cost: $31.50.

This handbook contains a vast collection of tables, equations, formulas, charts, and solutions to hydraulic engineering problems. Included are discussions of such topics as the properties of fluids, the flow of fluids through closed and open channels,

and various means of measuring the flow of water by the use of different types of meters. Conversion factors are provided along with information on the metric units in this well-indexed source which also contains many references to the literature of hydraulic engineering. For related works, please refer to the *Handbook of Applied Hydraulics.*

1393. Handbook of Hydrocarbons. Seymour W. Ferris. Academic Press, 1955. ISBN: 0-12-254050-5. Cost: $35.50.

1394. Handbook of Hypergeometric Integrals: Theory, Applications, Tables, Computer Programs. H. Exton. John Wiley and Sons, 1978, 316 pp. LC: 78-40120. ISBN: 0-470-26342-3. Cost: $59.95. Series: *Mathematics and Its Applications Series.*

This handbook has an excellent listing of over 700 integrals and over 50 computer programs in Fortran IV.

1395. Handbook of Hypnosis and Psychosomatic Medicine. G.D. Burrows; L. Dennerstein. Elsevier, 1980, 554 pp. ISBN: 0-444-80148-0. Cost: $83.00.

1396. Handbook of I.V. Additive Reviews. Donald E. Francke. Drug Intelligence Publications, 1973. ISBN: 0-914768-06-9. Cost: $3.75.

1397. Handbook of Illinois Stratigraphy. H.B. Willman et al. Illinois State Geological Survey, 1975, 261 pp. LC: 76-620818. Series: *Illinois State Geological Survey Bulletin,* No. 95.

Bibliography on pages 240-251; also includes index.

Handbook of Immunology in Aging *See* **CRC Handbook of Immunology in Aging.**

1398. Handbook of Industrial Chemistry. Emil Raymond Riegel; James A. Kent (ed). Van Nostrand Reinhold, 7th ed., 1974, 902 pp. LC: 73-14798. ISBN: 0-442-24347-2. Cost: $38.50.

First through fifth editions entitled *Industrial Chemistry;* sixth edition was entitled *Riegel's Industrial Chemistry.*

1399. Handbook of Industrial Control Computers. Thomas J. Harrison. Wiley-Interscience, 1972, 1,056 pp. LC: 72-68. ISBN: 0-471-35560-7. Cost: $59.95.

1400. Handbook of Industrial Electronic Circuits. John Markus; Vin Zeluff. McGraw-Hill, 1948, 272 pp.

1401. Handbook of Industrial Electroplating. Eric A. Ollard; E.B. Smith. American Elsevier, 3d ed., 1964, 400 pp. LC: 64-22352.

1402. Handbook of Industrial Energy Analysis. I. Boustead; G.F. Hancock. John Wiley and Sons, 1979, 422 pp. ISBN: 0-470-26492-6. Cost: $69.95.

This handbook shows the various methods of analysis that are required to make energy calculations to determine how much energy is consumed in different manufacturing processes. Part 2 of this handbook is a listing of tables showing the energy requirements for several industrial processes. Such topics as coke production, natural gas production, and how much energy is required to produce a particular fuel are discussed. In addition, such topics as naphtha cracking, energy requirements in transportation, energy requirements in recycling processes, and energy requirements in metals production are presented.

1403. Handbook of Industrial Engineering and Management. William Grant Ireson; Eugene L. Grant. Prentice-Hall, 2d ed., 1970, 907 pp. LC: 71-139954. ISBN: 0-13-378463-0. Cost: $32.50.

1404. Handbook of Industrial Engineering. Gavriel Salvendy. John Wiley and Sons, Inc, 1981, 2,128 pp. LC: 81-23059. ISBN: 0-471-05841-6. Cost: $64.95.

Such industrial engineering topics as ergonomics, manufacturing engineering, engineering economy, job design, and quality assurance are covered in this handbook. In addition, such topics as industrial products, operations research, and optimization in industrial engineering are discussed in this publication.

1405. Handbook of Industrial Fasteners. Trade and Technical Press (available from Renouf USA Inc), 1st ed., 1975, approx. 700 pp. ISBN: 0-85461-0626. Cost: $99.00.

Covers both mechanical and adhesive fastening techniques. Information is provided on the selection and application of different types of fasteners and joining methods. Such topics as machine screws, lock washers, self-tapping screws, circlips and retaining rings, clevises, pivot fasteners, rivots, staples, metal stitching, spring fasteners, strap fasteners, fabric fasteners, and different types of adhesive fasteners and chemical bonding techniques are presented in this handbook.

1406. Handbook of Industrial Fire Protection and Security. Trade and Technical Press. Renouf USA Inc, 1st ed., 1976, 600 pp. ISBN: 0-85461-059-6. Cost: $83.00.

This handbook covers topics related to fire protection, such as fire resistant materials, noncombustible materials, plastics in fires, electrical fires, hazardous liquids, and hazardous environments. In addition, such topics as fire extingushers, eye protection, fire alarms, fire detection systems, heat detectors, smoke detectors, and carbon dioxide installations are discussed. Emergency lighting and power, communication equipment, fire ventilation, and rescue equipment are also presented.

1407. Handbook of Industrial Gas Utilization: Engineering Principles. R. Pritchard; J.J. Guy; N.E. Connor. Van Nostrand Reinhold, 1977, 771 pp. LC: 77-22278. ISBN: 0-442-26635-9. Cost: $42.50.

Discusses theoretical aspects of gas uses and their application to such processes as air conditioning, metal melting, steam raising, and ceramic firing.

1408. Handbook of Industrial Hearing Conservation. C.E. Kline. Fairmont Press, 1976. ISBN: 0-915586-03-7. Cost: $24.50.

1409. Handbook of Industrial Infrared Analysis. Robert G. White. Plenum Press, 1964, 440 pp. LC: 64-7506. ISBN: 0-306-30174-1. Cost: $49.50.

1410. Handbook of Industrial Loss Prevention. Factory Mutual System. McGraw-Hill, 2d ed., 1967, 912 pp. LC: 66-23623. ISBN: 0-07-019888-8. Cost: $36.85.

1411. Handbook of Industrial Materials. Trade and Technical Press (available from Renouf USA Inc), 1st ed., 1977, 689 pp. ISBN: 0-85461-060-X. Cost: $98.00.

The materials in this handbook are grouped into categories such as ferrous materials, nonferrous metals and alloys, nonmetallic materials, plastics, and thermoplastics. Under each material there is a description and a listing of the material's corresponding properties. For example, under glass, there is a listing of the different types of glass and such properties as light transmission, sound transmission, thermal transmission, mechanical properties of glass, hardness of glass, modulus of elasticity, steady state thermal stresses, resistance to corrosion, viscosity and selection of glasses for specific purposes. Under glycerine, there is a description of the material and a listing of properties, such as vapor pressure, freezing point, refractive index, viscosity, latent heat of vaporization, specific heat, flash point, and heat of fusion. This is an excellent guide to the properties of different types of materials and should be very easy to use because of its encyclopedic format and its thorough indexing.

1412. Handbook of Industrial Metrology; A Reference Book on Principles, Techniques, and Instrumentation Design and Application for Physical Measurements in the Manufacturing Industries. John W. Greve (ed) (American Society of Tool and Manufacturing Engineers). Prentice-Hall, 1967, 492 pp. LC: 67-12084. ISBN: 0-13-378505-X. Cost: $22.95.

This handbook contains information on the different types of measurements that are made in manufacturing and production situations. It includes a discussion of the various types of measuring devices and tools that are used to make these precision measurements. The handbook contains a discussion of such measuring devices as gage blocks, sinebars, air gages, interferometers, density gages, jig transits, and devices used to make radiological measurements. Various tables, graphs, and equations are used to help convey some of the most complex concepts in metrology. In addition, the handbook discusses such topics as gear measurements and the measurement of different types of surface textures as well as screw thread gaging methods. The handbook includes an index to this wealth of information on measuring devices and techniques. For related works, please refer to such titles as the *Instrument Engineers' Handbook,* the *Handbook of Applied Instrumentation,* and *Standards and Practices for Instrumentation.*

1413. Handbook of Industrial Noise Control. L.L. Faulkner. Industrial Press, 1976, 584 pp. LC: 75-41315. ISBN: 0-8311-1110-0. Cost: $43.00.

This handbook is a collection of articles devoted primarily to the practical applications of engineering noise control. It discusses the different sources of noise such as vibration, machine element noise, combustion, fluid flow, hydraulic or process air supplies. A number of case histories are provided in order to see how the information in this handbook can be applied to specific noise problems. Many noise measuring devices are discussed in addition to noise reduction methods such as vibration control and damping materials. This handbook includes bibliograhical references and a thoroughly prepared index. For related works, please refer to the following titles: *Machinery's Handbook; ASTM Standards in Building Codes; The Lubrication Engineer's Manual; Standard Handbook for Mechanical Engineers; Shock and Vibration Handbook; Handbook of Environmental Civil Engineering; Environmental Engineers' Handbook; Handbook of Industrial Noise Management,* and *Building Construction Handbook.*

1414. Handbook of Industrial Noise Management. Richard K. Miller. Fairmont Press, 1976. ISBN: 0-915586-02-9. Cost: $24.95.

This handbook is concerned primarily with the management aspects of industrial noise control and the decisions that are necessary to provide a safe working environment for employees. The book contains information about acoustical materials, costs of various types of equipment and various tables that would help management make decisions about the particular types of equipment and materials to purchase. The handbook also mentions the many OSHA regulations and requirements that need to be met in order to provide a good working environment. In addition, such information as protective devices for the ears, hearing damage, and audiometric testing are mentioned. For related titles, please refer to such handbooks as the *Handbook of Industrial Noise Control Machinery's Handbook; ASTM Standards in Building Codes; Lubrication Engineer's Manual; Standard Handbook for Mechanical Engineers; Shock and Vibration Handbook; Handbook of Environmental Civil Engineering; Environmental Engineers' Handbook*, and *Building Construction Handbook*.

1415. Handbook of Industrial Pipework Engineering. Ernest Holmes. Halsted Press (also McGraw-Hill), 1974, 570 pp. LC: 73-21538. ISBN: 0-470-40769-7; 0-07-084427-5 (McGraw-Hill). Cost: $44.95.

1416. Handbook of Industrial Research Management. Carl Heyel. Van Nostrand Reinhold, 2d ed., 1968, 562 pp. LC: 68-25155. ISBN: 0-442-15613-8. Cost: $18.95.

1417. Handbook of Industrial Safety and Health. Trade and Technical Press, Ltd, 1980. ISBN: 0-85461-075-8. Cost: $108.00.

Covers such topics as electrical safety, lighting, laser safety, pressure vessels, boilers, noise and vibration control, hearing conservation, asbestos, ionizing radiation, gas detection, temperature and pressure monitoring. In addition, this publication covers eye protection, face and body shields, handling of materials, lifting equipment, transportation of toxic waste, fire, explosions, inflammable substances, and breathing and rescue apparatus.

1418. Handbook of Industrial Textile. E.R. Kaswell; Wellington Sears (ed). Textile Book Service, 1963. ISBN: 0-87245-160-7. Cost: $30.00.

1419. Handbook of Industrial Toxicology. E.R. Plunkett. Chemical Publishing Co, 2d ed., 1976, 537 pp. ISBN: 0-8206-0201-9. Cost: $40.00.

This handbook contains information on the toxicity of various industrial chemicals, drugs, and other types of compounds. It is arranged something like a dictionary, and each entry including such information as the chemical name and any synonyms that may exist. A brief description, threshold limit values and information on the types of poisoning from these chemicals are provided including symptoms, treatment, and various types of tests. An excellent bibliography has been prepared for the user who would like to make a further investigation into the various types of chemicals. For related works, please refer to *Dangerous Properties of Industrial Materials* by N. Irving Sax.

1420. Handbook of Industrial Wastes Pretreatment. Jon C. Dyer; Arnold S. Vernick; Howard D. Feiler.. Garland Press, 1981, 262 pp. LC: 79-25702. ISBN: 0-8240-7066-6. Cost: $32.50.

Presents the regulations associated with the Clean Water Act of 1977 and the General Pretreatment Regulation of 1978 and discusses standards and methods for getting in compliance with these reguations. Such topics as sludge disposal, wastewater treatment, industrial waste control, and elements of pretreatment are presented.

1421. Handbook of Infant Development. J.D. Osofsky. John Wiley and Sons, 1979, 937 pp. LC: 78-17605. ISBN: 0-471-65703-4. Cost: $37.50. Series: *Wiley Series on Personality Processes.*

Covers prenatal, perinatal, and neonatal development.

1422. Handbook of Infrared Radiation From Combustion Gases. C.B. Ludwig et al (NASA). NTIS, 1973, 486 pp. LC: 73-602407. SuDoc: NAS-1.21:3080. Cost: $6.00. Series: NASA SP-3080.

Includes bibliographical references.

1423. Handbook of Innovative Psychotherapies. R.J. Corsini. Wiley-Interscience, 1981, 1,016 pp. LC: 80-29062. ISBN: 0-471-06229-4. Cost: $40.00. Series: *Wiley Series on Personality Processes.*

Such topics as biofeedback, conditioned reflex, aqua-energetics, and autogenic training are discussed in this publication.

Handbook of Inorganic Electrochemistry. *See* **CRC Handbook Series in Inorganic Electrochemistry.**

1424. Handbook of Institutional Pharmacy Practice. Mickey C. Smith; T.R. Brown. Williams and Wilkins, 1979, 693 pp. LC: 78-16182. Cost: $27.95.

Discusses the administration and management of pharmacies in institutions such as hospitals. Information is provided on such topics as standards of practice and organization of services.

1425. Handbook of Instrumentation and Controls: A Practical Manual for the Mechanical Services. H. Kallen. McGraw-Hill, 1961, 750 pp. LC: 60-6886. ISBN: 0-07-033235-5. Cost: $55.00.

1426. Handbook of Instruments and Instrumentation. Trade and Technical Press. Renouf USA Inc, 1st ed., 1977, 650 pp. ISBN: 0-85461-064-2. Cost: $108.00.

Provides information on different types of measuring techniques and dimensioning devices. Such topics as linear measurement, angular measurement, measurement of temperature and pressure, computer-based measurement, and screw thread measurement are presented. In addition, such topics as microscopy, transducers, gears, digital meters, volt meters, ammeters, electrical meters, potentiometers, telemeters, oscilloscopes, signal generators, and photometers are discussed. Methods of measuring fluids, liquids, noise, migration, and time are also presented.

1427. Handbook of Integer Sequences. N.J. Sloans. Academic Press, 1974, 206 pp. LC: 72-82647. ISBN: 0-12-648550-X. Cost: $21.00.

1428. Handbook of Integrated Circuits, Equivalents, and Substitutes. Bernard Baruch Babani. Hayden, 1976. Cost: $2.95 paper copy.

1429. Handbook of Integrated Circuits. Harry E. Thomas. Prentice-Hall, 1971, 346 pp. LC: 77-146681. ISBN: 0-13-378687-0. Cost: $15.65.

1430. Handbook of Integrated Circuits: For Engineers and Technicians. John D. Lenk. Reston (division of Prentice-Hall), 1978, 480 pp. LC: 78-7260. ISBN: 0-8359-2744-X. Cost: $18.95.

Digital integrated circuits are presented in this handbook along with such topics as operational amplifiers, IC audio amplifiers, operational transconductance amplifiers, and effects of temperature extremes on integrated circuits.

1431. Handbook of Integrated-Circuit Operational Amplifiers. George B. Rutkowski. Prentice-Hall, 1975, 321 pp. LC: 74-13332. ISBN: 0-13-378703-6. Cost: $18.95.

1432. Handbook of Interactive Computer Terminals. Duane E. Sharp. Reston Publishing Co, 1977, 266 pp. LC: 77-1377. ISBN: 0-87909-331-5. Cost: $19.95.

Includes an index and bibliography.

1433. Handbook of Intermediary Metabolism of Aromatic Compounds. B.L. Goodwin. Halsted Press, 1976. LC: 76-47. ISBN: 0-470-15026-2. Cost: $75.00.

1434. Handbook of International Alloy Compositions and Designations. Vol.1 Titanium. Vol.2 Superalloys. NTIS (Metals and Ceramics Information Center), Vol.1, 1976; Vol.2, 1979, Vol.1, 241 pp.; Vol.2, 405 pp. Accession: Vol.1, N77-29297 AD-A036433; Vol.2, AD-A065 740. Cost: Vol.1, $31.00; Vol.2, $40.00.

Interrelates the alloy designations and standards of one country with the alloy designations, compositions, and standards of other countries for titanium materials.

1435. Handbook of International Food Regulatory Toxicology. Gaston Vetorrazzi. SP Medical and Scientific Books (division of Spectrum Publications Inc), 1980, 161 pp. LC: 79-24008. Cost: $30.00.

Discusses such topics as food additives, pesticides, and various other experimental findings associated with food toxicology.

1436. Handbook of Interstitial Brachytherapy. Basil Hilaris et al. Publishing Science Group, 1975. LC: 75-4043. ISBN: 0-88416-027-0. Cost: $18.00.

1437. Handbook of Investigative Hypnosis. Martin Reiser. Lehi Publishing Co, 1980, 275 pp. LC: 79-53215. ISBN: 0-0934486-00-X. Cost: $24.95.

Presents information on the usage of hypnosis as a technique of obtaining information from victims and witnesses. Presents hypnosis as a method of improving memory to be used in the context of the criminal justice system. Such topics as memory, amnesia, and drugs are discussed.

1438. Handbook of Iron Meteorites: Their History, Distribution, Composition and Structure. Vagn F. Bachwald. University of California Press, 1976, 3 Vols., 1,418 pp. LC: 74-27286. ISBN: 0-520-02934-8. Cost: $205.00.

Includes index and cited literature.

Handbook of Irrigation Technology. *See* **CRC Handbook of Irrigation Technology.**

1439. Handbook of Key Federal Regulations and Criteria for Multimedia Environmental Control. D.R. Greenwood; G.L. Kingsbury; J.G. Cleland. NTIS, 1979, 271 pp. Accession: PB 80-107998.

1440. Handbook of Kidney Nomenclature International Committee for Nomenclature and Nosology of Renal Disease. Little, Brown & Co, 1975. LC: 73-17665. ISBN: 0-316-41920-6. Cost: $17.50.

1441. Handbook of Laboratory Distillation. Erich Krell; E.C. Lumb (ed); C.G. Verver (trans). Elsevier, 1963, 561 pp. LC: 63-9918.

1442. Handbook of Laboratory Solutions. M.H. Gabb; W.E. Latcham. Chemical Publishing Co, 1968, 128 pp. LC: 68-810. ISBN: 0-8206-0055-5. Cost: $15.00.

Includes numerous recipes for chemical solutions. Covers solutions for titrations, biochemical solutions, and solutions in histology as well as physiological saline solutions and culture solutions.

1443. Handbook of Laboratory Unit Operations for Chemists and Chemical Engineers. Jan Pinkava; J. Bryant (trans). Gordon and Breach, 1970, 470 pp. LC: 74-122851. ISBN: 0-677-60600-1. Cost: $93.00.

1444. Handbook of Laplace Transformation. Floyd E. Nixon. Prentice-Hall, 2d ed., 1965, 260 pp. LC: 65-14937.

1445. Handbook of Large Scale Systems Engineering Applications. M.G. Singh; A. Titli. Elsevier, 1979, 564 pp. ISBN: 0-444-85283-2. Cost: $110.50.

Such large scale systems as water systems, transportation systems, telecommunications systems, power systems, and chemical systems are discussed in this handbook. This is actually a collection of invited papers by systems experts on various aspects of large scale systems.

1446. Handbook of Lattice Spacing and Structure of Metals and Alloys. William Burton Pearson. Pergamon, 1967, multivolume set, Vol. 2, 1,446 pp. LC: 57-14965 (G.B.). Cost: $75.00.

1447. Handbook of Legal Medicine. Alan Richards Moritz et al. Mosby, 4th ed., 1975. LC: 74-14826. ISBN: 0-8016-3508-X. Cost: $12.50.

1448. Handbook of Legumes of World Economic Importance. James A. Duke. Plenum Press, 1981, 350 pp. ISBN: 0-306-40406-0. Cost: $45.00.

1449. Handbook of Leprosy. W.H. Jopling. 2d ed., 1978. Cost: $11.95.

1450. Handbook of Linear Integrated Electronics For Research. T.D.S. Hamilton. McGraw-Hill, 1977, 469 pp. LC: 77-30037. ISBN: 0-07-084483-6. Cost: $24.50.

This book covers operational amplifiers, oscillators, optoelectronics, photodiodes, phototransistors, rectifiers, thyristors, and many other applications of linear integrated circuits.

1451. Handbook of Lipid Research. Arnis Kuksis. Plenum, 1978-, Vol.1, 440 pp (more volumes are planned in the future). LC: 77-25277. ISBN: Vol.1, 0-306-33581-6. Cost: Vol.1, $35.00.

Vol.1, entitled *Fatty Acids and Glycerides*. Vol.2, *Fat Soluble Vitamins*.

1452. Handbook of Liquid Crystals. H. Kelker; R. Hatz. Verlag Chemie, 1980, 917 pp. ISBN: 3-527-25481-1. Cost: $262.50; DM 420.

This is primarily a guide to the literature, through 1976, on liquid crystals. It includes over 800 bibliographic citations and over 30 tables on such topics as optical properties, liquid crystals, thermodynamic properties of liquid crystals, nuclear magnetic resonance, and the application of liquid crystals as solvents in gas-chromatography.

1453. Handbook of Livestock Management Techniques. R.A. Battaglia; V. Mayrose. Burgess Publishing Co, 595 pp. LC: 80-70003. ISBN: 0-8087-2957-8. Cost: $24.95.

1454. Handbook of Living Primates. J.R. Napier; P.H. Napier. Academic Press, 1967. ISBN: 0-12-513850-4. Cost: $24.75.

1455. Handbook of Logic Circuits. John Lenk. Reston Publishing Co, 1972, 480 pp. LC: 70-155947 (70-185947). ISBN: 0-87909-332-3. Cost: $18.95.

1456. Handbook of Loss Prevention. Springer-Verlag, 1978, 423 pp. ISBN: 3-540-07822-3. Cost: $39.50.

Discusses loss prevention as it relates to the management of machinery and technical plants. Loss prevention is presented in the context of such topics as turbogenerators, power cables, data processing equipment, steam generators, fluid flow machines, gas turbines, turbo compressors, blowers, centrifugal pumps, and cranes. In addition, such topics as furnaces, diesel engines, gears, and bearings are presented.

1457. Handbook of Machine Foundations. P. Srinivasulu; C.V. Vaidyanathan. McGraw-Hill, 1978, 238 pp. ISBN: 0-07-096611-7. Cost: $17.50.

Block foundations, vibration, and elastic bases are covered in this handbook as well as reciprocating machinery, hammers, fans, and blowers.

1458. Handbook of Mammographic X-ray Spectra. Thomas R. Fewell; Ralph E. Shuping. NTIS, 1978, 114 pp. Accession: PB-292 001. Tech Report: DHEW/PUB/FDA-79/8071; FDA/BRH-79/61. Cost: $3.50 microform; $11.00 paper copy.

This handbook is a compendium of experimentally acquired X-ray spectra obtained from 3 conventional mammographic X-ray tubes. Such X-ray tubes as tungsten, molybdenum-tungsten, and molybdenum targets are used in the study. Numerical tabulations of X-ray spectra produced by the mammographic tubes are listed in this publication.

1459. Handbook of Materials and Processes for Electronics. Charles A. Harper. McGraw-Hill, 1970, 1,200 pp. (various paging). LC: 76-95803. ISBN: 0-07-026673-5. Cost: $19.95.

This handbook contains information on the materials that are used to make electronic components and devices. Such topics as elastomers and various types of coatings and laminates are discussed along with materials such as ceramics and glasses. In addition, various ferrous and nonferrous metals are discussed in connection with their usage in electronic devices. Many definitions are provided to help the user better understand this subject, which is interdisciplinary in nature. An extensive index is provided which will help the user retrieve information on the above subjects in addition to such subjects as thin films and thick films. For related works, please refer to the *Standard Handbook for Electrical Engineers*; the *Electronics Engineers' Handbook*; the *Basic Electronic Instrument Handbook*; the *Handbook of Electronic Packaging*; the *Handbook of Modern Electronic Data; In-*

dustrial Electronics Handbook; the *Handbook of Semiconductor Electronics*; and the *Electrical Engineer's Reference Book*. Also refer to the *Handbook for Electronics Engineering Technicians; Electronics Handbook*; and *Electronic Conversions, Symbols and Formulas.*

1460. Handbook of Materials and Techniques for Vacuum Devices. Walter Heinrich Kohl. Reinhold, 1967, 623 pp.

1461. Handbook of Materials Testing Reactors and Associated Hot Laboratories in the European Community. Peter von der Hardt. D. Reidel, 1981, 160 pp. ISBN: 90-277-1347-2. Cost: $24.00.

This handbook presents information on experimental installations that can handle work with neutrons, radioisotope production, nuclear applications of neutron radiography, solid state research, nuclear physics research, and activation analysis. In addition, this handbook covers information on such topics as nuclear safety, fusion reactors, silicon doping, fission reactors, thermal reactors, and fast reactors.

1462. Handbook of Mathematical Calculations: For Science Students and Researchers. Karen Assaf; Said Assaf. Iowa State University Press, 1974, 350 pp. LC: 73-16451. ISBN: 0-8138-1135-X. Cost: $10.50.

1463. Handbook of Mathematical Economics. K.J. Arrow; M.D. Intriligator. Elsevier, 1981-82, 3 vols. ISBN: 0-444-86054-1. Cost: $150.00.

This 3-volume set is actually a review of the state of the art of mathematical economics and contains a collection of invited papers describing the current status and recent developments in various subfields of mathematical economics. For example, such topics as microeconomics, welfare economics, mathematical programing, probability, gain theory, competitive equilibrium, and economic planning are discussed in this compilation.

1464. Handbook of Mathematical Formulae for Engineers and Scientists. Barrie Raymond Bartlett; D.J. Fyfe. Denny Publications, Ltd, 1974, 216 pp. ISBN: 0-9503154-0-0. Cost: £2.25.

1465. Handbook of Mathematical Formulas. Hans-Jochen Bartsch. Academic Press, translated, 1974, 528 pp. LC: 73-2088. ISBN: 0-12-080050-0. Cost: $22.00.

1466. Handbook of Mathematical Functions with Formulas, Graphs, and Mathematical Tables. M. Abramowitz; I.A. Stegun. U.S. Government Printing Office (also available from Dover), 1964, 1,046 pp. ISBN: 0-486-61272-4 (Dover). Cost: $14.95 paper copy (Dover). Series: *U.S. National Bureau of Standards, Applied Mathematics Series*, No. 55.

1467. Handbook of Mathematical Functions. National Bureau of Standards. U.S. Government Printing Office, 1964, 1,061 pp. SuDoc: C13.32:55. Cost: $12.65. Series: NBS AMS55.

NOTE: This handbook includes many tables, formulas, graphs, and equations. There are many mathematical functions and bibliographies for further investigation.

1468. Handbook of Mathematical Logic. J. Barwise. Elsevier, 1977, 1,166 pp. ISBN: 0-7204-2285-X. Cost: $85.00. Series: *Studies in Logic*, Vol. 90.

Handbook of Mathematical Sciences. *See* CRC Handbook of Mathematical Sciences.

1469. Handbook of Mathematical Tables and Formulas. Richard Stevens Burington. McGraw-Hill, 5th ed., 1973, 500 pp. LC: 78-39634. ISBN: 0-07-009015-7. Date of First Edition: 1933. Cost: $10.50.

This one-volume ready-reference is similar to the *CRC Standard Mathematical Tables* in that it provides a basic collection of mathematical theorems, equations, formulas, and tables that are useful for basic mathematical calculations. Included is information on such subjects as differential equations, Laplace transforms, Fourier series, probability and statistics, trigonometric functions, and logarithms. An index is provided as well as a list of references for further investigation. For related works, please refer to the following titles: *A Table of Series and Products; CRC Handbook of Tables for Probability and Statistics; CRC Handbook of Tables for Probability Statistics; CRC Handbook of Tables for Mathematics; Eight-place Tables of Trigonometric Functions for Every Second of Arc; Engineering Mathematics Handbook*; and the various mathematical handbooks prepared by the U.S. National Bureau of Standards.

1470. Handbook of Mathematics. Lauwerens Kuipers; Reinier Timman. Pergamon Press, 1969, 782 pp. LC: 67-22830. ISBN: 0-08-011857-7 hard copy; 0-08-018996-2 paper copy. Cost: $52.00 hard copy; $21.00 paper copy.

1471. Handbook of Measurement and Control. Edward E. Herceg. Schaevitz Engineering, 1972, various paging. LC: 72-88883. Cost: $7.95.

The subtitle of this book is as follows: *An Authoritative Treatise on the Theory and Application of the LVDT.* The major product of Schaevitz Engineering is the Linear Variable Differential Transformer (LVDT). This book presents such topics as transducer elements, magnetic induction elements, piezoelectric elements, passive transducers, resistance strain gauges, and the history of the Linear Variable Differential Transducer. In addition, such topics as environmental effects on an LVDT, vibration and shock, installation of an LVDT, computer-based measurement, the DC-LVDT, and the measurement of force and fluid pressure are discussed.

1472. Handbook of Measurement and Evaluation in Rehabilitation. Brian Bolton (ed). University Park Press, 1976. ISBN: 0-8391-0861-3. Cost: $18.50.

1473. Handbook of Mechanical and Electrical Systems for Buildings. H.E. Bovay. McGraw-Hill, 1981, 864 pp. LC: 80-22666. ISBN: 0-07-006718-X. Cost: $49.50.

This handbook will be helpful in selecting mechanical and electrical systems for new and renovated structures. Such topics as boilers, air conditioning systems, refrigeration equipment, heating, humidity control, ventilation, air filtration, and odor control are discussed in this handbook. In addition, energy conservation, pumps, fans, vibration control, alarm systems, lighting, plumbing, and fire safety are also presented.

1474. Handbook of Mechanical Power Drives. Renouf USA Inc (available from International Ideas), 2d ed., 1977, 777 pp. ISBN: 0-85461-064-2. Cost: $96.00.

This book includes over 1,500 diagrams, tables, charts, and illustrations on mechanical power transmissions. Such topics as variable speed drives, electric motors, bearings, lubrication, belt drives, servomechanisms, torque limiters, cams, vibrations, and springs are covered. In addition, such subjects as piston seals, shaft seals, oil seals, shock absorbers, brakes, fluid drives, and universal joints are also discussed.

1475. Handbook of Mechanical Properties of Steels and Alloys During Plastic Deformation. A.V. Tretyakov; G.K. Trofimov; M.K. Guryanova. NTIS, 1972, 85 pp. Accession: N72-33538; AD-745246.

This handbook presents data on the changes in mechanical properties of steels and alloys during plastic deformation. Tables of actual resistances are given for deformation in a hot state relative to temperature, degree and rate of deformation. Formulas are also given for determining yield point, critical point, and relative elongation and hardness.

1476. Handbook of Mechanical Specifications for Buildings and Plants. R.H. Emerick. McGraw-Hill, 1966, 496 pp. LC: 66-16332. ISBN: 0-07-019313-4. Cost: $27.50.

1477. Handbook of Mechanical Wear, Frettage, Pitting, Cavitation, Corrosion. Charles Lipson; L.V. Colwell. University of Michigan Press, 1961, 469 pp. LC: 60-7931.

1478. Handbook of Medical Acupuncture. Frank Z. Warren. Van Nostrand Reinhold, 1976, 273 pp. ISBN: 0-442-29198-1. Cost: $15.00.

1479. Handbook of Medical Emergencies. Jack J. Kleid; Bruce H. Heckman. Medical Examination Publishing, Co, 1970. ISBN: 0-87488-635-X. Cost: $8.00.

1480. Handbook of Medical Laboratory Formulae. R.E. Silverton; M.J. Anderson. Butterworth, 1961, 676 pp.

Handbook of Medical Physics. *See* CRC Handbook of Medical Physics.

1481. Handbook of Medical Specialties. Henry Wechsler. Human Science Press, 1976. LC: 74-10951. ISBN: 0-87705-232-8. Cost: $16.95.

1482. Handbook of Medical Treatment. Milton John Chatton. Lange Medical Publishers, Inc, 14th ed., 1974, 640 pp. ISBN: 0-87041-044-X. Cost: $7.50.

1483. Handbook of Metal Ligand Heats and Related Thermodynamic Quantities. James J. Christensen et al. Dekker, 2d ed., 1975, 495 pp. LC: 75-4480. ISBN: 0-8247-6317-3. Cost: $65.00.

1484. Handbook of Metal Powders. Arnold R. Poster. Reinhold (also Chapman), 1966, 274 pp. LC: 66-20059.

This handbook is a vast collection of data on metal powders including physical and chemical properties of over 1,800 metal powders. In addition, a supplier's list is provided to help the user identify the availability of certain metal powders. The handbook also includes methods of testing metal powders, as well as an extensive annotated bibliography on pages 241-274. For related works, please refer to the *Metals Handbook*.

1485. Handbook of Metal Treatments and Testing. Robert B. Ross. John Wiley (Halsted), 1977, 467 pp. LC: 77-3929. ISBN: 0-470-99104-6. Cost: $37.95.

Covers casting, blasting processes, brazing, corrosion protection processes, fatigue resistance processes, and plating processes.

1486. Handbook of Meteorology. F. Berry; E. Bollay; N. Beers. McGraw-Hill, 1945, 1,116 pp. LC: 45-10426. ISBN: 0-07-005030-9. Cost: $55.00.

1487. Handbook of Methods of Applied Statistics. Indra Mohan Chakravarti et al. John Wiley and Sons, 1967, 2 Vols.. LC: 66-26737.

This handbook consists of 2 volumes entitled as follows: Vol.1, *Techniques of Computation, Descriptive Methods, and Statistical Inference*; Vol.2, *Planning of Surveys and Experiments*.

1488. Handbook of Microbiology. Morris Boris Jacobs; M.J. Gerstein. Van Nostrand Reinhold, 1960, 322 pp. LC: 60-15811.

1489. Handbook of Microcircuit Design and Application. David F. Stout; Milton Kaufman. McGraw-Hill, 1980, 576 pp. LC: 79-19999. ISBN: 0-07-061796-1. Cost: $32.50.

This handbook emphasizes microprocessors and microcomputers and presents information on circuit design with integrated circuits. In addition, such topics as LSI circuits, digital-to-analog converters, active filters, and nonlinear analog microcircuits are also presented.

1490. Handbook of Micromethods for the Biological Sciences. Georg Keleti; William H. Lederer. Van Nostrand Reinhold, 1974, 166 pp. LC: 73-12027. ISBN: 0-442-24290-5. Cost: $15.95.

1491. Handbook of Microprocessors, Microcomputers, and Minicomputers. John D. Lenk. Prentice-Hall, 1979, 404 pp. LC: 78-24307. ISBN: 0-13-380378-3. Cost: $20.95.

This handbook discusses such topics as different types of microprocessors and their corresponding peripheral equipment. For example, such equipment as the RCA COSMAC 1800 microprocessors, the Motorola M-6800 microcomputer system, the Intel MCS-48 microcomputers, the Mostek Z-80 microcomputer system, the Texas Instruments TM 9900 16-bit microprocessor family, and the Tektronic 8002 microporcessor system are presented in this handbook.

1492. Handbook of Microscopic Anatomy for the Health Sciences. Annabelle Cohen. Mosby (distributed in U.K. by Kimpton), 1975, 135 pp. LC: 74-13572. ISBN: 0-8016-1012-5. Cost: &£2.60 ($6.95).

1493. Handbook of Microscopical Technique for Workers in Animal and Plant Tissues. C.E. McClung; R. McClung Jones. Hafner, 3d ed., reprint of 1950 ed., 1964. ISBN: 0-02-849050-9. Cost: $26.25.

1494. Handbook of Microwave Ferrite Materials. Wilhelm H. Von Aulock. Academic Press, 1965, 518 pp. LC: 65-27086. ISBN: 0-12-723350-4. Cost: $51.00.

1495. Handbook of Microwave Measurements. Max Sucher; Jerome Fox. Halsted Press, 3d ed., 1975 (also 1963), 3 Vols., 1,144 pp. LC: 63-20838. ISBN: 0-470-83538-9. Date of First Edition: 1954. Cost: $59.50.

This 3-volume set contains information on the different types of techniques and equipment that are used to make microwave measurements. Many tables, charts, and examples are used to help the reader understand the various methods of measurement and testing. Such microwave measurements as power, attenuation, frequency, and dielectric constants are discussed. In addi-

tion, such topics as measurments with regard to antenna performance are discussed along with measurements that deal with parametric amplifiers and tunnel diode amplifiers. This handbook provides charts for transmission lines and an excellent cross-referenced index which helps to retrieve the data on microwave measurments. For related works, please refer to such titles as *Microwave Engineers' Handbook*; and *Instrument Engineers' Handbook*.

1496. Handbook of Microwave Techniques and Equipment. Harry E. Thomas. Prentice-Hall, 1972, 319 pp. LC: 72-2347. ISBN: 0-13-380329-5. Cost: $23.95.

1497. Handbook of Military Infrared Technology. William L. Wolfe (Office of Naval Research). U.S. Government Printing Office, 1965, 906 pp. LC: 65-62266.

1498. Handbook of Mineral Dressing: Ores and Industrial Minerals. Arthur F. Taggart. John Wiley and Sons, Inc, 2d ed., 1945, 1905 pp. (various paging). LC: 45-2507. ISBN: 0-471-84348-2. Date of First Edition: 1927. Cost: $55.00.

Although this handbook is somewhat dated, it provides a considerable amount of useful information on mineral dressing, production, distribution, and storage. The separation of solid crudes is discussed in addition to such preparations of raw materials as crushing, grinding, scrubbing, and tumbling. Air sizing, magnetic separation, electrostatic separators, various methods of flotation, and filtration are discussed. Information is provided on such subjects as conveyers, elevators, feeders, tailing disposal, and the design of ore-treatment plants. This handbook includes tables, flow sheets, mathematical formulas, definitions, abbreviations, and a very useful index. For related works, please refer to *Metals Handbook; ASME Handbook; Thermophysical Properties of Matter; Corrosion Guide; Handbook of Aluminum; Metals Reference Book; Fatigue Design Handbook; Handbook of Binary Phase Diagrams; Steel Designers' Manual; CRC Handbook of Materials Science; Engineering Alloys*, and the *Aerospace Structural Metals Handbook*.

1499. Handbook of Minerals. John Tomikel. Allegheny Press, 1968, 133 pp. LC: 68-8384.

1500. Handbook of Minimal Brain Dysfunctions: A Critical View. H.E. Rie; E.D. Rie. John Wiley and Sons, 1979, 800 pp. LC: 78-25656. ISBN: 0-471-02959-9. Cost: $39.95. Series: *Wiley Series on Personality Processes*.

This book would be useful for clinical psychology and behavioral science collections.

1501. Handbook of Mobile Home Inspection. Robert C. Smith. Cahners Publishing Company, Inc, 1975, 198 pp. ISBN: 0-8436-0155-8. Cost: $20.00.

Various reference standards and specifications are discussed in connection with mobile home inspection. This handbook covers such topics as plumbing, electrical systems, heating systems, cooling systems, fuel burning systems, and mobile home construction.

1502. Handbook of Modern Electrical Wiring. John Traister. Reston, 1979, 302 pp. LC: 78-27289. ISBN: 0-8359-2754-7. Cost: $16.95.

Discusses electrical wiring and the National Electrical Code. Wiring information is provided for such installations as trailer parks, mobile homes, lighting, branch circuits, and low voltage wiring.

1503. Handbook of Modern Electronic Data. Matthew Mandl. Reston Publishing Co, 1973, 274 pp. LC: 72-91116. ISBN: 0-87909-329-3. Cost: $18.95.

This handbook contains many tables, graphs, and equations for those users who are interested in basic electronic information associated with such topics as transistors and tubes, antennas, and lasers. In addition, information dealing with such topics as transmission lines, meter ranges, and vectors are presented along with discussions on various types of circuitry. An index is provided as well as information on SI units. For related works please refer to the *Electrical Engineer's Reference Book*; the *Electronics Engineers' Handbook*; the *Basic Electronic Instrument Handbook*; the *Handbook of Electronic Packaging*; the *Handbook of Materials and Processes for Electronics, Industrial Electronics Handbook*; the *Handbook of Semiconductor Electronics*; and the *Standard Handbook for Electrical Engineers*. Also refer to the *Handbook for Electronics Engineering Technicians; Electronics Handbook*, and *Electronic Conversions, Symbols and Formulas*.

1504. Handbook of Modern Keypunch Operations. William J. Keys; Carl H. Powell. Canfield Press (division of Harper-Row), 1970, 199 pp. LC: 73-95880. ISBN: 0-06-384500-8. Cost: $13.50 paper copy.

1505. Handbook of Modern Manufacturing Management. H.B. Maynard. McGraw-Hill, 1970, 1,100 pp. LC: 77-83271. ISBN: 0-07-041087-9. Cost: $34.65.

1506. Handbook of Modern Personality Theory. Raymond B. Cattell; Ralph Mason Dreger. Hemisphere Publishing Corp, 1977, 815 pp. ISBN: 0-470-15201-X. Cost: $37.50. Series: *Series in Clinical and Community Psychology*.

1507. Handbook of Modern Solid-State Amplifiers. John D. Lenk. Prentice-Hall, 1974, 414 pp. LC: 73-21834. ISBN: 0-13-380394-5. Cost: $18.95.

1508. Handbook of Moisture Determination and Control: Principles, Techniques, Applications. A. Pande. Dekker, 1974-75, 4 Vols. LC: 73-86820. ISBN: Vol.1,. 0-8247-6184-7; Vol.2, 0-8247-6185-5; Vol.3, 0-8247-6186-3; Vol.4, 0-8247-6201-0. Cost: Vol.1, $36.00; Vols.2, 3, and 4, $49.50 each.

1509. Handbook of Molecular Constants of Inorganic Compounds. K.S. Krasnov et al; J. Schmorak (trans). Israel Program for Scientific Translation (available in U.S. from International Scholarly Book Service), 1971, 275 pp. LC: 74-27650. ISBN: 0-706-50620-0.

1510. Handbook of Molecular Cytology. A. Lima-DeFaria. North-Holland (also Wiley-Interscience), 1969, 1,508 pp. LC: 69-18387. ISBN: 0-444-10413-5; 0-7204-7115-X (Wiley-Interscience). Cost: $122.00. Series: *Frontiers of Biology*, Vol. 15.

1511. Handbook of Multiphase Systems. Gad Hetsroni. McGraw-Hill (Hemisphere Publishing Corp), 1982, 1,492 pp. LC: 81-6790. ISBN: 0-07-028460-1. Cost: $64.50.

This handbook contains numerous articles on such topics as pool boiling, pre-burnout convective boiling, condensation, fluidized bed hydrodynamics, fluidized bed heat transfer, fluidized bed mass transfer, gas-particle separation, and liquid suspensions. In addition, such topics as conveying, 3-phase flow, drop phenomena, various types of gas-solid systems, interfacial phenomena,

and various pipeflow equations are included. Each chapter concludes with a long list of references, and the book is heavily illustrated with a variety of charts, tables, and correlations.

1512. Handbook of Municipal Administration and Engineering. William S. Foster. McGraw-Hill, 1978, 512 pp. (various paging). LC: 77-14128. ISBN: 0-07-021630-4. Cost: $24.50.

Covers snow and ice control, street cleaning, water distribution, refuse collection, traffic control, street lighting, and sewer maintenance.

1513. Handbook of Municipal Waste Management Systems. Barbara J. Stevens. Van Nostrand Reinhold, 1980, 336 pp. ISBN: 0-442-23362-0. Cost: $22.50.

This is a compilation of information on solid waste collection systems for 14 different cities in the United States. It compares such information as cost effectiveness, equipment, haulage practices, service level, and management practices.

1514. Handbook of Mutagenicity Test Procedures. B.J. Kilbey. Elsevier, 1977, 486 pp. ISBN: 0-444-41338-3. Cost: $86.75.

1515. Handbook of Natural Gas Engineering. D.L. Katz. McGraw-Hill, 1959, 802 pp. LC: 58-6686. ISBN: 0-07-033384-X. Cost: $64.50.

This handbook includes information on natural gas from its production to its eventual utilization. A considerable amount of information is provided on the properties of natural gas and the various hydrocarbons that are associated with it. Flow calculations are provided as well as compressibility information and the various sources of natural gas from gas wells and gas fields. A considerable amount of processing data, tables, and charts have been provided for the transportation of natural gas. In addition, information is also provided for the transportation of natural gas to the ultimate consumer, including such areas as pipelines, underground storage, and legal regulations by the government. A well-prepared index has been provided in addition to a list of references that will be useful in obtaining additional information.

1516. Handbook of Naturally Occurring Compounds. T.K. Devon; A.I. Scott. Academic Press, Vol.1, 1975; Vol.2, 1972, 2 Vols. LC: 76-187258. ISBN: Vol.1, 0-12-213601-2; Vol.2, 0-12-213602-0. Cost: Vol.1, $51.50; Vol.2, $42.00.

This handbook contains 2 volumes entitled as follows: Vol.1, *Acetogenins, Shikimates, and Carbohydrates;* Vol.2, *Terpenes.*

1517. Handbook of Neonatology. Rita G. Harper; Jing J. Yoon. Year Book Medical Publishers, 1974. ISBN: 0-8151-4149-1. Cost: $8.95.

1518. Handbook of Neurochemistry. Abel Lajtha. Plenum Press, 1969-72, 7-Vol. set. LC: 68-28097. ISBN: Vol.1, 0-306-37701-2; Vol.2, 0-306-37702-0; Vol.3, 0-306-37703-9; Vol.4, 0-306-37704-7; Vol.5a, 0-306-37705-5; Vol.5b, 0-306-37715-2; Vol.6, 0-306-37706-3; Vol.7, 0-306-37707-1,. Cost: $39.50 per Vol.

This handbook consists of 7 volumes with the following titles: Vol.1, *Chemical Architecture of the Nervous System;* Vol.2, *Structural Neurochemistry;* Vol.3, *Metabolic Reactions in the Nervous System;* Vol.4, *Control Mechanisms in the Nervous System;* Vols.5a and 5b, *Metabolic Turnover in the Nervous System;* Vol.6, *Alterations of Chemical Equilibrium in the Ner-*

vous System; Vol.7, *Pathological Chemistry of the Nervous System.*

1519. Handbook of Neurological Examination and Case Recording. Derek E. Denny-Brown. Harvard University Press, rev. ed., 1957. LC: 58-2654. ISBN: 0-674-37100-3. Cost: $3.95.

1520. Handbook of NMR Spectral Parameters. W. Brugel. Heyden and Son Inc, 3 Vols., 1,016 pp. ISBN: 0-88501-170X. Cost: $336.00.

Covers more than 7,500 compounds and correlated these compounds with chemical structure in detailed tables arranged by class. NMR spectral data are presented for such organic compounds as acetylenes, amino acids, benzenes, butanes, dioxanes, ethylenes, furans, propanes, quinolines, steroids, and flourinated compounds.

1521. Handbook of Noise and Vibration Control. Trade and Technical Press, Ltd, 4th ed., 1979, 893 pp. ISBN: 0-85461-073-1. Cost: $94.00.

Covers noise generation from electric motors, pumps, compressors, hydraulic systems, air distribution systems, diesel engines, and fans.

1522. Handbook of Noise Assessment. Daryl May. Van Nostrand Reinhold, 1978, 400 pp. LC: 77-27657. ISBN: 0-442-25197-1. Cost: $24.50.

1523. Handbook of Noise Control. C.M. Harris. McGraw-Hill, 2d ed., 1979, (various paging). LC: 78-6764. ISBN: 0-07-026814-2. Cost: $39.50.

The editors of this handbook have taken an interdisciplinary approach to the subject of noise control since noise is related to so many different subject areas. The handbook discusses various sources of noise such as trains and aircraft, automobiles, noise resulting from vibrations, noises resulting from the movement of fans and gears, and heating and ventilating equipment. In addition, this handbook provides information on the physical properties of the noise as well as the measurement of noise. Various means of noise reduction are treated including vibration damping and a discussion of acoustical materials used to reduce noise. Hearing loss and ear protectors are discussed along with various codes and ordinances associated with building requirements in an effort to control noise levels. For related works, please refer to such titles as: *The Handbook of Noise and Vibration Control, Handbook of Industrial Noise Control*, and the *Handbook of Industrial Noise Management.*

1524. Handbook of Noise Measurement. Arnold P.G. Peterson; Ervin E. Gross Jr. General Radio Corp, 7th ed., 1972.

1525. Handbook of Noise Ratings. Karl S. Pearsons; Ricarda L. Bennett. NTIS, 1974, 334 pp. Accession: N74-23275. Tech Report: NASA-CR-2376.

1526. Handbook of Non-Prescription Drugs. George B. Griffenhagen; Linda L. Hawkins. American Pharmaceutical Association, 6th ed., 1979, 488 pp. LC: 68-2177. ISBN: 0-917330-27-0. Cost: $20.00.

Includes bibliographies.

1527. Handbook of Non-topographic Photogrammetry. H.M. Karara. American Society of Photogrammetry, 1979, 206 pp. LC: 79-9270. ISBN: 0-937294-02-0. Cost: $20.00.

1528. Handbook of Nonparametric Statistics; Results for Two and Several Sample Problems, Symmetry and Extremes. John Edward Walsh. Van Nostrand Reinhold, 1965, Vol.3, 1968, 3 Vols. LC: 62-6555. ISBN: Vol.2, 0-442-09193-1; Vol.3, 0-442-39194-3. Cost: Vol.2, $22.50; Vol.3, $22.50.

1529. Handbook of Nonpoint Pollution: Sources and Management. Vladimir Novotny; Gordon Chesters. Van Nostrand Reinhold, 1981, 555 pp. LC: 81-1525. ISBN: 0-442-225633-6. Cost: $24.50. Series: *Van Nostrand Reinhold Environmental Engineering Series.*

This book deals primarily with pollution from storm water and runoff from both urban and rural areas. For example, such topics as runoff from croplands, strip minings, construction sites, erosion, sediment yields, and contamination of ground water are discussed. In addition, hydrologic considerations in precipitation and runoff, as well as acid rainfall data, are also presented.

1530. Handbook of Nuclear Data for Neutron Activation Analysis. A.I. Aliev et al; Baruch Benny (trans). Israel Program for Scientific Translation (also Halsted Press), 1970, 168 pp. LC: 72-176084. ISBN: 0-470-02260-4 (Halsted Press). Cost: $22.95.

Includes bibliographies.

1531. Handbook of Numerical and Statistical Techniques. J.H. Pollard. Cambridge University Press, 1977, 349 pp. LC: 76-27908. ISBN: 0-521-29750-8. Cost: $39.95.

1532. Handbook of Numerical Harmonic Analysis. Vladimir Ivanovich Krylov; L.G. Kruglikova; David Louvish (trans). Israel Program for Scientific Translation, 1969, 151 pp. LC: 70-8166.

1533. Handbook of Numerical Inversion of Laplace Transforms. V.I. Krylov; N.S. Skoblya. Davy, 1969, 293 pp. LC: 79-4487.

1534. Handbook of Numerical Matrix Inversion and Solution of Linear Equations. Joan Robinson Westlake. Wiley (available from Krieger), 1968, rep. 1975, 182 pp. LC: 74-26623. ISBN: 0-88275-225-1. Cost: $13.50.

1535. Handbook of Numerical Methods and Applications. L.G. Kelly. Addison-Wesley, 1967, 354 pp. LC: 67-23979.

1536. Handbook of Numerical Methods for the Solution of Algebraic and Transcendental Equations. Vladimir L'vovich Zaguskin; G.O. Harding (trans). Pergamon Press, 1961, 195 pp. LC: 61-9787.

Handbook of Nutritional Requirements in a Functional Context. *See* CRC Handbook of Nutritional Requirements in a Functional Context.

Handbook of Nutritive Value of Processed Food. *See* CRC Handbook of Nutritive Value of Processed Food.

1537. Handbook of Obstetric Emergencies. Richard H. Schwarz. Medical Examination Publishing, Inc, 1973. ISBN: 0-87488-634-1. Cost: $8.00.

1538. Handbook of Obstetrical and Gynecological Data. Robert C. Goodlin. Geron-X, 1972, 526 pp. LC: 78-188790. ISBN: 0-87672-032-7. Cost: $7.00.

1539. Handbook of Obstetrics and Gynecology. Ralph Criswell Benson; Laurel V. Schaubert (ed). Lange Medical Publishers, 6th ed., 1977. ISBN: 0-87041-143-8. Cost: $9.50.

1540. Handbook of Occupational Safety and Health. Frank E. McElroy. National Safety Council, 1975. LC: 75-11314. ISBN: 0-87912-036-3. Cost: $6.00 paper copy.

1541. Handbook of Ocean and Underwater Engineering. John J. Meyers; Carl H. Holm; Raymond F. McAllister. McGraw-Hill, 1969, 800 pp. (various paging). LC: 67-27280. ISBN: 0-07-044245-2. Cost: $46.50.

This handbook is a remarkable collection of information, tables, data, facts, and figures associated with ocean engineering and marine sciences. Any engineer involved with the design of equipment or structures used in ocean environments will find this handbook of interest, for it provides information on such topics as the properties of fresh water and salt water, characteristics of tides and waves, the corrosive effects of sea water exposure, wind and wave forces on fixed floating structures, the transmission of sound, light, and radiowaves through sea water. This handbook is concerned with the different types of equipment utilized in ocean environments such as winches, underwater welding equipment, diving equipment, and equipment used to lay cables underneath the ocean. In addition, it covers such topics as materials suitable for use in ocean environments, safety operations on ships, as well as radio communications, and forecasting weather. The handbook includes an appendix listing various sources of information on ocean technology, and it provides an index as well as a listing of conversion factors. and the *CRC Handbook of Marine Science*.

1542. Handbook of Oceanographic Engineering Materials. S.C. Dexter. John Wiley and Sons, 1979, 250 pp. LC: 78-26196. ISBN: 0-471-049506. Cost: $33.50. Series: *Ocean Engineering Series*.

This handbook provides information on different types of materials that are used in oceanographic environments. Mechanical and physical properties are given for various metals and alloys, including aluminum alloys, copper alloys, nickel-based alloys, stainless steels, titanium alloys, magnesium, zinc, and lead. In addition, such nonmetallic materials as polymers, rubbers, elastomers, concrete, ceramics, glass, and wood are discussed. Corrosion resistance, cost of materials, and availability of materials are also mentioned.

1543. Handbook of Oceanographic Tables, 1966. Eugene L. Bialek (Naval Oceanographic Office). Government Printing Office, 1967, 427 pp. LC: 67-61991. SuDoc: D203.22/3:68.

Replaces H.O. publication 614, *Planning Oceanographic Data.*

1544. Handbook of Ocular Pharmacology. Marvin B. Smith. Publishing Sciences Group, 1974, 156 pp. LC: 74-75353. ISBN: 0-88416-018-1. Cost: $15.00.

1545. Handbook of Ocular Toxicity. Marvin B. Smith. Publishing Sciences Group, 1976. LC: 75-12027. ISBN: 0-88416-114-5. Cost: $20.00.

1546. Handbook of Operational Amplifier Circuit Design. David F. Stout; Milton Kaufman. McGraw-Hill, 1976, 434 pp. LC: 76-3491. ISBN: 0-07-061797-X. Cost: $24.50.

This handbook is a collection of all types of circuits that are associated with operational amplifiers. Schematics and design equations are provided for such circuits as bandpass filters, demodulators, oscillators, power circuits, and waveform generators. Many tables, illustrations, and examples are provided to make this handbook easy to use. For related works, please refer to the *Electronics Engineers' Handbook*.

1547. Handbook of Operations Research. Joseph J. Moder; Salah E. Elmaghraby. Van Nostrand Reinhold, 1978, 2-Vol. set, 622 pp. LC: 77-21580. ISBN: Vol.1, 0-442-24595-5; Vol.2, 0-442-24596-3. Cost: $34.50 per Vol.

This 2-volume set is entitled: Vol.1, *Foundation and Fundamentals*; Vol.2, *Models and Applications*. This handbook covers all types of operations research and includes articles on such topics as stochastic processes, mathematical programing, linear programing, dynamic programing, queuing, and optimal control.

1548. Handbook of Ophthalmologic Emergencies. George M. Gombos. Medical Examination Publishing, Inc, 1973. ISBN: 0-87488-633-3. Cost: $8.00.

1549. Handbook of Optical Holography. H.J. Caufield. Academic Press, 1979, 638 pp. LC: 79-51672. ISBN: 0-12-165350-1. Cost: $63.00.

This handbook is a collection of 31 articles on various topics related to holography. For example, such topics as Fresnel holography, Fraunhofer holograms, Fourier holography, reflection holograms, multiplexed holograms, color holograms, and polarization holograms. In addition, such topics as solid state lasers, gas lasers, multiple-image generation, spectroscopy, and photogrammetry are also discussed.

1550. Handbook of Optics. Walter G. Driscoll; William Vaughn (ed) (Optical Society of America). McGraw-Hill, 1978, 1,184 pp. LC: 77-22123. ISBN: 0-07-047710-8. Cost: $55.00.

Discusses such topics as fiber optics, photosensitive materials, colorimetry, coatings, and filters.

1551. Handbook of Oral Diagnosis and Treatment Planning. Donald L. McElroy. Krieger, reprint of 1969 ed., 1974. LC: 69-14460. ISBN: 0-88275-126-3. Cost: $9.50.

1552. Handbook of Oral Pharmacology. Alvin F. Gardner. Publishing Sciences Group, 1976. LC: 75-12026. ISBN: 0-88416-115-3. Cost: $20.00.

1553. Handbook of Organic Analysis: Qualitative and Quantitative. Hans Thacher Clarke; B. Haynes (rev). Crane-Russak Co, 5th ed., 1975, 291 pp. LC: 75-330409. ISBN: 0-8448-0518-1 hard copy; 0-8448-0662-5 paper copy; Arnold edition (British), 0-7131-2460-1 hard copy; 0-7131-2490-3 paper copy. Cost: $35.00 hard copy; $20.50 paper copy Arnold edition (British) £12 hard copy, £4.95 paper copy.

1554. Handbook of Organic Industrial Solvents. Alliance of American Insurers. Alliance of American Insurers, 5th ed., 1980, 120 pp. LC: 79-91726. Cost: $3.00.

Provides information on the problems associated with industrial solvents and particularly addresses the safety and loss control fields dealing with the explosion hazards associated with organic solvents.

1555. Handbook of Organic Reagents in Inorganic Analysis. Zavis Holzbecher. Halsted Press, translated, 1976, 734 pp. LC: 75-34459. ISBN: 0-470-01396-6. Cost: $55.00.

Includes bibliographies and index.

1556. Handbook of Organic Structural Analysis. Yasuhide Yukawa. Benjamin Co, Inc, 1965, 765 pp. LC: 66-3844.

1557. Handbook of Organic Waste Conversion. Michael W.M. Bewick. Van Nostrand Reinhold, 1980, 419 pp. LC: 79-18568. ISBN: 0-442-20679-8. Cost: $24.50. Series: *Van Nostrand Reinhold Environmental Engineering Series.*

This handbook deals with the conversion and utilization of organic waste produced by municipalities and industrial facilities. Such topics as sewage sludge, agricultural residues, food processing, and antibiotic production are presented. In addition, several reprocessing techniques are included for animal waste and crop residues, municipal waste, distilling by-products, fermentation waste materials, cellulose waste, cannery waste, slaughterhouse waste, and fish waste. A chemical analysis of each type of waste is provided and a discussion is presented as to its suitability for use as fertilizer or foodstuffs. Fifteen papers by different authors are included in this handbook and each paper has a listing of references as well as excellent tables and illustrations.

1558. Handbook of Organometallic Compounds. Nobue Hagihara et al. W. A. Benjamin, 1968, 1,044 pp. LC: 68-31355. ISBN: 0-8503-3780-6. Cost: $45.00.

Bibliography on pp. ix-x.

1559. Handbook of Organosilicon Compounds: Advances Since 1961. Vladimir Bazant et al. Dekker, translated, 1973, 4 Vols., Vol.1, 762 pp.; Vol.2, 624 pp.; Vol.3, 736 pp.; Vol.4, 1,008 pp. LC: 74-21879. ISBN: Vol.1, 0-8247-6259-2; Vol.2, 0-8247-6267-3; Vol.3, 0-8247-6268-1; Vol.4, 0-8247-6269-X. Cost: $99.75 per Vol.

1560. Handbook of Orthodontics. Robert E. Moyers. Year Book Medical Publishers, 3d ed., 1973, 757 pp. ISBN: 0-8151-6002-X. Cost: $25.50.

1561. Handbook of Orthopaedic Surgery. Alfred Rives Shands et al. Mosby (also available from Kimpton), 7th ed., 1967, 572 pp. LC: 67-13419.

1562. Handbook of Orthoptic Principles. G.T. Cashell; I.N. Durran. Longman, 3d ed., 1975. LC: 74-78383. ISBN: 0-443-01216-4. Cost: $8.00.

Covers such topics as heterophoria, binocular single vision, and causes and effect of squint.

1563. Handbook of Oscilloscope Waveform Analysis and Applications. Miles Ritter-Sanders Jr (i.e., Robert Gordon Middleton). Reston Publishing Co, 1977, 200 pp. LC: 76-41259. ISBN: 0-87909-337-4. Cost: $15.95.

1564. Handbook of Oscilloscopes: Theory and Application. John D. Lenk. Prentice-Hall, 1968, 212 pp. LC: 68-20858. ISBN: 0-13-380543-3. Cost: $15.95.

1565. Handbook of Package Design Research. W. Stern. John Wiley and Sons, 1981, 600 pp. LC: 80-39935. ISBN: 0-471-05901-3. Cost: $40.00.

This handbook deals with product packaging, primarily as a means of enhancing the saleability of products.

1566. Handbook of Package Engineering. J.R. Hanlon. McGraw-Hill, 1971, 544 pp. LC: 70-124138. ISBN: 0-07-025993-3. Cost: $36.50.

Covers physical and chemical properties of packaging materials.

1567. Handbook of Package Materials. Stanley Sacharow. AVI, 1976, 243 pp. ISBN: 0-87055-207-4. Cost: $26.50 ($29.00 outside U.S.).

This handbook is a very comprehensive treatment of the usage of different types of packaging materials. Such topics as the usage of glass, metals, and plastics as packaging materials; the usage of corrugated fiberboard and paperboard for packaging materials; the usage of paper as a flexible packaging material; and the usage of films and aluminum foil as flexible packaging materials are discussed. In addition, this book also describes laminates as a packaging material, and presents information on the various properties of packaging materials and their environmental and ecological impact with regard to disposability. These packaging materials are primarily used for food packaging.

1568. Handbook of Paleontological Techniques. Bernhard Kummel; D.M. Raup. Freeman (prepared under the auspices of the Paleontological Society), 1965, 852 pp. LC: 64-7537.

1569. Handbook of Palynology: Morphology, Taxonomy, Ecology: An Introduction to the Study of Pollen Grains and Spores. Gunnar Erdtman. Munksgaard (also Hafner), 1969, 486 pp. LC: 76-432483. ISBN: 0-02-84425-4 (Hafner). Cost: $31.75.

1570. Handbook of Paper Science. H.F. Rance. Elsevier, 1980, Vol.1, 298 pp. ISBN: Vol.1, 0-444-41778-8. Cost: Vol.1, $78.00.

Vol.1 of this handbook is entitled *The Raw Materials and Processing of Papermaking*. This handbook series deals with such topics as the science and technology of papermaking, the properties of paper, and the usage of paper. Vol.1 covers such topics as cellulose fibers and water interaction, the preparation and treatment of cellulose fibers for papermaking, the properties of cellulose fibers, and web consolidation. In addition, such topics as fiber beating, flocculation, pressing, drying, and bonding are discussed. Different types of wood sources and wood properties are covered in this handbook and numerous references are provided at the end of each chapter.

1571. Handbook of Paperboard and Board. Robert Robin Higham. Beekman Publishers, Inc, 1970-71, 2 Vols.. ISBN: 0-8464-0469-9. Cost: $50.00 per set.

This handbook, also available from Business Books, Ltd, London, contains 2 volumes entitled as follows: Vol.1, *Manufacturing Technology*; Vol.2, *Technology of Conversion and Usage*.

1572. Handbook of Papermaking: The Technology of Pulp, Paper and Board Manufacture. Robert R. Higham. Beekman Publishers, Inc, 2d ed., 1968, 400 pp. Cost: $29.95.

1573. Handbook of Parapsychology. Benjamin B. Wolman. Van Nostrand Reinhold, 1977, 870 pp. ISBN: 0-442-29576-6. Cost: $35.00.

1574. Handbook of Pediatric Anesthesia David J. Steward. Churchill Livingstone, 1979. ISBN: 0-443-08019-4. Cost: $15.00.

This handbook provides information on anesthesia for specific procedures. Such procedures as neurosurgery, neuroradiology, ophthamology, plastic surgery, cardiac surgery, urologic surgery, orthopedic surgery, and otorhinolaryngology and thoracoabdominal surgery are covered in this handbook.

1575. Handbook of Pediatric Cardiology. Louis Gerome Krovetz et al. Harper, 1969, 388 pp. LC: 68-57703.

1576. Handbook of Pediatric Ophthalmology. S. Feman; R. Reinecke. Grune and Stratton (available from Stechert Macmillan), 1978, 208 pp. ISBN: 0-8090-1115-5. Cost: $19.50.

1577. Handbook of Pediatric Primary Care. M.P. Chow et al. John Wiley and Sons, 1979, 1,084 pp. LC: 78-19731. ISBN: 0-471-01771-X. Cost: $19.50.

This handbook covers such topics as nutrition of children, parent-child interactions, family planning, medication, learning disabilities, and basic health care of children and adolescents. The book is primarily for pediatric nursing students.

1578. Handbook of Pediatric Radiology. Herbert J. Kaufmann. Masson Publishing USA, Edited and adapted from the French version by J. Lefebvre, et al, 780 pp.

NOTE: Vol.1 is entitled: *Pulmonary and Cardiovascular Systems.*

1579. Handbook of Pediatric Surgery. Jack H. Hertzler; Medo Mirza. Year Book Medical Publishers, 1974. ISBN: 0-8151-4400-8. Cost: $7.95 paper.

1580. Handbook of Pediatric Surgical Emergencies. Diller Groff. Medical Examination Publishing Co, 1975. ISBN: 0-87488-642-2. Cost: $10.00.

1581. Handbook of Pediatrics. Henry K. Silver et al. Lange Medical, 12th ed., 1977, 705 pp. ISBN: 0-87041-064-4. Cost: $9.00.

1582. Handbook of Pest Control. Arnold Mallis. Pest Control Technology, 6th ed., 1982, 1,200 pp. Cost: $65.00.

This handbook is a highly respected book on control techniques of various types of insects and pests. Numerous topics are presented, including a discussion of such pests as silverfish, cockroaches, crickets, termites, earwigs, rats, mice, fungi, lice, bedbugs, moths, beetles, ants, bees, wasps, fleas, gnats, mosquitoes, spiders, mites, and ticks. In addition, control techniques are presented, including a discussion of the chemicals used, fumigation methods, and the different types of equipment needed.

1583. Handbook of Pest Control; The Behavior, Life History, and Control of Household Pests. Arnold Mallis. MacNair-Dorland, 4th ed., 1964, 1,148 pp. LC: 64-2249.

Handbook of Pest Management in Agriculture. *See* **CRC Handbook of Pest Management in Agriculture.**

Handbook of Pharmaceutical Analysis. *See* **Handbuch der Arzneimittle-Analytik.**

1584. Handbook of Pharmaceutical and Clinical Measurements and Analysis. Thomas. Reston, 1977. Cost: $29.95.

1585. Handbook of Pharmacology; The Actions and Uses of Drugs. Windsor Cooper Cutting. Appleton (ACC), 5th ed., 1972, 659 pp. LC: 71-160806. ISBN: 0-390-25095-3 (0-8385-3622-0). Cost: $13.95.

1586. Handbook of Phase Diagrams of Silicate Systems, Vol. 1, Binary Systems; Vol. 2, Metal-Oxygen Compounds in Silicate Systems. N.A. Toropov; V.P. Barzakovskii; V.V. Lapin; N.N. Kurtseva. NTIS (Russian translation for the National Bureau of Standards), 2d rev. ed., 1969, trans. 1972; Vol. 1, 1972, Vol. 1, 731 pp. Accession: Vol. 1, TT71-50040; Vol. 2, TT71-50041. Cost: Vol. 1, $21.50; Vol. 2, $12.50.

1587. Handbook of Photochemistry. Steven L. Murov. Marcel Dekker, 1973, 272 pp. LC: 73-89496. ISBN: 0-8247-6164-2. Cost: $43.50.

Handbook of Photography and Reprography: Materials and Processes. *See* **Nebletts' Handbook of Photography and Reprography: Materials and Processes.**

1588. Handbook of Phycological Methods; Culture Method and Growth Measurements; sponsored by the Phycological Society of America. Janet Ruth Stein. Cambridge Universty Press, 1973, 448 pp. LC: 73-79496. ISBN: 0-521-20049-0. Cost: $39.50.

1589. Handbook of Physical Calculations. Jan J. Tuma. McGraw-Hill, 1976, 384 pp. LC: 75-23347. ISBN: 0-07-065438-7. Cost: $19.50.

1590. Handbook of Physical Constants. Sydney P. Clark Jr. Geological Society, rev. ed., 1966, 587 pp. LC: 66-19814. ISBN: 0-8137-1097-9. Cost: $7.00. Series: *GSA Memoir*, 97.

1591. Handbook of Physical Distribution Management. Felix Wentworth. Beekman Publishers, Inc (also from Gower Press), 2d ed., 1976, 534 pp. ISBN: 0-7161-0257-9. Cost: $36.00 (£10.50).

1592. Handbook of Physical Medicine and Rehabilitation. Frank Hammond Krusen et al. Saunders, 2d ed., 1971, 920 pp. LC: 77-139429. ISBN: 0-7216-5571-8. Cost: $25.00.

Handbook of Physical Properties of Rocks. *See* **CRC Handbook of Physical Properties of Rocks.**

1593. Handbook of Physical Therapy. Robert Shestack. Spring Publications, 3d ed., 1977. ISBN: 0-8261-0173-9.

1594. Handbook of Physics. E.U. Condon; Hugh Odishaw. McGraw-Hill, 2d ed., 1967, 1,690 pp. (various paging). LC: 66-20002. ISBN: 0-07-012403-5. Date of First Edition: 1958. Cost: $59.50.

This handbook includes information on all the basic fields of physics including nuclear and atomic physics, optics, heat relationships, electricity, mechanics, and magnetism. A special section on mathematics is included for the convenience of the user, and the index has been improved over the first edition. In addition, such topics as relativity, solid state physics, and plasma physics are discussed at length. For related works, please refer to such titles as the *CRC Handbook of Chemistry and Physics*; the *CRC Handbook of Tables for Applied Engineering Science*; and the *American Institute of Physics Handbook*.

Handbook of Physiology in Aging. *See* **CRC Handbook of Physiology in Aging.**

1595. Handbook of Physiology; A Critical, Comprehensive Presentation of Physiological Knowledge and Concepts. American Physiological Society; H.W. Magoun; John Field et al. Williams and Wilkins, 1959-, multivolume set, 7 sections in 21 vols. LC: 60-4587. ISBN: Vol.1, 0-683-05404-X. Cost: Consult publisher for price.

The *Handbook of Physiology* consists of 7 sections, each section having a number of volumes. The sections are entitled as follows: Section 1: *Neurophysiology*, in 3 vols.; Section 2: *Circulation*, in 3 vols.; Section 3: *Respiration*, in 2 vols.; Section 4: *Adaptation to the Environment*, in 1 vol.; Section 5: *Adipose Tissue*, in 1 vol.; Section 6: *The Alimentary Canal*, in 5 vols.; Section 7: Endocrinology, *in 6 vols.*

1596. Handbook of Plant Virus Infections: Comparative Diagnosis. Edouard Kurstak. Elsevier, 1981, 943 pp. ISBN: 0-444-80309-2. Cost: $164.50.

1597. Handbook of Plasma Instabilities. Ferdinand F. Cap. Academic Press, 1976, 2 Vols. (Vol.3 in prep.). LC: 76-4075. ISBN: Vol.1, 0-12-159101-8; Vol.2, 0-12-159102-6. Cost: Vol.1, $36.50; Vol.2, $41.50.

1598. Handbook of Plastic Fillers and Reinforcements for Plastics. Harry S. Katz; John V. Milewski. Van Nostrand Reinhold, 1977, 800 pp. ISBN: 0-442-25372-9. Cost: $49.50.

1599. Handbook of Plastic Optics, With Emphasis on Injection-Molded Optics. Precision Lens, Cincinnati, 1973, 105 pp. LC: 75-126126.

1600. Handbook of Plastic Product Design Engineering. Sidney Levy; J. Harry DuBois. Van Nostrand Reinhold, 1977, 332 pp. ISBN: 0-442-24764-8. Cost: $27.50.

This handbook covers such topics as the physical structure and molecular configuration of polymer materials, as well as the geometric design considerations for improving the effective stiffness of plastic structures. In addition, mathematical formulas are provided to describe plastic structures under various static and dynamic loads.

1601. Handbook of Plastics and Elastomers. Charles A. Harper. McGraw-Hill, 1975, 1,024 pp. (various paging). LC: 75-9790. ISBN: 0-07-026681-6. Cost: $51.50.

Includes bibliographies and index. Covers laminates, reinforced plastics, fibers, foams, liquid resin systems, coatings, and adhesives.

1602. Handbook of Plastics in Electronics. Dan Grzegorczyk; George Feineman. Reston Publishing Co, 1974, 444 pp. LC: 73-19641. ISBN: 0-87909-338-2. Cost: $21.95.

1603. Handbook of Plastics Test Methods. G.C. Ives. CRC Press, 1971. LC: 76-173786. Cost: $35.00.

1604. Handbook of Poisoning. Robert H. Dreisbach. Lange Medical Publishers, 10th ed., 1980, 578 pp. LC: 79-92918. Cost: $9.00.

1605. Handbook of Poisoning: Diagnosis and Treatment. Robert H. Dreisbach. Lange Medical Publishers, 9th ed., 1977. Cost: $9.00.

1606. Handbook of Pollution Control Management. Herbert F. Lund. Prentice-Hall, Inc, 1978, 444 pp. LC: 77-17330. ISBN: 0-13-380634-0. Cost: $39.95.

A unique aspect of this handbook is the usage of the critical path sequence format designed to help make the proper decisions with regard to the installation and usage of pollution control equipment. This handbook includes approximately 22 papers in the field of environmental engineering. Such topics as stack emission monitoring, solid waste control, water effluent control, air emission control systems, noise control, and cost estimating are discussed in this handbook.

1607. Handbook of Polyolefin Fibres. James G. Cook. Textile Books Service (also available from Merrow Pub. Co.), 1967, 602 pp. LC: 67-81317 (GB). ISBN: 0-87245-058-9. Cost: $15.00.

1608. Handbook of Powder Metallurgy. Henry H. Hausner. Chemical Publishing Co, 1973, 494 pp. Cost: $56.50.

More than 900 tables and graphs on such topics as powder mixing, compaction of powders, hot pressing, refractory metal compounds, electrical contact materials, porous materials, friction and antifriction materials, and dispersion strengthening.

1609. Handbook of Powder Technology. J.C. Williams; T. Allen. Elsevier, 3 Vols. (Vol. 4 in preparation). LC: 79-26589.

This 4-volume set is entitled as follows: Vol.1, *Particle Size Enlargement*. C.E. Capes. 1980, 192 pp. $71.75; Vol.2, *Fundamentals of Gas-Particle Flow*. B. Rudingers. 1980, 143 pp. $53.75; Vol.3, *Solid-Gas Separation*. L. Svarovsky. 1981, 123 pp. $46.25; Vol.4, *Dust Explosions*. P. Field. This is a continuing series of handbooks that will ultimately consist of 20 to 25 volumes. Currently, Vols.1-3 are available and Vol.4 is in preparation. This multivolume handbook covers such topics as gas-particle heat transfer, momentum transfer, electrostatic precipitators, dry separators, conglomeration of powders, and various types of particle filters.

1610. Handbook of Power Drives. Trade and Technical Press, Ltd, 1974, 527 pp. LC: 74-174928. ISBN: 0-85461-051-0.

1611. Handbook of Practical Electronic Tests and Measurements. John D. Lenk. Prentice-Hall, 1969, 302 pp. ISBN: 0-13-380626-X. Cost: $18.95.

1612. Handbook of Practical Microcomputer Troubleshooting. John Lenk. Reston Publications, 1979, 432 pp. LC: 79-1357. ISBN: 0-8359-2757-1. Cost: $18.95.

This book includes information on testing and servicing microcomputers. Such test equipment as logic analyzers, including the HP 1610A and the TEK 7D01, are presented in this handbook.

1613. Handbook of Practical Organic Microanalysis: Recommended Methods for Determining Elements and Groups. S. Bance. John Wiley and Sons, 1981, 206 pp. LC: 80-40145. ISBN: 0-470-26972-3. Cost: $64.95. Series: *Ellis Horwood Series in Analytical Chemistry.*

Presents reliable procedures for organic microanalysis.

1614. Handbook of Practical Pharmacology. Sheila A. Ryan; Bruce D. Claton. Mosby, 1977. LC: 76-41194. ISBN: 0-8016-0978-X. Cost: $7.00 paper.

1615. Handbook of Practical Solid-State Troubleshooting. John D. Lenk. Prentice-Hall, 1972, 310 pp. LC: 75-167631. ISBN: 0-13-380642-1. Cost: $18.95.

A practical guide to circuit and component failures. Covers test equipment and servicing of digital electronic equipment.

1616. Handbook of Practical Urology. Richard Clark Hirschhorn. Lea and Febiger (also available from Kimpton), 1965, 222 pp. LC: 65-19431.

1617. Handbook of Pre-Natal Pediatrics for Obstetricians and Pediatricians. Gifford F. Batstone et al. Lippincott, 1971. LC: 73-166202. ISBN: 0-397-58087-8. Cost: $10.00.

1618. Handbook of Precision Engineering. A. Davidson. McGraw-Hill, 1971-74, 10 Vols. LC: 75-137567. ISBN: Vol. 1, 0-07-015455-4; Vol 2, 0-07-015456-2; Vol. 3, 0-07-015461-9; Vol. 4, 0-07-015462-7; Vol. 5, 0-07-015472-4; Vol. 6, 0-07-015473-2; Vol. 7, 0-07-015476-7; Fol. 8, 0-07-015476-7; Vol. 8, 0-07-015477-5; Vol. 9, 0-07-015501; Vol. 10, 0-07-015493-7. Cost: $29.50 each volume.

This multivolume set includes information on such subjects as properties of glass, ceramics, and plastics; working and shaping nonmetallic materials; making permanent joints in precision products; and an analysis of accurate dimensioning and interchangability of parts. In addition, this comprehensive work includes information on the techniques of chemical and physical fabrication, such as thin-film technology, etching processes, and electrical discharge ultrasonics. Methods of heat treating are discussed in connection with the preparation of surfaces, as well as as improving mechanical properties of the material.

1619. Handbook of Precision Sheet, Strip, and Foil. Hamilton B. Bowman. American Society for Metals, 1980, 312 pp. Cost: $76.00.

This handbook presents information on thin flat-rolled metal and emphasizes mainly stainless steel, titanium, and nickel-based alloys.

1620. Handbook of Preparative Inorganic Chemistry. George Brauer (Scripta Technica, Inc, trans). Academic Press, 2d ed., Vol. 1, 1963; Vol. 2, 1965, 2 Vols. LC: 63-14307. ISBN: Vol.1, 0-12-126601-X; Vol.2, 0-12-126602-8. Cost: $55.00 per Vol.

1621. Handbook of Pressure-Sensitive Adhesive Technology. Don Satas. Van Nostrand Reinhold, 1981, 680 pp. ISBN: 0-442-25724-4. Cost: $35.00.

Presents information on the properties of pressure-sensitive adhesives, such as peeling resistance and bonding strength. Covers information on the adhesive application industry, pressure-sensitive products, and raw materials that are used to make pressure-sensitive materials.

1622. Handbook of Preventive Medicine and Public Health. Murray Grant. Lea and Febiger, 1967, 242 pp. LC: 67-13884.

1623. Handbook of Probability and Statistics with Tables. R.S. Burington; D.C. May Jr. McGraw-Hill, 2d ed., 1970, 416 pp. LC: 68-55264. ISBN: 0-07-009030-0. Cost: $19.50.

1624. Handbook of Process Stream Analysis. Kenneth J. Clevett. Halsted Press, 1974, 544 pp. LC: 73-14416. ISBN: 0-470-16048-9. Cost: $52.95.

1625. Handbook of Property Conservation. Factory Mutual Engineering. Factory Mutual Engineering, 2d ed., 1979, 240 pp. Cost: $1.50.

This handbook is basically a guide to the protection of property and presents information on loss control. The latest techniques in loss control are presented along with such new equipment as rack storage sprinklers and flood shields.

1626. Handbook of Protein Sequence Analysis: A Compilation of Amino Acid Sequences of Proteins. L.R. Croft. John Wiley and Sons, 2d ed., 1980, 628 pp. LC: 79-41487. ISBN: 0-471-27703-7. Cost: $105.00.

Includes information on over 800 protein structures up to the end of 1978. Covers such topics as the purification of peptides, sequencing of amino acids, and the usage of mass spectrometry to the sequence analysis of peptides and proteins.

1627. Handbook of Proton Ionization Heats and Related Thermodynamic Quantities. J.J. Christensen et al. John Wiley and Sons, 1976, 269 pp. LC: 76-16511. ISBN: 0-471-01991-7. Cost: $29.95.

1628. Handbook of Psychiatric Emergencies. Andrew E. Slaby; Julian Lieb; Laurence R. Tancredi. Medical Examination Publishing Co, 1975, 191 pp. ISBN: 0-87488-645-7. Cost: $10.00.

1629. Handbook of Psychiatry. Phillip Solomon; Vernon D. Patch. Lange Medical Publishers, 4th ed., 1977. Cost: $11.00.

1630. Handbook of Psychobiology. Michael S. Gazzaniga (ed); Colin Blakemore (ed). Academic Press, 1975. ISBN: 0-12-278656-4. Cost: $29.50.

1631. Handbook of Psychonomics. J.A. Michon; E.G.J. Eijkam; L.F.W. DeKlerk. Elsevier, 1979, Vol.1, 654 pp.; Vol.2, 658 pp. ISBN: Vol.1, 0-444-85109-7; Vol.2, 0-444-85194-1; 2-Vol. set, 0-444-85201-8. Cost: Vol.1, $124.50; Vol.2, $131.75; 2-Vol. set, $244.00.

1632. Handbook of Psychopharmacology. L.L. Iversen; S.D. Iversen. Plenum Press, 1975-, 14 Vols. available. LC: 75-6851. ISBN: Vol.1, 0-306-38921-5. Cost: Vols. range from $22.50 to $39.50.

This handbook currently consists of 14 volumes entitled as follows: Vol.1, *Biochemical Principles and Techniques in Neuropharmacology*; Vol.2, *Principles of Receptor Research*; Vol.3, *Biochemistry of Biogenic Amines*; Vol.4, *Amino Acid Neurotransmitters*; Vol.5, *Synaptic Modulators*; Vol.6, *Biogenic Amine Receptors*, Vol.7, *Animal Psychopharmacology*; Vol.8, *Drugs, Neurotransmitters and Behavior*; Vol.9, in preparation;

Vol.10, *Neuroleptics and Schizophrenia*; Vol.11, *Stimulants*; Vol. 12, *Drugs of Abuse*; Vol.13, *Biology of Mood and Antianxiety Drugs*; and Vol.14, *Affective Disorders: Drug Actions in Animals and Man*.

1633. Handbook of Psychophysiology. N.S. Greenfield; R.A. Sternbach. Holt, Rinehart, and Winston, 1973. LC: 72-158478. ISBN: 0-03-086656-1. Cost: $25.95.

1634. Handbook of Psychotherapy and Behavior Change. S.L. Garfield; A.E. Bergin. John Wiley and Sons, 2d ed., 1978, 1,024 pp. LC: 78-8426. ISBN: 0-471-29178-1. Date of First Edition: 1971. Cost: $35.00.

1635. Handbook of Pulp and Paper Technology. Kenneth W. Britt. Van Nostrand Reinhold, 2d ed., 1970, 724 pp. LC: 71-129018. ISBN: 0-442-15645-6. Cost: $36.50.

1636. Handbook of Pulse-Digital Devices for Communication and Data Processing. Harry Elliot Thomas. Prentice-Hall, 1970, 335 pp. LC: 72-76878. ISBN: 0-13-380675-8. Cost: $16.95.

1637. Handbook of PVC Pipe. Unie-Bell Plastic Pipe Association. Unie-Bell Plastic Pipe Association. Cost: $20.00.

Contains over 300 pages with 120 illustrations, charts, figures, and tables on PVC piping systems. The emphasis in this handbook is on municipal water mains, water distribution, and sanitary sewer systems.

1638. Handbook of Pyrotechnics. Karl O. Brauer. Chemical Publishing Co, 1974, 412 pp. ISBN: 0-8206-0220-5. Cost: $37.00.

Covers property of explosives and applications of explosive methods in manufacturing processes. Includes information on release mechanisms, missile systems, quality assurance testing, explosive forming, and piston and bellows devices.

1639. Handbook of Radar Engineering Fundamentals. V. Ya. Tsylov (Army Foreign Science Technology Center, trans). NTIS, 1970, 698 pp. Accession: AD-716660. Tech Report: FSTC-HT-23-1114-70.

1640. Handbook of Radar Measurement. David K. Barton; Harold R. Ward. Prentice-Hall, 1969, 426 pp. LC: 76-77553. ISBN: 0-13-380683-9.

Handbook of Radiation Doses in Nuclear Medicine and Diagnostic X-ray. *See* CRC Handbook of Radiation Doses in Nuclear Medicine and Diagnostic X-ray.

Handbook of Radiation Measurement and Protection. See *CRC Handbook of Radiation Measurement and Protection.*

1641. Handbook of Radioactivity Measurements Procedures. National Council on Radiation Protection & Measurement. NCRP Publications, 1978, 506 ppp. LC: 78-54276. Cost: $11.00. Series: NCRP Report No. 58.

This report covers such topics as radiation detectors, measurement of radioactive decay, and the identification of radionuclides in the environment. An appendix includes radioactive decay data for approximately 250 radionuclides and also includes statistical data of radioactive decay.

1642. Handbook of Radiochemical Analytical Methods. Frederick B. Johns (National Environmental Research Center, Las Vegas, NV). NTIS, 1975, 149 pp. Accession: PB-240621. Tech Report: EPA/680/4-75-001. Cost: $2.25 microfiche; $5.75 paper copy.

1643. Handbook of Radiochemical Exercises. Andrei Nikolaevich Nesmeianov et al; E. Kloczko (trans). Macmillian (also Pergamon), 1965, 448 pp. LC: 62-9182.

1644. Handbook of Radioimmunoassay, Part 1. Guy E. Abraham (ed). Marcel Dekker, 1977, 832 pp. ISBN: 0-8247-6453-6. Cost: $97.50. Series: *Clinical and Biochemical Analysis Series*, Vol. 5.

1645. Handbook of Radiological Protection. Great Britain Radioactive Substances Advisory Committee. Her Majesty's Stationary Office, 1971-, Vol.1, 192 pp., continuation. LC: 73-502043. ISBN: Vol.1, 0-11-260079-8. Cost: £2.25.

1646. Handbook of Range Distributions for Energetic Ions in All Elements. U Littmark; J.F. Ziegler. Pergamon, 1980, 490 pp. LC: 79-27825. ISBN: 0-08-023879-3. Cost: $72.00. Series: *The Stopping and Ranges of Ions in Matter*, 6.

1647. Handbook of Rectifier Circuits. Graham John Scoles. John Wiley and Sons, 1980, 238 pp. LC: 79-41814. ISBN: 0-470-26950-2. Cost: $69.95. Series: *Ellis Horwood Series in Electrical and Electronic Engineering.*

This handbook covers a wide range of rectifier circuits and includes such information as overload protection, gate-control by thyristors, and the switching of transformers. In addition, such topics as 3-phase rectifier bridges, voltage multiplying bridges, double input circuits, hammer circuits, and anvil circuits are presented.

1648. Handbook of Refractory Compounds. Grigorii Valentinovich Samsonov; Igo'Markovich Vinitskii; Kenneth Shaw (trans)ii+i. Plenum, 1980, 555 pp. LC: 79-17968. ISBN: 0-306-65181-5. Cost: $75.00.

This handbook covers phase diagrams of some binary systems, as well as various properties of refractory compounds, including magnetic properties, thermal properties, thermodynamic properties, electric properties, optical properties, mechanical properties, chemical properties, and refractory properties.

1649. Handbook of Refrigerating Engineering. W.R. Woolrich. AVI Publishing Co, 4th ed., 2 Vols., Vol.1, 1965; Vol.2, 1966, Vol.1, 460 pp.; Vol.2, 434 pp. LC: 65-12890. ISBN: Vol.1, 0-87055-054-3; Vol.2, 0-87055-055-1. Cost: $29.00 per Vol.

The 4th ed., of this handbook is divided into 2 vols., entitled: Vol.1, *Fundamentals*; Vol.2, *Applications*. This handbook contains a contains a considerable amount of information on some of the thermodynamic properties of refrigerants; and various charts, formulas, equations, and data in connection with the fundamentals of regfrigeration and food processing. For example charts are provided on the thermodynamic properties of azeotrope refrigerant 500. (For a comprehensive listing of refrigerants and their properties, please refer to the *ASHRAE Handbook of Fundamentals*.) The *Handbook of Refrigerating Engineering* also includes information on the applications of basic principles of refrigeration in processing perishable foods, transportation, storage, and refrigeration of perishable foods and related products. Bibliographic references are provided for further investigation.

1650. Handbook of Reinforced Plastics of the Society of the Plastics Industry, Inc. Samuel S. Oleesky; J.G. Mohr. Reinhold, 1964, 640 pp. LC: 64-15205.

1651. Handbook of Relay Switching Technique. J.T. Appels; B.H. Geels; R.H. Bathgate (trans). Springer-Verlag (also available from Cleaver-Hume), 1966, 321 pp. LC: 66-23051. ISBN: 0-387-91005-0. Cost: $14.90.

1652. Handbook of Remote Control and Automation Techniques. John E. Cunningham. TAB Books, 1978, 292 pp. LC: 78-11475. ISBN: 0-8306-9848-5. Cost: $12.95.

Numerous topics are presented in this handbook, including tone-operated systems, time-controlled systems, ultrasonic systems, light-beam systems, electron motors, and solenoids. In addition, such topics as radio-controlled systems, closed-looped control systems, and computer-based systems are discussed.

1653. Handbook of Research and Development Forms and Formats. Robert D. Carlsen. Prentice-Hall, 1978, 573 pp. Cost: $45.00.

1654. Handbook of Rigging: In Construction and Industrial Operations. W. Rossnagel. McGraw-Hill, 3d ed., 1964, 375 pp. LC: 63-19770. ISBN: 0-07-053940-5. Cost: $26.50.

1655. Handbook of Rockets and Guided Missiles. N.J. Bowman.. Perastadion Press, 2d ed., 1963, 1,008 pp.

1656. Handbook of Rotating Electric Machinery. Donald V. Richardson. Reston, 1980, 636 pp. LC: 79-26216. ISBN: 0-8359-2759-8. Cost: $29.95.

This is a useful training manual for those interested in electric motors. Such topics as direct current generators, direct current motor control, alternating current motors, synchronous alternators, transformers, polyphase induction motors, and single-phase AC motors are discussed. A bibliography and index are provided for the convenience of the user.

1657. Handbook of Rubber Silicones and Compounds. Wilfred Lynch. Van Nostrand Reinhold, 1977, 350 pp. ISBN: 0-442-24962-4. Cost: $25.00.

1658. Handbook of Satellites and Space Vehicles. Robert Paul Haviland; C.M. House. Van Nostrand Reinhold, 1965, 457 pp. LC: 65-7525.

1659. Handbook of Seagrass Biology; An Ecosystem Perspective. Ronald C. Phillips. Garland STPM Press, 1980, 450 pp. LC: 77-90470. ISBN: 0-8240-7025-9. Cost: $42.50.

This handbook covers such topics as the taxonomy of seagrasses, remote sensing of seagrass beds, seagrass epiphytes, cellular carbohydrate composition of seagrasses, and various types of animals associated with seagrasses.

1660. Handbook of Selected Organ Doses for Projections Common in Diagnostic Radiology. Marvin Rosenstein. NTIS, 1976. Accession: PB-257-482. Tech Report: DHEW/PUBL/FDA-76/8031; FDA/BRH-76/99. Cost: $3.00 microform; $4.00 paper copy.

1661. Handbook of Selected Properties of Air- and Water-Reactive Materials. Jack R. Gibson; Geanne D. Weber (Naval Ammunition Depot, Carne, Indiana). NTIS, 1966-69, 225 pp. Accession: AD-688422, N69-35282. Tech Report: NAD-CR-RDTR-144.

Such properties as melting point, boiling point, vapor pressure, thermodynamic properties and flammability are listed. In addition, solubility, toxicity, and molecular weight are included for each compound.

1662. Handbook of Semiconductor Circuits. Armed Forces Supply Support Center; Gernsback Library Staff. TAB Books (originally published by the U.S. Government Printing Office), 1967, 448 pp. ISBN: 0-8306-7030-0. Cost: $8.95.

Original title was *Military Standardization Handbook, Selected Semi-Conductor Circuits.*

1663. Handbook of Semiconductor Electronics: A Practical Manual Covering the Physics, Technology, and Circuit Applications of Transistors, Diodes, and Photocells. Lloyd P. Hunter. McGraw-Hill, 3d ed., 1970, 1,100 pp. (various paging). LC: 69-17183. ISBN: 0-07-031305-9. Date of First Edition: 1956. Cost: $46.50.

This handbook will be of interest to those engineers who are concerned with the characteristics, various applications, and fabrication of different types of semiconductors. The handbook includes background information on the physics of different types of semiconductors including transistors, diodes, and photocells. Information is provided on how semiconductors are made, the materials that are used, and the various processes associated with the design of semiconductors. Much information is provided on the various circuits that utilize semiconductors including such circuits as amplifiers, oscillators, microwave circuits, and switching circuits. An excellent bibliography and index is included for the user's convenience. For related works, please refer to the *Electrical Engineer's Reference Book*, the *Electronics Engineers' Handbook*, the *Basic Electronic Instrument Handbook*, the *Handbook of Electronic Packaging*, the *Handbook of Modern Electronic Data, Industrial Electronics Handbook*, and the *Standard Handbook for Electrical Engineers*. Also refer to the *Handbook for Electronics Engineering Techncians, Electronics Handbook, Electronic Conversions, Symbols, and Formulas.*

1664. Handbook of Sensory Physiology. H. Autrum et al. Springer-Verlag, 1971-, 8 Vols. including many parts. ISBN: Vol.1, 0-387-05144-9. Cost: By subscription.

1665. Handbook of Separation Techniques for Chemical Engineers. Philip A. Schweitzer. McGraw-Hill, 1979, 1,120 pp. LC: 79-4096. ISBN: 0-07-055790-X. Cost: $42.50.

Separation techniques are discussed for all types of mixtures ranging from liquid-liquid mixtures, gas mixtures, solid-liquid mixtures, solid mixtures, gas-solid mixtures, to liquids with dissolved solids. Such topics as continuous distillation, batch distillation, solvent recovery, extraction, ion exchange separations, dialysis, electrodialysis, and parametric pumping are discussed. In addition, such techniques as membrane filtration, evaporation, foam separation, crystallization from solutions, hydrometallurgical extraction, gas-phase absorption, centrifugation, and hydrocyclone separation methods are presented.

1666. Handbook of Series for Scientists and Engineers. Visvaldis Mangulis. Academic Press, 1965, 134 pp. LC: 65-26418. ISBN: 0-12-468850-0. Cost: $26.50.

1667. Handbook of Sexology. J. Money; H. Musaph. Excerpta Medica, 1977, 1,400 pp. ISBN: 90-219-2104-9. Cost: $79.95.

1668. Handbook of Silicone Rubber Fabrication. Wilfred Lynch. Van Nostrand Reinhold, 1978, 257 pp. LC: 77-10986. ISBN: 0-442-24962-4. Cost: $25.50.

Covers continuous transfer molding, automatic insert molding, and the molding of liquid rubber by the injection molding. Also, treats heat vulcanizing and extrusion die design.

1669. Handbook of Simplified Electrical Wiring Design. John D. Lenk. Prentice-Hall, 1975, 416 pp. LC: 76-86598. ISBN: 0-13-381723-7. Cost: $18.95.

1670. Handbook of Simplified Solid-State Circuit Design. John D. Lenk. Prentice-Hall, 2d ed., 1978, 429 pp. LC: 77-23555. ISBN: 0-13-381715-6. Cost: $18.95.

1671. Handbook of Simplified Television Service. John Lenk. Prentice-Hall, 1978, 432 pp. ISBN: 0-13-381772-5. Cost: $18.95.

This is a helpful guide to the use of test equipment for troubleshooting television circuit and component failures. This handbook would be useful for all types of TVs, whether they be color, black and white, solid state, or contain integrated circuits.

1672. Handbook of Sludge Handling Processes. Gordon L. Culp. Garland Press, 1979, 238 pp. LC: 78-20644. ISBN: 0-8240-7062-3. Cost: $22.50.

This handbook discusses such topics as sludge stabilization, sludge thickening, sludge dewatering, incineration, sludge disposal, and the transportation of sludge.

1673. Handbook of Soil Mechanics, Vol. 1, Soil Physics. Vol. 2, Soil Testing. Arpad Kezdi; I. Lazanyi (trans). Elsevier Sci, Vol.1, 1974; Vol.2, 1980, Vol.1, 294 pp.; Vol.2, 260 pp. LC: 73-85224. ISBN: 0-444-99890-X. Cost: Vol.1, $70.75; Vol.2, $70.75.

Covers soil testing, load bearing capacity of soils, ground water, earthwork, and soil exploration.

1674. Handbook of Solar and Wind Energy. Floyd Hickok. Cahners, 1975, 125 pp. LC: 75-22495. ISBN: 0-8436-0159-0. Cost: $20.00.

1675. Handbook of Solar Simulation for Thermal Vacuum Testing. John S. Griffith. Institute of Environmental Sciences, 1968, various paging. LC: 68-6392.

Includes bibliographies.

1676. Handbook of Solid Waste Disposal: Material and Energy Recovery. Joseph L. Pavoni et al. Van Nostrand Reinhold, 1975, 560 pp. LC: 74-26777. ISBN: 0-442-23027-3. Cost: $31.50.

This handbook includes information on such topics as sanitary landfills, compost methods, and incineration processes. Many bibliographic references are provided at the end of each chapter for further investigation, and an index is provided for the user's convenience. In addition, this handbook contains information on the recovery of wastes, the recycling of wastes, and energy recovery systems such as methane production and the use of heat from wastes. For related works, please refer to such titles as the *Environmental Engineers' Handbook; Pollution Engineering Practice Handbook; Handbook of Environmental Civil Engineer-*ing; *Air Pollution; Air Pollution Engineering Manual; Betz Handbook of Industrial Water Conditioning,* and the *CRC Handbook of Environmental Control.*

1677. Handbook of Solid Waste Management. David G. Wilson. Van Nostrand Reinhold, 1977, 752 pp. LC: 76-20505. ISBN: 0-442-29550-2. Cost: $42.50.

1678. Handbook of Solid-Liquid Equilibria in Systems of Anhydrous Inorganic Salts. N.K. Voskresenskay (ed) (U.S. Atomic Energy Commission). Israel Program for Scientific Translation (NTIS), 1970, 2 Vols. LC: 75-607293. Tech Report: ACE TR-6983/1.

1679. Handbook of Solid-State Devices: Characteristics and Applications. Michael Thomason. Reston, 1979, 264 pp. LC: 78-15658. ISBN: 0-8359-2761-X. Cost: $17.95.

This handbook covers all different types of solid state devices, including diodes, transistors, thyristors, integrated circuits, microprocessors, and optoelectronic devices. A glossary is provided for the convenience of the user.

1680. Handbook of Solid-State Troubleshooting. Hershal Gardner. Reston, 1976, 318 pp. LC: 76-832. ISBN: 0-87909-339-0. Cost: $17.95.

1681. Handbook of Solvents. Leopold Scheflan; Morris B. Jacobs. Krieger, reprint of 1953 edition, 1974, 728 pp. ISBN: 0-88275-130-1. Date of First Edition: 1953. Cost: $36.00.

1682. Handbook of Sound and Vibration Parameters. General Dynamics. NTIS, 1971, 210 pp. Accession: AD-A071 837. Tech Report: U443-78-072. Cost: $3.50 microform; $17.00.

Discusses such topics as mechanical vibrating systems, acoustics, fluid-liquid interactions, and various methods of frequency analysis.

1683. Handbook of Soviet Adhesives. Howard Jaffe et al (Army Foreign Science and Technology Center). NTIS, 1974, 245 pp. Accession: AD-780713. Tech Report: FSTC-HB-01-100-74.

Presents an analysis of Soviet conventional and high-performance adhesives and tabular data on the properties and compositions of adhesive material. Also covers glue welding technology.

1684. Handbook of Soviet Alloy Compositions. Marshall J. Wahll; Roy F. Frontani (Battelle Columbus Laboratory, Ohio Metals and Ceramics Information Center). NTIS, Feb. 1975, 321 pp. Accession: AD-A009 118. Tech Report: MCIC-HB-05. Cost: $25.00 microfiche; $25.00 paper copy.

This handbook is a computer-generated listing of data on chemical compositions, designations, and other information on a wide variety of metals and alloys that are used in the Soviet Union. The 25 tables and index of alloys are computer-generated. An explanation of the Soviet system of alloy and metal designations is provided in a general section as well as in an introduction at the beginning of each table. First Supplement issued April 1976, 207 pp., AD-A013699, Report No.: MCIC-HB-10551.

1685. Handbook of Soviet Lunar and Planetary Exploration. Nicholas L. Johnson. American Astronautical Society (available from Univelt, Inc), 1979, 276 pp. ISBN: 0-87703-130-4; 0-87703-131-2, paper. Cost: $35.00; $25.00, paper.

This is a detailed accounting of the Soviet unmanned space program from its beginning to late 1979. The exploration of the moon, Venus, and Mars is discussed. Various Soviet flights, including the Luna, the Kosmos, Zond, and Venera are presented.

1686. Handbook of Soviet Manned Space Flight. Nicholas L. Johnson. Univelt, 1980, 461 pp. ISBN: 0-87703-115-0. Cost: $45.00; $35.00 paper. Series: *Science and Technology*, Vol 48.

This is a companion volume to the *Handbook of Soviet Lunar and Planetary Exploration.*

1687. Handbook of Soviet Reinforced Plastics. Charles A. Petschke (Army Foreign Science and Technology Center). NTIS, 1973, 400 pp. Accession: AD-766 311/5. Tech Report: FSTC-CW-01-102-73.

1688. Handbook of Soviet Space Science Research. G.E. Wukelic (Battelle Memorial Institute). Gordon and Breach, 1968, 505 pp. LC: 67-30479. ISBN: 0-677-11770-1. Cost: $78.75.

1689. Handbook of Specialty Elements in Architecture. Andrew Alpern. McGraw-Hill, 1981, 448 pp. LC: 81-8417. ISBN: 0-07-001360-8. Cost: $42.50.

This handbook covers such topics as exterior plantings, trees and plants for interior design, pools and fountains, exterior lighting, signs, flagpoles, and recreational areas associated with structures. This handbook also covers such topics as designing structures for the handicapped.

1690. Handbook of Specific Losses in Flow Systems. Robert P. Benedict; Nicola A. Carlucci. IFI/Plenum Press, 1966, 193 pp. LC: 65-25129. ISBN: 0-306-65122-X. Cost: $32.50.

Handbook of Spectrophotometric Data of Drugs. *See* **CRC Handbook of Spectrophotometric Data of Drugs.**

1691. Handbook of Speech Pathology and Audiology. Lee E. Travis. Prentice-Hall, 1971. LC: 74-146360. ISBN: 0-13-381764-4. Cost: $34.95.

1692. Handbook of Stack Sampling and Analysis. Richard J. Prowals; Lorin J. Zaner; Karl Sporek. Technomic Publishing Co, 1978, 441 pp. LC: 78-19914. ISBN: 0-87762-233-7. Cost: $45.00.

Over 692 air contaminants are listed in this handbook with their corresponding stack sampling methods. The emphasis of this book is on air contaminants emitted from stationary sources. Such comtaminants as ethane, ethyl butyl ketone, ferban, hydrogen perozide, nicotine, naptha, octane dimethyl sulfide, turpentine, vinyl toluene, yttrium, and zirconium are discussed.

1693. Handbook of Stainless Steels. D. Peckner; I.M. Bernstein. McGraw-Hill, 1977, 928 pp. (various paging). LC: 76-54266. ISBN: 0-07-049147-X. Cost: $52.50.

Discusses wrought and cast steels, including physical characteristics, also corrosion, machining cast parts, and powder metallurgy.

1694. Handbook of Statistical Distribution. J.K. Patel et al. Marcel Dekker, Inc, 1976, 320 pp. ISBN: 0-8247-6362-9. Cost: $24.50. Series: *Statistics: Textbooks and Monographs Series*, Vol 20.

1695. Handbook of Statistics. P.R. Krishnaiah. North-Holland, 1980.

The first 2 Volumes are entitled: Vol.1, *Analysis of Variance*, 1,000 pp., ISBN: 0-444-85335-9, $134.25. Vol.2, *Classification, Pattern Recognition, and Reduction of Dimension*. This handbook series has been designed to disseminate information on statistical methodology. It is eventually planned for 14 volumes. Vol.1 contains such topics as tests for normality, univariant and multivariant populations, Anova models, Manova models, growth curve models, dynamic programing, Bayesian inference, and quadratic unbiased estimation.

1696. Handbook of Steel Construction. Canadian Institute of Steel Construction, 1967. LC: 68-118852.

1697. Handbook of Steel Drainage and Highway Construction Products. Highway Task Force for Committee of Galvanized Sheet Producers; Committee of Hot-Rolled and Cold-Rolled Sheet and Strip Producers. American Iron and Steel Institute, 2d ed., 1971, 348 pp. LC: 78-174344. Date of First Edition: 1967.

Contains information on storm drainage, aerial conduits, hydraulics, culverts, tunnels, bridge railing, retaining walls, guardrails, and signs.

1698. Handbook of Stock Gears Featuring Inch and Metric Sizes, Technical Section. Catalog 75. Stock Drive Products, Inc (available from NTIS), 1975, 59 pp. Accession: PB-242149. Cost: $2.25 microfiche.

1699. Handbook of Stock Gears. Stock Drive Products, Inc, 1975. Cost: $2.95.

1700. Handbook of Stopping Cross-Sections for Energetic Ions in All Elements. J.F. Ziegler. Pergamon Press, 1980, 432 pp. Cost: $50.00. Series: *The Stopping and Ranges of Ions in Matter*, 5.

1701. Handbook of Strata-Bound and Stratiform Ore Deposits. K.H. Wolf. Elsevier, 1976, 7 Vols. Cost: $478.00 set.

This 7-volume set is entitled as follows: Vol.1, *Classifications and Historical Studies*, 338 pp., ISBN: 0-444-41401-0, $80.50. Vol.2, *Geochemical Studies*, 364 pp., ISBN: 0-444-41402-9, $80.50. Vol.3, *Supergene and Surficial Ore Deposits: Textures and Fabrics*, 354 pp., ISBN: 0-444-41403-7, $80.50. Vol.4, *Tectonics and Metamorphism—Indexes Vols. 1-4*, 430 pp., ISBN: 0-444-41404-5, $80.50. Vol.5, *Regional Studies*, 320 pp., ISBN: 0-444-41405-3, $80.50. Vol.6, *Cu, Zn, Pb, and Ag Deposits*, 586 pp., ISBN: 0-444-41406-1. $129.95. Vol.7, *Au, U, Fe, Mn, Hg, Sb, W, and P Deposits*, 656 pp., ISBN: 0-444-41407-X, $129.25. This 7-volume set contains substantial information on economic geology. It is noted that additional volumes are planned for the future, namely Volumes 8-10. Geologic information is provided for most ore deposits, including such information as metamorphic, sedimentary, and petrological data.

1702. Handbook of Stress and Strength; Design and Material Applications. Charles Lipson; R.C. Juvinall. Macmillan, 1963, 447 pp. LC: 63-10398.

1703. Handbook of Stress-Intensity Factors: Stress-Intensity Factor Solutions and Formulas for Reference. G.C. Sih. Institute of Fracture and Solid Mechanics, 1973, various paging.

This book includes a substantial amount of information on fracture mechanics and represents a catalog of stress-intensity fracture solutions. This reference work includes information on the concept of stress intensity factors to structural members under complex loading conditions. Such numerical calculations as integral transforms, asymptotic expansions, boundary collocation, and finite differences are described in this book. A wide variety of boundary-value crack problems are addressed, such as cracks in specimens of finite size, plates with stiffeners, and two-dimensional linear elastostatic crack problems. In addition, solutions to straight and circularly-shaped cracks in isotropic elasticity are also included.

1704. Handbook of Structural Design. Irvine Ernest Morris. Van Nostrand Reinhold, 1963, 822 pp. LC: 62-10728. ISBN: 0-442-12104-0.

1705. Handbook of Structural Stability. Column Research Committee of Japan. Corona Publishing Co. Ltd, 1971, various paging. LC: 72-185735.

This excellent Japanese handbook is a collection of data, formulas, charts, graphs, and design curves that will provide useful assistance in performing structural stability calculations. This handbook contains information on such subjects as buckling of straight members, different types of plates and shells, arches, and frames. Information is provided on various loading conditions, and different geometrical shapes for each one of the above-mentioned structural entities. In addition, data are provided on such subjects as trusses, boundary conditions, axial compression, bending, shear, torsion, external, and internal pressure. Although the subject index is omitted, the handbook has an excellent and very detailed table of contents, and extensive bibliographies are provided for further investigation. For related works, please refer to the following titles: *Structural Engineering Handbook; Metals Handbook; Steel Designers' Manual; Handbook of Heavy Construction; Handbook of Environmental Civil Engineering; Civil Engineer's Reference Book; Civil Engineering Handbook,* and the *Standard Handbook for Civil Engineers.*

1706. Handbook of Subsurface Geology. Carl Allphin Moore. Harper, 1963, 235 pp. LC: 64-10009.

1707. Handbook of Sugars. W. Ray Junk; Harry M. Pancoast. AVI Publishing Co, 2d ed., 1980, 400 pp. ISBN: 0-87055-348-8. Cost: $45.00.

Covers properties of sugars, bulk handling, corn syrups, blends, and standards for processing. Includes more than 100 tables of data.

1708. Handbook of Superalloys: International Alloy Compositions and Designations Series. M.J. Wahll et al. Van Nostrand Reinhold, 1979, 408 pp. ISBN: 0-442-86097-8 (Van Nostrand); 0-935470-01-8 (Battelle). Cost: $40.00.

This handbook helps to standardize the superalloy nomenclature and designations by listing approximately 200 superalloys that have been produced worldwide. The composition of the superalloys is compared along with the names, designations, and standards. In addition, physical and mechanical properties of many of the superalloys are listed. The 4 indexes help the user approach the desired superalloy by country, company, standard number, or standard designation. Also available from Battelle Columbus Laboratory, and Ohio Metals and Ceramics Information Center.

1709. Handbook of Surface Preparation. Richard C. Snogren. Palmerton Publications, 1974, 577 pp. LC: 73-91040.

Essentially, this handbook deals with the subject of adhesive bonding and includes information and data on such areas as organic coatings, metal conditioners, solvent degreasing, ultrasonic cleaning, sealants for around joints, and various other types of chemical treatments used to prepare surfaces. Various detailed data are provided on numerous metals and nonmetals, including aluminum, steel, magnesium, titanium, and different types of plastics.

1710. Handbook of Surfaces and Interfaces. Leonard Dobrzynski. Garland Press, Vols.1-2, 1978; Vol.3, 1979. LC: Vol.1, 77-24776; Vol.2, 77-24776; Vol.3, 78-27392. ISBN: Vol.1, 0-8240-9857-9; Vol.2, 0-8240-9856-0; Vol.3, 0-8240-9855-2. Cost: $52.50 per Vol.

This handbook is on surface physics and represents a comprehensive treatment of surface phonons and polaritons. Such topics as photoemission, diffusion of crystal surfaces, surface tension, acoustic surface waves, low-energy electron diffraction, surface crystallography, auger electron spectroscopy, and inelastic scattering of electrons are discussed in this handbook. In addition, such topics as chemisorption, surface states of tetrahedral semiconductors, and field emission spectroscopy are treated in this 3-volume set.

1711. Handbook of Surgery. John Long Wilson. Lange Medical Publishers, 5th ed., 1973, 877 pp. LC: 78-180540. ISBN: 0-87041-112-8. Cost: $7.00.

1712. Handbook of Surgical Diathermy. John Phillimore Mitchel; G.M. Lumb. Williams and Wilkins, 1966, 100 pp. LC: 66-66781.

1713. Handbook of Survey Notekeeping. F.W. Pafford. John Wiley and Sons, 1962, 140 pp. LC: 62-17467. ISBN: 0-471-65751-4. Cost: $15.95.

1714. Handbook of Swimming Pool Construction, Maintenance, and Sanitation. Frank L. Cross Jr; W.W. Cameron. Technomic Publishing Co, 1974, 136 pp. LC: 74-76524. ISBN: 0-87762-135-7. Cost: $15.00.

Covers filter systems, disinfection, and winterizing. Includes over 60 data tables.

1715. Handbook of Synchronous Drive Components. Stock Drive Products, Inc, 1974, 356 pp. ISBN: 0-686-05807-0. Cost: $3.95.

1716. Handbook of System and Product Safety. Willie Hammer. Prentice-Hall, 1972, 368 pp. LC: 72-2683. ISBN: 0-13-382226-5. Cost: $32.95.

1717. Handbook of Systematic Botany. Subhash Chandra Datta. Asia Publishing House, 2d ed., 1970, 562 pp. ISBN: 0-210-27082-9. Cost: $9.50.

1718. Handbook of Systems Analysis. J.E. Bingham. Halsted Press (division of John Wiley and Sons), 2d ed., 1978. LC: 77-28954. ISBN: 0-470-99129-1. Date of First Edition: 1972. Cost: $18.95.

Covers such topics as data processing, data communications, data management, input methods, data security, and decision tables.

1719. Handbook of Tables and Formulas for Home Construction Estimating. Paul Thomas. Prentice-Hall, Inc, 1979, 228 pp. ISBN: 0-13-382200-1. Cost: $34.95.

This handbook would be very useful for the home construction industry or for estimating the cost of property losses. Numerous formulas and tables are provided for practical kinds of construction situations and many examples are described in this handbook.

Handbook of Tables for Mathematics. *See* **CRC Handbook of Mathematical Sciences.**

1720. Handbook of Taste and Odor Control Experiences in the U.S. and Canada. American Water Works Association. American Water Works Association, 1976, 106 pp. (various paging). LC: 76-368120.

1721. Handbook of Techniques in High-Pressure Research and Engineering. Daniil Semenovich Tsiklis. Plenum Press, based on 3d rev. ed. (trans. from the Russian), 518 pp. LC: 68-14854. ISBN: 0-306-30324-8. Cost: $49.50.

1722. Handbook of Telemetry and Remote Control. E. Gruenberg. McGraw-Hill, 1967, 1,344 pp. LC: 65-24524. ISBN: 0-07-025075-8. Cost: $49.50.

1723. Handbook of Teratology. James G. Wilson; Clarke F. Fraser. Plenum Press, 1977-, 4 Vols. ISBN: Vol.1, 0-306-36241-4; Vol.2, 0-306-36242-2; Vol.3, 0-306-36243-0; Vol.4, 0-306-36244-9. Cost: Vol.1, $45.00; Vol.2, $39.00; Vol.3, $39.00; Vol.4, $39.00.

The *Handbook of Teratology* consists of 4 volumes entitled as follows: Vol.1, *General Principles and Etiology*; Vol.2, *Mechanisms and Pathogenesis*; Vol.3, *Comparative, Maternal, and Epidemiologic Aspects*; and Vol.4, *Research Procedures and Data Analysis*.

1724. Handbook of Textile Fibers, Dyes, and Finishes. Howard L. Needles. Garland STPM Press, 1981, 170 pp. LC: 79-23188. ISBN: 0-8240-7046-1. Cost: $27.50.

This handbook is a compilation of information on textile fibers, including such fibers as cellulosic fibers, cellulose ester fibers, polyamide fibers, polyester fibers, vinyl fibers, acrylic fibers, protein fibers, and metallic fibers. Such data as the physical properties, chemical properties, and structural properties of these fibers are included. In addition, information is provided on major cleaning agents for maintenance of textile fibers.

1725. Handbook of Textile Fibers. James Gordon Cook. Merrow Publishing Co, Ltd, 4th ed., 1968, 2 Vols. LC: 72-376353.

The *Handbook of Textile Fibers* consists of 2 volumes with the following titles: Vol.1, *Natural Fibers*; Vol.2, *Man-made Fibers*.

1726. Handbook of Textile Finishing. Archibald J. Hall. Textile Book Service, 3d ed., 1966, 453 pp. LC: 65-8760.

1727. Handbook of Textile Testing and Quality Control. E.B. Grover; D.S. Hamby. Wiley-Interscience, 1960, 622 pp. LC: 60-11026. ISBN: 0-470-32901-7. Cost: $47.00.

The main emphasis in this handbook is on cotton and various synthetic fibers. This handbook is concerned with different types of tests that can be performed on these fibers and fabrics which will limit the number of defects and improve quality control. Various tests are performed on these fibers and include tests associated with the strength of the fiber of yarn, twist testing, evenness testing, and fabric strength. Information is also provided on grading of fabrics, and many appendices are included in addition to an index. For related works, please refer to such titles as *The Textile Handbook*, the *Wood Handbook*, the *Cotton Ginners Handbook*, and the *American Cotton Handbook*.

1728. Handbook of Textiles. Ann Margaret Collier. Pergamon Press, 1970, 258 pp. ISBN: 0-08-015548-0.

1729. Handbook of the Analytical Chemistry of Rare Elements. A.I. Busev. Halsted Press, 1972. LC: 75-104379. ISBN: 0-470-12620-5. Cost: $29.95.

1730. Handbook of the Atomic Elements. Richard A. Williams. Philosophical Library, Inc, 1970, 124 pp. LC: 78-92089. ISBN: 0-8022-2340-0. Cost: $6.00.

1731. Handbook of the Atomic Energy Industry S. Jefferson. Newnes, 1958, various paging. LC: 59-20682.

1732. Handbook of the Biology of Aging. Caleb E. Finch; Leonard Hayflick. Van Nostrand Reinhold, 1976, 700 pp. LC: 76-52755. ISBN: 0-442-20796-4. Cost: $32.50.

Includes information on such topics as nutrition, pathobiology, abnormal cell growth, molecular genetics, metabolism, and cell longevity.

1733. Handbook of the British Astronomical Association. British Astronomical Association. British Astronomical Association, annual, 1921-.

1734. Handbook of the Crayfishes of Ontario. Denton W. Crocker; David W. Barr. University of Toronto Press, 1967, 158 pp. LC: 68-79080. ISBN: 0-8020-1816-5. Cost: $15.00.

This handbook covers the natural history and biology of 9 different species of crayfishes that are found in Ontario. A close examination is made of the literature on crayfish.

1735. Handbook of the Elements. Samuel Ruben. Sams, 2d ed., 1967. LC: 67-23714. ISBN: 0-672-20563-7 paper copy; 0-672-20419-3 wall chart. Cost: $3.50 paper copy; $5.95 wall chart.

1736. Handbook of the Engineering Sciences. James H. Potter. Van Nostrand Reinhold, 1967, 2 Vols, Vol.1, 1,347 pp.; Vol.2, 1,428 pp. LC: 67-4632. ISBN: Vol.2, 0-442-06612-0. Cost: Vol.2, $37.50.

Vol.1 is entitled *The Basic Sciences*; Vol.2 is entitled *The Applied Sciences*. This 2-volume set is a massive collection of basic engineering information which is useful for practicing engineers or graduate students as a ready-reference source. As the titles of each volume suggest, the first volume includes such information as chemistry, mathematics, and physics, while the second volume covers such topics as material science, electronics, nuclear engineering, and heat and mass transfer. Many bibliographic references are provided throughout, and there are sections on writing engineering reports as well as retrieving information on engineering, and an extensive index has been prepared. For related works, please refer to the *Handbook of Tables for Applied Engineering Science*, and the *Handbook of Engineering Fundamentals*.

1737. Handbook of the European Iron and Steel Works. Heinman, 5th ed., 1966, 1,442 pp. LC: 63-32146.

1738. Handbook of the Hypothalamus. Peter J. Morgane; Jaak Panksepp. Marcel Dekker, Inc, 1979-, 3 Vols.

The 3 volumes in this set are entitled as follows: Vol.1, *Anatomy of the Hypothalamus*, ISBN: 0-8247-6834-5, $95.00. Vol.2, *Physiology of the Hypothalamus*, ISBN: 0-8247-6881-7, $145.00. Vol.3, *Behavior Studies of the Hypothalamus*, Part A and B, ISBN: Part A, 0-8247-6904-X; Part B, 0-8247-6905-8, $93.50 A; $93.50 B. This is a very comprehensive treatise on the various aspects of the hypothalamus, including articles from such related fields as the anatomy, physiology, biochemistry, endocrinology, and pharmacology of the hypothalamus. Numerous articles are included on such topics as the rodent hypothalamus, electrophysiological studies of the hypothalamus, neurophysiological studies of the hypothalamus, and behavioral studies of the hypothalamus.

1739. Handbook of the Metallurgy of Tin. Dmitrii Vasilevich Beliaev; J.J. Cornish (trans). Pergamon Press, 1963, 122 pp. LC: 62-9696.

1740. Handbook of the Normal Distribution. Jagdish K. Patel; Campbell Read. Dekker, 1982, 352 pp. ISBN: 0-8247-1541-1. Cost: $35.00.

This handbook deals with the usage of distributions and statistics. This handbook includes such topics as bivariate distribution, and various algorithms that are used for computer calculations, as well as numerous references to the literature on normal distribution. In addition, such processes as the Wiener process and the Gaussian process are discussed.

1741. Handbook of the Nutritional Contents of Food. U.S. Department of Agriculture. Dover, reprint, 1975. LC: 75-2616. ISBN: 0-486-21342-0. Cost: $4.00.

NOTE:This book was originally entitled *Composition of Foods*.

1742. Handbook of the Optical Thermal and Mechanical Properties of Six Polycrystalline Dielectric Materials. D.P. Dewitt (Thermophysical and Electronic Properties Information Analysis Center). NTIS, 1972, 247 pp. Accession: N72-33713. Tech Report: NASA-CR-114500. Cost: $9.50.

1743. Handbook of the Physiochemical Properties of the Elements. Grigorii Valentinovich Samsonov. Plenum Press (also Oldbourne Press), trans. ed., 1968, 941 pp. LC: 67-10536. ISBN: 0-306-65126-2. Cost: $75.00.

1744. Handbook of the Poisson Distribution. F.A. Haight. John Wiley and Sons, 1967, 168 pp. LC: 66-28750.

1745. Handbook of the Practice of Anesthesia. John Robert Stalker Shields. Mosby (Kimpton), 1963, 203 pp. LC: 63-14263.

1746. Handbook of the Psychology of Aging. James E. Birren; K. Warner Schaie. Van Nostrand Reinhold, 1977, 740 pp. ISBN: 0-442-20797-2. Cost: $32.50.

Covers behavior genetics, psychophysiology, and how aging affects personality.

1747. Handbook of the Psychology of Obesity. Benjamin B. Wolman. Van Nostrand, 1981, 336 pp. ISBN: 0-442-22609-8. Cost: $24.50.

This handbook addresses the topic of gaining weight and the psychology of overeating. In addition, such topics as depression and poor interrelationships with other people and its effect on obesity and eating habits are discussed.

1748. Handbook of the Rare Elements. M.A. Filiand; E.I. Semenova; Micheal E. Alferieff (trans). Boston Technical Press, 1968, 3 vols. LC: 68-4329.

Vol.1 is entitled *Trace Elements and Light Elements*; Vol.2, *Refractory Elements*; and Vol.3, *Radioactive Elements*. Also available from MacDonald and Co, 8201 Hunting Hill Lane, McLean, Va 22101.

1749. Handbook of Theoretical Activity Diagrams Depicting Chemical Equilibria in Geologic Systems Involving an Aqueous Phase at One ATM and Zero Degrees to 300 Degrees Centigrade. Harold Charles Helgeson et al. Freeman, Cooper and Co, 1969, 253 pp. LC: 73-97467. ISBN: 0-87735-331-X. Cost: $7.50.

1750. Handbook of Thermal Expansion Tables. George R. Bechtell. Pacific Coast Publishers, 1968. ISBN: 0-87465-015-1. Cost: $4.95.

1751. Handbook of Thermal Insulation Design Economics for Pipes and Equipment. William C. Turner; John F. Malloy. Krieger Publishing Co, 1980, 432 pp. LC: 79-13584. ISBN: 0-88275-837-3. Cost: $55.00.

This handbook will help to determine the proper thickness of insulation that would be required for various pipes and equipment. In addition, such topics as fire protection, and temperature control are presented. Energy loss of bare and insulated piping is discussed and numerous examples are provided to determine the economic thickness of insulation for pipes, tubings, and fittings.

1752. Handbook of Thermionic Properties: Electronic Work Functions and Richardson Constants of Elements and Compounds. V.S. Fomenko. IFI/Plenum Press, translated from Russian, 1966, 151 pp. LC: 65-23385. ISBN: 0-306-65117-3. Cost: $42.50.

1753. Handbook of Thermochemical Data for Compounds and Aqueous Species. H.E. Barnes; R.V. Scheuerman. John Wiley and Sons, 1978, 176 pp. LC: 77-20244. ISBN: 0-471-03238-7. Cost: $24.95.

Presents values for heat of formation, free energy of formation, enthalpy, and entropy of ions over the temperature range of 25° to 300°C.

1754. Handbook of Thermodynamic Constants of Inorganic and Orgnaic Compounds. M. Khand Karapet'Yants. Halsted Press, trans. ed., 1968, 461 pp. LC: 72-122509. ISBN: 0-470-45850-X. Cost: $34.95.

1755. Handbook of Thermodynamic Data. G.B. Naumov et al. NTIS, translated from the Russian for the U.S. Geological Survey, 1974, 373 pp. Accession: PB-226722. Tech Report: USGS-WRD-74-001. Cost: $7.75.

1756. Handbook of Thermodynamic Properties of Geological Substances. G.B. Naumov; B.N. Reyence; I.L. Khooakouski. NTIS, Russian translation for the National

Bureau of Standards, 1971, trans. 1974. Accession: TT73-53051. Cost: $9.25.

1757. Handbook of Thermodynamic Tables and Charts. Kuzman Raznjevic. McGraw-Hill (Hemisphere), 1976, 400 pp. ISBN: 0-07-051270-1. Cost: $35.00.

This handbook includes various properties of different types of gases, vapors, solids, and liquids. Some of the properties that are given in these tables are thermal conductivities, emissivities, specific heats, melting points, densities, and viscosities of various substances. This handbook includes such information as conversion factors, SI Units, and various properties of refrigerants. Many charts are included, and an index is provided. For related works, please refer to the *CRC Handbook of Tables for Applied Engineering Science.*

1758. Handbook of Thermophysical Properties of High Temperature Solid Materials. Y.S. Touloukian. Macmillan Co, 1967, 6 Vols., approx. 8,479 pp. ISBN: 0-02-42105. Date of First Edition: 1960 by the Armour Research Foundation as *Handbook of Thermophysical Properties of Solid Materials,* (WADC TR-58 476). Cost: $250.00 complete set.

Although this handbook consists of 6 vols. it actually includes 9 books since vols., 2, 4, and 6 each have 2 parts. The properties coverered include vapor pressure, electrical resistivity, thermal conductivity, thermal diffusivity, heat of vaporization, melting point, heat of fusion, heat of sublimation, density, specific heat at constant pressure, and several thermal radiative properties. The materials that are included are ferrous and nonferrous alloys, oxides, oxygen compounds, polymers, glasses and ceramic materials. Most of these materials were included because their melting points are above 1,000°F.

1759. Handbook of Thermophysical Properties of Solid Materials. A. Goldsmith et al (Armour Research Foundation). Pergamon Press (Macmillan), 1961-63, 5 Vols. LC: 61-11362.

Titles of Vols. I-V are as follows: Vol. I, *Elements*; Vol. II *Alloys*; Vol. III *Ceramics*; Vol. IV *Cermets, Intermetallics, Polymerics and Composites*; Vol. V *Appendix.*

1760. Handbook of Thermoplastic Elastomers. Benjamin M. Walker. Van Nostrand Reinhold, 1979, 345 pp. ISBN: 0-442-29163-9. Cost: $24.95.

This handbook covers such thermoplastic elastomers as block polymers, polyolefins, polyesters, and polyurethane elastomers. A considerable amount of information is provided on the uses, composition, chemical properties, structural properties, aging properties, mechanical properties, electrical properties, and compounding characteristics of elastomers.

1761. Handbook of Thick Film Hybrid Microelectronics: A Practical Sourcebook for Designers, Fabricators and Users. Charles A. Harper. McGraw-Hill, 1974, 1,024 pp. (various paging). LC: 74-2460. ISBN: 0-07-026680-8. Cost: $39.50.

Handbook of Thin Film Technology. *See* **Handbook for Thin Film Technology.**

1762. Handbook of Total Parenteral Nutrition. John Palmer Grant. Saunders, 1980, 197 pp. LC: 79-64592. Cost: $17.50.

1763. Handbook of Toxic and Haxardous Chemicals. Marshall Sittig. Noyes Data Corp, 1981, 729 pp. LC: 81-4950. ISBN: 0-8155-0841-7. Cost: $64.00.

Covers data on approximately 600 toxic and hazardous chemicals. Includes information on chemical composition, safety data, and health as it relates to hazardous waste.

1764. Handbook of Toxic Chemicals. O.N. Vrokofevrmy (Foreign Science and Technology Center, trans). NTIS, 2d ed., 1972, 129 pp. Accession: AD-746649. Tech Report: FSTC-HT-23-1069-72.

1765. Handbook of Toxicity of Pesticides to Wildlife. U.S. Sport Fisheries and Wildlife Bureau. U.S. Government Printing Office, 1970.

1766. Handbook of Toxicology. Committee on the Handbook of Biological Data, Division of Biology and Agriculture, National Academy of Sciences. Saunders, 1956-59, 5 vols. LC: 56-6976 rev. 2.

This *Handbook of Toxicology* consists of 5 volumes entitled as follows: Vol.1, *Acute Toxicities of Solids, Liquids, and Gases to Laboratory Animals*; Vol.2, *Antibiotics*; Vol.3, *Insecticides*; Vol.4, *Tranquilizers*; Vol.5, *Fungicides.*

1767. Handbook of Traction, Casting, and Splinting Techniques. Royce E. Lewis. Lippincott, 1977. LC: 77-10326. ISBN: 0-397-50380-6. Cost: $17.75.

1768. Handbook of Transducers for Electronic Measuring Systems. Harry N. Norton. Prentice-Hall, Inc, 1969, 704 pp. LC: 76-76879. ISBN: 0-13-382242-7. Cost: $35.00.

This handbook contains information on transducers as sensing elements in electronic measuring devices. It is design-oriented, and it discusses transducers in terms of the particular measurement to be made. For example, measurement of such phenomena as humidity, flow of fluids, nuclear radiation, acceleration, and pressure are discussed. In addition, devices used to measure sound, speed, and temperature also receive treatment. Various bibliographies on the different types of measuring devices have been provided as well as an index and a glossary of terms. For related works, please refer to such titles as the *Instrument Engineers' Handbook,* and the *Handbook of Applied Instrumentation.*

1769. Handbook of Transistor Circuits. Allan Herbert Lytel. Sams, 1963, 224 pp. LC: 63-20742.

1770. Handbook of Transistors, Semiconductors, Instruments, and Microelectronics. Harry Thomas. Prentice-Hall, 1968, 453 pp. LC: 69-11205. ISBN: 0-13-382283-4. Cost: $26.50.

Handbook of Transportation and Marketing in Agriculture. *See* **CRC Handbook of Transportation and Marketing in Agriculture.**

1771. Handbook of Treatment of Acute Poisoning. Edward Horton Bensley; G.E. Joron. Williams and Wilkins, 3d ed., 1963, 227 pp. LC: 63-6426.

1772. Handbook of Treatment. H.W. Proctor (ed); V.S. Bryne (ed). University Park, 1976. ISBN: 0-8391-0944-X. Cost: $24.50.

1773. Handbook of Trees for the Midwest. Pamela Sue Stava. Kendall-Hunt, 1978. ISBN: 0-8403-1851-0. Cost: $14.95.

1774. Handbook of Trees, Shrubs and Roses. Walter G. Hazlewood. Tri-Ocean Books, rev. 2d ed., 1969, 271 pp. LC: 70-396322.

1775. Handbook of Trigonometric Functions: Including Haversines and Introducing Doversines. Leon Kennedy. Iowa State University Press, 1961, 397 pp. LC: 61-17423. ISBN: 0-8138-0750-6. Cost: $7.50.

1776. Handbook of Tropical and Subtropical Horticulture. Ernest Mortensen; Ervin T. Bullard. NTIS, 1970, 197 pp. Accession: PB-286437. Cost: $3.50 microform; $15.50 paper copy.

Major fruit, nut, and tree crops are discussed, including information on pruning, fertilizing, budding, disease, and insect control. Temperature requirements, soil, and cultivation information is also presented.

1777. Handbook of Tropical Forage Grasses. B. Ira Judd. Garland STPM Press, 1981, 126 pp. LC: 78-456. ISBN: 0-8240-7080-X.

This handbook covers different types of grasses that grow in the Southwest and tropical areas, including such grasses as Guinea grass, Napier grass, molasses grass, and tropical carpet grass. In addition, such topics as cultivation, diseases that affect grasses, and morphology are discussed in this handbook.

1778. Handbook of Turbulence. Walter Frost; Trevor H. Moulden. Plenum, 1977, 2 Vols., 498 pp. LC: 77-23781. ISBN: Vol.1, 0-306-31004-X. Cost: Vol.1, $49.50, (Vol.2 is forthcoming).

Vol.1 is entitled *Fundamentals and Applications*; Vol.2 is entitled *Modeling and Measurement*. This handbook covers such subjects as fluid mechanics, turbulent flow, laminar boundary layers, and presents models of turbulence.

1779. Handbook of U.S. Colorants for Foods, Drugs, and Cosmetics. Daniel M. Marmion. John Wiley and Sons, 1979, 350 pp. LC: 78-10949. ISBN: 0-471-04684-1. Cost: $31.00.

This handbook covers such topics as the properties of colorants, specifications, regulation of colorants, and a brief history of their usage. In addition, information is presented on colorant analysis and means of chemical identification.

1780. Handbook of Ultrasonic Testing. British Railways Board, Department of the Chief Engineering, London. British Railways Board, 1966.

1781. Handbook of Ultraviolet and Visible Absorption Spectra of Organic Compounds. Kenzo Kirayama. IFI/Plenum Press, 1967, 642 pp. LC: 66-24948. ISBN: 0-306-65123-8. Cost: $55.00.

1782. Handbook of Ultraviolet Methods. Robert Gordon White. Plenum Press, 1965, 373 pp. LC: 64-23240. ISBN: 0-306-30202-0. Cost: $35.00.

1783. Handbook of Underwater Calculations. Wayne C. Tucker. Cornell Maritime, 1980, 182 pp. Cost: $10.00.

Presents information on methods used to conduct underwater calculations.

1784. Handbook of Underwater Exploration. Bill S. Wilkes. Stein and Day, 1975. LC: 77-159558. ISBN: 0-8128-1807-5. Cost: $2.45.

1785. Handbook of Unit Operations. David A. Blackadder; R.M. Nedderman. Academic Press, 1971, 284 pp. LC: 71-129783. ISBN: 0-12-102950-6. Cost: $38.00.

1786. Handbook of Universal Conversion Factors. Steven Gerolde. Petroleum Publishing Co, 1971, 276 pp. LC: 71-164900. ISBN: 0-87814-005-0.

1787. Handbook of Urban Landscape. C. Tandy. Architectural Press, 1972, 275 pp. LC: 72-194527. Cost: £6.00.

1788. Handbook of Urban Transport. UITP (Union Internationale des Transports Publics), 4th ed., 1979, 2 Vols., 1,100 pp. Cost: $110.00.

This handbook provides statistical data on over 600 cities throughout the world. Such mass transportation systems as motorbuses, tramways, charter buses, and light railroad systems are discussed. Metropolitan roadways and rapid transit systems as presented along with such topics as fare collection, tariffs, and extensive technical tables.

1789. Handbook of Urinalysis and Urinary Sediment. Neil A. Kurtzman; P.W. Rogers. C.C. Thomas, 1974, 103 pp. LC: 73-7856. ISBN: 0-398-02918-0. Cost: $8.50.

1790. Handbook of Urological Endoscopy. J.G. Gow; H.H. Hopkins. Churchill-Livingstone, 1978, 128 pp. ISBN: 0-443-01419-1.

Information is presented on the endoscope as an instrument for medical exams. Such topics as diathermy and the examination of the urinary tract, prostate, urethra, and bladder are discussed in this handbook.

1791. Handbook of Utilization of Aquatic Plants. E.C.S. Little. Food and Agriculture Organization, 1979, 176 pp. Tech Report: FAO Fisheries Technical Paper No. 187. Cost: $9.75.

This handbook discusses the usage of aquatic plants for fuels, soil improvement, water purification, and for building materials. An annotated list is provided for over 250 publications dealing with the utilization of aquatic plants.

1792. Handbook of Vacuum Physics. A.H. Beck. Pergamon Press, 1964-65, 3 volumes - Vol.1 has 3 parts; Vol.2 has 6 parts; Vol.3 has 3 parts. LC: 63-21443. ISBN: Vol.1, parts 1-3, 0-08-010425-8; Vol.2, part 1, 0-08-010888-1; Vol.2, parts 2-3, 0-08-011469-5; Vol.2, parts 4-6, 0-08-012440-2; Vol.3, parts 1-3, 0-08-011051-7. Cost: Vol.1, parts 1-3 $11.50; Vol.2, part 1 $10.25; Vol.2, parts 2-3 $12.75; Vol.2, parts 4-6 $13.25; Vol.3, parts 1-3 $11.50.

The *Handbook of Vacuum Physics* consists of 3 volumes entitled as follows: Vol.1, *Gases and Vacua*; Vol.2, *Physical Electronics*; and Vol.3, *Technology*.

1793. Handbook of Valves. Philip A. Schweitzer. Industrial Press, 1972, 258 pp. LC: 72-5835. ISBN: 0-8311-3026-1. Cost: $22.00.

This handbook contains information on different types of valves that are used for different purposes. This handbook will be useful to those engineers involved with the selection of valves, because information about the design of valves, specifications, installation, and operation of valves is included. Many types of valves are discussed in this handbook including plug valves, piston check valves, solenoid valves, cryogenic valves, and pressure relief valves. An index is provided in addition to a section of definitions and abbreviations. For related works, please refer to the *Equivalent Valves Reference Manual.*

1794. Handbook of Vapor Degreasing. American Society for Testing and Materials, 1976, 60 pp. ISBN: 0-686-52055-6. Cost: $6.00. Series: ASTM, STP 310A.

A guide to the selection of degreasing solvents. ASTM Publication Code: 04-310010-15.

1795. Handbook of Vapor Pressures and Heats of Vaporization of Hydrocarbons and Related Compounds. B.J. Zwolinski; R.C. Wilhoit. American Petroleum Institute, 1971, 329 pp. ISBN: 085501-2511. Cost: $49.50.

This is a compilation of data on vapor-liquid equilibrium phenomena of organic compounds that are used in the refining of petroleum and in the manufacture of petrochemicals. The data are listed in 6 sections, including the following: tables of vapor pressures, tables for boiling points, heats of vaporization of hydrocarbons, sulfur compounds, and an index of compounds.

1796. Handbook of Variables for Environmental Impact Assessment. Larry W. Canter; Loren G. Hill. Ann Arbor Science Publishers, 1979, 203 pp. LC: 79-89718. ISBN: 0-250-40321-8. Cost: $25.00.

This is basically a guide for the preparation of environmental impact statements. A discussion is presented on 62 different environmental variables that would be applicable in preparing any type of environmental impact statement.

1797. Handbook of Vegetation Science. Robert H. Whitaker (ed). The Hague, Junk, 1971-.

Multivolume set covering such topics as environment, water resources, landscape management, forestry and floristic analysis.

1798. Handbook of Ventilation for Contaminant Control. Henry J. McDermott. Ann Arbor Science Publishers, 1976, 376 pp. LC: 76-22253. ISBN: 0-250-40139-8. Cost: 29.50.

This handbook includes many tables, figures, and design data on the evaluation, slection, and installation of ventilation systems for industrial plants. Information is provided on such subjects as air flow, fans, hood selection, ducts, and threshold limit values. Recommended OSHA standards are included as well as information on employee exposure to inadequate ventilation. Many bibliographic references are provided for further investigation, and cost information and examples are provided to help engineers save money with regard to ventilation decisions. For related works, please refer to the following titles: *Dangerous Properties of Industrial Materials; CRC Handbook of Laboratory Safety; ASHRAE Handbook and Product Directory; Air Pollution Engineering Manual; Air Pollution; IHVE Guidebook and Handbook,* and *Handbook of Heating, Ventilating, and Air Conditioning.*

1799. Handbook of Veterinary Drugs; A Compendium for Research and Clinical Use. Irving S. Rossoff. Springer Publications, 1975, 730 pp. LC: 73-88322. ISBN: 0-8261-1530-6. Cost: $42.50.

1800. Handbook of Veterinary Pharmacy, Toxicology, and Pharmacotherapeutic Index. R.K. Mehta. International Publications Service, 1974. Cost: $7.50.

1801. Handbook of Veterinary Procedures and Emergency Treatment. Robert Warren Kirk; Stephen I. Bistner. Saunders, 3d ed., 1981. LC: 80-50294. ISBN: 0-7216-5475-4. Cost: $26.50.

1802. Handbook of Vitamins, Minerals, and Hormones. Roman J. Kutsky. Van Nostrand Reinhold, 2d ed., 1981. ISBN: 0-442-24557-2. Cost: $24.50.

1803. Handbook of Wastewater Collection and Treatment. Muhammad Anis Al-Layla; Shamin Ahmad; E. Joe Middlebrooks. Garland STPM Press, 1980, 504 pp. LC: 79-7515. ISBN: 0-8240-7124-7. Cost: $47.50.

This handbook covers such topics as the design of sewers and pumping stations, diffusion and absorption of gas and liquids, pressures on underground pipes, and various different types of wastewater treatment methods. In addition, such topics as sludge treatment and disposal, solid waste disposal, sedimentation, and wastewater examination are presented.

1804. Handbook of Wastewater Treatment Processes. Arnold S. Vernick; Elwood C. Walker. Marcel Dekker, 1981, 272 pp. ISBN: 0-8247-1652-3. Cost: $38.50. Series: *Pollution Engineering and Technology Series,* Vol. 18.

Covers such topics as wastewater collection and disposal, fixed film processes, suspended growth processes, nitrogen removal, phosphorus removal, and other types of physical-chemical processes of wastewater treatment. In addition, sludge disposal is presented with a discussion on transport, conversion processes, thickening, and dewatering.

1805. Handbook of Water Purification. Walter Lorch. McGraw-Hill, 1982, 768 pp. ISBN: 0-07-084555-7. Cost: $89.95.

This handbook discusses purification processes and various case histories of water purification.

1806. Handbook of Water Quality Management Planning. Joseph L. Pavoni. Van Nostrand Reinhold, 1977, 419 pp. LC: 77-21601. ISBN: 0-442-23282-9. Cost: $31.50. Series: *Van Nostrand Reinhold Environmental Engineering Series.*

This handbook addresses the existing legal structure involved in water quality planning as well as the technical aspects of the planning process.

1807. Handbook of Water Resources and Pollution Control. Harry W. Gehm; Jacob I. Bregman. Van Nostrad Reinhold, 1976, 846 pp. LC: 75-29050. ISBN: 0-442-21041-8. Cost: $44.50.

Includes bibliographical references and index.

1808. Handbook of Waterfowl Behavior. Paul A. Johnsgard. Comstock Editions Inc, 1965, 392 pp. LC: 65-15717. ISBN: 0-8014-0207-7. Cost: $29.50.

1809. Handbook of Watersoluble Gums and Resins. Robert L. Davidson. McGraw-Hill, 1980, 640 pp. ISBN: 0-07-015471-6. Cost: $39.50.

This handbook covers 23 major varieties of commercial gums and resins, including such varieties as gum Ghatti, gum Arabic, locust bean gum, pectins, polyacrylamide, polyethylene glycol, polyvinyl alcohol, alginates, carboxymethycellulose, and xanthan gum. Many properties are listed, such as solution viscosities, rheology, viscosity blending curves, toxicology, and environmental impact.

1810. Handbook of Welded Steel Pipe. Armco Drainage and Metal Products, Inc, 1950, 399 pp. Date of First Edition: 1950.

The history and development of fabricated pipe is presented along with such topics as the flow of water, oil, gas, and air in welded pipelines. In addition, the design of welded steel pipe is discussed, along with the protection and installation of welded steel pipe. Numerous specifications, data, and tables are listed in the appendix.

1811. Handbook of Well Log Analysis: For Oil and Gas Formation Evaluation. Sylvain J. Pirson. Prentice-Hall, 1963, 326 pp. LC: 63-9396. ISBN: 0-13-382804-2. Cost: $29.50.

1812. Handbook of Wiring, Cabling, and Interconnecting for Electronics. Charles A. Harper. McGraw-Hill, 1972, 1,152 pp. LC: 72-4069. ISBN: 0-07-026674-3. Cost: $32.50.

This handbook deals with various types of electrical and electronic wiring, connectors, and termination systems that would be of interest to those engineeers involved with electrical or electronic design systems. Such topics as microelectronics, coaxial cables, printed circuits, and wiring for high voltage systems are discussed. In addition, such subjects as microstrips and magnet wire are treated, and bibliographical references are included. An index and glossary have been prepared for the user's convenience. For related information, please refer to the *American Electricians' Handbook*, the *National Electrical Code*, the *NFPA Handbook of the National Fire Code*, and the *IES Lighting Handbook*. Also refer to the *Standard Handbook for Electrical Engineers; Handbook for Electronics Engineering Technicians; Electronics Engineers' Handbook; Electrical Engineer's Reference Book; Electronic Conversions, Symbols and Formulas; Handbook of Materials and Processes for Electronics*, and *Handbook of Electronic Packaging*.

1813. Handbook of World Salt Resources. Stanley J. Lefond. Plenum Press, 1969, 384 pp. LC: 68-13391. ISBN: 0-306-30315-9. Cost: $42.50. Series: *Monographs in Geoscience.*

1814. Handbook of X-Ray Analysis of Polycrystalline Materials. Lev Iosifovich Mirkin. IFI/Plenum Press, translated from Russian, 1964, 731 pp. LC: 64-23250. ISBN: 0-306-65109-2. Cost: $75.00.

1815. Handbook of X-Ray and Microprobe Data. R.D. Dewey et al. Pergamon Press, 1968, 364 pp. Series: *Progress in Nuclear Energy*, 9.

1816. Handbook of X-Ray and Ultraviolet Photoelectron Spectroscopy. D. Briggs. Heyden, 1977, 400 pp. ISBN: 0-85501-208-0. Cost: $86.50.

This handbook deals with spectrometer design, sample preparation, and interpretation of spectra. Such topics as energy calibration, Auger lines, ultraviolet photoemission, and vacuum ultraviolet photoelectron spectroscopy are discussed in this handbook.

1817. Handbook of X-Rays: For Diffraction, Emission, Absorption and Microscopy. E.F. Kaelble. McGraw-Hill, 1967, 1,044 pp. LC: 63-23535. ISBN: 0-07-033200-2. Cost: $54.50.

1818. Handbook of Zoo Medicine. Heinz-Georg Klos; Ernst M. Lang. Van Nostrand Reinhold, 432 pp. ISBN: 0-442-21367-0. Cost: $49.50.

Such topics as infectious disease, parasites, nutritional diseases, artificial breeding, feeding, and surgery of zoo animals are discussed in this handbook. In addition, such topics as capturing of zoo animals, restraint, and handling of different types of zoo animals are presented. This handbook also lists different types of drugs by trade names, chemical names, and manufacturers in the appendices.

1819. Handbook on Aerosols. Richard Dennis. NTIS, 2d ed., 1976, 142 pp. LC: 75-33965. ISBN: 0-87079-024-2. Tech Report: TID-26608. Date of First Edition: 1950. Cost: $6.00.

This handbook was prepared for the Division of Biomedical and Environmental Research and Division of Reactor Research and Development, U.S. Energy Research and Development Administration. It includes information on air pollution and particle size determination. It includes bibliographic references and index.

1820. Handbook on Air Pollution Control. Frank L. Cross Jr. Technomic Publishing Co, 1973, 171 pp. LC: 72-85083. ISBN: 0-87762-097-0. Cost: $15.00.

Covers collection devices and physical and chemical characteristics of air pollution.

1821. Handbook on Alcoholism and Its Treatment. W. Poley; G. Lea; B. Vibe.. John Wiley and Sons, 1979, 300 pp. LC: 78-13435. ISBN: 0-470-26523-X. Cost: $14.95.

This handbook includes information on various types of treatment methods for alcoholism. It also includes information on research topics in alcoholism and an appendix listing additional sources of information on this topic.

1822. Handbook on Astronaut Crew Motion Disturbances for Control System Design. M. Conlon Kullas. NTIS, 1979, 196 pp. Accession: N79-27237. Tech Report: NASA-RP-1225. Cost: $3.50 microform; $15.50 hard copy.

Various motion experiments are summarized in this handbook.

1823. Handbook on Calibration of Radiation Protection Monitoring Instruments. International Atomic Energy Agency (available from UNIPUB), 1971, 93 pp. ISBN: 92-0-125071-1. Cost: $9.50. Series: *Technical Report Series*, No. 133.

This handbook deals with the reliability and accuracy of radiation measuring instruments. It is a detailed discussion of calibration methods for radiation monitoring equipment.

1824. Handbook on Continuous Casting. E. Herrman. Aluminum Verlag (distributed by Heyden), 1980, 742 pp. ISBN: 3-87017-134-0. Cost: $735.00.

This handbook deals with the latest techniques in continuous casting of steel and nonferrous metals. It is a compilation and listing of all the inventions and patent documentation which appeared between 1958 and 1979 and includes over 3,700 illustrated patents with descriptions.

1825. Handbook on Diseases of Children, Including Dietetics and the Common Fevers. Norman Bruce Williamson. Williams and Wilkins, 9th rev. ed., 1964, 524 pp. LC: 65-7367.

1826. Handbook on Drug and Alcohol Abuse: The Biomedical Aspects. Frederick G. Hofmann; Adele D. Hofmann. Oxford University Press, 1975. ISBN: 0-19-501844-3. Cost: $11.95.

1827. Handbook on Electroplating, Polishing, Bronzing, and Lacquering. W. Canning and Co. W. Canning and Co, 20th ed, 1966, 786 pp.

1828. Handbook on Environmental Management. Carlos J. Hilado. Technomic Publishing Co.

Includes 2 volumes entitled as follows: Vol.1, *Fundamentals*, 1972, 113 pp., $5.00; Vol.2, *Food for Mankind*, 1973, 165 pp., $5.00. A considerable amount of statistical information is provided on such topics as food, water, air, population, and energy. The second volume is devoted entirely to the topic of food and such related topics as land usage, cultivation, food from the sea, fertilizers, irrigation, pest control, and food processing.

1829. Handbook on Ferromagnetic Materials. E.P. Wohlfarth. Elsevier, 1981. ISBN: Vol.1, 0-444-85311-1; Vol.2, 0-444-85312-X. Cost: $102.50 per Vol.

This 2-volume set includes a substantial amount of information on ferromagnetic materials.

1830. Handbook on Hospital Solid Waste Management. Frank L. Cross Jr; George Noble. Technomic, 1973, 107 pp. LC: 72-86164. ISBN: 0-87762-096-2. Cost: $15.00.

Covers such systems as carts, chutes, elevators, pneumatic and hydraulic systems, incinerators and pulp compactors.

1831. Handbook on Hospital-Associated Infections. Dieter Groschel. Marcel Dekker, Inc, 1978-.

This multivolume handbook currently includes 3 volumes entitled as follows: Vol.1, *Occurence, Diagnoses, and Sources of Hospital-Associated Infections*, 1978, 138 pp., ISBN: 0-8247-6724-1, $18.50. Vol.2, *Hospital-Associated Infections in the Compromised Host*, Gerald P. Bodey, 1979, 280 pp., ISBN: 0-8247-6785-3, $32.75.; Vol.3, *Hospital-Associated Infections in the General Hospital Population and Specific Measures of Control*, Dieter Groschel, 1979, 208 pp., ISBN: 0-8247-6815-9, $25.00. This series includes information on such topics as nosocomial infections, the acquisition and spread of hospital-associated infections, control measures, isolation procedures, nursing techniques, and general antimicrobial procedures. This series will eventaully include 10 volumes.

1832. Handbook on Incineration. Frank L. Cross Jr. Technomic Publishing Co, 1972, 64 pp. LC: 72-77083. ISBN: 0-87762-086-5. Cost: $12.00.

This handbook discusses such topics as combustion, incineratior design, maintenance of incinerators, air pollution control, emissions, resource recovery, and agricultural waste disposal.

1833. Handbook on Injectable Drugs. Lawrence A. Trissel. American Society of Hospital Pharmacists, 2d ed., 1980, 612 pp. LC: 80-80210. Cost: $20.00.

Such information as dosage, pH, and stability are addressed in this handbook. The topic of physicochemical principles of interactions of injectable drugs is also presented.

1834. Handbook on Marine Environmental Engineering. Frank L. Cross Jr. Technomic Publishing Co, 1974, 186 pp. LC: 74-80459. ISBN: 0-87762-145-4. Cost: $35.00.

The emphasis of this handbook is on the environmental factors that should be considered in connection with marine operations. Various articles have been presented on such topics as oil spills, various types of marine pollution including solid waste and air pollution, and bulk loading operations. Information has also been included on environmental impact statements in connection with marine operations, as well as safety and health information in connection with OSHA recommendations. For related works, please refer to such titles as the *Handbook of Ocean and Underwater Engineering* and *Oil Spill Prevention and Removal Handbook*.

1835. Handbook on Material and Energy Balance Calculations in Metallurgical Processes. H. Alan Fine; Gordon H. Geiger. Metallurgical Society of AIME, 1981, 572 pp. Cost: $25.00. Series: *Metallurgical Textbook Series*.

Covers such topics as stoichiometry, sampling methods, measurements, material balances, thermochemistry, and energy balances. Tabular enthalpy data are presented for 65 elements, including oxides, nitrides, halides, and carbides. A bibliography of related material is also presented.

1836. Handbook on Materials for Superconducting Machinery. J.E. Campbell; E.A. Eldridge; J.K. Thompson (Battelle Columbus Laboratory, Ohio Metals and Ceramics Information Center). NTIS, Nov. 1974, 597 pp. Accession: AD/A-002 698. Tech Report: MCIC-HB-04. Cost: $35.00 microfiche; $35.00 paper copy.

This handbook includes information on the mechanical, thermal, magnetic, and electrical properties of the following structural materials: Al and Al Alloys, Cu alloys, Ni alloys, Ti alloys, and steels. The equipment to be designed with these materials would be exposed to cryogenic temperatures, and the compilation include data in the temperature range of 0-300° K.

1837. Handbook on Materials for Superconducting Machinery. K.R. Hanby (Metals and Ceramics Information Center). NTIS, 1st supplement, 1975, 372 pp. Accession: ADA-023 228. Tech Report: MCIC-HB-04S1. Cost: $20.00.

Supplement No. 1 to report dated November 1974, ADA-002 698.

1838. Handbook on Materials for Superconducting Machinery; Includes data sheet for first and second supplements Nov. 1, 1975 and Jan. 1977. Metals and Ceramics Information Center (available from NTIS), 2d supplement, 1977, 858 pp. Accession: ADA-035926. Cost: $60.00.

Supplement No. 2 to report dated November 1974, ADA-002-698

1839. Handbook on Mechanical Properties of Rocks. V.S. Vutukuri et al. Trans Tech Publications, 1974-78, 4 Vols. LC: 74-82971. ISBN: 0-87849-031-0 set. Cost: $65.00 per Vol.; $248.00 set.

Covers tensile strength of rock, sheer strength, compressive strength, and strength of rock under triaxial and biaxial stresses. Includes dynamic elastic constants and static elastic constants. Over 292 tables and 860 figures.

1840. Handbook on Nuclear Activation Cross-Sections: Neutron, Photon, and Charged-Particle Nuclear Reaction Cross-Section Data. D. Brune; J.J. Schmidt. International Atomic Energy Agency (available from Unipub), 1974, 558 pp. LC: 76-64669. ISBN: 92-0-135074-0. Tech Report: 156. Cost: $26.00. Series: *Technical Report Series*, No. 156.

Contains data on cross-sections for thermal, epithermal, and fast neutron induced nuclear reactions. Useful in activation analysis and radioisotope production.

1841. Handbook on Radio Frequency Interference. Frederick Research Corp. Frederick Research Corp, 1962, 4 Vols.

1842. Handbook on Semiconductors. T.S. Moss. North-Holland, 1980, 4 Vols.

This handbook includes 4 volumes entitled as follows: Vol.1, *Band Theory and Transport Properties*, 546 pp., ISBN: 0-444-85346-4. Vol.2, *Optical Properties of Solids*, 648 pp., ISBN: 0-444-85273-5, $122.00. Vol.3, *Materials, Properties, and Preparation*, 870 pp., ISBN: 0-444-85274-3, $170.75. Vol.4, *Device Physics*, 900 pp., ISBN: 0-444-85347-2, $183.00. This 4-volume set includes an abundance of information on such topics as crystal structures, lattice vibrations, impurities in semiconductors, transport in inversion layers, conductivity, and scattering in semiconductors. In addition, such topics as electron-phonon interaction, different methods of preparation of semiconductors, MOS transistors, light-emitting diodes, semiconductor lasers, solar cells, infrared detectors, and nuclear radiation detectors are also presented in this handbook.

1843. Handbook on Structural Steel Work: Properties and Safe Laods. British Construction Steel Work Assoc, 2d ed, 1964, 376 pp.

1844. Handbook on Synchronous Drive Components Featuring Inch and Metric Sizes, Catalog 74. Technical Section. Ferdinand Freudenstein; Hitoshi Tanaka. Stock Drive Products, 1974, 157 pp. Accession: PB-233206.

1845. Handbook on the Care and Management of Laboratory Animals. Universities Federation for Animal Welfare. Longman, 5th ed., 1976. ISBN: 0-443-01404-3. Cost: $42.50.

1846. Handbook on the Laboratory Mouse. Charles G. Crispens Jr. C. C. Thomas, 1975. ISBN: 0-398-03403-6. Cost: $9.75.

1847. Handbook on the Organization of Veterinary Service. Parts I and II. A.D. Tretyakov. NTIS, 1975, Part I, 210 pp., Part II, 157 pp. Accession: Part I, JPRS-66341-1; Part II, JPRS-66341-2. Cost: Part I, $3.50 microform, $15.00 bound copy; Part II, $3.50 microform, $17.00 bound copy.

Covers topics related to veterinary service, such as compensation, material benefits, and employee relations.

1848. Handbook on the Physics and Chemistry of Rare Earths. K.A. Gschneidner; L. Eyring. Elsevier, Vol.1, 1978; Vol.2, 1979; Vol.3, 1979; Vol.4, 1979, Vol.1, 918 pp.; Vol.2, 634 pp.; Vol.3, 678 pp.; Vol.4, 616 pp. LC: 78-12371. ISBN: 0-444-85022-8. Cost: Vol.1, $146.50; Vol.2, $97.75; Vol.3, $107.50; Vol.4, $95.25.

This Handbook includes 4-volumes entitled as follows: Vol.1, Metals; *Vol.2*, Alloys and Intermetallics; *Vol.3*, Non-metallic Compounds-1; *Vol.4*, Non-metallic Compounds-2; *This work covers such topics as electronic structure of rare earth metals, low-temperature heat capacity of the rare earth metals, magnetic structure and inelastic scattering of rare earth metals, the elastic and mechanical properties of the rare earth metals, and high-pressure studies of rare earth metals, alloys, and compounds. In addition, such topics as diffusion in rare metals, diffusion in rare earth metals, superconductivity in rare earth metals, magnetic properties of intermagnetic compounds of rare earth metals, nuclear magnetic resonance and the Mössbauer effect of rare earth metals and the valence changes in compounds are presented.*

1849. Handbook on the Principles of Hydrology. Donald M. Gray. Water Information Center, 1975. LC: 73-82157. ISBN: 0-912394-07-2. Cost: $20.00.

This handbook deals with the study of water resources and provides information on such topics as ground water, sedimentation, precipitation, run-off, and the chemical and physical properties of water. Canadian water data have been included along with information on the basic principles of hydrology. For related works, please refer to the *Handbook of Applied Hydrology*.

1850. Handbook on the Toxicology of Metals. L. Friberg; G.F. Nordberg. Elsevier, 1979, 710 pp. ISBN: 0-444-80075-1. Cost: $117.00.

This handbook includes a considerable amount of information on metals and heavy metals and their toxicology on plants and animals.

1851. Handbook on the Ultrasonic Examination of Welds. International Institute of Welding. Institute of Welding, 1977, 44 pp. Cost: £6.75.

This handbook deals with the nondestructive testing technique of ultrasonic testing of welds.

1852. Handbook on the Unified Code for Structural Concrete (CP110: 1972). S.C. Bate et al. Cement and Concrete Association, London, 1972, 153 pp. LC: 74-181712 (GB 73-05754). ISBN: 0-7210-0889-5. Cost: $12.50.

This is a companion volume to the code CP110.

1853. Handbook on Thermophysical Properties of Oxygen. H.M. Roder et al. NTIS, 1973. Accession: B73-10187. Tech Report: LEWIS-11962.

This handbook has been compiled by the Cryogenic Data Center of the National Bureau of Standards. It presents such information as the physical properties and heat transfer data for oxygen as well as thermodynamic functions. This handbook includes primarily information for low temperatures but does include some data for room temperatures.

1854. Handbook on Torsional Vibration. E.J. Nestorides.. Cambridge University Press, 1958, 636 pp. ISBN: 0-521-04326-3. Cost: $72.50.

This handbook was sponsored by the British Internal Combustion Engine Research Association, and has considerable data have been obtained from the BICERA Laboratories as well as member institutions. This handbook includes a discussion of the different types of instruments used to measure torsional vibration as well as different types of dampig devices and detuners used to limit orreduce vibrations. Many formulas, examples, tables, and test procedures are provided to explain the various problems associated with torsional vibration. Such topics as moments of inertia, stiffness of different engine components or machinery components, and torque are discussed. Metric units and British Units have been included throughout, and an index has been prepared. For related titles please refer to *Shock and Vibration Handbook* and the *SAE Handbook*.

1855. Handbook on Wave Analysis and Forecasting. Unipub, 1976, 87 pp. Cost: $20.00. Series: WMO 446.

Handbook Series in Clinical Laboratory Science. *See* **CRC Handbook Series in Clinical Laboratory Science.**

Handbook Series in Marine Sciences. *See* **CRC Handbook of Marine Science.**

Handbook Series in Nutrition and Food. *See* **CRC Handbook of Nutrition and Food.**

Handbook Series in Zoonoses. *See* **CRC Handbook of Zoonoses.**

1856. Handbook, Butane-Propane Gases. William W. Clark. Butane-Propane News, Inc, 5th ed., 1973. LC: 75-319876. Date of First Edition: 1932.

1857. Handbook-Index of Hawaii Groundwater and Water Resources Data Extracted from Reports of the Water Resources Research Center, Univerity of Hawaii. Vol. 1. John F. Mink. NTIS, 1977, 115 pp. Accession: PB-281 794. Tech Report: TR-113. Cost: $3.50 microfiche; $11.00 hard copy.

These groundwater data are listed in 6 major categories, including aquafier characteristics, physical properties of basalt, hydraulic and physical properties of soils, regional geophysical features, geochemistry, and radioactive dating.

1858. Handbook: Ductile Iron Pipe, Cast Iron Pipe. Cast Iron Pipe Research Association, 5th ed., 1978, 448 pp. LC: 75-405-82. ISBN: 0-916-442-01-2. Date of First Edition: 1967.

This handbook is a collection of information on cast and ductile iron pipe for the development of piping design and specifications. Covers such topics as gray cast iron pipe, ductile iron pipe, various standards for cast iron pipe, and the metallurgy of both gray and ductile iron pipe. In addition, such information as expansive soil loads, frost penetration loads, rounding tolerances, and pressure-head conversion is presented. Various ANSI and AWWA standards are included for the user's convenience.

1859. Handbuch Der Allgemeinen Patholgie. F. Buchner et al. Springer-Verlag, 1955-, multivolume set. LC: 56-2297.

NOTES: Vol.6, Part 5, *Tumors 1: Morphology, Epidemiology, Immunology*, 1974, 870 pp., ISBN: 0-387-06813-9, $182.50. Vol.6, Part 8, *Transplantation*, 1977, 1,070 pp., ISBN: 3-540-07751-0, $187.00.

1860. Handbuch der Arzneimittel-Analytik. Siegfried Ebel. Verlag Chemie, 1977, 417 pp. LC: 77-482219. ISBN: 3-527-25679-2.

Deals with drug analysis and identification and quantitative determination of approximately 600 different drugs. Various analytical methods are discussed, including chromatographic, titrametric, colorimetric, and spectrometric.

1861. Handbuch der Bauphysik. Hans W. Bobran. Vieweg, 4th rev. ed., 1979, 344 pp. Cost: DM 89.

This handbook includes information on building physics, such as sound insulation, room acoustics, heat insulation, and humidity control.

1862. Handbuch Der Experimentellen Pharmakologie. New Series. W. Heubner et al. Springer-Verlag, continuing series, Vol.1, 1935-, multivolume set, each Vol. having several parts.

NOTE: In 3 languages—English, French, and German. This is a comprehensive treatise rather than a handbook as such.

1863. Handbuch der Laplace-Transformation. Gustav Doetsch. Birkhauser Boston, Inc, 1971-73. Series: *Mathematische Reihe Series*, Nos. 14, 15, and 19.

This 3-volume set is entitled as follows: Vol.1, *Theorie der Laplace Transformation*, 581 pp., ISBN: 3-7643-0083-3, $73.50. Vol.2, *Anwendungen der Laplace Transformation, 1, Abteilung*, 436 pp., ISBN: 3-7643-0653-X, $56.80. Vol.3, *Anwendungen der Laplace Transformation, 2, Abteilung*, 300 pp., ISBN: 3-7643-0674-2, $38.80. This is a compilation of information on Laplace transforms in 3 volumes.

1864. Handbuch der Medizinischen Radiologie (Encyclopedia of Medical Radiology). Diethelm L. Herausgeber. Springer (Available from Otto Harrassawitz), 1977-, multivolume set. ISBN: Vol.6, Part 5, 3-540-08312-X. Cost: $162.80, $128.00 by subscription.

This multivolume set includes a substantial amount of information on medical radiology, dealing with such topics as roentgen diagnosis, bone tumors, and diseases of the skeletal system.

1865. Handbuch Der Mikroscopischen Anatomie Des Menschen. A. Benninghoff et al. Springer-Verlag, 1927-; Vol.1, 1929, multivolume set. LC: 55-37658. ISBN: 0-387-07456-2. Cost: $55.80.

NOTE: Part 3: *Chromosomes in Mitosis and Interphase*, 1976, 182 pp.

1866. Handbuch der Photometrischen Analyse Organischer Verbindungen (Handbook of Photometrical Analysis of Organic Compounds.) B. Kakac; Z.J. Vejdelek. Verlag Chemie, 1974, 1st supplement, 1977, Vol.1, 718 pp.; Vol.2, 598 pp.; 1st supplement, 464 pp. ISBN: 3-527-25441-2; 1st supplement, 3-527-25707-1. Cost: Vols.1 and 2, DM 360 (available only as set); 1st supplement, DM 138.

This German handbook includes over 970 tables with data on organic compounds and their color reactions with organic and inorganic reagents. This book will be of interest to biochemists, food chemists, pharmaceutical chemists, and medical students.

1867. Handbuch Der Physik. S. Flugge. Springer, 1955, multivolume set. LC: 56-2942.

A very comprehensive treatise on physics.

1868. Handbuch Der Spurenanalyse. Die Anreicherung Und Bestimmung Von Spurenelementen Unter Anwendung Chemischer Und Mikrobiologischer Verfahren. O.G. Koch (ed); G.A. Koch-Dedic (ed). Springer-Verlag, 2d ed., 1973, 2 Vols.

1869. Handbuch der Textilhilfsmittel. A. Chwala. Verlag Chemie International, 1977, 1,158 pp. ISBN: 3-527-25367-X. Cost: $223.90.

The translated title of this book is *Handbook of Textile Auxiliaries*. Although this handbook is written in German, it includes a considerable amount of information on the textile industry with over 316 figures and 175 tables describing data and information necessary for producing marketable textiles from raw materials.

1870. Handbuch des Kathodischen Korrosionsschutzes. W. Baeckmann; W. Schwenk. Verlag Chemie, 2d enlarged ed., 1980, 472 pp. ISBN: 3-527-25859-0. Cost: DM 198.

This handbook includes over 200 illustrations and 600 tables on cathodic corrosion prevention.

1871. Handbuch zum Filterentwurf: Handbook of Filter Design. Rudolf Saal. AEG-Telefunzen, 1979, 663 pp. Cost: DM 220.

Over 25 different examples of filter design problems are worked out and provided for ease of solving filter design problems.

1872. Handling Guide for Potentially Hazardous Commodities. A.D. Baskin. Railway Systems and Management Association, Commodity Safety System, 1972.

1873. Handy Matrices of Unit Conversion Factors for Biology and Mechanics. C.J. Pennycuick. Halsted Press, 1975, 47 pp. LC: 74-26762. ISBN: 0-470-67948-4. Cost: $4.95.

A tabulation of conversion factors including different units such as membrane tension, space, mass, time, metabolic rate, and force.

Harmful Chemicals—The Toxicological and Medical Establishment of MAK-Values. *See* **Gesundheitsschadliche Arbeitsstoffe Toxikologisch-Arbeitsmedizinische Begrundung von MAK-Werten.**

1874. Hayden's Complete Tube Caddy: Tube Substitution Handbook. Herman A. Middleton. Hayden, 24th ed., 1979, 128 pp. ISBN: 0-8104-0809-0. Cost: $5.50 paper.

This handbook covers over 6,700 tubes and over 66,000 substitutions.

1875. Hazardous and Toxic Effects of Industrial Chemicals. Marshall Sittig. Noyes Data, 1979, 460 pp. LC: 78-70739. ISBN: 0-8155-0731-3. Cost: $42.00.

This handbook is an alphabetically arranged guide to over 250 chemicals that have been described as hazardous and toxic. Each listing includes a description of the chemical, permissable exposure limits, harmful effects, special tests, different names for the chemical, and possibility of occupational exposure.

1876. Hazardous Materials Guide: Shipping, Materials Handling, and Transportation. J.J. Keller and Associates, Inc, 1977, loose-leaf, over 900 pp. Cost: $79.00.

Covers the shipping and labeling, loading and storage, and authorized containers for hazardous materials.

1877. Hazardous Materials Handbook. James H. Meidl. Glencoe Press, 1972, 349 pp. LC: 72-177437. ISBN: 0-02-476370-5. Cost: $7.95.

1878. Hazardous Materials Table. 172.101 American Trucking Associations, 57 pp. Cost: $2.50.

This publication provides information on the transportation of hazardous materials and wastes.

1879. Hazards of Chemical Rockets and Propellants Handbook. Chemical Propulsion Information Agency. NTIS. Accession: Vol.1, AD-889-763; Vol.2, AD-870-258; Vol.3, AD-870-259. Cost: $25.00.

1880. Health Handbook. G.K. Chacko. North-Holland, 1979, 1,104 pp. ISBN: 0-444-85254-9. Cost: $146.25.

This handbook has been prepared by over 111 internationally famous experts on the topic of cure of illness and care of health. It covers such topics as environmental management, including air pollution, water pollution, solid waste pollution, genetics management, and aging control. It also covers such topics as computers and their application to the study of heart disease, pattern recognition, developing medical databases, and multiphasic screening.

1881. Hearing Aid Handbook of Measurement 1979. U.S. Government Printing Office, 1979, 323 pp. SuDoc: VA 1. 22:11-68. Cost: $5.50.

This is a guide to the performance measurements of hearing aids. It covers such hearing aid topics as compression, high-pass, CROS, BICROS, and head-worn bone conduction.

1882. Hearing-Loss Handbook. Richard Rosenthal. St. Martin's Press, Inc, 1975. LC: 75-9496. ISBN: 0-312-36540-3. Cost: $8.95.

1883. Heat and Mass Transfer Data Book. C.P. Kothandaraman; S. Subramanyan. Halsted Press, 3d ed., 1978, 149 pp. LC: 77-7346. ISBN: 0-470-99078-3. Cost: $7.95.

Includes Psychometric Charts and covers property values of numerous materials at different temperatures. Data are provided for conduction, radiation, and convection.

1884. Heat Exchanger Design Handbook. E.U. Schlünder (ed). Hemisphere Publishing Corp, 1982, 5 Vol. loose-leaf service with updates, 2,080. ISBN: 0-89116-125-2.

The individual parts are entitled as follows: Part 1, *Heat Exchanger Theory*; Vol.2, *Fluid Mechanics and Heat Transfer*; Vol.3, *Thermal and Hydraulic Design of Heat Exchangers*; Vol.4, *Mechanical Design of Heat Exchangers*; Vol.5, *Physical Properties*. This is a collection of over 100 articles on such topics as hydraulic design of heat exchangers, thermal design of heat exchangers, mechanical design of heat exchangers, fluid mechanics, and heat transfer as it relates to heat exchangers.

1885. Heat Pumps: Design and Applications. A Practical Handbook for Plant Managers, Engineers, Architects, and Designers. David Anthony Reay; D.B.A. MacMichael. Pergamon, 1979, 302 pp. LC: 79-40622. ISBN: 0-08-022716-3. Cost: $47.00.

This practical handbook provides information on the theory of heat pumps, their design, and various types of design problems. In addition, information is provided on the application of heat pumps in industry, commercial, and municipal buildings. A comprehensive bibliography is provided along with a section on conversion factors and energy equivalents. There is a section on refrigerant properties in the appendix.

1886. Heat Transfer and Fluid Flow Data Book. R. Hosmer Norris. General Electric Co, 1970-, 2 Vols., loose-leaf service. ISBN: 0-931690-02-1. Cost: 2 vols, $375.00 plus $95.00 for annual updating service.

This handbook deals with the traditional topics of heat transfer and fluid flow, and includes, among other things, a listing of properties of gases, liquids, and solids. Such properties as thermal conductivity, density, specific heat, viscosity, thermal expansion, and emissivity are included. In addition, fluid flow information is provided for different types of ducts and fans. Such topics as heat exchangers, forced and free convection, radiation and condensation are also discussed in this 2-volume set. Many bibliographic references are given in this handbook as sources for the original data. For related works, please refer to the *Handbook of Fluid Dynamics* and the *Handbook of Heat Transfer.*

1887. Heat Transfer in Microelectronic Equipment: A Practical Guide. J.H. Seely. Marcel Dekker, Inc, 1972, 360 pp. ISBN: 0-8247-1611-6. Cost: $32.50.

1888. Heating Handbook: A Manual of Standards, Codes and Methods. R.H. Emerick. McGraw-Hill, 1964, 522 pp. LC: 63-16465. ISBN: 0-07-019312-6. Cost: $29.50.

1889. Heating, Ventilating and Air Conditioning Systems Estimating Manual. A.M. Khashab. McGraw-Hill, 1977, 311 pp. LC: 77-4208. ISBN: 0-07-034535-X. Cost: $24.95.

1890. Heavy Oil Gasification. Arnold H. Pelofsky. Marcel Dekker, Inc, 1977, 176 pp. ISBN: 0-8247-6638-5. Cost: $24.75. Series: *Energy, Power, and Environment: A Series of Reference Books and Textbooks*, Vol. 1.

This is an engineering handbook on such topics as pyrolysis, partial oxidation, refinery processes, and basic fundamentals of oil gasification.

1891. Heavy Water: Thermophysical Properties. Y.Z. Kazavchinskii. NTIS, 1971, 270 pp. Accession: N71-38757. Tech Report: TT70-50094.

This represents a collection of thermophysical properties of heavy water, including nuclear properties, parameters of the equation of state, and isotopic analysis.

1892. Heilbron's Dictionary of Organic Compounds. J. Buckingham (ed, 5th ed). Chapman and Hall (available in U.S. through Methuen), 5th ed., 1982, 7 Vols., (over 7,500 pp.). ISBN: 0-412-17000-3. Date of First Edition: 1934. Cost: $1,950.00.

This 7-volume set consists of an alphabetical listing of over 50,000 organic compounds providing such information as structural physical properties of a compound and its derivatives together with alternative names and bibliographic references. In addition, such information as CAS Registry Number, molecular formula, molecular weight, melting point, boiling point, freezing point, spectroscopic data, solvent of recrystallization, solubility, density, optical rotation, refractive index, and hazard information is also presented. Access is provided by the 4 indexes, including name index, formula index, heteroatom index, and CAS Registry Number index, as well as by the main alphabetical listing on entries.

Helium and Neon—Gas Solubilities. *See* Solubility Data Series, Vol. 1.

1893. High Temperature Properties and Decomposition of Inorganic Salts, Part 2. Carbonates. K.H. Stern; E.L. Weise (National Bureau of Standards). NTIS, 1969. Cost: $4.00. Series: NSRDS-NBS-30.

1894. High Temperature Properties and Decomposition of Inorganic Salts, Part I. Sulfates. K.H. Stern; E.L. Weise (National Bureau of Standards). Office of Standard Reference Data, 1966. Cost: $0.85. Series: NSRDS-NBS-7.

1895. Highly Hazardous Materials Spills and Emergency Planning. J.E. Zajic; W.A. Himmelman. Marcel Dekker, Inc, 1978, 240 pp. ISBN: 0-8247-6622-9. Cost: $29.75. Series: *Hazardous and Toxic Substances: A Series of Reference Books and Textbooks*, Vol. 1..

This book presents case histories of many different types of spills which have occurred in the United States and Canada. Reference information is provided on the transportation of hazardous materials, containment and treatment techniques in the event of a spill, and various types of environmental effects of spills. A classification of hazardous materials is provided.

1896. Highly Refractory Elements and Compounds. S.N. Bashlykov et al. Izdate'stvo Metallurgiia, 1969, 378 pp. Accession: A70-37470.

The text of this book is written in Russian. It is a Soviet reference book on highly refractory elements and compounds including such properties as electrical, optical, nuclear, thermophysical, thermodynamic, crystalline, and chemical.

1897. Highway Capacity Manual. Highway Research Board. Transportation Research Board, 1965, 397 pp. Series: *HRB Special Report*, No. 87.

1898. Highway Engineering Handbook. Kenneth B. Woods; Donald S. Berry; William H. Goetz. McGraw-Hill, 1960, 1,696 pp., (various paging). LC: 58-59682. ISBN: 0-07-071735-4. Cost: $75.00.

This handbook gathers together an abundance of information dealing with highway design, location, construction, and maintenance. Among the topics discussed are soil mechanics, materials used in highway construction, drainage problems associated with highway construction, and the use of photogrammetry as an aid in the selection of highway routes. This handbook includes many references to the literature as source materials, and an index is provided. For related works, please refer to the *Highway Capacity Manual; The Handbook of Highway Engineering; Transportation and Traffic Engineer Handbook*; and *Highway Surveying Tables.*

1899. Highway Surveying Tables. Lefax Publishing Co., loose-leaf binding. ISBN: 0-685-14142-X. Cost: $8.50. Series: *Lefax Technical Manual*, No. 798.

1900. Histology; A Color Atlas of Cytology, Histology, and Microscopic Anatomy. Frithjoi Hammersen. Urban and Schwarzenberg, 1980, 229 pp. Cost: $29.00.

This is a reference book on tissue and cell morphology. Provides illustrations of organs and structures on a microscopic level as well as an ultramicroscopic view of finer tissue elements.

1901. Holography Handbook: Making Holograms the Easy Way. Fred Unterscher; Jeanne Hansen; Bob Schlesinger. Ross Books, 1980, 408 pp. ISBN: 0-89496-017-2. Cost: $16.95.

This book describes how to create holograms, the equipment necessary, and how to make and display different types of holograms. It also describes how to construct optical tables.

1902. Home Plumbing Handbook. Charles N. McConnell. Audel, 2d ed., 1978. LC: 78-59370. ISBN: 0-672-23321-5. Cost: $7.95 paper.

This is a practical guide to plumbing problems that occur in residential dwellings.

1903. Homeowner's Handbook of Plumbing and Repair. Kendall Webster Sessions. John Wiley and Sons, 1978, 421 pp. LC: 77-21333. ISBN: 0-471-02550-X. Cost: $23.95.

Covers such topics as water distribution, piping, valves, piping insulation, sewage disposal, and different tools that are used in plumbing repair.

1904. Honour's Energy and Environmental Handbook. Walter W. Honour. Technomic, 1978, 150 pp. Cost: 18.00.

This is primarily a dictionary that includes over 2,000 terms and abbreviations on pollution control, energy, and environmental problems. In addition, tables are provided on sewage flow, criteria for air pollutants, energy equivalency tables, and conversion factors for pollution-related problems.

1905. Horse Power Tables for Agitator Impellers. L.T. Advani. Gulf Publishing Co, 1976.

The primary purpose of this handbook is to assist engineers in calculating horsepower absorbed and the design of agitator impellers for various types and sizes of turbines. As a result, data are presented for different turbine designs and diameters. It is the hope of the editor that this collection of tables will save the engineer time in his/her horsepower calculations.

1906. Hospital Engineering Handbook. American Hospital Association. American Hospital Association, 1974, 318 pp. LC: 73-77521. ISBN: 0-87258-125-X. Cost: $16.25.

1907. Hospital Planning Handbook. Rex W. Allen; Ilona Van Karolyi. Renouf USA Ltd, 1976, 242 pp. LC: 75-30599. ISBN: 0-471-02319-1. Cost: $22.00.

1908. Hospital-Medical Automation Handbook. Automated Education Center.. Management Information Service. Cost: $55.00; updating service $45.00.

1909. House Physician's Handbook. C. Allan Birch. Longman, Inc, 4th ed., 1977. ISBN: 0-443-01536-8. Cost: $8.50 paper.

1910. Household and Automotive Chemical Specialties: Recent Formulations. Ernest W. Flick. Noyes Data Corp, 1979, 390 pp. LC: 79-84429. ISBN: 0-8155-0751-8. Cost: $32.00.

Provides formulations for household chemicals and automotive chemicals.

1911. How to Become the Successful Construction Contractor. Taylor F. Winslow. Craftsman Book Co, 2 Vols. LC: 75-19189.

This 2-volume set is entitled as follows: Vol.1, *Plans, Specs, and Building*, 1976, 450 pp., ISBN: 0-910460-14-0, $11.75. Vol.2 *Estimating, Sales, and Managment*, 1976, 496 pp., ISBN: 0-910460-15-9. $12.50. This 2-volume set includes over 1,000 figures and tables on how to use plans and specifications and how to handle such construction materials as steel, concrete, masonry, and drywall. A considerable amount of information is pro-vided on how to compile estimates for excavation, carpentry, and concrete, including personhour tables and forms.

1912. Huebner's Machine Tool Specs. Harlan B. Reeves; James B. King. Huebner Publications, 1980, 4 Vols., 4,650 pp. Cost: $332.00 set.

This 4-volume set is entitled as follows: Vol.1, *Boring, Drilling, Gear Making and Grinding Machines*; Vol.2, *Machining Centers Through Spark Erosion Machines*; Vol.3, *Threading Through Turning Machines*, Vol.4, *Metal Forming Machine Tools*. This 4-volume set includes specifications and photographs of most of the machine tools that are sold in the United States.

1913. Human Brain in Figures and Tables: A Quantitative Handbook. Samuel M. Blinkov; Ilya I. Glezer. Basic Books, 1968, 482 pp. LC: 66-12890. ISBN: 0-465-03115-3. Cost: $26.50.

1914. Human Factors Design Handbook: Information and Guidelines for the Design of Systems, Facilities, Equipment, and Products for Human Use. Wesley E. Woodson. McGraw-Hill, 1981, 1,072 pp. LC: 80-13299. ISBN: 0-07-071765-6. Cost: $75.00.

This handbook deals with designing facilities, systems and buildings based on the needs, limitations, and expectations of the users, whether they be children, the elderly, or the handicapped. The book covers architecture, furnishings, equipment, machinery, tools, and environmental control systems.

1915. Human Health and Disease. Philip L. Altman; Dorothy D. Katz. Federation of American Societies for Experimental Biology, 1977. LC: 76-53166. ISBN: 0-913822-11-6. Cost: $45.00.

1916. Human Oocytes and Their Chromosomes: An Atlas. Uebele-Kalhardt, B-M (available from Otto Harrassowitz), 1978, approx. 100 pp. ISBN: 3-540-08879-2. Cost: $30.50.

This book includes 73 micrographs.

1917. Hybrid Microcircuit Reliability Data. Reliability Analysis Center. NTIS, 1977, 409 pp. Accession: AD-A040 303. Cost: $50.00.

Supersedes Report No. RAC-HMRD-0175, AD-A0110 058

1918. Hydraulic Engineering Handbook. Lefax Publishing Co, loose-leaf binding. ISBN: 0-685-14145-4. Cost: $8.00. Series: *Lefax Technical Manual*, No. 789.

1919. Hydraulic Handbook. Editors of *Hydraulic and Pneumatic Power* (available from Trade and Technical Press, Ltd), 7th ed., 1979, 792 pp. ISBN: 0-85461-074-X. Cost: $94.00.

Covers oil hydraulic power transmission, hydrostatics, hydrodynamics, pressure vessels, pump selection, and cylinders.

1920. Hydraulic Institute Engineering Data Book. Hydraulic Institute, 1980, 204 pp. Cost: $28.50.

This book replaces the following title: *The Hydraulic Institute Pipe Friction Manual*. This data book provides information on fluids for hydraulic systems and on the pumping and transfer of fluids.

1921. Hydraulic Standards, Lexicon, and Data. Institute for Power Systems. State Mutual Books, 1979, 200 pp. ISBN: 0-85461-005-7. Cost: $35.00.

This book contains information on such topics as power pumps, hydraulic motors, valves, hydraulic transmissions, intensifiers, cylinders, and specifications for hydraulic cylinder tubes. Numerous tables are included covering such topics as pressures, cylinders, swell and linear expansion, orifices, seal application, compatibility of fluids, pumps, accumulators, cylinder selection, and seal application.

1922. Hydraulic Technical Data. Trade and Technical Press, Ltd, Vol.3, 1971; Vol.4, 1977, Vol.1, 160 pp.; Vol.2, 200 pp.; Vol.3, 117 pp., Vol.4, 120 pp. ISBN: Vol.1, 0-85461-003-0; Vol.2, 0-85461-004-9; Vol.3, 0-85461-047-2; Vol.4, 0-85461-066-9. Cost: Vol.1, &£3.60; Vol.2, &£4.30; Vol.3, &£3.20; Vol.4, $19.95.

1923. Hydrocarbons for Fuel: 75 Years of Materials Research as NBS—Their Physical and Chemical Properties. G.T. Armstrong. NTIS, 1976, 25 pp. LC: 75-619428. Accession: N76-33352; PB 253665. Tech Report: NBS-SP-434. Cost: $3.50 hard copy.

This publication includes an historical review of the past 75 years of materials research at The National Bureau of Standards. This publication addresses the 3 major classes of natural hydrocarbonaceous fuels including natural gas, petroleum, and coal.

1924. Hydroelectric Handbook. William P. Creager; Joel D. Justin. John Wiley and Sons, 2d ed., 1950, 1,151 pp. LC: 50-6298. Date of First Edition: 1927.

This handbook has a great deal of information on different types of dams and ways to produce electric power from water flow. Such dams as steel dams, timber dams, earth dams, rockfill dams, buttressed concrete dams, and arched dams are included in this reference work. Information is provided on such subjects as turbines, different types of hydro plants, spillways, flumes and canals, transmission lines, and hydroelectric generators. In addition to an index at the end of the text, bibliographies are provided at the end of most of the sections which will help the user locate additional information, if necessary. For related works in the field of energy technology, please refer to the following titles: *Energy Reference Handbook; Keystone Coal Industry Manual; Gas Engineer's Handbook; Handbook of Natural Gas Engineering; Hydrogen: Its Technology and Implications; Reactor Handbook; Engineering Tables for Energy Operators; Petroleum Production Handbook; Nuclear Engineering Handbook; Handbook of Environmental Civil Engineering; Standard Handbook for Civil Engineers; Civil Engineers' Reference Book; Civil Engineering Handbook*, and the *Handbook of Applied Hydrology*.

1925. Hydrogen Stopping Powers and Ranges in All Elements. H.H. Andersen; James F. Ziegler. Pergamon Press, 1977, 324 pp. LC: 77-3068. ISBN: 0-08-021605-6. Cost: $40.00. Series: *The Stopping and Ranges of Ions in Matter*, Vol. 3.

This publication includes 7 tables and 265 charts of data displaying the hydrogen-stopping powers and ranges in all elements and for projectile energies between O KeV and 20 MeV.

1926. Hydrogen Technology Survey: Thermophysical Properties. R.D. McCarty. NTIS, 1975, 535 pp. Accession: N76-11297. Tech Report: NASA-SP-3089. Cost: $13.00.

The thermophysical properties of liquid and gaseous hydrogen are presented along with thermodynamic functions, transport properties, and physical properties. Low-temperature is emphasized.

1927. Hydrogen: Its Technology and Implications. Kenneth E. Cox; Kenneth D. Williamson. CRC Press, 1976-1977, 5 vols, each volume has approximately 200 pp. LC: 76-47677. ISBN: Vol.1, 0-8493-5121-9; Vol.2, 0-8493-5122-7; Vol.3, 0-8493-5123-5; Vol.4, 0-8493-5124-3; Vol.5, 0-8493-5125-1. Cost: Vol.1, $59.95; Vol.2, $44.95; Vol.3, 79.95; Vol.4, $69.95; Vol.5, $44.95.

This 5-vol. set is not actually a handbook, but it is included because it represents a comprehensive collection of information on hydrogen. The 5 vols. are entitled: Vol.1, *Hydrogen Production Technology*; Vol.2, *Transmission and Storage of Hydrogen*; Vol.3, *Hydrogen Properties*; Vol.4, *Utilization of Hydrogen*; Vol.5, *Implication of Hydrogen Energy*. This reference work includes information on such subjects as sources of hydrogen from natural gas, coal, nuclear and solar energy, and from water splitting. These vols. also contain information on the various costs associated with production, transmission, and use of hydrogen, as well as the many physical constants and properties of hydrogen which are tabulated in a collection of graphs and tables. A list of bibliographic references is provided, and an index has been prepared.

1928. Hydrogenation of Ethylene on Metallic Catalysts. J. Horiuti; K. Miyahara (National Bureau of Standards). U.S. Government Printing Office, 1968. SuDoc: C13.48:13. Cost: $3.00. Series: NSRDS-NBS-13.

1929. Hydrological Atlas of the Federal Republic of Germany. Reiner Keller. Available from Otto Harrassowitz, 1978, text, 370 pp.; atlas, 109 pp. ISBN: Text, 3-7646-1714-4; atlas, 3-7646-1715-2.

This is a German-English atlas consisting of 73 maps and 106 tables describing the hydrological characteristics of the Federal Republic of Germany.

1930. Hypergeometric and Legendre Functions with Applications to Integral Equations of Potential Theory. C. Snow (National Bureau of Standards). NTIS, 1952, 427 pp. Accession: PB-175815. Cost: $11.75. Series: NBS AMS19.

NOTE: Supersedes NBS MT15.

1931. Hypergeometric and Legendre Functions with Applications to Integral Equations of Potential Theory. Chester Snow. National Bureau of Standards (available from NTIS), 1942, 319 pp. Accession: PB-175815. Cost: $11.75. Series: NBS MT15.

NOTE: Superseded by NBS AMS19.

1932. I.A.T. Manual of Laboratory Animal Practice and Techniques. Douglas J. Short (ed). Lockwood, 2d ed., 1969, 492 pp. LC: 70-360830.

1933. I.B.I. Guide: World's Most Complete History of Bearing Number Alternatives; A Computerized Interchange Anti-Friction Ball and Roller Bearing Numbers for Ground Equipment. International Bearing Interchange. Interchange, Inc, 6th ed, 1979, 1,070 pp. LC: 78-70758. ISBN: 0-910966-07-0. Cost: $70.00.

1934. IC Master Herman Publishing Co, 1980, 1,250 pp. ISBN: 0-686-64340-2. Cost: $75.00.

Covers nearly 100 manufacturers of integrated circuits. Annual publication.

1935. IC Schematic Sourcemaster. K.W. Sessions. John Wiley and Sons, 1978, 560 pp. LC: 77-1304. ISBN: 0-471-02623-9. Cost: $38.50.

This publication contains 1,500 different schematic diagrams which demonstrate the usage of integrated circuits.

1936. Ice Cream Service Handbook. Wendell S. Arbuckle. AVI Publishing Co, 1976, 125 pp. ISBN: 0-87055-211-2. Cost: $7.50.

Provides information on handling, serving, and marketing ice cream and other types of frozen dairy products primarily from the retail point of view. Includes such topics as sanitation, recordkeeping, recipes, customer relations, and advertising.

1937. Identification and Analysis of Plastics. J. Haslam; H.A. Willis; D.C.M. Squirrell. Heyden, 2d ed., 1979, 758 pp. ISBN: 0-85501-193-9. Cost: $56.00.

Covers such topics as the identification and analysis of different types of polymers such as vinyl resins, ester polymers, nylon, polyolefins, fluorocarbon polymers, rubbers, thermosetting resins, and natural resins. Numerous tests are discussed such as burning, heating, and solubility tests, as well as the interpretation of infrared spectra, polymers, and plasticisers.

1938. Identification of Organic Compounds: A Students' Text Using Semimicro Techniques. Nicholas D. Cheronis; John B. Entrikin. Wiley-Interscience, 1963, 477 pp. ISBN: 0-470-15279-6. Cost: $29.50.

1939. IEEE Guide to the Collection and Presentation of Electrical, Electronic and Sensing Component Reliability Data for Nuclear Power Generating Stations. The Institute of Electrical and Electronics Engineers (distributed by John Wiley), 1977, 543 pp. LC: 77-77870. Cost: $41.50. Series: IEEE Std. 500-1977.

This is a collection of information in tabular form on electrical and electronic components that are used in nuclear power generating stations. Such components as transformers, relays, circuit breakers, and batteries are discussed. This is actually a standard with the designation IEEE Std 500-1977.

1940. IEEE National Electrical Safety Code, 1981 Edition. John Wiley and Sons Inc, 1981. ISBN: 0-471-09263-0. Cost: $14.50.

Includes standards of safety that are recommended by the Institute of Electrical and Electronics Engineers and the U.S. National Bureau of Standards.

1941. IEEE Standard Dictionary of Electrical and Electronics Terms. Wiley-Interscience, 2d ed., 1977, 896 pp. LC: 77-92333. ISBN: 0-471-04264-1. Date of First Edition: 1972. Cost: $37.50.

Although this is primarily a dictionary including 20,254 definitions, it also includes a listing of over 10,000 acronyms. Numerous diagrams are provided to help further explain some of the definitions. Each definition is followed by a number which corresponds to the original source document in the bibliography where the definition was obtained.

1942. IES Lighting Handbook: The Standard Lighting Guide. John E. Kaufman; Jack F. Christensen. Illuminating Engineering Society, 5th ed., 1972, various paging. LC: 77-186864. ISBN: 0-87995-000-5. Date of First Edition: 1949. Cost: $37.50.

This handbook gathers together a considerable amount of information on the use of light, the different sources of light, the applications of lighting to various types of buildings and industries, and various calculations with regard to light. It includes basic physical data on light and information on devices used to measure light sources. Lighting applications information is provided for such uses as lighting for sports events, lighting along highways, lighting at airports and underwater lighting. In addition, an index is provided as well as a listing of conversion factors and a collection of definitions of lighting terms.

1943. IFM Handbook of Practical Energy Management. Ronnie L. Highnote. Institute for Management, Inc, 1979, loose-leaf, 208 pp. LC: 77-93207.

1944. IHVE Guide. Institution of Heating and Ventilating Engineers, 4th ed, 1970, 3 Vols., (various paging). Date of First Edition: 1940.

This guide includes such subjects as corrosion, vibration and noise control, waste disposal systems, fire protection systems, weather data, moisture problems, solar data, and combustion systems. Information on properties of saturated steam, humid air, and other types of data such as heat transfer, fluid flow, and fuels is provided. Recommended BSI standards on governmental regulations are included in this handbook, which also provides an index and many bibliographic references to the literature.

1945. IITRI Fracture Handbook: Failure Analysis of Metallic Materials by Scanning Electron Microscopy. S. Bhattachanya et al. IIT Research Institute, 1978, 570 pp. LC: 78-78273. ISBN: 0-915802-11-20.

This handbook contains over 2,000 photographs produced by scanning electron microscopy. It also provides descriptions of alloy processing, microstructure and test conditions, and interpretation of the fracture features.

1946. Illumination, Color, and Contrast Tables: Naturally Illuminated Objects. M.R. Nagel et al. Academic Press, 1978, 494 ppp. ISBN: 0-12-513750-8. Cost: $52.50.

1947. Illustrated Handbook in Local Anaesthesia. Ejnar Eriksson; Anton Doberl; Victor Goldman (trans). Saunders, 2d ed., 1980, 159 pp. LC: 79-65870. Cost: $37.50.

Provides information on local anesthesia with many illustrations and photographs.

1948. Illustrated Handbook of Electronic Tables, Symbols, Measurements and Values. Raymond Ludwig. Parker (Prentice-Hall), 1977, 352 pp. ISBN: 0-13-450973-0. Cost: $17.95.

Covers troubleshooting electronic components and troubleshooting through circuit design analysis. Includes formulas for problems involving impedence, transformers, conductance decibels, compasitors, antennas, and AC and DC applications.

1949. Index Bergeyana. Robert E. Buchanan et al. Williams and Wilkins, 1966.

This is a companion volume to *Bergey's Manual of Determinative Bacteriology*.

1950. Index of Chemicals Used for the Treatment of Metal Surfaces. H. Benninghoff. Elsevier, 1974, 630 pp. ISBN: 0-444-41075-9. Cost: $134.25.

This index is in German, English, and French.

1951. Index of European Potato Varieties. Hermann Stegemann; Volkmar Loeschke. Verlag Parey Berlin.

Over 74 figures and 4 tables are provided in this index. Identification is by electrophoretic spectra. Genetic data is also provided.

1952. Index of Mathematical Tables. A. Fletcher et al. Addison-Wesley Publishing Co, 2d ed., 1962, 2 Vols. Date of First Edition: 1946.

1953. Index of Vibrational Spectra of Inorganic and Organometallic Compounds. Norman Neill Greenwood; E. J.F. Ross. Butterworth, Vol.1, 1976, 3 Vols. LC: 72-133910. ISBN: 0-408-70351-2. Cost: Vol.1, $89.95.

This is an index of all the research done on the vibrational spectra of inorganic and organometallic molecules from 1935 to 1966.

1954. Index on Mass Spectral Data. American Society for Testing and Materials. American Society for Testing and Materials, 1969, 632 pp. Tech Report: PCN: 10-011000-39 (order number). Cost: $50.00. Series: ASTM AMD11.

This index contains approximately 8,000 mass spectra of various chemical compounds.

1955. Index to Binary Phase Collections. William G. Moffatt. General Electric Co, 1979, 370 pp. LC: 79-25561. ISBN: 0-931690-12-9. Cost: $85.00.

This is a guide to binary phase metal alloy information and basically represents a merger of indexes to over 40 different collections on binary phase information.

1956. Induction Heating Handbook. E.J. Davies; P.G. Simpson. McGraw-Hill, 1979, 460 pp. LC: 78-7279. ISBN: 0-07-084515-8. Cost: $35.00.

This handbook covers such topics as heat transfer, electromagnetic theory, surface heating, metal melting, motor alternators, and radiofrequency.

1957. Industrial Air Pollution Handbook. A. Parker. McGraw-Hill, 1978, 657 pp. LC: 77-30126. ISBN: 0-07-084486-0. Cost: $35.00.

Industrial Chemicals. *See* Faith, Keyes, and Clark's Industrial Chemicals.

Industrial Chemistry. *See* Handbook of Industrial Chemistry.

1958. Industrial Electronics Handbook. William D. Cockrell. McGraw-Hill, 1958, 1,376 pp. (various paging). LC: 57-8620. ISBN: 07-011536-2.

This handbook discusses a number of subjects from electronic instrumentation to the different types of components used to make electronic devices. AC and DC power supplies are discussed as well as different types of circuits and their applications. A section is provided for the convenience of the user. For related works, please refer to the *Electrical Engineer's Reference Book*; the *Electronics Engineers' Handbook*; the *Basic Electronic Instrument Handbook*; the *Handbook of Electronic Packaging*; the *Handbook of Materials and Processes for Electronics*; the *Handbook of Modern Electronic Data*; the *Handbook of Semiconductor Electronics*; and the *Standard Handbook for Electrical Engineers*. Also refer to the *Handbook for Electronics Engineering Technicians; Electronics Handbook*; and *Electronic Conversions, Symbols, and Formulas*.

1959. Industrial Energy Conservation: A Handbook for Engineers and Managers. D.A. Reay. Pergamon Press, 2d ed., 1979, 375 pp. LC: 78-40982. ISBN: 0-08-023273-6. Cost: $52.00.

Covers such topics as energy storage, waste heat recovery, energy uses, and costs.

1960. Industrial Engineering Handbook. H.B. Maynard. McGraw-Hill, 3d ed., 1971, 1,984 pp. (various paging). LC: 77-128017. ISBN: 0-07-041084-4. Date of First Edition: 1956. Cost: $49.95.

This handbook includes a collection of data and information that industrial engineers will find useful in helping to solve problems related to industrial engineering. Such topics as inventory control, quality control, reliability engineering, and human factors engineering are discussed. In addition, such subjects as time and motion studies, value engineering, and computer applications are presented. Information is also provided on estimating costs, and an index has been included. For related works, please refer to the *Scheduling Handbook; Management Handbook for Plant Engineers*; and *Handbook of Industrial Engineering and Management*.

1961. Industrial Engineering Tables. Samuel Eilon. Reinhold, 1962, 232 pp. LC: 61-14613.

1962. Industrial Fasteners Handbook. International Ideas (Trade and Technical Press). ISBN: 0-85461-062-6. Cost: $ 115.00.

1963. Industrial Fire Hazards Handbook. Gordon P. McKinnon; Paul S. Tasner; Mary L. Hill. National Fire Protection Association, 1st ed., 1979, 933 pp. LC: 79-66427. ISBN: 0-87765-155-8. Cost: $30.00. Series: NFPA Spp-57.

This handbook deals with fire hazards in such major industries as textile manufacturing, rubber producing, furniture manufacturing, pulp and paper milling, printing, painting, asphalt industry, shipyards, and dry-cleaning plants. In addition, such fire hazards as chemical processes, boiler furnaces, welding, spray finishing, and dipping and coating processes are discussed. Fire hazards associated with solvent extraction, radioactive materials, grinding and milling operations, molten salt baths, and refrigeration systems are also presented.

1964. Industrial Hazard and Safety Handbook. Ralph W. King; John Magid. Butterworths, 1979, 793 pp. LC: 78-40474. ISBN: 0-408-00304-9. Cost: $74.95.

This handbook covers such topics as chemical hazards, radiation hazards, fire hazards, and hazards associated with machines, cranes, lifting tackle, hand tools, and portable power tools. In addition, such topics as falling objects, protective clothing, noise and hearing in the working environment, toxicology, and fire protection are also addressed.

1965. Industrial Instrument Servicing Handbook. G.C. Carroll. McGraw-Hill, 1960, 886 pp. LC: 59-8532. ISBN: 0-07-010146-9. Cost: $34.95.

1966. Industrial Lubrication: A Practical Handbook for Lubrication and Production Engineers. Michael Billett. Pergamon Press, 1979, 136 pp. LC: 79-40526. ISBN: 0-08-024232-4. Cost: $20.00.

Numerous types of lubricants and oils are discussed in this handbook.

1967. Industrial Motor Users' Handbook of Insulation for Rewinds. L.J. Rejda; Kris Neville. Elsevier, 1977, 408 pp. LC: 76-26949. ISBN: 0-444-00191-3. Cost: $37.25.

This handbook includes bibliogrpahic references and indexes. It is essentially about insulation and windings for electric motors.

1968. Industrial Noise and Hearing Conservation. Julian B. Olishifski; Earl R. Harford. National Safety Council, 1975. LC: 75-11313. ISBN: 0-87912-085-1. Cost: $24.00.

1969. Industrial Noise Control Handbook. Paul N. Cheremisinoff; Peter P. Cheremisinoff. Ann Arbor Science Publishers Inc, 1977, 350 pp. LC: 76-46023. ISBN: 0-250-40144-4. Cost: $29.50.

Covers silencers and suppressor systems vibration control, noise reduction, with glass, lead, hydrodynamic control of valve noise, ventilating system noise control, and audiometric testing.

1970. Industrial Noise Manual. American Industrial Hygiene Association, 1966, 171 pp.

1971. Industrial Pollution Control Handbook. Herbert F. Lund. McGraw-Hill, 1971, 864 pp. (various paging). LC: 70-101164. ISBN: 0-07-039095-9. Cost: $46.50.

1972. Industrial Power Systems Handbook. D. Beeman. McGraw-Hill, 1955, 948 pp. LC: 54-10629. ISBN: 0-07-004301-9. Cost: $45.50.

1973. Industrial Products Master Catalog. Herman Publishing, annual.

This catalog was formerly entitled *Electronic Distributor's Master.* This annual publication is basically a directory of electronic components and equipment.

1974. Industrial Safety Handbook. William Handley. McGraw-Hill, 2d ed., 1977, 480 pp. LC: 76-54350. ISBN: 0-07-08441-X. Date of First Edition: 1969. Cost: $27.50.

1975. Industrial Solvents Handbook 1977. I. Mellan. Noyes Data Corp, 2d ed., 1977, 567 pp. LC: 78-104728. ISBN: 0-8155-0651-1. Cost: $39.00.

Includes 975 tables on the physical properties of most solvents and on the solubilities of numerous materials. Covers ketones, acids, amines, esters, ethers, aldehydes, phenols, glycol ethers, halogenated hydrocarbons, and organic sulfur compounds.

1976. Industrial Wastewater Management Handbook. Hardam Singh Azad. McGraw-Hill, 1976, 608 pp. LC: 76-17884. ISBN: 0-07-002661-0. Cost: $35.75.

This handbook contains information on industrial water treatment and water pollution control in several different industries including food processing, power generation, metals, petroleum, chemical and pulp and paper manufacturing. This book provides information on recommended guidelines, standards, and legislation with regard to waste water discharges. Water monitoring systems are discussed along with waste water treatment equipment including equipment for activated sludge and aeration filtering and sludge disposal. Bibliographic references are provided throughout the handbook, an index, and a list of vendors for monitoring equipment has been prepared. For related works, please refer to such titles as *Betz Handbook of Industrial Water Conditioning; CRC Handbook of Environmental Control; Pollution Engineering Practice Handbook; Environmental Engineers' Handbook,* the *Handbook of Environmental Civil Engineering,* and *Water Treatment Handbook.*

1977. Industrialized Builders Handbook: Techniques of Component and Modular Fabrication. R.J. Lytle et al. Structures Publishing Co, 1971. LC: 70-147978. ISBN: 0-912336-01-3. Cost: $24.95.

1978. Infrared and Ultraviolet Spectra of Some Compounds of Pharmaceutical Interest. Association of Official Analytical Chemists, rev. ed., 1972, 278 pp. Cost: $10.00.

This is a compilation of absorption spectra of compounds of pharmaceutical interest. In addition, infrared spectra of antibiotics is also provided. More than 800 compounds are listed.

1979. Infrared Band Handbook. Herman A. Szymanski; Ronald E. Erickson. IFI/Plenum, rev. ed., 1970, 2 Vols., 1,491 pp. LC: 62-15543. ISBN: 0-306-65138-6. Cost: $150.00.

1980. Infrared Handbook. William L. Wolfe; G.J. Zissis. Environmental Research Institute, 1978, 1,736 pp. LC: 77-90786. ISBN: 0-9603590-0-1. SuDoc: D210.6/2:In 3/2. Cost: $25.00.

This volume replaces the *1965 Handbook of Military Infrared Technology.* It covers such topics as atmospheric scattering, optical materials, detectors, imaging tubes, photographic film, displays, radiometry, and tracking systems.

1981. Infrared Radiation: A Handbook for Applications. M.A. Bramson; William Wolfe trans. Plenum Press, 1968, 623 pp. LC: 66-26812. ISBN: 0-306-30274-8. Cost: $55.00.

1982. Infrared Spectra of Selected Chemical Compounds. R. Mecke; F. Langenbucher. Heyden. ISBN: 0-85501-023-1 microfilm; 0-85501-028-2 microfiche. Cost: $240.00 each.

This is collection of 1,879 different infrared spectra of various chemical compounds. Such information as concentration, capillary thickness, temperature, purity, and band table are presented.

1983. Injured Hand: A Handbook for General and Clinical Practice. H.R. Mittelbach. Springer-Verlag, 1979, 320 pp. ISBN: 0-387-90365-8. Cost: $19.80.

Provides information on hand surgery, traumatology, and orthopedics.

1984. Inorganic Compounds with Unusual Properties, I and II. R. Bruce King. American Chemical Society, Vol.1, 1976; Vol.2, 1979. LC: Vol.1, 76-13505; Vol.2, 78-31900. ISBN: Vol.1, 0-8412-0281-8; Vol.2, 0-8412-0429-2. Cost: Vol.1, $48.50; Vol.2, $43.50. Series: Vol.1, *Advances in Chemistry Series:* No. 150; Vol.2, *Advances in Chemistry Series:* No. 173.

The compounds covered are metal alkyks, metal alkoxides, metal alkylamides, metal chelates, polyboranes, polyphosphines, and macrocyclic derivatives.

1985. Inorganic Ligands. E. Hogfeldt. Pergamon Press, 1980, 1,200 pp. ISBN: 0-08-020959-9. Cost: $150.00. Series: *IUPAC Chemical Data Series*, No. 21.

1986. Inspection of Processed Photographic Record Films for Aging Blemishes. National Bureau of Standards. American National Standards Institute, Jan. 24, 1964. Accession: ANSI PH 1.28-1973. Series: NBS Handbook 96.

1987. Instrument Engineers' Handbook. Bela G. Liptak. Chilton Book Co, 1969-72, 2 Vols. and supplement; Vol.1, 1,194 pp.; Vol.2, 1,600 pp. LC: 73-80445. ISBN: 0-8019-5947-0. Cost: $80.00 per set.

Vol.1, is entitled, *Process Measurement*; Vol.2, *Process Control*. This handbook contains information on a multitude of different devices used to make scientific measurments. The devices are grouped by the type of measurement to be made and include such topics as temperature, pressure, process control, and weight measurments. This 2-volumne set discusses such instruments as pressure relief valves, strain gages, flow meters, and thermocouples. An excellent index has been provided, and conversion tables have been included for the convenience of the user. For related titles, please refer to such handbooks as the *Handbook of Applied Instrumentation; The Handbook of Biomedical Instrumentation and Measurements; The Handbook of Industrial Metrology; The Handbook of Microwave Measurements; The Handbook of Transducers for Electronic Measuring Systems*, and *Standards and Practices for Instrumentation*.

1988. Instrument Maintenance Manager's Sourcebook. R.A. Denoux. Instrument Soceity of America, 1974, 480 pp. ISBN: 0-87664-253-9. Cost: $30.00.

1989. Instrument Pilot Handbook: A Reference Manual and Exam Guide. Courtney L. Flatau; Jerome F. Mitchell. Van Nostrand, 1980, 372 pp. LC: 79-93. ISBN: 0-442-22411-7. Cost: $24.95.

This handbook is designed for private and commercial aviation pilots who are seeking an instrument rating of a fixed-wing aircraft. Over 350 illustrations are provided as well as 400 different questions with their corresponding answers. Navigation information is provided along with flight instruments, navigation calculations, and flight procedures.

Instrument Society of America. *See* **ISA.**

1990. Instrumentation and Control Systems Engineering Handbook. *Instrumentation Technology Magazine*, eds (available from TAB Books), 1979, 432 pp. LC: 78-11391. ISBN: 0-8306-9867-1. Cost: $22.95.

This is a good reference guide for information on instrumentation and controls.

1991. Integrals of Airy Functions. National Bureau of Standards. NTIS, 1958, 28 pp. Accession: PB-251904. Cost: $4.00. Series: NBS AMS52.

1992. Integrated Atlas of Gastric Diseases. K. Krentz; H. Lamm (trans). Georg Thieme (distributed in U.S. by PSG Publishing, Littleton, MA), 1976, 203 pp. ISBN: 3-13-522901-7. Cost: DM 120.

This publication contains 370 figures with 188 in color.

1993. Integrated Circuits Reference Book. Association Internationale Pro Electron. A.E. Klauwer, Antwerp, 1976-77, 2 Vols., Vol.1, 240 pp.; Vol.2, 224 pp. Cost: Vol.1, $25.00; Vol.2, $24.50.

Vol.1 is entitled: *Analogue Circuits*. Vol.2 is entitled: *Digital Circuits*. This handbook contains reference data on integrated circuits made in Western Europe.

1994. Integrated Circuits: A User's Handbook. Michael Cirovic. Reston, 1977, 368 pp. LC: 77-9022. ISBN: 0-87909-356-0. Cost: $19.95.

Covers operational amplifiers, voltage regulators, timers, and function generators.

1995. Interface Integrated Circuits Handbook. National Semiconductor Corp, 464 pp. Cost: $4.00 softcover.

1996. Intergrated Circuits Handbook. Robert E. Gibson. Boston Technical Publishers, Inc, 1966, 308 pp.

1997. Internal Dynamics Manual, Appendix. North American Rockwell (NTIS), 1970, 597 pp. Accession: N71-32061 AD-723841. Tech Report: NR68H-434-APP. Cost: $0.95 microform; $6.00 hard copy.

This manual consists of tabulations for calculating internal combustion engine thermodynamics.

1998. International and Metric Units of Measurement. Marvin H. Green. Chemical Publishing Co, 1973. ISBN: 0-8206-0069-5. Cost: $13.75.

International Bearing Interchange. *See* **IBI.**

1999. International Bibliography of Alternative Energy Sources. G. Hutton; M. Rostron. Nichols Publishing Company, 1979, 180 pp. LC: 78-53105. ISBN: 0-89397-036-0. Cost: $60.00.

This is a collection of articles listing over 15,000 references to books, journals, and papers on alternative energy sources.

2000. International Classification of Diseases. American Hospital Association, 9th rev. ed., 1978, 3-Vol. set, Vol.1, 1,140 pp.; Vol.2, 910 pp., Vol.3, 464 pp. Cost: $45.50 set.

This 3-volume set consists of an international classification of diseases. Vol.1 is a tabular lists of diseases. Vol.2 is an alphabetical lists of diseases. Vol.3 is an index to procedures.

2001. International Classification of Health Problems in Primary Care. American Hospital Association, 1975, 119 pp. ISBN: 0-87258-177-2. Cost: $5.00.

This book was designed for use in the primary care setting for the purpose of classifying diagnostic entities.

2002. International Cloud Atlas. Unipub, rev. ed., 1956; reprinted, 1969, 210 pp. ISBN: 0-685-68975-1. Date of First Edition: 1956. Cost: $62.00.

This is a reference on clouds providing explanations for optical phenomena for all different types of precipitation. Includes 72 plates, 30 of which are in color.

2003. International Countermeasures Handbook. Harry F. Eustace (ed). Franklin Watts, Inc, 5th ed., 1979, 530 pp. ISBN: 531-03918-8. Cost: $50.00.

This book contains a substantial amount of information on electronic warfare, including hardware systems, simulations, tactical warning technology, and budgetary information for the 3 branches of the military service.

2004. International Critical Tables of Numerical Data, Physics, Chemistry and Technology. Edward W. Washburn. McGraw-Hill, 1926-33, multivolume set, 7 vols plus index.

The *International Critical Tables* is considered by many to be the classical tabulation of physical constants in existence today. Although this 7-volume set was prepared between the years 1926 and 1933, it still includes information that is unavailable in any other tabulation. The data are presented in the form of tables, charts, graphs, and equations, and contains such information as physical and chemical properties of various chemical substances including solubility, viscosity, refractivity, dielectric properties, surface tension, and toxicology. In addition, such topics as radioactivity, data on the solar system, explosives, metallurgy, phase-equilibrium data, acoustics, and magnetism are discussed. All sources of information are well documented and a one-volume index has been very thoroughly prepared. For related works, please refer to such titles as the *Chemical Engineers' Handbook*, the *CRC Handbook of Chemistry and Physics*, the *CRC Handbook of Tables for Applied Engineering Science*, the *Handbook of Physics*, and *Lange's Handbook of Chemistry*.

2005. International Frequency List. International Frequency Registration Board (Geneva). International Telecommunications Union, 7th ed, 1974, supplements issued periodically.

2006. International Handbook of Coal Petrology: Supplement to the 2nd Edition. International Committee for Coal Petrology, 1971, 247 leaves.

This supplement contains definitions of coal related topics. Includes index and bibliographies.

2007. International Handbook of Contemporary Developments in Architecture. Warren Sanderson. Greenwood Press, 1981, 600 pp. LC: 80-24794. ISBN: 0-313-21439-5. Cost: $75.00.

This book includes articles by 35 leading architects on contemporary developments in architecture. Such topics as preservation, restoration, and urban planning are discussed, along with essays on architectural development in 32 different countries. Bibliographies and indexes of architects and buildings are provided along with more than 250 illustrations.

2008. International Handbook of Liquid-Crystal Displays. Martin Tobias (Omniscience, Ltd). Scholium International, Inc, 1974. Cost: $75.00.

2009. International Handbook of Medical Science. D. Horrobin. University Park Press, 1972. ISBN: 0-8391-0573-8. Cost: $17.50.

2010. International Handbook on Ageing. E. Palmore. Macmillan Press, Ltd, 1980, 592 pp. ISBN: 0-33-27828-3. Cost: $72.60.

Covers such topics as gerontology and demographic information.

2011. International Journal of Thermophysics: Journal of Thermophysical Properties, Thermal Physics, and Its Ap-plications. Ared Cezairliyan. Plenum, Vol.1, 1980-; 4 issues. Cost: $58.00 per year.

This is obviously not a handbook, but it is an excellent source of information on thermophysical properties data and thermophysics.

2012. International Metallics Materials Cross Reference. Daniel L. Potts. General Electric Co, 1979, 108 pp. ISBN: 0-931690-09-9. Cost: $34.50.

Such metallic materials as carbon steels, tool steels, stainless steels, aluminum and aluminum alloys, and copper alloys are included. This collection presents complete chemical specifications and is a useful guide for identifying materials in various national standards which are roughly equivalent to those of U.S. specifications.

2013. International Meteorological Tables. World Meteorological Organization. Unipub. Cost: $11.00 paper copy.

2014. International Oceanographic Tables. UNESCO; National Institute of Oceanographers. Unipub, 2 Vols., Vol.1, 1966; Vol.2, 1974, 141 pp. ISBN: Vol.1, 92-3-000906-7; Vol.2, 92-3-001044-8. Cost: Vol.1, $12.25; Vol.2, $15.75.

2015. International Operations Handbook for Measurement of Background Atmospheric Pollution. Unipub, 1979, 110 pp. ISBN: 92-63-10491-3. Cost: $30.00.

This handbook includes such information as turbidity measurements, sampling methods, analysis of precipitation, and methods of measuring carbon dioxide in the atmosphere. In addition, methods and standard procedures are provided for measuring such pollutants as ozone, sulfur, carbon monoxide, ammonia, nitrogen oxides, halocarbons, pesticides, hydrometals, and PCB's.

2016. International Plastics Flammability Handbook. Troitzsch. Carl Hanser Verlag, 1982, approx. 700 pp. ISBN: 0-02-949770-1. Cost: $90.00.

Includes principles and testing methods for the flammability of plastics.

International Seal Interchange. *See* **ISI.**

2017. International Statistical Handbook of Urban Public Transport. Lee H. Rogers. Union Internationale Des Transports Publics (UITP), 1979, 2 Vols., 1,116 pp.

This 2-volume work includes such information as statistical data on runways and rapid transit systems, electric buses, tramways, inclines, and public elevators. Substantial highway information is included along with numerous tables on such topics as vehicle capacities, vehicle length, vehicle width, speed, acceleration, vehicle weights, verticle gradients, and equivalents in petroleum fuel consumption.

2018. International Tables for X-Ray Crystallography. International Union of Crystallography. Kynoch Press, 2d ed., 1965-1968, 3-Vol. set.

The volumes are entitled as follows: Vol.1, *Symmetry Groups*; Vol.2, *Mathematical Tables*; Vol.3, *Physical and Chemical Tables*.

2019. International Thermodynamic Tables of the Fluid State. S. Angus. Pergamon Press, 1972-, 7 vols.

Each volume in this multivolume series follows a similar format and includes such information as equations of state, zero-pressure properties, zero-pressure tables, single-phase properties, proerties of the saturation curve, and properties of the melting curve. In addition, such information as entropy, enthalpy, isobaric heat capacity, compression factor, and fugacity/pressure ratio are provided in this comprehensive work.

2020. International Trademark Design. Peter Wildbur. Van Nostrand Reinhold, 136 pp. Cost: $16.95.

This is an illustrated collection of designs from all over the world showing different forms of trademarks and how to design trademarks. Trademarks are listed under such categories as environment, industry, products, resources, and services.

2021. Interpretation of Diagnostic Tests: A Handbook of Synopsis of Laboratory Medicine. Jacques B. Wallach. Little, Brown, 2d ed., 1974, 529 pp. ISBN: 0-316-92043-6. Cost: $8.95.

2022. Interpreting Graphs and Tables. Peter H. Selby. John Wiley and Sons, 1976, 204 pp. LC: 75-25761. ISBN: 0-471-77559-2. Cost: $4.95. Series: *Wiley Self-Teaching Guides.*

This handy self-teaching guide should help the user to understand graphs and tables. Includes information on such topics as area graphs, curvilinear graphs, and bar graphs.

2023. Interstate Motor Carrier Forms Manual. Harold C. Nelson. J.J. Keller and Associates, rev. ed., 1979, looseleaf. LC: 76-7194. ISBN: 0-686-16913-1. Cost: $69.00.

This manual provides samples of forms that are used in numerous types of regulatory agencies, such as samples of federal agency forms and samples of individual state agency forms. State forms are provided for uniform registration certificates and permits.

2024. Intravenous Medications: A Handbook for Nurses and Other Allied Personnel. Betty L. Gahart. C.V. Mosby, 2d ed., 1977, 224 pp. ISBN: 0-8016-1718-9. Cost: $7.75.

2025. Introduction to the Theory of Stochastic Processes Depending on a Continuous Parameter. H.B. Mann (National Bureau of Standards). NTIS, 1953, 45 pp. Accession: PB-175816. Cost: $4.00. Series: NBS AMS24.

2026. Inventor's Handbook. Terrence W. Fenner; James L. Everett. Chemical Publishing Co, 1968, 309 pp. LC: 73-5567. ISBN: 0-8206-0070-9. Cost: $17.00.

This handbook provides information on inventions, including how to patent and market new inventions.

2027. Investigation and Compilation of the Thermodynamic Properties of High Temperature Chemical Species. D.R. Stull; H. Prophet. NTIS, 1968, 7 pp. Accession: AD-837 620. Tech Report: T0009-20-68. Cost: $3.50 microform; $5.00 hard copy.

This is actually a supplement to the *JANAF Thermochemical Tables* and includes approximately 66 revised tables.

2028. Ion Beam Handbook for Material Analysis. James W. Mayer; E. Rimini. Academic Press, 1977, 488 pp. LC: 77-24538. ISBN: 0-12-480860-3. Cost: $31.50.

The subtitle of this book is as follows: *A Compilation of Tables, Graphs and Formulas for Ion Beam Analysis using the Methods of Backscattering, Channeling, Nuclear Reactions and X-Ray*

Emission Based on Catania Working Data, 1974. This handbook contains information on different methods of materials analysis by ion beams. Such topics as backscattering spectrometry, neutron cross section, and ion-induced X-ray emissions are discussed.

2029. Ion Exchange Equilibrium Constants. Y. Marcus; D.G. Howery. Pergamon Press, 1975, 42 pp. ISBN: 0-08-020992-0f. Cost: $10.00. Series: *IUPAC Chemical Data Series*, Vol. 10.

2030. Ionisation Constants of Organic Acids in Aqueous Solution. E.P. Serjeant; B. Dempsey. Pergamon Press, 1979, 998 pp. LC: 78-40988. ISBN: 0-08-022339-7. Cost: $190.00. Series: *IUPAC Chemical Data Series*, Vol. 23.

Data is summarized in this compilation for over 4,500 different acids that have been reported in the literature from 1956 to the end of 1970.

2031. Ionization Potentials and Ionization Limits Derived from the Analyses of Optical Spectra. C.E. Moore (National Bureau of Standards). U.S. Government Printing Office, 1970. SuDoc: C13.48:34. Cost: $1.65. Series: NSRDS-NBS-34.

2032. Ionization Potentials, Appearance Potentials and Heats of Formation of Gaseous Positive Ions. J.L. Franklin; J.G. Dillard; H.M. Rosenstock; J.T. Herron; K. Draxl; F.H. Field (National Bureau of Standards). U.S. Government Printing Office, 1969. SuDoc: C13.48:26. Cost: $6.20. Series: NSRDS-NBS-26.

2033. IP Standards for Petroleum and Its Products. Heyden, 1981, 2 Vols., 1,600 pp. ISBN: 0-85501-338-9. Cost: $127.50.

Presents over 300 methods for analysis and testing of petroleum and its products.

2034. Iron Castings Handbook: Including Data on Gray, Ductile, White and High Alloy Irons. Charles F. Walton. Iron Castings Society, 1981, 860 pp. LC: 70-165866. Cost: $27.50.

This handbook, which includes over 862 illustrations, includes such information as the physical and mechanical properties of cast iron and cast alloys. Heat treatment data are provided along with information about welding and hardness data. Various tests are discussed including nondestructive testing and tensile testing. Thermal expansion, magnetic properties, specific heat, thermal conductivity, and corrosion resistance are among the properties listed. An index is included along with a listing of definitions of iron casting terms. For related works please refer to such titles as *Metals Handbook; Metals Reference Book*, and the *ASME Handbook.*

2035. ISA Control Valve Compendium. Instrument Society of America, 1973, 320 pp. LC: 73-77184. ISBN: 0-87664-235-0. Cost: $40.00.

This handbook includes a listing of specifications and manufacturer's product data for over 1,000 different control valves. The control valves are compared, and advantages and disadvantages are listed. Actuators and valve accessories are also listed along with addresses of the different valve manufacturers. A handy index to retrieve the data is also included. For related works, please refer to *Equivalent Valves Reference Manual; ISA Handbook of Control Valves*, and the *Handbook of Valves.*

2036. ISA Handbook of Control Valves. Instrument Society of America. Instrument Society of America, 2d ed., 1976, 533 pp. LC: 73-163121. ISBN: 87664-234-2. Cost: $35.00.

This handbook represents a collection of design data, tables, and charts on control valves and information to assist the technician or engineer in solving control valves problems. The handbook presents such topics as flow characteristics, cavitation, installation, selection, and noise associated with control valves. Included are a subject index, an index for the tabulated data, and a section on terminology associated with control valve technology. For related works, please refer to the *ISA Control Valve Compendium; Equivalent Valves Reference Manual*, and the *Handbook of Valves.*

2037. ISA Transducer Compendium. G.F. Harvey. Instrument Society of America, 2d ed., Part 1, 212 pp., Part 2 323 pp., Part 3, 277 pp. LC: Part 3, 68-57392. ISBN: Part 1, 87664-096-X; Part 2, 87664-123-0; Part 3, 87664-124-9. Date of First Edition: Part 1, 1969; Part 2, 1970; Part 3, 1977. Cost: Part 1, $27.50; Parts 1 & 2, $52.50; Parts 1, 2, and 3 $72.00.

This is a compendium of performance data tables on over 13,000 transducers. Performance data are provided for such variables as temperature, heat flux, nuclear radiation, penetrating radiation, electromagnetic radiation, magnetic field strength, humidity, and moisture. Performance data are also provided for transducers that measure pressure, flow, sound, force, torque, motion, and dimension.

2038. ISI Guide. International Seal Interchange. Interchange, Inc, 3d ed., 1979. LC: 78-70760. ISBN: 0-916966-2. Cost: $50.00.

2039. ISO Standards Handbook. American National Standards Institute, Inc.

This set currently consists of 3 handbooks entitled as follows: *ISO Standards Handbook 1—Information Transfer*, 1977, 525 pp., $46.20. *ISO Standards Handbook 2—Units of Measurement*, 1979, 240 pp., $20.00. *ISO Standards Handbook 3—Statistical Methods*, 1979, 300 pp., $25.20. These volumes represent compilations of standards on given topics. For example, the *Information Transfer* volume includes a compilation of 56 selected international standards dealing with such topics as bibliographic references, documents, bibliographic control, information systems, libraries, and mechanization.

2040. Isodose Atlas for Use in Radiotherapy. G. Nemeth; H. Kuttig. Martinus Nijhoff, 1981, 272 pp. ISBN: 90-247-2476. Cost: $54.50. Series: *Series in Radiology*, 5.

This is a reference book on the topic of irradiation treatment of tumors. Dosage distribution information is provided for treatment of tumors in various locations of the body.

2041. Isophotometric Atlas of Comets. W. Hogner; N. Richter. Springer-Verlag, 2 parts, 1979, Loose-leaf. ISBN: Part 1, 0-387-09171-8; Part 2, 0-387-09172-6. Cost: Part 1, $72.60; Part 2, $64.90.

This 2-part loose-leaf service provides material on the topic of the physics of comets.

2042. IUPAC Compendium of Analytical Nomenclature *The Orange Book.* H.M.N.H. Irving. Pergamon Press, 1978, 228 pp. LC: 77-8949. ISBN: 0-08-022008-8 hard copy; 0-08-022347-8 flexi-cover. Cost: $25.00 hard copy; $15.00 flexi-cover.

This is a collection of recommendations for terminology in various areas of analytical chemistry such as contamination phenomena, precipitation from aqueous solution, automatic analysis, thermal analysis, mass spectrometry, and other types of chemical analysis.

2043. JANAF Handbook. Rocket Exhaust Plume Technology. Johns Hopkins University Chemical Propulsion Information Agency, May 1975, 253 pp. Accession: AD-A013-427. Tech Report: CPIA-Pub-263.

This loose-leaf handbook is designed primarily to provide data which will help solve common plume technology problems. Many subjects are covered including fluid dynamics, chemical composition, multiphase flow, turbulence, and phase studies as they all relate to rocket exhaust plume-technology.

2044. JANAF Thermochemical Tables, 2nd ed D.R. Stull; H. Prophet. U.S. National Bureau of Standards (U.S. Government Printing Office), 2d ed., 1971, 1,141 pp. SuDoc: C13.48 No. 37. Cost: $9.75. Series: *National Bureau of Standards*, 37.

This handbook is widely recognized as one of the most authoritive collections of thermophysical data. Although it was originally prepared by the Dow Chemical Company for the Chemical Rocket Propulsion Industry, the handbook has become recognized as on of the best sources for accurate thermochemical data. Many of the properties of chemicals are listed including heat information, enthalpy, and entropy. Methods of calculations are presented as well as definitions of terminology and indexes to the tables. For related works, please refer to the following titles: *CRC Handbook of Tables for Applied Engineering; Handbook of Thermodynamic Tables and Charts; Tables on the Thermophysical Properties of Matter*; as well as the *International Critical Tables.*

2045. Jane's World Aircraft Recognition Handbook. Derek Wood. Jane's, 1979, 512 pp. Cost: &£6.95.

Primarily for the purpose of identifying airplanes. Numerous illustrations are provided as well as an index to different types of planes.

2046. Kempe's Engineers Yearbook for 1981. C.E. Prockter. Morgan-Grampian Book Publishers, Ltd, 86th ed., 1981, 2 Vols., annual publication. Cost: £27.00.

This yearbook is really a handbook which consists of information on such topics as gears, springs, water treatment, cranes, transportation systems, properties, and specifications of metals and fuel injection systems. This 2-volume set, with its wide diversification of data, would be of value to any practicing engineer or library. An index is provided in the second volume.

2047. Kempe's Metrication Handbook. Jack Peach. Morgan-Grampian, 1972, 79 pp. LC: 73-156481.

Free with *Kempe's Engineering Yearbook.*

2048. Kent's Mechanical Engineers' Handbook: Vol. 1, Design and Production; Vol. 2, Power. Colin Carmichael. John Wiley and Sons, 12th ed., 1950, 2 Vols. (various paging). LC: 50-8435. ISBN: Vol.1, 0-471-46959-9; Vol.2, 0-471-46992-0. Date of First Edition: 1895. Cost: $34.50 per Vol. Series: *Wiley Engineering Handbook Series.*

This 2-volume set includes a considerable amount of basic engineering information on the properties of metals and nonmetals, the purpose of which is to provide information to those engineers who are involved with the design and manufacture of different types of machinery and equipment. Such topics as vibration and lubrication, and joining techniques are discussed. In addition, information is provided on heating, ventilating, air conditioning, and refrigeration as well as boilers and turbines and the various fuels that are used in power-producing processes. An excellent index is included, and many bibliographic references have been supplied throughout the handbook so that the user can obtain additional information. For related works, please refer to the *Mechanical Engineer's Reference Book* and the *Standard Handbook for Mechanical Engineers*.

2049. Kentucky Mine Safety Training Program: Trainee Handbook. Kentucky Bureau of Vocational Education. Kentucky Bureau of Vocational Education, 2 manuals.

The 2 training manuals in this program are entitled as follows: *Underground*, 185 pp., Accession No.: C80-428. *Surface*, 134 pp. Accession No.: C80-427. These two manuals, both the training manual for surface mining and the training manual for underground mining, present information on mine safety and the practical aspects of learning safe procedures when mining coal.

2050. Kentucky Solar Energy Handbook. Blaine F. Parker. Kentucky Department of Energy, 1979, 182 pp.

This handbook includes 11 articles on such topics as solar heating systems, solar collectors, solar energy storage in water, solar energy storage in rock, temperature control of solar energy, and solar water heating systems.

2051. Keystone Coal Industry Manual, 1982. Mining Information Services. McGraw-Hill, 1982, 1,466 pp. LC: 20-20634. ISBN: 07-608825-0. Date of First Edition: 1918. Cost: $99.50.

This manual is an annual publication containing a wealth of information on the coal industry. Included are such subjects as statistics on various aspects of coal production, reserve consumption, coal seams, transportation of coal, fly ash, slurry pipe lines, and various reports on the latest coal research programs. The state-by-state listing of coal mines, companies, and coal seams will be of interest. In additition, information is available on such subjects as fuel conversion factors, coal gasification and liquefaction, scrubbers, exporting and listing of various assiciations dealing with coal information. This thumb-indexed publication includes an index in the front of the book as a detailed table of contents.

2052. Kinetic Data on Gas Phase Unimolecular Reactions. S.W. Benson; H.E. O'Neil (National Bureau of Standards). NTIS, 1970. Accession: PB-191956. Cost: $16.25. Series: NSRDS-NBS-21.

2053. Kinetic Parameters of Electrode Reactions of Metallic Compounds. R. Tamamushi. Pergamon Press, rev. and enlarged ed. 1975, 190 pp. ISBN: 0-08-020991-2. Cost: $34.00. Series: *IUPAC Chemical Data Series*, Vol. 11.

This publication, which is sponsored by the International Union of Pure and Applied Chemistry, provides information on Kinetic Parameters of electrode reactions of atomic compounds.

2054. Knott's Handbook for Vegetable Growers. O.A. Lorenz; D.N. Maynard. John Wiley and Sons, 2d ed., 1980, 390 pp. LC: 79-26840. ISBN: 0-471-05322-8. Cost: $16.50.

This reference book provides practical information on seeding, fertilizing, soil analysis, pest control, and weed control.

2055. Kronecker Product Tables. A.P. Cracknell; B.L. Davies. Plenum Publishing Corp, 1979, 4 Vols. ISBN: 0-306-65175-0. Cost: $325.00.

This 4-volume set is entitled as follows: Vol.1, *General Introduction and Tables of Irreducible Representations of Space Groups.* 600 pp. Vol.2, *Wave Vector Selection Rules and Reductions of Kronecker Products for Irreducible Representations of Orthorhombic and Cubic Space Groups.* 900 pp. Vol.3, *Wave Vector Selection Rules and Reductions of Kronecker Products for Irreducible Representations of Triclinic, Monoclinic, Tetragonal, Trigonal, and Hexagonal Space Groups.* 400 pp. Vol.4, *Symmetrised Powers of Irreducible Representations of Space Groups.* 400 pp. This 4-volume set provides information that will help determine the selection of crystalline materials.

2056. Krypton, Xenon, and Radon—Gas Solubilities. H.L. Clever. ISBN: 0-08-022352-4.

See Solubility Data Series, Vol. 2.

2057. Laboratory Handbook for Oil and Fat Analysis. Leslie V. Cocks; C. Van Rede. Academic Press, 1966, 419 pp. LC: 65-27886. ISBN: 0-12-178550-5. Cost: $66.00.

2058. Laboratory Handbook of Chromatographic and Allied Methods. O. Mikes. Halsted Press, 1979, 764 pp. LC: 78-40595. ISBN: 0-470-26399-7. Cost: $149.95. Series: *Analytical Chemistry Series.*

Includes information on adsorption column chromatography, ion exchange chromatography, gel chromatography, thin-layer chromatography, gas chromatography, and counter-current distribution. This compilation is basically a review of the literature between 1962 and 1976. Tables of chromatographic data and laboratory procedures are included in the compilation.

2059. Laboratory Handbook of Chromatographic Methods. Otakar Mikes. Van Nostrand-Reinhold, 1961, 399 pp. LC: 62-25895. ISBN: 0-442-05364-9. Cost: $22.50.

2060. Laboratory Handbook of Medical Mycology. Michael R. McGinnis. Academic Press, 1980, 661 pp. ISBN: 0-12-482850-7. Cost: $55.00.

This book emphasizes the identification of fungi and identification methods are based primarily on modern concepts of conidiogenesis and sporogenesis. Such topics as yeast identification, mould identification, culture collection, bioassay procedures, and fungal pathogens are discussed.

2061. Laboratory Handbook of Methods of Food Analysis. R. Lees. Leonard Hill Books (London), 2d ed., 1971, 192 pp. LC: 72-115563.

Sampling methods are discussed in addition to testing and laboratory procedures with regard to food analysis.

2062. Laboratory Handbook of Paper and Thin-Layer Chromatography. J. Gasparic; J. Churacek. Halsted Press, 1978, 362 pp. LC: 77-14168. ISBN: 0-470-99298-0. Cost: $74.95.

This book discusses analytical chemistry as it relates to paper chromatography and thin-layer chromatography. Such organic compounds as alkaloids, vitamins, antibiotics, and dyes are discussed in connection with chromatographic analysis.

2063. Laboratory Handbook of Petrographic Techniques. Charles S. Hutchison. Wiley-Interscience, 1974, 544 pp. LC: 73-17336. ISBN: 0-471-42550-8. Cost: $44.95.

Includes thin-section preparation, photomicrography, mineral separation, powder methods of X-ray diffraction, refractive index, and spectrophotometry techniques.

2064. Laboratory Handbook. N.L. Parr. D. Van Nostrand Co, Inc (also by George Newnes, Ltd, Great Britain), 1963, 1,161 pp. LC: 65-7092.

2065. Laboratory Primate Handbook. Robert A. Whiteny Jr et al. Academic Press, Inc, 1973. ISBN: 0-12-747450-1. Cost: $14.25.

2066. Landolt-Börnstein Numerical Data and Functional Relationships in Science and Technology, New Series. K.H. Hellwege. Springer-Verlag, New Series, 6th ed, 1961-, multivolume set. LC: 62-53136. ISBN: 3-540-02715-7 (Vol.1, Group 1). Cost: Consult publisher for price.

A very comprehensive collection of data primarily in the field of chemistry, but also including nuclear physics, astronomy and astrophysics, solid state physics, atomic and molecular physics.

2067. Lange's Handbook of Chemistry. John A. Dean. McGraw-Hill, 12th ed., 1978, 1,600 pp. LC: 78-5335. ISBN: 0-07-016191-7. Date of First Edition: 1934. Cost: $28.50.

Lange's Handbook of Chemistry is an excellent one-volume reference book containing many and varied tabulations of data on chemistry. This handbook is noted specifically for including extensive tables on the physical constants for organic and inorganic compounds. Many topics are discussed including spectroscopy, vitamins and hormones, and separation methods. An extensive table of nuclides as well as conversion tables have been provided. An excellent subject index is included. For related works, please refer to the *CRC Handbook of Chemistry and Physics*; the *Chemical Engineers' Handbook*; and the *International Critical Tables*.

2068. Laser Experimenter's Handbook. Frank G. McAleese. TAB Books, 1979, 210 pp. LC: 79-17465. ISBN: 0-8306-9770-5. Cost: $9.95.

Includes information on electromagnetic wave theory, light wave theory, mirrors, lenses, phasors, and masers.

2069. Laser Handbook. F.T. Arecchi; E.O. Schultz-Dubois. Elsevier (North Holland), Vol.1 and 2, 1972; Vol.3, 1979, 3 Vols., Vol.1, 1,027 pp.; Vol.2, 921 pp.; Vol.3, 878 pp. LC: 73-146191. ISBN: Vols.1 and 2, 0-7204-0213-1; Vol.3, 0-444-85271-9. Cost: Vols.1 and 2, $232.50; Vol.3, $162.75.

Covers such topics as laser beam quality, wavelength regions, tunability, and increased power and efficiency. Isotope separation, pulsed holorgrpahy, picosecond spectroscopy, inertial confinement, and laser induced chemical reactions are also discussed.

2070. Laser Parameter Measurement Handbook. H.G. Heard. John Wiley and Sons, 1968, 489 pp. LC: 67-31181. ISBN: 0-471-36665-X. Cost: $21.00.

2071. Laser Safety Handbook. Alex Mallow; Leon Chabot. Van Nostrand Reinhold Company, 1978, 353 pp. LC: 78-1484. ISBN: 0-442-25092-4. Cost: $24.50.

The hazards of lasers are discussed along with protective standards and the usage of lasers in a safe and proper manner. In addition, the biological effects of laser radiation are discussed as well as laser protective eyewear and various types of laser control measures.

2072. LASL PHERMEX Data. Charles L. Mader et al. Los Alamos Scientific Laboratory, 1980, 2 Vols. LC: 79-66580. ISBN: V.1, 0-520-04009-0; V.2, 0-520-04010-4. Cost: V.1, $57.00; V.2, $47.50. Series: *Los Alamos Series on Dynamic Material Properties*.

PHERMEX stands for Pulse High Energy Radiographic Machine Emitting X-Rays. It includes a collection of high-energy electron accelerated data in the fields of shock waves, hydrodynamics, and detonations. Vol.2 in this series describes the dynamic fracture of materials and particle-velocity flow patterns of detonation products.

2073. LASL Shock Hugoniot Data. Stanley P. Marsh. Los Alamos Scientific Laboratory, 1980, 658 pp. LC: 79-65760. ISBN: 0-520-04008-2. Cost: $45.50. Series: *Los Alamos Series on Dynamic Material Properties*.

Presents information on over 5,000 unclassified experiments for determining the locus of single-shocked states at high pressures. The data is grouped by elements, alloys, minerals, compounds, rocks, mixtures of minerals, elastics, synthetics, woods, liquids, aqueous solutions, high explosives, and propellants.

2074. Lead Telluride-Tin Telluride Data Sheets. M. Neuberger. NTIS, 1970, 215 pp. Accession: N70-29473; AD-701075. Tech Report: EPIC-DS-164.

Various electronic properties of lead telluride, tin telluride, and lead-tin-tellurium systems are included in this publication.

2075. Leakage Testing Handbook. A.J. Bialous et al (General Electric Co). NTIS, 1969, 624 pp. Accession: N69-38843. Tech Report: NASA-CR-106139.

2076. Lectures on Modular Forms. Joseph Lehner (National Bureau of Standards). NTIS, 1969, 77 pp. Accession: PB-188790. Cost: $5.00. Series: NBS AMS61.

2077. Letter Symbols for Chemical Engineering Plus S.I. Conversion Tables. American Institute of Chemical Engineers. American Institute of Chemical Engineers, 1979, 12 pp. Cost: $3.00.

Includes letter symbols for chemical engineering as well as a handy guide for using the Systeme International (SI).

2078. Leybold Vacuum Handbook. R. Jaeckel; K. Diels. Pergamon Press (translated from the 2nd German edition), 1966, 360 pp. LC: 67-2345. ISBN: 0-08-010830-X. Cost: $22.50.

2079. Life History Data on Centrarchid Fishes of the United States and Canada. Kenneth Dixon Carlander. Iowa State University Press, 1976, 431 pp. LC: 77-4176. ISBN: 0-8138-0670-4. Cost: $18.00.

This is a compilation and index of literature on the centrarchid fishes in North America. Such topics as habitat, food habits, population estimates, and age-weight-length relationships are presented.

2080. Life in and Around the Salt Marshes: A Handbook of Plant and Animal Life in and Around the Temperate Atlantic Coastal Marshes. Michael J. Ustin. Crowell (also available from Apollo Eds. in paperback), 1972, 144 pp. ISBN: 0-8152-0329-2. Cost: $2.95 paper copy.

2081. Life Safety Code Handbook. John A. Sharry. National Fire Protection Association, 1978, 475 pp. Cost: $14.50.

This handbook is a guide to the 1976 Life Safety Code and addresses such topics as building design, construction, renovation, and inspection. Over 125 diagrams are used to help explain the Life Safety Code.

2082. Light Metals Handbook. George A. Pagonis. Van Nostrand Reinhold, 1954, 2-Vol. set. LC: 54-6279.

2083. Lightning and its Spectrum: An Atlas of Photographs. Leon E. Salanave. University of Arizona, 1980, 136 pp. LC: 80-18882. ISBN: 0-8165-0374-5. Cost: $25.00.

This is an atlas of photographs including over 100 photographs of the optic phenomena produced by lightning.

2084. Linear IC Applications Handbook. George Burbridge Clayton. TAB Books, 1975, 269 pp. LC: 77-71. ISBN: 0-8306-7938-3. Cost: $9.95.

2085. Linear Integrated Circuit DATA Book. Derivation and Tabulation Associates, Inc. DATA, Inc, continuation issued semiannually, pages vary per issue. Cost: $38.00 per year by subscription.

2086. Lineman's and Cableman's Handbook. Edwin B. Kurtz; Thomas M. Shoemaker. McGraw-Hill, 6th ed., 1981, 768 pp. LC: 73-19940. ISBN: 0-07-035678-5. Cost: $49.50.

2087. Liquid Chromatographic Data Compilation. American Society for Testing and Materials. American Society for Testing and Materials, 1975, 186 pp. LC: 75-18418. ISBN: 0-12-169001-6.

This publication is number AMD 41 of the ASTM *Atomic and Molecular Data Series* and consists of 4 tables. Table I is a listing of compounds by name and key separation parameters. Table II is a list of compound formulas, names and abstract numbers. Table III is a list by compound type, and Table IV is a listing of the detailed abstract with abstract numbers from Tables I,II, and III. Consequently, this compilation contains name, formula, and compound indexes, and listings of abstracts. The primary function of this compilation is to provide information regarding the separation of one compound from another.

2088. Liquid-Liquid Equilibrium and Extraction: A Literature Source Book. Elsevier North-Holland Publishing Company, 1980, 1,252 pp. ISBN: 0-444-41909-8. Cost: $183.00. Series: *Physical Sciences Data*, 7.

This is a 2-part work. Part A contains compounds from A-CM and Part B from CM100-Z. This is basically a literature reference guide for chemical equilibrium data published between 1900 and early 1980. Over 8,500 articles have been located reporting information on over 8,000 compounds.

2089. Liquid-Liquid Equilibrium Data Collection. J.M. Sorensen; W. Arlt. DECHEMA, 1979-, 3 Parts, Part 1, 622 pp.; Part 2, 625 pp.; Part 3, 605 pp. ISBN: Part 1, 3-921-567-17-3; Part 2, 3-921-567-18-1; Part 3, 3-921-567-19-x. Cost: $99.00 set. Series: *Chemistry Data Series*, Vol. 5, Parts 1,2, and 3.

The 3 parts are entitled as follows: Part 1 - *Binary Systems* 622 pp., ISBN: 3-921-567-17-3; Part 2 - *Ternary Systems* 625 pp., ISBN: 3-921-567-18-1; Part 3 - *Ternary and Quaternary Systems* 605 pp., ISBN: 3-921-567-19-X. This collection of information consists of physical and thermodynamic property data of numerous chemical compounds and mixtures primarily in the fluid state. Such properties as heat capacity, entropy data, phase equilibrium, interfacial tension data, and transport data are provided. It also includes solubility data of more than 200 binary, ternary, and quaternary mixtures of organic liquids.

2090. List of Conversion Factors for Atomic Impurities to PPM by Weight. A. Cornu; R. Massot. Heyden, 1968, 142 pp. ISBN: 0-85501-003-7. Cost: $35.00.

Eighty-three tables are provided which help to convert analytical results from spectroscopic and other techniques from a number of atoms per million to a number of parts per million. The publications is divided into Part A, which enables conversion from atomic impurities in pure elements to parts per million, and Part B, which enables conversion from impurities in inorganic compounds to parts per million by weight.

2091. Local Anesthesia in Dentistry: Dental Practitioner Handbook No. 14 G.L. Howe; F.I.H. Whitehead. Year Book Medical Publishers, 1975, 98 pp. ISBN: 0-8151-4716-3. Cost: $9.75.

2092. Local Stresses in Cylindrical Shells. B. Fred Forman. Pressure Vessel Handbook Publishing, Inc, 1975, 126 pp. LC: 75-26248.

This collection of data addresses the problem of local stresses on cylindrical shells, such as attached supports, piping, and various appurtenances. The data is presented for the most commonly used diameters and shell thicknesses.

2093. Low Temperature Mechanical Properties of Copper and Selected Copper Alloys—A Compilation from the Literature. R.P. Mikesell; R.P. Reed. U.S. Government Printing Office, 1967, 165 pp. Accession: N68-14721. Tech Report: NBS-MONOGRAPH-101. Cost: $2.75.

This compilation provides such information as creep analysis, stress analysis, temperature gradients, and data on copper in low temperature environments.

2094. Lubrication Engineers Manual. Charles A. Baily; Joseph S. Aarons. United States Steel Corp, 1971, 460 pp.

Among the topics included in this handbook are gear and bearing maintenance and the testing of lubricants and lubricant additives. The subject of hydraulics receives treatment as well as viscosity of lubricants. There is a listing of lubrication terminology and definitions as well as an index to the handbook. For related works, please refer to the *Standard Handbook of Lubrication Engineering* and the *Tribology Handbook*.

2095. Lubrication Handbook for the Space Industry. Midwest Research Institute, Kansas City, MO (available from NTIS), 1978, 497 pp. Accession: N79-19346. Tech Report: NASA-CR-161109. Cost: $3.50 microform; $33.50 hard copy.

Presents information on solid and liquid lubricants that are used in the space industry. Chemical and physical property data of more than 250 solid lubricants, bonded solid lubricants, dispersions, and composites are presented. In addition, data are presented on 250 liquid lubricants, greases, oils, fluids, and compounds.

2096. Lubrication Handbook for Use in the Space Industry. Part A: Solid Lubricants. Part B: Liquid Lubricants. Mahlon E. Campbell; Mason B. Thompson. NTIS, 1972, 412 pp. Accession: N73-32365. Tech Report: NASA-CR-130490.

A compilation of chemical and physical properties of more than 250 lubricants.

2097. Lyon's Valve Designer's Handbook. Jerry L. Lyons. Van Nostrand Reinhold, 1982, 882 pp. LC: 80-20030. ISBN: 0-442-24963-2. Cost: $62.50.

This handbook is a collection of articles by experts on valve design. A comprehensive set of references is provided along with numerous line drawings and photographs dealing with valve design.

2098. Lysosomes: A Laboratory Handbook. J.T. Dingle. Elsevier, 2d ed., 1977, 323 pp. LC: 77-3863. ISBN: 0-7204-0627-7. Cost: $73.25.

2099. M.S.I.-L.S.I. Memories D.A.T.A. Book. DATA, Inc, continuation issued semiannually, pages vary per issue. Cost: $31.50 per year by subscription.

2100. Machine Construction Hydraulics: A Reference Manual. T.M. Bashta (trans) (Wright Patterson Air Force Base). NTIS, 1973, 1,028 pp. Accession: AD-772652. Tech Report: FTD-HC-23-1343-72.

2101. Machine Shop, Machine Repairs Handbook. Superintendent of Documents. NTIS, 1966 (rep.1977), 205 pp. SuDoc: D 209.14:m 18/966/rep. S/N: 008-050-00134. Cost: $4.00.

This is a listing providing time data for making machine shop repairs.

2102. Machinery World Standards Mutual Speedy Finder, Vol.3. Media International Promotions, 1,100 pp. Cost: $95.00. Series: *World Standards Mutual Speedy Finder*, Vol. 3.

This is a listing of over 14,800 machinery standards from 5 different countries including the United States, Japan, the United Kingdom, West Germany, and France. There are 3 basic indexes including the product-parts name index, the standards number index, and the key word index. All types of machinery standards are presented including machinery standards for such areas as the aircraft industry, railway engineering, shipbuilding, welding, automotive engineering, and all types of mechanical engineering.

2103. Machinery's Handbook Guide to the Use of Tables and Formulas. John M. Amiss; Franklin D. Jones. Industrial Press, 21st ed., 1980, 224 pp. ISBN: 0-8311-1131-3. Cost: $8.00.

Formerly entitled *The Use of Handbook Tables and Formulas.* This book is designed as a companion volume to the 21st edition of *Machinery's Handbook*.

2104. Machinery's Handbook: A Reference Book for the Mechanical Engineer, Draftsman, Toolmaker and Machinist. Erik Oberg; Franklin D. Jones; Holbrook Horton; Paul B. Schubert (trans). Industrial Press, Inc, 21st ed, 1975, 2,482 pp. LC: 14-2166. ISBN: 0-8311-1129-1. Date of First Edition: 1914. Cost: $36.00.

It should be mentioned at the outset that a companion volume to *Machinery's Handbook* is available. *Machinery's Handbook Guide to the Use of Tables and Formulas*, 21st ed., 1980, 224 pp., $8.00 by Franklin D. Jones and John M. Amiss. The *Machinery's Handbook* is a classical reference book that has been through 21 edition and has sold approximately 2 million copies to date. This handbook provides information on metal working, the machining of metals, the tools that are used in working with metals, and all the technical data that are necessary to run a successful machine shop. Information is provided on gears, bearings, springs, pulleys, belts, chains, taps and dyes, fastening devices such as bolts, nuts, nails, rivets, and screws. This handbook is very well-indexed in additioned to having a thumb-index for the convienence of the user.

2105. Machinery's Mathematical Tables. Holbrook L. Horton. Industrial Press, 4th ed., 1975, 284 pp. ISBN: 0-8311-1086-4. Cost: $9.00.

This information has been selected and compiled from the tabulated data of *Machinery's Handbook*. It also includes SI Metric practice and conversion tables.

2106. Machining Data Handbook. Machinability Data Center Technical Staff. Metcut Research Associates, Inc (also NTIS), 1981, 2 Vols. Date of First Edition: 1966. Cost: $150.00.

This publication supplies a considerable amount of information on such subjects as gears, threading and tapping, grinding metals, the tools that are used in machining and how these tools can be used to their best advantage, drilling and honing, and milling of metals. It also includes information on the problems associated with machine tool vibration, as well as turning techniques during the machining process. Information on drilling is provided including cooling systems and the many different types of drills and bits that can be used. Various standards are mentioned in this handbook in connection with machining, and a materials index is provided in addition to an overall index for the entire handbook.

2107. Machinists' Ready Reference. C. Weingardner. Prakken Publications, Inc, rev. ed., 1977, spiral binding. LC: 77-85648. ISBN: 0-911168-31-1. Cost: $5.95.

2108. Magnetic Field of Cylindrical Coils and of Annular Coils. C. Snow (National Bureau of Standards). NTIS, 1953, 29 pp. Accession: PB-251900. Cost: $4.00. Series: NBS AMS38.

2109. Maintainability Engineering Design Notebook, Revision II, and Cost of Maintainability. Lyle R. Greenman (Martin Marietta Aerospace). NTIS, final report, Jan, 1975, 3 Vols.,, Vol.1, 378 pp.; Vol.2, 311 pp.; Vol.3, 463 pp. Accession: Vol.1, AD-A009043; Vol.2, AD-A009044; Vol.3, AD-A009045. Tech Report: Vols.1, 2 and 3, RADC-TR-74-308. Cost: Vol.1, $2.25 microfiche, $10.75 paper copy; $2.25 microfiche, $9.75 paper copy; Vol.3, $2.25 microfiche, $12.00 paper copy.

This handbook is primarily about Air Force ground electronic systems and equipment and their maintainability requirements and design. Mathematical models are used to show their application to maintainability tasks. The handbook contains a description of time phasing of tasks, guidelines, procedures, and methodology of each task. The principles in this 3-volume set can be applied to a much wider area of maintainability problems than just ground electronic systems since the material is presented in a general fashion. Vol.2 contains information on a revision of the Discard-at-Failure Maintenance (DAFM) mathematical model that was presented in an earlier report, AD-842-300. Demonstration plans are covered in Vol.3.

2110. Maintainability Engineering Handbook. U.S. Naval Ordinance Systems Command. U.S. Government Printing Office, 1970. LC: 72-88394.

2111. Maintenance Engineering Handbook. Lindley R. Higgins; L.C. Morrow. McGraw-Hill, 3d ed., 1977, 1,478 pp. LC: 76-46969. ISBN: 0-07-028755-4. Date of First Edition: 1957. Cost: $49.50.

1st and 2nd edition entered under L. C. Morrow. This handbook covers a wide range of subjects including sanitation; painting of buildings; maintenance of heating, ventilating, air conditioning equipment; elevators; and built-up roofs. Information is provided on the maintenance of various types of mechanical and electrical equipment as well as a presentation on different types of lubricants and ways of reducing vibration propblems. An index has been included to help the user retrieve the information on maintenance engineering. For related works, please refer to *Ground Maintenance Handbook; Plant Engineering Handbook* and *Management Handbook for Plant Engineers.*

2112. Making, Shaping and Treating of Steel. Harold E. McGannon (ed). U. S. Steel Corp, 9th ed., 1971, 1,420 pp. Cost: $25.00.

2113. Management Handbook for Plant Engineers. Bernard T. Lewis. McGraw-Hill, 1977, 736 pp. LC: 75-12936. ISBN: 0-07-037530-5. Cost: $36.50.

This handbook includes information on plant engineering ranging from safety and security to cost reduction measures and applications of data processing systems to plant management. Budgeting, planning, and personnel management have been included as well as cost reduction measures for utilities management. An index is provided. For related works, please refer to such titles as *Industrial Engineering Handbook; Plant Engineering Handbook*; and *Production Handbook.*

2114. Management Handbook: Operating Guidelines, Techniques and Practices. P. Mali. John Wiley and Sons, 1981, 1,536 pp. LC: 80-20514. ISBN: 0-471-05263-9. Cost: $37.50.

This large volume is an excellent reference tool that comprehensively covers the major topics in the practice of management.

2115. Manometer Tables. Instrument Society of America. Instrument Society of America, 1978, 31 pp. ISBN: 87664-325-X. Cost: $8.00.

Covers information on techniques used in measuring fluid pressure. Tables are provided for a large number of liquids and these tables of pressures are indicated by heights of columns at various temperatures.

2116. Manual for MOS Users. John D. Lenk. Reston Publishing Co, 1975, 340 pp. LC: 74-23929. ISBN: 0-87909-478-8. Cost: $18.50.

2117. Manual of Adverse Drug Reactions. John Parry Griffin; P.F. D'Art. Wright, 2d ed., 1979, 402 pp. Cost: £9.00.

Provides information on the various adverse interactions of different types of drugs.

2118. Manual of Hazardous Chemical Reactions Reported to be Potentially Hazardous. National Fire Protection Association. National Fire Protection Association, 4th ed., 1971. Cost: $3.25.

2119. Manual of Lower Gastrointestinal Surgery. C.E. Welch; L.W. Ottinger; J.P. Welch. Springer (also available from Otto Harrassowitz), 1979, 250 pp. ISBN: 3-540-90205-8.

This manual includes approximately 150 figures which graphically illustrate lower gastrointestinal surgery.

2120. Manual of Mathematics. G.A. Korn; T.M. Korn. McGraw-Hill, 1967, 391 pp. LC: 66-23624. ISBN: 0-07-035362-X. Cost: $9.95.

Manual of Mineralogy. *See* **Dana's Manual of Mineralogy.**

2121. Manual of New Mineral Names. P.G. Embrey; J.P. Fuller. Oxford University Press, Inc, 1980, 467 pp. LC: 80-40204. Cost: £24.00.

This is an alphabetical listing of new mineral names with references to the original papers and related literature.

Manual of Nuclear Medicine Procedures. *See* **CRC Manual of Nuclear Medicine Procedures**

2122. Manual of Radioactivity Procedures. National Bureau of Standards. NCRP Publications, Nov. 20, 1961. Cost: $3.00. Series: NBS Handbook 80.

Available from NCRP Publications, PO Box 30175, Washington, DC 20014 as NCRP Report 28.

2123. Manual of Steel Construction. American Institute of Steel Construction. American Institute of Steel Construction, 7th ed, 1970 and 1973, various paging. Date of First Edition: 1926. Cost: $18.00.

This authoritative manual on steel construction provides information on many high-strength steels as well as design and detail information for many steels. Various tables, formulas, equations, and data are presented on such topics as allowable loads and stress on different types of beams and columns, welding information, and codes and specifications related to structural steel. This handbook includes an index, conversion factors, and trigonometric information that would be useful in making structural steel calculations. A handy thumb index has been provided for the convenience of the user. For related titles, please refer to such handbooks as *Steel Designers' Manual; Structural Engineering Handbook; Welding Handbook; Handbook of Structural Stability*; and *Metals Handbook.*

2124. Manual of Symbols and Terminology for Physicochemical Quantities and Units. D.H Whiffen. Pergamon, rev. ed., 1979, 44 pp. LC: 78-41015. ISBN: 0-08-022386-9. Cost: $10.25.

Provides information such as recommended names and symbols for quantities in chemistry and physics, recommended mathematical symbols, symbols for spectroscopy, electric potential differences, electromotive forces, and electrode potentials, and definitions relating to these topics. The definitions and symbols provided have been recommended by the International Union of Pure and Applied Chemistry.

2125. Manual on Aeroelasticity: Vol. 1, Introductory Survey and Structural Aspects; Vol. 2, Aerodynamic Aspects; Vol. 3, Predictions of Aerolastic Phenomena; Vol. 4, Experimental Methods; Vol. 5, Factural Information on Flutter Characteristics; Vol. 6, Collected Tables and Graphs. R. Mazet; E.C. Pike. North Atlantic Treaty Organization: Advisory Group for Aerospace Research and Development, Vols 1-5, 1968 (Mazet); Vol. 6, 1971 (Pike), 6 Vols., various paging. LC: 62-884.

2126. Manual on Aircraft Loads J. Taylor (ed). Pergamon Press, 1965, 350 pp. LC: 64-24959.

2127. Manual on International Oceanographic Data Exchange. UNESCO. Unipub, 3d ed., 1973.

2128. Manual on Radiation Dosimetry. Niels W. Holm (ed); Roger J. Berry (ed). Dekker, 1970, 472 pp. LC: 73-84775. ISBN: 0-8247-1307-9. Cost: $38.25.

2129. Manual on Radiation Protection in Hospitals and General Practice. K. Koren; A.H. Wuehrmann. World Health Organization, Vol.1, 1974; Vol.2, 1975; Vol.3, 1975; Vol.4, 1977, 52 pp. LC: 76-355956. ISBN: Vol.1, 92-4-154038-9; Vol.2, 92-4-154039-7; Vol.3, 92-4-154040-0. Cost: Vol.1, $9.80; Vol.2, $9.60; Vol.3, $7.20.

This 4-volume set includes information on radiation, exposure, and protective safety measures. Vol.1, is entitled *Basic Protection Requirements*; Vol.2, *Unsealed Sources*; Vol.3, *X-ray Diagnosis*; Vol.4, *Radiation Protection in Dentistry*.

2130. Manual on Uniform Traffic Control Devices for Streets and Highways. U.S. Government Printing Office, 1971. S/N: 5001-0021.

This manual was prepared by the U.S. Department of Transportation, Federal Highway Administration. It provides information related to the design and use of traffic control devices in the United States.

2131. Manual on Water. ASTM, 4th ed., 1978, 472 pp. LC: 78-5159. ISBN: 0-686-52119-6. Cost: $28.50.

Provides testing information for water as recommended by the American Society for Testing and Materials.

2132. Manuel d'Horticulture Tropicale et Sub-tropicale (Handbook of Tropical and Subtropical Horticulture). Ernest Mortensen; Ervin T. Bullard. Agency for International Development, Office of Science and Technology, 1970, 283 pp. Accession: PB-286-437. Cost: $3.50 microform; $21.50 paper copy.

Text is in French. Covers such topics as pruning, fertilizing, budding, disease control, and insect control for fruit and nut trees. In addition, numerous vegetables are discussed in this manual with corresponding soil and cultivation requirements.

2133. Manufacturing Engineer's Manual: American Machinists Reference Book Sheets. Rupert LeGrand. McGraw-Hill, 1971, 317 pp. LC: 73-167557. ISBN: 0-07-037066-4. Cost: $26.50.

2134. Manufacturing Management and Engineering Handbook. Bruno A. Moski. Prentice-Hall, Inc, 1977, 385 pp. LC: 76-48082. ISBN: 0-13-555755-0. Cost: $29.95.

2135. Manufacturing Planning and Estimating Handbook. F.W. Wilson (American Society of Manufacturing Engineers). McGraw-Hill, 1963, 849 pp. LC: 62-16761. ISBN: 0-07-001536-8. Cost: $41.50.

The subtitle of this book is *A Comprehensive Work on the Techniques for Analysing the Methods of Manufacturing a Product and Estimating its Manufacturing Costs*. Handbook covers machine and tool replacement, linear programing, tool design and selection, production scheduling, and tolerance charts.

2136. Marine Electronics Handbook. Leo G. Sands. TAB Books, 1973. LC: 72-97209. ISBN: 0-8306-3638-2. Cost: $7.95.

Primarily pertaining to marine radios and marine communications.

2137. Marine Environmental Engineering Handbook. Frank L. Cross Jr. Technomic Publishing Co, Inc, 1974, 186 pp. LC: 74-80459. ISBN: 0-87762-245-4. Cost: $35.00.

This handbook describes such pollution sources as dockside facilities emissions, oil spills, shipstack emissions, dredging, dredge spoil, solid waste disposal, and tank washings. Environmental control technology is discussed in connection with vessels, shipyards, and terminal dock loading facilities.

2138. Marks' Standard Handbook for Mechanical Engineers. Theodore Baumeister; Lionel S. Marks. McGraw-Hill, 8th ed., 1978, 1,950 pp. (various paging). LC: 16-12915. ISBN: 0-07-004123-7. Date of First Edition: 1916. Cost: $49.50.

Commonly referred to as *Marks'*, after the original editor Lionel S. Marks, this handbook covers such topics as non destructive testing, power and energy materials, fuels, HVAC, pumps, compressors, and fluid mechanics. The reader is also referred to the *Mechanical Engineer's Reference Book* and to *Kent's Mechanical Engineer's Handbook*.

2139. Masonry Estimating Handbook. Michael F. Kenny. Van Nostrand Reinhold, 1973, 136 pp. LC: 72-97069. ISBN: 0-442-12162-8. Cost: $19.95.

2140. Mass and Abundance Tables for Use in Mass Spectrometry. J.H. Beynon; A.E. Williams. Elsevier-North Holland, 1963, 570 pp. ISBN: 0-444-40044-3. Date of First Edition: 1970. Cost: $92.75.

These tables will be useful in analytical chemistry for the purposes of identifying chemical structures and determining mixtures, and for use in quantitative elemental analysis.

2141. Mass Spectral Correlations. Fred W. McLafferty. American Chemical Society, 1963, 117 pp. LC: 63-17704. ISBN: 0-8412-0041-6. Cost: $11.00. Series: *Advances in Chemistry*, No. 40.

Approximately 4,000 spectra are listed by mass/charge ratios of fragment ions with their cooresponding structures.

2142. Mass Spectrometry of Pesticides and Pollutants. Stephen Safe; Otto Hutzinger. Chemical Rubber Co, 1973, 220 pp. LC: 73-83415. ISBN: 0-8493-5033-6. Cost: $52.95.

Mass spectra and fragmentation patterns are presented for such chemicals as halogenated pollutants, polycyclic aromatic hydrocarbons, chlorinated benzene isomers, chlorinated phenols, carbamates, ureas, anilides, triazines, and uracils.

2143. Master Guide to Electronic Circuits. Thomas McConnell Adams. TAB Books, 1980, 616 pp. LC: 79-25412. ISBN: 0-8306-9971-6. Cost: $19.95.

This guide covers such circuits as amplifier circuits, rectifier circuits, transistor circuits, radio circuits, and oscillator circuits.

2144. Master Handbook of 1001 More Practical Electronic Circuits. Michael L. Fair. TAB Books, 1st ed., 1979, 698 pp. LC: 78-10709. ISBN: 0-8306-8804-8. Cost: $17.95.

Numerous circuits have been diagrammed in this handbook including such circuits as battery chargers, servo motor circuits, logarithmic amplifiers, solid state switches, detectors, smoke detectors, video amplifiers, photo-activated circuits, various types of filters, chopper circuits, audio amplifiers, waveform amplifiers, oscillators, timers, multiplexers, and transmitter and receiver circuits. An index has been provided for those who have difficulty finding the circuit they want within the classified listing of circuits.

2145. Master Handbook of Acoustics. F. Alton Everest. TAB Books, 1981, 352 pp. LC: 81-9212. ISBN: 0-8306-0008-6 hardcover; 0-8306-1296-3 paper. Cost: $18.95 hardcover; $12.95 paper.

This book is written primarily for the hi-fi buff and those interested in an elementary handbook on acoustics. Approximately 222 illustrations are provided as well as numerous references throughout the text.

2146. Master Handbook of Digital Logic Applications. William L. Hunter. TAB Books, 1976, 392 pp. LC: 76-24788. ISBN: 0-8306-6874-8. Cost: $12.95.

2147. Master Handbook of Electronic Tables and Formulas. Martin Clifford. TAB Books, 3d ed., 1980, 322 pp. ISBN: 0-8306-9943-0 hardcover; 0-8306-1225-4 paper. Cost: $15.95 hardcover; $8.95 paper.

Numerous electronic tables and formulas have been compiled in this handbook for the convenience of those interested in electrical and electronics engineering and design.

2148. Master Handbook of IC Circuits. Thomas R. Powers. TAB Books, 1981, 532 pp. Cost: $22.95 hardcover; $14.95 paper.

This handbook includes over 932 circuits which utilize integrated circuits.

2149. Master Index to Materials and Properties. Y.S. Touloukian; C.Y. Ho. Plenum Data Co, 1979, 178 pp. LC: 79-11021. ISBN: 0-306-67092-5. Cost: $75.00. Series: *Thermophysical Properties of Matter: the TPRC Data Series*, Index Volume.

This volume is actually the index volume to the 13-vol set, *Thermophysical Properties of Matter*. The *Master Index* indexes over 6,362 individual materials and properties reported in the *Data Series*.

2150. Master List of Nonstellar Optical Astronomical Objects. Robert S. Dixon; George Sonneborn. Ohio State University Press, 1980, 835 pp. ISBN: 0-8142-0250-0. Cost: $30.00.

This is a compendium of approximately 185,000 nonstellar optical astronomical objects.

2151. Master Tables for Electromagnetic Depth Sounding Interpretation. Rajni K. Verma. Plenum, 1980, 473 pp. LC: 79-27044. ISBN: 0-306-65188-2. Cost: $75.00.

This compendium includes multi-layer master curves for frequency electromagnetic sounding. This collection of tables will be useful for petroleum and geothermal resource exploration.

2152. Master Transitor/IC Substitution Handbook. TAB Books, 1977, 518 pp. LC: 76-45065. ISBN: 0-8306-5970-6. Cost: $11.95 vinyl.

2153. Master Tube Substitution Handbook. TAB Books, 1977, 322 pp. LC: 76-24782. ISBN: 0-8306-6870-5. Cost: $8.95.

2154. Material Properties Handbook. Royal Aeronautical Society. NATO-Advisory Group for Aerospace Research and Development, 1959-66, 4-Vol. set.

The chief emphasis in this handbook is in ths aerospace and aeronautical sciences since it was prepared by the Royal Aeronautical Society for NATO's Advisory Group for Aerospace Research and Development (AGARD). This handbook is very well organized into a list of properties listing such basic categories as physical properties, static properties, fatigue properties, short-time properties at elevated temperatures, recovered properties after exposure to elevated temperature, and properties under load at elevated temperatures. The 4 vols. are entitled: Vol.1, *Aluminum Alloys*; Vol.2, *Steels*; Vol.3, *Magnesium, Nickel, and Titanium Alloys*; Vol.4, *Heat Resistant Alloys*. Each volume contains extensive footnotes and references.

2155. Materials and Compounding Ingredients for Rubber. Richard S. Walker. Herman Publishing Co, published annually, 500 pp. ISBN: 0-89047-021-9. Cost: $47.50.

Provides information on rubber compounding materials, latex compounding materials, and the physical and chemical properties of natural and synthetic rubbers, latexes, and reclaims.

2156. Materials Data Book for Engineers and Scientists. Earl R. Parker. McGraw-Hill, 1967, 398 pp. LC: 67-28303.

2157. Materials Data Handbook: Aluminum Alloy 2219. R.F. Muraca; J.S. Whittick. NTIS, 1972, 149 pp. Accession: N72-30461. Tech Report: NASA-CR-123777. Cost: $9.50 hard copy.

Physical and mechanical properties are provided at cryogenic, ambient, and elevated temperatures. Corrosion, environmental effects, fabrication, and joining techniques are also discussed in this handbook.

2158. Materials Data Handbook: Aluminum Alloy 5456. R.F. Muraca; J.S. Whittick. NTIS, 1972, 102 pp. Accession: N72-30457. Tech Report: NASA-CR-123827. Cost: $7.25 hard copy.

Property information is provided for aluminum alloy 5456. Information is presented on corrosion, environmental effects, fabrication, and joining techniques for aluminum alloy 5456. In addition, physical and mechanical properties data are presented at cryogenic, ambient, and elevated temperatures.

2159. Materials Data Handbook: Aluminum Alloy 6061. R.F. Muraca; J.S. Whittick. NTIS, Rev. ed., 1972, 114 pp. Accession: N72-30458. Tech Report: NASA-CR-123772. Cost: $7.75 hard copy.

Information is presented on aluminum alloy 6061 covering physical and mechanical properties of this alloy at cryogenic, ambient, and elevated temperatures. In addition, information is provided on the corrosion of this alloy as well as its environmental effects, fabrication, and joining techniques.

2160. Materials Data Handbook: Aluminum Alloy 7075. R.F. Muraca; J.S. Whittick. NTIS, rev. ed., 1972, 141 pp. Accession: N72-30459. Tech Report: NASA-CR-123773. Cost: $9.25 hard copy.

A summary of property information is provided for aluminum alloy 7075 including physical and mechanical properties of this alloy at cryogenic, ambient, and elevated temperatures. In addition, such topics as corrosion, envrionmental effects, fabrication, metallurgy, and joining techniques of this alloy are also discussed.

2161. Materials Data Handbook: Inconel Alloy 718. R.F. Muraca; J.S. Whittick. NTIS, 1972, 115 pp. Accession: N72-30460. Tech Report: NASA-CR-123774. Cost: $7.75 hard copy.

Data is provided on inconel alloy 718, including the physical and mechanical properties at various temperatures ranging from cryogenic and ambient to elevated temperatures. In addition, such topics as the metallurgy, corrosion, environmental effects, joining techniques, and fabrication of this alloy are also discussed.

2162. Materials Data Handbook: Stainless Steel Alloy A-286. R.F. Muraca; J.S. Whittick. NTIS, 1972, 109 pp. Accession: N72-30463. Tech Report: NASA-CR-123776. Cost: $7.50 hard copy.

Information is provided on stainless steel alloy A-286 including the physical and mechanical properties at cryogenic, ambient, and elevated temperatures. In addition, such topics as the metallurgy of stainless steel alloy A-286 as well as corrosion and its environmental effects, bonding, and fabrication are also presented.

2163. Materials Data Handbook: Stainless Steel Type 301. R.F. Muraca; J.S. Whittick. NTIS, 1972, 106 pp. Accession: N72-30462. Tech Report: NASA-CR-123780. Cost: $7.50 hard copy.

A summary of the materials information for stainless steel 301 is presented, including information on the physical and mechanical properties at cryogenic, ambient, and elevated temperatures. In addition, such topics as corrosion, environmental effects, fabrication, bonding, and the metallurgy of stainless steel type 301 are presented.

2164. Materials Data Handbook: Titanium 6A1-4V. R.F. Muraca; J.S. Whittick. NTIS, 1972, 124 pp. Accession: N72-30464. Tech Report: NASA-CR-123775. Cost: $8.25 hard copy.

A summary of materials properties for titanium 6A1-4V alloy is presented including such information as the physical and mechanical properties at cryogenic, ambient, and elevated temperatures. In addition, such topics as metallurgy, corrosion, environmental effects, bonding, and fabrication are also discussed for titanium 6A1-4V alloy.

2165. Materials Data Handbooks on Stainless Steels. R.F. Muraca; J.S. Whittick. NTIS, 1973. Accession: B73-10397.

Information is provided on stainless steel alloy A-286 and type 301. Both mechanical and physical property data are provided for cryogenic, ambient, and elevated temperatures.

2166. Materials Data Handbooks Prepared for Aluminum Alloys 2014, 2219, and 5456, and Stainless Steel Alloy 301. NTIS, 1967. Accession: B67-10089.

Information is provided on structural aluminum alloys 2014, 2219, and 5456, as well as structural stainless steel alloy 301. Mechanical and physical property data are presented as well as design tables.

2167. Materials Handbook: An Encyclopedia for Managers, Technical Professionals, Purchasing and Production Managers, Technicians, Supervisors and Foremen. George S. Brady; Henry R. Clauser. McGraw-Hill, 11th ed., 1977, 1,011 pp. LC: 29-1603. ISBN: 0-07-007069-5. Date of First Edition: 1929. Cost: $39.50.

As the title suggests, this book is arranged in encyclopedia format and provides brief discussions of each listed material. All types of metals, alloys, chemical compounds, and nonmetallic materials are included. For example, such topics as grease, gum, gasoline, protein, and different types of wood are presented. Tables of physical properties and a comprehensive index are included. For related works, please refer to such titles as *Material Properties Handbook, Encyclopedia/Handbook of Materials, Parts, and Finishes, Materials Handling Handbook, Handbook of Engineering Materials, Dangerous Properties of Industrial Materials, CRC Handbook of Materials Science,* and the *Handbook of Materials and Processes for Electronics.*

2168. Materials Handling Handbook. Harold A. Bolz. John Wiley and Sons, 1958, 1,740 pp (various pagings). LC: 57-11291. ISBN: 0-8260-1175-6. Cost: $53.50.

This handbook was jointly sponsored by the American Society of Mechanical Engineers and the American Materials Handling Society. It covers information on such subjects as transportation of materials, storage of materials, manufcturing of materials, packaging, warehousng, and different types of processing materials. This handbook also includes information on the different types of transporation used to haul materials. Different types of conveyors are discussed as well as the different types of transporation terminals. Containers receive treatment as does the problem of handling materials of different shapes, ranging from long stock to flat stock to bags and sacks. An impressive list of contributors have combined to provide a well-written handbook which includes definitions, terminology, and a thoroughly prepared index for easy access to the information.

2169. Materials Safety Data Sheets. General Electric Co. General Electric Co, updating service. Cost: $260.00; 1 year update service $90.00.

This loose-leaf service presents information on over 370 different materials that are commonly used in industrial substances. Such topics as ingredients, hazards, material identification, reactivity data, and disposal procedures are discussed.

2170. Materials Selector. Reinhold Publishing Co, revised annually by subscription, approx. 500 pp.

Provides information for materials evaluation and selection, including metals, nonmetallics, forms, and finishes.

2171. Mathematical Handbook for Scientists and Engineers. G.A. Korn; T.M. Korn. McGraw-Hill, 2d ed., 1968, 1,129 pp. LC: 67-16304. ISBN: 0-07-035370-0. Cost: $46.50.

2172. Mathematical Handbook of Formulas and Tables. Murray Ralph Spiegel. McGraw-Hill, 1968, 271 pp. LC: 68-55425. ISBN: 0-07-060224-7. Cost: $5.95.

2173. Mathematical Tables and Formulas. Robert D. Carmichael; Edwin R. Smith. Dover, 1931, 269 pp. ISBN: 0-486-60111-0. Cost: $3.25.

2174. Mathematical Tables Project. Work Projects Administration of the Federal Works Agency. Columbia University Press, 1939-44, 40 Vols.

2175. Mathematical Tables. Lefax Publishing Co. Lefax Publishing Co, loose-leaf binding. ISBN: 0-865-14155-1. Cost: $8.50. Series: *Lefax Technical Manuals*, No. 785.

2176. Mathematical Tables. National Physical Laboratory (Great Britain). Her Majesty's Stationery Office (London), Vol.1, 1956.

2177. Matheson Gas Data Book. William Braker; Allen L. Mossman. Matheson Gas Products, 5th ed., 1971, 574 pp.

Provides information on compressed gases and gas regulators.

2178. Matrix Representations of Groups. Morris Newman (National Bureau of Standards). U.S. Government Printing Office, 1968, 79 pp. SuDoc: C13.32:60. Cost: $1.45. Series: NBS AMS60.

2179. Maximum Permissible Body Burdens and Maximum Permissible Concentrations of Radionuclides in Air and in Water for Occupational Exposure. National Bureau of Standards. NCRP Publications, June 5, 1959. Cost: $3.00. Series: NBS Handbook 69.

Available from NCRP Publications, PO Box 30175, Washington, DC 20014 as NCRP Report 22. Supersedes NBS Handbook 52.

2180. MC6800 Microprocessor Application Manual. Motorola Semiconductor Products. Motorola, Inc, 1975, (various paging).

This applications manual provides such information as MC6800 family elements, memory allocation, hardware requirements, logic operations, program control operations, and input/output techniques. In addition, such topics as RAM memory, floppy disks, PROM codes, and terminal keyboard instructions are included.

2181. McCance and Widdowson's The Composition of Foods. A.A. Paul; D.A.T. Southgate. Elsevier/North-Holland, 4th ed., 1978, 418 pp. ISBN: 0-444-80027-1. Cost: $66.50. Series: *MCR Special Report*, No. 297.

This is a compilation of tables of the chemical composition of foods. Arrangement is by food groups such as fruits, nuts, dairy products, cereals, etc. Such information as inorganic constituents, vitamins, amino acids, fatty acid composition, energy values, cholesterol, and iodine are listed.

2182. McGraw-Hill Construction Business Handbook: A Practical Guide to Accounting, Credit, Finance, Insurance, and Law for the Construction Industry. Robert F. Cushman. McGraw-Hill, 1978, 760 pp. LC: 77-13349. ISBN: 0-07-014982-8. Cost: $32.50.

This work is primarily designed for CPA's involved in the construction industry.

2183. McGraw-Hill/CINDAS Data Series on Material Properties. U.S. Touloukian; C.Y. Ho (eds). McGraw-Hill, 1981-.

This new series represents a very ambitious joint project between McGraw-Hill and the Center for Numerical Information and Data Analysis and Synthesis (CINDAS) at Purdue University. Four volumes are currently available and it is expected that eventually this set will comprise of at least 42 volumes dealing with all different types of materials and their properties.

2184. McGraw-Hill's Compilation of Data Communications Standards. Harold C. Folts; Harry R. Karp. McGraw-Hill, 1978, 1,133 pp. LC: 78-17191. ISBN: 0-07-099782-9. Cost: $165.00.

This is a compilation of standards on data communications, computers, information processing systems. digital interfaces, and analog interfaces and peripheral equipment. Such organizations as The Consultive Committee for International Telephone and Telegraph (CCITT), International Organization for Standardization (ISO), American National Standards Institute (ANSI), the Electronic Industries Association (EIA), and the Federal Telecommunications Standards Committee (FTSC) have standards that are included in this book.

2185. McGraw-Hill's National Electrical Code Handbook. Joseplh F. McPartland. McGraw-Hill, 17th ed., 1981, 1,162 pp. ISBN: 0-07-045693-3. Date of First Edition: 1932. Cost: $24.50.

This handbook is designed to be used with the NFPA National Electrical Code. This handbook helps to explain the code and provides interpretation for rules that are frequently difficult to understand. Numerous topics are covered, such as circuit breakers, motor starters, switches, relays, contactors, and transformers used in control circuits.

2186. Measurement of Absorbed Dose of Neutrons and of Mixtures of Neutrons and Gamma Rays. National Bureau of Standards. NCRP Publications. Cost: $2.00. Series: NBS Handbook 75.

Available from NCRP Publications, PO Box 30175, Washington, DC 20014 as NCRP Report 25.

2187. Measurement of Neutron Flux and Spectra for Physical and Biological Applications. National Bureau of Standards. National Bureau of Standards, July 15, 1960. SuDoc: C13.11:72. Cost: $1.30. Series: NBS Handbook 72.

2188. Measurement Uncertainty Handbook. R.B. Abernethy et al. Instrument Society of America, rev. ed., 1980, 174 pp. ISBN: 0-87664-483-3. Cost: $12.00.

Although the first edition of this handbook was referring primarily to uncertainty in gas turbine measurements, this new title clearly has wide applications in the measurement of error in many fields. The Monte Carlo Model and Method is primarily stressed in this handbook.

2189. Meat Handbook. Albert Levie. AVI Publishing Co, 4th ed, 1979, 354 pp. ISBN: 0-87055-315-1. Cost: $19.50.

This handbook provides information on the meat processing industry including the slaughtering process, meat inspection, and grading.

2190. Mechanic's Handbook on Deep Drilling. E.A. Palashkin. NTIS, translated from the Russian (Joint Publications Research Service), 1975, 157 pp. Accession: JPRS-65115. Cost: $6.25 hard copy; $2.25 microfiche.

2191. Mechanical and Corrosion Properties. David J. Fisher. Trans Tech Publications, 1975-, multi-volume series. LC: 76-640376.

Formerly entitled *Mechanical Properties*. Has been through numerous editorial changes and is currently being published in 2 series entitled as follows: Series A — *Key Engineering Materials*, and Series B — *Single Crystal Properties*. Each series has 2 volumes per year and the subscription cost is $92.00/series. This 20-volume set provides extended abstracts in the form of graphs and tables on the mechanical and corrosion properties of materials. The set includes approximately 20,000 evaluated reviews and data compilations. Numerous materials are covered, including carbides, nitrides, borides, oxides, halides, glasses, and semiconductors. In addition, various alloys are discussed along with carbon steels, low-alloy steels, high-alloy steels, and pure iron. Such properties as elasticity, fracture, creep, fatigue, deformation, and environmental effects are discussed.

2192. Mechanical and Electrical Cost Data. Robert S. Godfrey. R.S. Means Co, 1982, 5th annual ed, 450 pp. Cost: $32.50.

Cost data is provided on labor, equipment, and materials used in heating, ventilation, electrical installations, air conditioning, plumbing, and fire extinguishing installations.

2193. Mechanical and Physical Properties of the British Standard En Steels. J. Woolman; R.A. Mottram. Macmillan (and Pergamon Press), 1964-69, 3-Vol. set, approx. 1,500 pp. ISBN: Vol.1, 0-08-010835-0; Vol.2, 0-08-011167-X; Vol.3, 0-08-012787-8. Cost: $45.00 per Vol.

This 3-volume set includes information on steels covered in British specification B.S.970. Such properties as torsion, hardness, tensile strength, yield strength, proof stress, and fatigue are covered.

2194. Mechanical Design and Systems Handbook. H.A. Rothbart. McGraw-Hill, 1965, 1,648 pp. LC: 62-21118. ISBN: 0-07-054019-5. Date of First Edition: 1965. Cost: $54.50.

Includes 1,513 illustrations covering such topics as vibrations, wear, noise, couplings, hydraulic components, pneumatic components, power transmissions, cams, seals, and joining techniques.

2195. Mechanical Engineer's Reference Book. A. Parrish. Butterworth and Co (also available from CRC Press, Inc), 11th ed, 1973, approximately 1,800 pp (various pagings). LC: 73-177366. ISBN: 0-408-00083-X. Date of First Edition: 1946. Cost: $59.95 (CRC Press, Inc).

It should be pointed out to all users of this handbook that the data are in SI units only. Consequently, this is an invaluable source of information on various mechanical standards and dimensions where the metric conversions have already been made. This handbook covers a wide range of engineering subjects including the mechanical properties of ferrous and nonferrous materials, screw threads, steam tables, lubricants, pipes, flanges, valves, and pumps. Literature references and bibliographies are provided in addition to an excellent index, which will help to retrieve this information. The reader is also referred to the *Standard Handbook for Mechanical Engineers*, and to *Kent's Mechanical Engineers' Handbook*.

2196. Mechanical Engineer's Reference Tables. Z. Elizanowski. Chemical Publishing Co, 1966.

Mechanical Engineers' Handbook. *See* **Kent's Mechanical Engineer's Handbook.**

2197. Mechanical Estimators' Handbook. Charles S. Wood. Craftsman Book Co, 1971, 288 pp. ISBN: 0-910460-64-7. Cost: $8.95.

2198. Mechanical Press Handbook. Harold R. Daniels. Herman Pub, 3d rev. and enlarged ed., 1975, 320 pp. LC: 71-79187. ISBN: 0-89047-036-7. Cost: $27.50.

A reference work for metal and plastics fabricators and die designers. Includes press types and capacities, fly wheels and counterbalances, and electrical press controls.

2199. Mechanical Properties of Industrial Nonferrous Metals at Low Temperatures. S.I. Gudkov. NTIS, 1972, 353 pp. Accession: N72-24568. Tech Report: JPRS-55861. Cost: $19.75 hard copy.

Data is provided for the thermophysical and mechanical properties of nonferrous metals and alloys in the temperature ranges between 20°C and −269°C. Such physical properties as heat expansion, thermal conductivity, and heat capacity are provided. Such mechanical properties as strength, plasticity, impact strength, and fatigue are provided.

2200. Mechanical Properties of Steel at Low Temperatures. S.I. Gudkov. NTIS, Vol.1, 1969; Vol.2, 1969, Vol.1, 376 pp; Vol.2, 158 pp. Accession: Vol.1, N70-11807 AD-693538; Vol.2, N70-11806 AD-693278. Tech Report: Vol.1, FTD-HT-23-624-68; Vol.2, FTD-HT-23-624-68.

This 2-volume set was originally published in Russian and translated by the Air Force Systems Command, at Wright Patterson Air Force Base.

2201. Medical and Biologic Effects of Environmental Pollutants Platinum-group Metals. NAS-NRC Committee on Medical and Biological Effects of Environmental Pollution (available from NAS-NRC), 1977, 138 pp. ISBN: 0-309-02640-7. Accession: N78-29659 PB-279464. Cost: $3.50 microform; $12.50 hard copy.

The environmental effects of pollution from platinum-group metals including iradium, osmium, paladium, platinum, rhodium, and ruthenium are discussed.

2202. Medical Aspects of Radiation Accidents: A Handbook for Physicians, Health Physicists, and Industrial Hygienists. E. Saenger (for the Atomic Energy Commission). U.S. Government Printing Office, 1963, 357 pp.

2203. Medical Device Industry. A Handbook. Food and Drug Administration. Food and Drug Administration, 1974, 86 pp. Accession: PB-248 273. Tech Report: ACPE-2; FDA/APE-76-5. Cost: $3.50 microform; $9.50 hard copy.

This is a statistical analysis of the medical device industry between the years 1958 and 1972. Statistical information is provided on the 12 largest medical device manufacturers. Such topics as surgical appliances, X-ray apparatus, orthopedic equipment, prosthetic devices, and dental equipment are presented.

2204. Medical Handbook to End All Medical Handbooks. L. Pheasant. British Book Centre, 1976. ISBN: 0-8277-0462-3. Cost: $10.00.

2205. Medical Microbiology: A Guide to the Laboratory Diagnosis and Control of Infection. Robert Cruickshank et al. Williams and Wilkins, 11th ed., 1965, 1,067 pp. LC: 65-6413.

NOTE: Previously entitled: *Handbook of Bacteriology.*

2206. Medical Mycology Handbook. M.C. Campbell; J.L. Stewart. John Wiley and Sons, 1980. LC: 80-11935. ISBN: 0-471-04728-7. Cost: $25.00.

Presents information on different types of fungi that have medical significance.

2207. Medical X-Ray Protection Up to Three Million Volts. National Bureau of Standards. NCRP Publications. Series: NBS Handbook 76.

Available from NCRP Publications, PO Box 30175, Washington, DC 20014 as NCRP Reports 33 and 34, respectively. $3.00 and $4.00, respectively.

2208. Merck Index: An Encyclopedia of Chemicals and Drugs. Merck and Co, Inc. Merck and Co, Inc, 9th ed., 1976, approx. 2,000 pp. Cost: $21.50.

This volume is an alphabetical listing of over 10,000 different chemicals and drugs providing such information as chemical formula, structural formula, medical uses, and toxicity. Much information has been tabulated, and access can be gained through the regular alphabetical listing or through the exhaustive cross index which provides synonyms, generic and other alternative names for the various chemicals and drugs. For related works, please refer to the *Physicians' Desk Reference to Pharmauceutical Specialities and Biologicals.*

2209. Merck Manual of Diagnosis and Therapy. Merck, Sharpe and Dohme Research Laboratories. Merck, Sharpe and Dohme Research Laboratories, 13th ed., 1977, 2,100 pp. Date of First Edition: 1899. Cost: $9.75.

NOTE: Thumb-indexed. This contains a presentation of disease entities and their corresponding diagnosis and treatment information. Pathology, etiology, and physiological background information is provided.

2210. Merck Veterinary Manual. Merck and Co, Inc. Merck and Co, Inc, 4th ed., 1973, 1,618 pp. LC: 73-174840. ISBN: 0-911910-51-4. Cost: $13.25.

2211. Metabolism; Biological Handbook. P.L. Altman; D.S. Dittmer. Federation of American Societies for Experimental Biology, 1968.

2212. Metal Cutting Tool Handbook. Metal Cutting Tool Institute, 6th ed, 1969, 933 pp.

2213. Metal Finishing: Guidebook Directory. Nathaniel Hall. Metals and Plastics Publications, Inc, 49th ed., 1981, annual 1,008 pp.

Published as part of an annual subscription to the monthly publication *Metal Finishing*. Covers mechanical surface preparation, chemical surface preparation, vapor degreasing, different solutions, and stripping metallic coatings.

2214. Metal Progress Databook. American Society for Metals, annual publication, 1976.

2215. Metallic Materials and Elements for Flight Vehicle Structures. Wright Patterson Air Force Base. U.S. Government Printing Office, revised annually, 1971, approx. 900 pp. Tech Report: MIL-HDBK-5B.

Includes information on approximately 150 different metallic materials used in military flight structures. Mechanical properties are included such as tensile yield strength, ultimate tensile strength, modulus of rigidity, yield strength in shear, and stress-rupture properties.

2216. Metallic Materials Data Handbook. Engineering Sciences Data Unit. Engineering Sciences Data Unit, 1981, currently 2 Vols. (8 Vols. are eventually planned). Cost: $210.00. Series: *U.K. Ministry of Defense, Defense Aviation Publication*, AVP-932.

This handbook includes information on strength data of metallic materials for the aerospace engineering industry. For example, information is supplied for aerospace aluminum, magnesium, titanium alloys, heat-resisting steel, and corrosion-resisting steel. Such physical properties as density, thermal conductivity, addition, such properties as tensile strength, compression, shear, torsion, bearing properties, modulus of elasticity, fracture toughness, fatigue, crack propagation, and creep properties are also included.

2217. Metallic Materials Specification Handbook. Robert B. Ross. E. and F.N. Spon Ltd, 3d ed., 1980, 790 pp. ISBN: 0-419-11360-6. Cost: £37.50.

Covers such topics as chemical composition of metals, mechanical properties, uses of metals, suppliers, machinability, corrosion protection, and other types of information on metals. Such properties as ultimate tensile strength, elongation, and numerous types of specifications are provided for different types of metals.

2218. Metallic Materials-Specification Handbook. Robert B. Ross. E. and F.N. Spon, Ltd (also Halsted Press), 3d ed., 1980, 805 pp. LC: 79-40761. Cost: $89.95.

Provides information on the composition and properties of many different metallic materials. Data are given in metric units. Information such as corrosion protection, thermal treatment, machinability, and weldability of the different metallic materials is included. Over 50,000 trade names are included. Also covers ultimate tensile strength.

2219. Metallic Shifts in NMR—A Review of the Theory and Comprehensive Critical Data Compilation of Metallic Materials. G.C. Carter. Pergamon, 4 vols. ISBN: 0-08-021143-7. Cost: $300.00. Series: *Progress in Materials Science*, Vol. 20, 1977.

2220. Metals Handbook. Taylor Lyman (Vols.1-8); Howard E. Boyer (Vols.9-10). American Society for Metals, 8th ed., Vol.1, 1961; Vol.2, 1964; Vol.3, 1967; Vol.4, 1969; Vol.5, 1970; Vol.6, 1971; Vol.7, 1972; Vol.8, 1973; Vol.9, 1974; Vol.10, 1975; Vol. 11, 1976, Vol.1, 1,300 pp.; Vol.2, 708 pp.; Vol.3, 552 pp.; Vol.4, 528 pp.; Vol.5, 472 pp.; Vol.6, 734 pp.; Vol.7, 366 pp.; Vol.8, 466 pp.; Vol.9, 499 pp.; Vol.10, 604 pp. LC: 27-12046; Vol.1, 9th ed, 78-14934. ISBN: Vol. 1 (9th ed) 0-87170-007-7. Date of First Edition: 1929. Cost: Vols.1,6,9 $47.50; Vols.2,3,4,7,8 $45.00; Vol.5, $40.00; Vol.10, $50.00; Vol.11 $49.00; 11-Vol. set $495.00.

The *Metals Handbook* is a very comprehensive collection of data on metals technology. This handbook includes properties of pure metals as well as properties of alloys. It contains information on such subjects as electroplating, heat treating of ferrous and nonferrous metals, threading and tapping, and drilling of metals. Information on binary and ternary alloys is included along with data on the crystal structure of metals and a detailed discussion of metal fractures along with 1,816 fractographs. This refernce work is considered to be a major handbook in the field of metallurgy for it provides data, tables, and formulas for almost every conceivable aspect of metal technology.

2221. Metals in Mercury—Solubilites of Solids. Z. Galus; C. Guminski. Pergamon Press, 1979, 350 pp. ISBN: 0-08-023921-8. Cost: $100.00. Series: *Solubility Data Series*, Vol. 12.

Solubility data information is provided for metals and mercury.

2222. Metals Reference Book. C.J. Smithells. Butterworths, 5th ed., 1976, 1,150 pp. LC: 67-23290 (4th ed). ISBN: 0-408-70627-9. Date of First Edition: 1949. Cost: $140.00.

The 5th ed., of the *Metals Reference Book* includes a considerable amount of information, in the form of tables and diagrams, on the subject of metallurgy and metal physics. Of particular interest to engineers and metallurgists is the information on the various properties of metals including electrical, radiating, magnetic, mechanical, elastic, thermo-electric, and general physical properties of most of the metals that are in use today. In addition this handbook supplies information on such subjects as lubricants that are used with metals, various fuels including metallurgical coke and coal, corrosion, and masers and lasers. Bibliographic references are provided to direct the reader to articles for further study, and an excellent index has been prepared to help the user retrieve data relating to metals and technology.

2223. Metals: Thermal and Mechanical Data. S. Allard. Pergamon Press, 1969, 252 pp. ISBN: 0-08-006588-0. Cost: $88.00.

2224. Metalworking Handbook: Principles and Procedures. Jeannette T. Adams. Arco Publishing Co, 1977 (copyright 1976), 480 pp. LC: 75-23577. ISBN: 0-668-03857-8. Cost: $12.95.

2225. Meteorites: Classification and Properties. J.T. Wasson. Springer-Verlag, 1974. LC: 74-4896. ISBN: 0-387-06744-2. Cost: $34.90. Series: *Minerals, Rocks and Inorganic Materials Series*, Vol. 10.

2226. Meteorological Data Catalogue. J.R. Nicholson. East-West Center Press (University of Hawaii Press), 1969, 59 pp. LC: 69-17881. ISBN: 0-8248-0082-6. Cost: $12.00.

NOTE: International Indian Ocean Expedition Meteorological Monographs: No. 3.

2227. Methods of Assessment of Absorbed Dose in Clinical Use of Radionuclides. ICRU Publications, 1979. Cost: $11.00. Series: *ICRU Report*, 32.

This publication provides physical data on radionuclides and procedures to calculate absorbed dosage of various radiopharmaceuticals.

2228. Methods of Evaluating Radiological Equipment and Materials. National Bureau of Standards. ICRU Publishers. Cost: $2.50. Series: NBS Handbook 89.

Available from ICRU Publications, PO Box 30165, Washington, DC 20014 as ICRU Report 10f.

2229. Metric and Other Conversion Tables. G.E.D. Lewis. Construction Press, Ltd, 1973, 90 pp. ISBN: 0-582-42940-4. Cost: £2.35 paper copy.

2230. Metric Conversion Handbook. M.H. Green. Chemical Publishing Co, 1978. ISBN: 0-8206-0229-9. Cost: $18.50.

An explanation of the Systeme International is provided along with conversion tables. Units of measurement are provided for different categories such as electricity and magnetism, energy, force, length, mass, pressure, temperature, time, and velocity.

2231. Metric Conversion Tables and Factors. Machinery's Handbook, eds. Industrial Press, 1973, 35 pp. ISBN: 0-8311-1103-8. Cost: $4.50.

Contains over 180 conversion factors.

2232. Metric Handbook for Reinforced Concrete. C.H. Garthwaite. Butterworths, 1969, 64 pp. ISBN: 0-408-06800-0. Cost: $3.95.

Metric dimensioning and conversion factors are provided on the topic of reinforced concrete.

2233. Metric Manual. Lawrence D. Pedde et al. U.S. Department of the Interior, 1978, 278 pp.

Metric conversion information is provided for numerous applications, such as pipe dimensions, metric screw threads, electrical conductors, soil classification, coarse aggregates, and other areas of contract specification. Examples of engineering problems are presented for such areas as vibration, mass, pump power, boom deflection, fluid dynamics, water utilization, and irrigation applications. Numerous figures and tables are provided for the convenience of the user.

2234. Metric Manual: Development, Regulations, Tables, Comparisons, Definitions for Metrication in U.S.A. Patricia Laux. Keller, 1974, 362 pp., loose-leaf service. LC: 74-31864. ISBN: 0-934674-29-9. Cost: $35.00.

2235. Metric System Guide. John J. Keller. Keller, 1974-, 5-Vol. set, loose-leaf service. LC: 75-25158. Cost: $59.00 per Vol.

This guide has 5 volumes entitled as follows: Vol.1, *Metrication in the United States*; Vol.2, *Legislation and Regulatory Activities*; Vol.3, *Metric Units Edition*; Vol.4, *References Sources*; and Vol.5, *Definitions and Terminology*.

2236. Metrication for Engineers. Ernst Wolff. Society for Manufacturing Engineers, 1974, 110 pp.

The International System of Units is presented and various types of measurements are discussed, such as those of area, volume, liquid mass, angles, and temperature. In addition, information on metric drawings is presented as well as metric dimensioning of threads and gears.

2237. Metrication Handbook. J.J. Keller. J.J. Keller and Associates, Inc, 1974, 142 pp. ISBN: 0-934674-32-9. Cost: $15.00.

This is a general overall view of metrication in the United States with a listing of specialized conversion factors, such as units of length, volume, mass, and area. A listing of reference material is provided as well as sources of additional information and organizations specializing in the metric system.

2238. Metrics for Mechanics and Other Practical People. Douglas F.C. Richard. Dos Reals Publishing, 2d ed., 1975, 171 pp. LC: 75-11122. ISBN: 0-915004-02-X. Cost: $6.00.

This handy little volume provides a useful source for metric information on different types and units of measurement. Such measurements as pressure, temperature, and threaded components are included. This tabulation of metric information provides the kind of metric units that are needed in practical situations, and the emphasis is on quick look-up data.

2239. Meyler's Side Effects of Drugs. M.N.G. Dukes. Elsevier, 9th ed., 1980, 860 pp. ISBN: 0-444-90102-7. Cost: $122.00.

Presents information on the side effects of drugs and drug interactions.

2240. Microbiological Transformation of Steroids: A Handbook. William Charney; Herschel Herzog. Academic Press, 1968. ISBN: 0-12-169950-1. Cost: $49.50.

2241. Microcircuit Device Reliability—Linear/Interface Data. Thomas E. Paquette (Reliability Analysis Center). NTIS, 1975, 304 pp. Accession: AD-A019 433. Tech Report: RAC-MDR-2. Cost: $50.00.

See also Accession No.: AD-A014 537

2242. Microcircuit Device Reliability—Memory/L.S.I. Data. Henry C. Rickers (Reliability Analysis Center). NTIS, 1976, 215 pp. Accession: AD-A023-277. Tech Report: RAC-MDR-3. Cost: $50.00.

2243. Microcircuit Device Reliability-Digital Generic Data. Roy C. Walker (Reliability Analysis Center). NTIS, 1975, 260 pp. Accession: AD-A014537. Tech Report: RAC-MDR-1. Cost: $50.00.

2244. Microcomputer Board Data Manual. Dave Bursky. Hayden, 1978, 128 pp. ISBN: 0-8104-0898-8. Cost: $7.95.

Tables are provided which summarize the microcomputer board specifications of over 30 different manufacturers. Peripheral equipment, support products, and software are also summarized in the second section of this manual.

2245. Microcomputer D.A.T.A. Book. Derivation and Tabulation Associates, Inc. Derivation and Tabulation Associates, Inc, semiannual continuation, pages vary per issue. Cost: $54.50 per year.

2246. Microcomputer Handbook. Charles J. Sippl. Van Nostrand Reinhold, 1977, 454 pp. LC: 76-53791. ISBN: 0-442-80324-9. Cost: $19.95.

2247. Microelectronic Device Data Handbook. ARINC Research Corp. NTIS, 1966, 327 pp. Accession: N72-73275. Tech Report: NASA-NPC-275-1.

Vol.2 is entitled *Manufacturer and Specific Device Information,* 1968, 163 pp., NASA-CR-1111, Accession No., N68-29526.

2248. Micrographics Handbook. Charles Smith. Artech House, Inc, 1978, 300 pp. LC: 78-2561. ISBN: 0-89006-061-4. Cost: $29.00.

Primarily devoted to the study of microfilm and its related formats and indexing systems.

2249. Microprocessor Applications Handbook. David F. Stout. McGraw-Hill, 1982, 448 pp. LC: 81-11787. ISBN: 0-07-061798-8. Cost: $35.00.

Microprocessors are discussed in connection with digital filters, slow-scan television systems, programmable video games, controlled color t.v. receivers, and other applications of microprocessors.

2250. Microprocessor Data Manual. Dave Bursky. Hayden, 1978, 128 pp. ISBN: 0-8104-5114-X. Cost: $7.95.

Provides specifications for microprocessors, microcomputers, and bit slices available from over 30 manufacturers. In addition, support circuits, computer architecture, and software information is also provided.

2251. Microscope Technique: A Comprehensive Handbook for General and Applied Microscopy. W. Burrells. Halsted, 1977, 574 pp. LC: 77-26687. ISBN: 0-470-99376-6. Cost: $27.50.

Various laboratory methods are presented, including staining, section cutting, counting particles, and microprojection. In addition, such topics as electron microscopy, photomicrography, microscope eyepieces, microscope lamps, and the proton microscope are presented.

2252. Microwave Engineer's Handbook. Theodore S. Saad. Artech House, Inc, 1971, Vol.1, 192 pp.; Vol.2, 209 pp. LC: 76-168891. ISBN: 0-89006-004-5. Date of First Edition: 1971. Cost: $43.00 per set.

This 2-volume set contains an excellent collection of tables on wave guides in addition to many other types of microwave information presented in the format of tables, graphs, charts, equations, and formulas. Among the topics discussed in this handbook are such subjects as microstrips, filters, coaxial cables, and transmission lines. Antenna information is also presented along with microwave tubes and various state of the art reports. For related works, please refer to such titles as *The Handbook of Microwave Measurements* and *Microwave Tube D.A.T.A. Book.*

2253. Microwave Spectral Tables Polyatomic Molecules without Internal Rotation. M.S. Cord; J.D. Petersen; M.S. Lojko; R.H. Haas (National Bureau of Standards). NTIS, 1968. Accession: COM 74-10795. Cost: $11.75. Series: NBS Monograph 70, Vol. IV.

2254. Microwave Spectral Tables, Diatomic Molecules. P.F. Wacker; M. Mizushima; J.D. Peterson; J.R. Ballard (National Bureau of Standards). NTIS, 1964. Accession: PB-168072. Cost: $6.75. Series: NBS Monograph 70, Vol. 1.

2255. Microwave Spectral Tables, Line Strengths of Asymmetric Rotors. P.F. Wacker; M.R. Pratto (National Bureau of Standards). NTIS, 1964. Accession: PB-180714. Cost: $10.50. Series: NBS Monograph 70, Vol. II.

2256. Microwave Spectral Tables, Polyatomic Molecules with Internal Rotation. P.L. Wacker; M.S. Cord; D.G. Burkhard; J.D. Petersen; R.F. Kukol (National Bureau of

Standards). NTIS, 1969. Accession: COM 75-10794. Cost: $9.25. Series: NBS Monograph 70, Vol. III.

2257. Microwave Spectral Tables, Spectral Line Listing. M.S. Cord; M.S. Lojko; J.D. Petersen (National Bureau of Standards). NTIS, 1968. Accession: COM 74-10796. Cost: $13.00. Series: NBS Monograph 70, Vol. V.

2258. Microwave Tube D.A.T.A. Book. DATA, Inc, semiannual continuation, pages vary per issue. Cost: $33.50 per year by subscription.

2259. Midwest Farm Planning Handbook. Sidney C. James. Iowa State University Press, 3d ed., 1975. ISBN: 0-8138-1101-5. Date of First Edition: 1949. Cost: $9.50 paper.

2260. Mineral Facts and Problems, 1975 Edition. U.S. Bureau of Mines. Superintendent of Documents, U.S. Government Printing Office, 1975 ed., 1976, 1,266 pp. ISBN: 0082-9129. Cost: $17.00. Series: *U.S. Bureau of Mines Bulletin,* No. 667.

Over 80 articles on different minerals and elements. This guide covers current developments for mineral commodities and forecasts and projections to 1985 and 2000.

2261. Mineral Names: What Do They Mean? Richard Mitchell. Van Nostrand, 1979, 256 pp. ISBN: 0-442-24593-9. Cost: $13.95.

Over 2,600 different mineral names are listed with their corresponding derivation and historical information. A bibliography is included for further information on minerals.

2262. Mineral Tables: Hand-Specimen Properties of Fifteen Hundred Minerals. Richard V. Dietrich. McGraw-Hill, 1969. ISBN: 0-07-016895-4. Cost: $8.95.

Provides such information as chemical composition, color and hardness data.

2263. Minerals Yearbook 1980. U.S. Bureau of Mines. U.S. Government Printing Office, Annual, 1973; Vol.1, 1975; Vols. 2 & 3, 1976. LC: 33-26551. SuDoc: I28.37. Date of First Edition: 1933.

This 3-volume set is entitled as follows: Vol.1, *Metals and Minerals,* 950 pp.; Vol.2, *Area Reports: Domestic,* 603 pp.; Vol.3, *Area Reports: International,* 1,367 pp. This 3-volume set is an annual publication of the U.S. Bureau of Mines, and it contains information on the mineral industry for the United States and for over 130 other countries. Statistical data are provided on production of various metals, minerals, and fuels, such as a state-by-state summary of the annual production and cosumption of coal. For related works, please refer to such publications as the *Keystone Coal Industry Manual, Mining Engineers' Handbook,* and the *Handbook of Mineral Dressing.*

2264. Mining Cable Engineering Handbook. Anaconda Company, Wire and Cable Division. Anaconda Company, 1977, 168 pp.

Provides detailed specifications and data on different wire and cable used in mining operations.

2265. Mining Engineers' Handbook. Robert Peele. John Wiley and Sons, 3d ed., 1941, 2 Vols., 2,442, pp. LC: 41-11922. ISBN: 0-471-67716-7. Date of First Edition: 1918. Cost: $62.50 per set.

Although this handbook is somewhat dated, it still represents a valuable source of information on mining technology. Information is provided on such subjects as the usage of explosives, equipment used in drilling and boring, drilling of tunnels, ventilation of mines including air distribution and fans, and the transport of ores and personnel in mines. In addition, information is provided on the surveying of mines and the area around mines, ore sampling, the preparation of anthracite and bituminous coal, and compressed air. Some basic engineering information is presented including background information on mechanics and thermodynamics, structural design, and electrical engineering in relation to mining. Various engineering tables are presented including conversion factors. An index has been prepared. For related works, please refer to such titles as the *Minerals Yearbook, Keystone Coal Industry Manual, Petroleum Production Handbook,* and the *Handbook of Mineral Dressing.*

2266. Minolta Systems Handbook. Joseph Cooper. Prentice-Hall, 2d. ed, 1979, 637 pp. ISBN: 0-13-584581-5. Cost: $34.95.

This manual provides the latest information on Minolta cameras and their accessories. Numerous photographs, diagrams, and tables help to provide information on basic photographic principles and techniques. In addition, there is an extensive directory of Minolta lenses included.

2267. Miscellaneous Physical Tables: Planck's Radiation Functions, and Electronic Functions. National Bureau of Standards. National Bureau of Standards, 1941, 58 pp. (out of print). Series: NBS MT17.

NOTE: Part 1—tables of Planck's Radiation Functions, reprinted from the *Journal of the Optical Society of America,* February, 1940. Part 2—"Electronic Functions" superseded by NBS Circular 571, *Electronic Physics Tables,* 1956, 83 pp.

2268. Missile Engineering Handbook. Carl William Besserer. Van Nostrand Reinhold (available from Kreiger), 1958, 600 pp. LC: 58-9433. ISBN: 0-442-00720-5. Cost: $24.50.

2269. Mixing and Excess Thermodynamic Properties: A Literature Source Book. J. Wisniak; A. Tamir. Elsevier, 1978, 936 pp. ISBN: 0-444-41687-0. Cost: $134.25. Series: *Physical Sciences Data,* 1.

This is a systematic compilation of articles published between 1900 and early 1977 on mixing and excess properties. Approximately 6,000 articles are listed on such topics as metallurgical systems, nonelectrolyte solutions, and electrolyte solutions. Descriptions are provided on principle theories and equations for correlating the data.

2270. Modern Drafting Practices and Standards Manual. General Electric Co, 1976 and updates, various paging. ISBN: 0-931690-01-3. Cost: $95.00 manual; $40.00 year updating service.

This is a comprehensive treatment on modern drafting practices. Numerous illustrations are provided that describe proper drafting procedures. Such topics as dimensions and tolerances, electrical drafting principles, and indications of surface textures are discussed. Over 6,000 abbreviations are included in this drafting manual, as well as approximately 150 pages of graphic symbols.

2271. Modern Drug Encyclopedia and Therapeutic Index. Arthur J. Lewis. Yorke Medical Books, 14th ed., 1977, 1,000 pp. ISBN: 0-914316-07-9. Cost: $37.00.

2272. Modern Electronic Circuits Reference Manual. John Markus. McGraw-Hill, 1980, 1,238 pp. LC: 79-22096. ISBN: 0-07-040446-1. Cost: $44.50.

Includes over 3,600 diagrams of electronic circuits on a multitude of topics. Some examples of the cirucits included are as follows: intercom circuits, keyboard circuits, lamp control circuits, memory circuits, music circuits, noise circuits, oscillator circuits, photography circuits, amplifier circuits, automotive circuits, antenna circuits, cathode-ray circuits, digital clock circuits, filter circuits, fire alarm circuits, servo circuits, siren circuits, switching circuits, tape recorder circuits, telephone circuits, television circuits, temperature control circuits, and timer circuits. Approximately half of the circuits use integrated circuits.

2273. Modern Electroplating. V.I. Lainer Israel Program for Scientific Translations (trans). NTIS, 1970, 312 pp. Accession: N70-41744. Tech Report: TT-69-55082.

2274. Modern Food Analysis. Frank Leslie Hart; Harry J. Fisher. Springer-Verlag, 1971. LC: 72-83661. ISBN: 0-387-05126-0. Cost: $67.20.

This handbook is similar to the *Laboratory Handbook of Methods of Food Analysis* in that it provides information on sampling methods, standard test methods and analysis of the composition of food.

2275. Modern Formulas for Statics and Dynamics: A Stress and Strain Approach. Walter D. Pilkey; Pin Yu Chang. McGraw-Hill, 1978, 418 pp. LC: 77-15093. ISBN: 0-07-049998-5. Cost: $21.00.

This is a compilation of formulas that are used in the analysis of beams, shafts, plates, and shells. Tables are provided for finding stresses and deformations in structural members.

2276. Modern Oscilloscope Handbook. Douglas Bapton. Prentice-Hall, 1979, 214 pp. LC: 78-14455. ISBN: 0-8359-4582-0. Cost: $18.95.

This handbook covers oscilloscope operation and applications. Such topics as oscilloscope types, waveform analysis, data domain analysis, wave shaping, and ignition waveform analysis are discussed.

2277. Mollier Enthalpy-Entropy Diagram for Water, Steam and Ice in SI Units From 0.000 01 to 30,000 bar and from -60 and 1400 C. F. Bosnjakovic; V. Renz; P. Burow. Hemisphere Publishing, 1976, chart (40 x 29 in). ISBN: 0-89116-025-6. Cost: $4.95.

This is actually a chart rather than a handbook. It is a 2-color chart measuring 40 x 29 inches. Includes instructions for use and nomenclature.

2278. Molten Salts Handbook. George J. Janz. Academic Press, 1967, 588 pp. LC: 66-30087. ISBN: 0-12-380445-0. Cost: $68.00.

2279. Molten Salts: Vol. 1. Electrical Conductance, Density, and Viscosity Data. G.J. Janz; F.W. Dampier; G.R. Lakshminarayanan; P.K. Lorenz; R.P.T. Tomkins (National Bureau of Standards). NTIS, 1968. Cost: $6.00. Series: NSRDS-NBS-15.

2280. Molten Salts: Vol. 2, Section 1. Electrochemistry of Molten Salts: Gibbs Free Energies and Excess Free Energies from Equilibrium-Type Cells. Section 2, Surface Tension Data. G.J. Janz; C.G.M. Dijkhuis (Section 1); G.J. Janz; G.R. Lakshminarayanan; R.D.T. Tomkins; J. Wong (Section 2). NTIS, 1969. Cost: $5.50. Series: NSRDS-NBS-28.

2281. Molybdenum: Physico-chemico Properties of Its Compounds and Alloys. L. Brewer et al. International Atomic Energy Agency, 1980. Cost: DM 143. Series: *Atomic Energy Review*: Special Issue, No. 7, 1980.

Includes such information as the crystal structure and density data of molybdenum, molybdenum halides, and chalcogenides. In addition, diffusion information is provided as well as phase diagrams of molybdenum and its compounds and alloys.

2282. Moments, Shears, and Reactions for Continuous Highway Bridges. American Institute of Steel Construction. American Institute of Steel Construction, rev., 1966, various paging. Date of First Edition: 1959.

Structural information is provided for highway bridges.

Mono- and Disaccharides in Water—Solubilities of Solids. *See* **Solubility Data Series, Volume 8.**

2283. Monte Carlo Method. A.S. Householder (ed) (National Bureau of Standards). NTIS, 1951, 42 pp. Accession: PB-184887. Cost: $4.00. Series: NBS AMS12.

2284. More Than Dispensing: A Handbook on Providing Pharmaceutical Services to Long Term Care Facilities. American Pharmaceutical Association. American Pharmaceutical Association, 1980, 119 pp. LC: 80-65958. Cost: $20.00.

Such topics as drug interactions, the use of nonprescription drugs, therapeutic drugs, and pharmaceutical service are presented.

2285. Motor Application and Maintenance Handbook. R. Smeaton. McGraw-Hill, 1969, 656 pp. LC: 68-26316. ISBN: 0-07-058438-9. Cost: $37.50.

Covers motor load data and characteristics of electric motors.

2286. Moving, Rigging Handbook. Superintendent of Documents, 1962, (rep. 1977), 52 pp. SuDoc: D209.14:M 86. Cost: $2.00.

Time Data are provided for moving and rigging.

2287. Multilingual Compendia of Plant Diseases. Paul R. Miller; Hazel L. Pollard. American Phytopathological Society, 1977, Vol.1, 519 pp.; Vol.2, 446 pp. LC: 75-46932. ISBN: Vol.1, 0-89054-018-7; Vol.2, 0-89054-020-9. Cost: $35.00 per Vol.; $65.00 per set.

This 2-volume set is entitled as follows: Vol. 1, *Fungi and Bacteria*, and Vol. 2, *Viruses and Nematodes*. Each Plant disease is listed along with the Latin name of the host, the Latin name of the pathogen, and synonyms for the pathogen. In addition, disease descriptions are provided in 4 languages, English, Interlingua, French, and Spanish.

2288. Multiple Light Scattering: Tables, Formulas and Applications. H.C. van de Hulst. Academic Press, 1980, various paging. LC: 79-51687. ISBN: Vol.1, 0-12-710701-0; Vol.2, 0-12-710702-9. Cost: Vol.1, $33.50; Vol.2, $46.50; $70.00 set.

This 2-volume set provides information on optical formulas and tables with regard to multiple light scattering.

2289. Multiplet Table of Astrophysical Interest Revised Edition, Part I. Table of Multiplets; Part II. Finding List of all Lines in the Table of Multiplets. C.E. Moore (National Bureau of Standards). NTIS, rev. ed., 1972. Accession: COM 72-50439. Cost: $9.00. Series: NSRDS-NBS-40.

2290. MUMPS Language Standard. National Bureau of Standards. U.S. Government Printing Office. SuDoc: C13.11:118. Cost: $2.70. Series: NBS Handbook 118.

2291. Municipal Refuse Disposal. Public Administration Service. American Public Works Association, 3d ed., 1970, 538 pp. LC: 76-126639. Date of First Edition: 1966. Cost: $15.00.

2292. Muskeg Engineering Handbook. Ivan C. MacFarlane. University of Toronto Press, 1969, 297 ppp. LC: 78-447167. ISBN: 0-8020-1595-6. Cost: $20.00.

This handbook includes information on the various problems that are associated with peat bogs, such as road construction, drainage, and trafficability.

2293. Mycotoxic Fungi, Mycotoxins, Mycotoxicoses: An Encyclopedic Handbook. T.D. Wyllie; L.G. Morehouse. Marcel Dekker, 1977-78, 3 Vols. ISBN: Vol.1, 0-8247-6550-8; Vol.2, 0-8247-6551-3; Vol.3, 0-8247-6552-4. Cost: Vol.1, $75.00; Vol.2, $75.00; Vol.3, $45.00; $171.50 set.

Vol.1, *Mycotoxic Fungi and Chemistry of Mycotoxins*; Vol.2, *Mycotoxicoses of Domestic and Laboratory Animals, Poultry, and Aquatic Invertebrates and Vertebrates*; Vol.3, *Mycotoxicosis of Man and Plants: Mycotoxin Control and Regulatory Practices*. This 3-volume handbook deals with the topic of toxic fungi as it relates to animals, humans, and plants.

2294. Mycotoxins Mass Spectral Data Bank. Association of Official Analytical Chemists. Association of Official Analytical Chemists, 1978, 60 pp. Cost: $14.00.

This compilation includes 104 mass spectra of mycotoxins compiled by the U.S. Food and Drug Administration.

2295. Mysterious Universe: A Handbook of Astronomical Anomalies. William R. Corliss. Sourcebook Project, 1979, 710 pp. LC: 78-65616. ISBN: 0-915554-05-4. Cost: $15.95.

Provides information on astronomical curiosities and mysteries dealing with the universe.

2296. Mössbauer Effect Data Index. John G. Stevens; Virginia E. Stevens. Plenum Publishing Co, Vol.1, 1975; Vol.2, 1971; Vol.3, 1972; Vol.4, 1972; Vol.5, 1973; Vol.6, 1975; Vol.7, 1975; Vol.8, 1976; Vol.9, 1978, Vol.1, 522 pp; Vol.2, 292 pp; Vol.3, 382 pp; Vol.4, 430 pp; Vol.5, 489 pp; Vol.6, 496 pp; Vol.7, 398 pp; Vol.8, 445 pp; Vol.9, 367 pp. LC: Vol.2, 76-146429; Vol.3, 76-146429; Vol.4, 76-146429; Vol.5, 76-146429; Vol.7, 76-146429. ISBN: Vol.1, 0-306-65162-9; Vol.2, 0-306-65140-8; Vol.3, 0-306-65141-6; Vol.4, 0-306-65142-4; Vol.5, 0-306-65145-9; Vol.6, 0-306-65144-0; Vol.70-306-65145-9; Vol.8, 0-306-65146-7; Vol.9, 0-306-65149-1. Cost: $95.00 each volume.

This is an annual index to the Mössbauer effect literature. It includes information on equipment and supplies for Mössbauer spectroscopy, tables for Mössbauer spectroscopy, and information on Mössbauer isotopes.

2297. NACE Coatings and Linings Handbook. National Association of Corrosion Engineers. National Association of Corrosion Engineers, 1980. Cost: $62.50; $50.00 members.

This handbook deals with the topic of corrosion protection of such facilities as tanks, concrete surfaces, petroleum producing equipment, industrial equipment, and marine and offshore facilities. The newest types of coatings are covered in this handbook, such as coal tar, chlorinated rubber, epoxy, urethane, vinyl, alkyd, phenolic resins, and zinc coatings. Such topics as acid-proof vessel construction, thin-film baked coatings, abrasion-resistant testing, and application of cathodic protection are discussed.

2298. NACE Corrosion Engineers' Reference Book. R.S. Treseder. National Association of Corrosion Engineers, 1980, 233 pp. LC: 79-67175. Cost: $30.00.

This compilation is a comprehensive collection of information on the topic of corrosion and protection. Tables are included which list the physical properties of elements, gases, and liquids. Includes such information as polarization curves, galvanic series, standard electrodes, critical velocities in seawater, stress corrosion cracking, caustic soda service chart, and maximum temperature for scaling. Composition and mechanical properties of carbon steels, stainless steels, magnesium alloys, titanium alloys, cobalt alloys, and aluminum alloys are presented. Properties of nonmetallic materials and coatings, including graphite, silicon, carbide, glass, silica and hydraulic cements are presented.

2299. NACE Surface Preparation Handbook. National Association of Corrosion Engineers. National Association of Corrosion Engineers, 1980. Cost: $17.00; $15.00 members.

Surface preparation is discussed in terms of such structures as concrete surfaces, tanks, petroleum equipment, offshore facilities, marine facilities, and various types of industrial equipment. Sandblasting, water blasting, and visual standards of surface preparation are discussed.

2300. NALCO Water Handbook. Frank N. Kemmer (ed). McGraw-Hill, 1979, 799 pp. LC: 78-25825. Cost: $28.50.

Emphasizes water treatment methods, municipal uses of water, and industrial uses of water. Such types of industrial uses of water as the aluminum industry, automotive industry, food processing industry, textile industry, petroleum industry, and the chemical industry are discussed. Such topics as water gauging, sampling methods, ion exchange, degasification, corrosion control, coagulation, and flocculation and evaporation are presented.

2301. National Construction Estimator. Craftsman Book Co, 1981, 288 pp. ISBN: 0-910460-29-9. Cost: $10.00.

Information on material costs, labor costs, and methods of reducing equipment expenditures is presented.

2302. National Electric Code Reference Book. J.D. Garland. Prentice-Hall, Inc, 2d ed., 1979, 612 pp. LC: 78-12447. ISBN: 0-13-609313-2. Cost: $18.95.

2303. National Electrical Code Handbook. Joseph Ross. National Fire Protection Assocation, 2d ed., 1981, 1,000 pp. Cost: $19.50.

Includes a complete text of the 1981 National Electrical Code, as well as extensive commentary, diagrams, and photographs which help to interpret and explain the code. Recommended practices and standards are provided on such topics as wiring, lightning arrestors, transformers, generators, storage batteries, X-ray equipment, fire protection, elevators, and data processing systems.

2304. National Electrical Safety Code. American National Standards Institute. American National Standards Institute. Cost: $9.75. Series: ANSI-C2-1981.

2305. National Emissions Data System (NEDS) Control Device Workbook. William M. Vatavuk. NTIS, 1973, 222 pp. Accession: PB-258 529. Tech Report: APTD-1570. Cost: $3.50 microform; $17.00 hard copy.

This publication is a supplement to *Guide for Compiling a Comprehensive Emission Inventory*, Report No. APTD1135. Information is provided on different types of pollutant control devices and methods that are most commonly used by industries collecting air pollution data.

2306. National Fire Codes; A Compilation of NFPA Codes, Standards, Recommended Practices, and Manuals. National Fire Protection Association, 1981, 16 Vols. LC: 38-27236. ISBN: 0-87765-140-X.

As the subtitle states, this 16-volume set is a comprehensive collection of recommended practices and standards with regard to fire protection and safety. Information is supplied on sprinkling systems, fire extinguishers, handling and shipping explosive materials and chemicals, and the proper installation of electrical wiring. In addition, standards and recommended practices are provided for such structures as airports, mobile homes, boats, aircraft, etc. Information is also provided on such subjects as fire retardant materials, fires related to nuclear reactors, dangerous chemical reactions, and fire safety with regard to doors, windows, and fireplaces. For related works, please refer to such titles as the *Handbook of Flammability Regulations*; *Fire Protection Handbook*; the *Flammability Handbook for Plastics*; and *Flammability Test Methods Handbook*.

2307. National Formulary. American Pharmaceutical Association. Mack Publications, 14th ed., 1975. LC: 55-4116. Cost: $24.00.

2308. National Handbook of Conservation Practices. Superintendent of Documents. Superintendent of Documents, 1978, loose-leaf; 278 pp. SuDoc: A57.6/2:C76/2/978. Cost: $6.00.

This service provides information on national standards and specifications regarding soil and water conservation programs.

2309. National Handbook of Recommended Methods for Water-Data Acquisition. Office of Water Data Coordination. Office of Water Data Coordination, Geological Survey, U.S. Department of the Interior, 1977, various paging, loose-leaf.

Helps to document methodologies that are used by collectors of hydrologic data. Numerous collection methods are discussed in connection with the following topics: sediment, groundwater, surface water, microbiological organisms, radioactive constituents in water, inorganic chemicals in water, organic chemicals in water, snow and ice. In addition, such topics as evaporation, transpiration, hydrometeorological observations, and drainage basin characteristics are also discussed.

2310. National Plumbing Code Handbook. V. Manas. McGraw-Hill, 1957, 503 pp. LC: 56-6965. ISBN: 0-07-039850-X. Cost: $24.30.

2311. National Repair and Remodeling Estimator. Craftsman Book Company, 1980, 200 pp. ISBN: 0-910460-78-7. Cost: $10.25.

Material costs and labor figures are provided for different types of repair and remodeling estimating. Such topics as concrete work, roofing, and various types of installations are discussed.

2312. National Soils Handbook: Soil Survey, Cartography, and Land Inventory and Monitoring. Kenneth E. Grant. Soil Conservation Service, United States Department of Agriculture, 1974, various paging; loose-leaf service.

This handbook provides policy information and procedure information for carrying out the soil survey program of the Soils Conservation Service. Information is provided on soil mapping and soil classification.

2313. National Space Science Data Handbook of Correlative Data. J.H. King. NTIS, 1971, 210 pp. Accession: N71-35168. Tech Report: NASA-TM-X-67294.

2314. National Standard Petroleum Oil Tables, Circular 410. National Bureau of Standards. American Society for Testing and Materials. Cost: $12.75. Series: NBS Circular 410.

NOTE: Information in this circular has been incorporated in the ASTM-IP *Petroleum Measurement Tables* issued by ASTM.

2315. National Standard Reference Data System Plan of Operation. E.L. Brady; M.B. Wallenstein (National Bureau of Standards). U.S. Government Printing Office, 1964. SuDoc: C13.48:1. Cost: $0.55. Series: NSRDS-NBS-1.

2316. Natural Gas Users' Handbook. Stephen A. Herman et al. Bureau of National Affairs, Inc. Cost: $12.50.

This handbook provides resource information on natural gas.

2317. Natural Trigonometric Functions to Seven Decimal Places for Every Ten Seconds of Arc. Howard C. Ives. Wiley-Interscience, 2d ed., 1942, 368 pp. ISBN: 0-471-43065-X. Cost: $31.50.

2318. Naval Reactor Physics Handbook. U.S. Atomic Energy Commision, Division of Reactor Development (available from U.S. Government Printing Office), 1959-.

2319. NBS Alloy Data Center: Permuted Materials Index. G.C. Carter; D.J. Kahan; L.H. Bennett; J.R. Cuthill; R.C. Dobbyn. Institute for Materials Research, National Bureau of Standards, 1971, 683 pp. LC: 76-607785.

This is an index to approximately 10,000 research papers on the physical properties of various metals and alloys. This index covers papers on nuclear magnetic resonance, Knight Shifts, Soft X-ray spectroscopy papers, Soft X-ray adsorption papers, and papers dealing with magnetic susceptibility, density of states, quantum description of solids, specific heats, and the Mössbauer effect. The literature covered is primarily since 1940, although in some cases it goes back as early as 1929, and continues through approximately 1969 to 1970.

2320. Near Infrared Photographic Sky Survey: A Field Index. George S. Rossano; Eric R. Craine. Pachart Publishing House, 1980, 202 pp. LC: 80-81185. ISBN: 0-912918-11-X. Cost: $38.00. Series: *Astronomy and Astrophysics*, 8.

This index is the result of a computer sort of the contents of 16 astronomical catalogs. The output of the catalog sort is included, as well as a table of planned field centers.

2321. Neblette's Handbook of Photography and Reprography: Materials and Processes. John M. Sturge. Van Nostrand Reinhold, 7th ed., 1977, 592 pp. ISBN: 0-442-25948-4. Cost: $52.50.

2322. Neutron Activation Analysis Tables. J.C. Leclerc; A. Cornu; A. Ginier-Gillet. Heyden and Son, Inc, 1974, 72 pp. ISBN: 0-85501-085-1. Cost: $38.50.

These tables provide such information as the saturation sensitivities of isotopes, lists of each radionuclide, and the isotopes that interfere in the gamma energy spectrum. In addition, such information as thorium and radium disintegration series, the half-lives of radionuclides, and radionuclide energies are included.

2323. Neutron Activation Tables. Gerhard Erdtmann. Verlag Chemie International, 1976, 146 pp. ISBN: 3-527-25693-8. Cost: $45.90. Series: *Topical Presentations in Nuclear Chemistry*, Vol 6.

Includes such information as cross-section values of thermal, epithermal, and fast reactor neutrons. In addition, such information as decay rates of the activation products and data tables on radionuclides and the natural isotopes are provided.

2324. Neutron Cross-Sections. D.J. Hughes; R.B. Schwartz. Brookhaven National Laboratory, Atomic Energy Commission, 2d ed., 1958, multivolume set. Date of First Edition: 1960.

This collection of data are often referred to as BNL-325. Included are such data as thermal cross-sections, resonance parameters, cross-section curves, absorption, and scattering. References are provided to the original source material. Cross-section data are provided for the various elements as well as their isotopes. For related works, please refer to such titles as *Reactor Physics Constants; Reactor Shielding Design Manual*; the *CRC Handbook of Radioactive Nuclides*; the *Naval Reactor Physics Handbook*; and the *Reactor Handbook*.

2325. Neutron Data of Structural Materials for Fast Reactors. K.H. Bockhoff. Pergamon Press, Inc, 1979, 824 pp. LC: 78-41017. ISBN: 0-08-023424-0. Cost: $120.00.

This collection of information provides data on such topics as evaluation of natural molybdenum between 5KeV and 5MeV, neutron sensitivity of capture gamma ray detectors, and coherent optical and statistical model calculations of neutron cross sections for Mo isotopes.

2326. New American Machinists' Handbook. R. LeGrand. McGraw-Hill, 1955, 1,572 pp. LC: 55-8908. ISBN: 0-07-037065-6. Cost: 39.50.

2327. New Cosmetic Formulary. H. Bennett. H. Bennett, 1970, 156 pp. Cost: $22.00.

Provides information on how to manufacture perfumes, soaps, and various other types of cosmetics. Such topics as lotions, makeup, and shampoos are discussed in terms of their chemical composition.

2328. New Dictionary and Handbook of Aerospace: With Special Section on the Moon and Lunar Flight. Robert W. Marks ed. Praeger and Bantam Books, 1969, 531 pp. LC: 79-94768. Series: *Bantam Science and Mathematics*.

Although the major portion of this book is a dictionary, it does contain many tables and star charts.

2329. New Handbook of Prescription Drugs: Official Names, Prices, and Sources for Patient and Doctor. Richard Burack. Ballantine Books, Inc, Rev. ed., 1975. ISBN: 0-24341-2-195. Cost: $1.95 paper.

2330. New Physical and Chemical Properties of Metals of Very High Purity. Centre National de la Recherche Scientifique. Gordon and Breach, 1965, 502 pp. ISBN: 0-677-10060-4. Cost: $104.00.

Numerous physical and chemical properties are listed for different pure metals.

2331. New Physical, Mechanical and Chemical Properties of Very High Purity Iron. Centre National de la Recherche Scientifique. Gordon and Breach, 438 pp. ISBN: 0-677-30730-6. Cost: $93.00.

Includes information on the physical, mechanical, and chemical properties of pure iron.

2332. New Table of Indefinite Integrals, Computer Processed. Melvin Klerer; Fred Grossman. Dover (also Peter Smith), 1971. ISBN: 0-8446-0168-3. Cost: $6.00.

2333. Newburger's Manual of Cosmetic Analysis. Alan J. Senzel. Association of Analytical Chemists, 2d ed., 1977, 150 pp. Cost: $12.00.

This manual provides information and methods for chemically analyzing cosmetics. Such methods as chromatographic techniques, including gas chromatography, and spectroscopy techniques are discussed. Numerous types of cosmetics are discussed such as lipsticks, nail lacquers, hair fixatives, cold wave solutions, antiperspirants, deodorants, sunscreens, and hair dyes. In addition, such topics as cosmetic microbiology, perfumes, fragrances, lotions, various types of creams, preservatives, and toothpastes are discussed in terms of chemical analysis.

2334. Newnes Engineer's Reference Book. F.J. Camm; A.T. Camm (rev). Newnes, 9th ed., 1960, 2,067 pp. LC: 61-3865.

2335. Newnes Radio and Electronics Engineer's Pocket Book. Electronics Today International. Butterworths Publishing, Inc, 15th ed., 1978. ISBN: 0-408-00314-6. Cost: $8.50.

Includes information calculations, formulas, and equations that are used in the field of radio and electronics.

2336. Nineteen Seventy Nine Dodge Manual for Building Construction Pricing and Scheduling. McGraw-Hill, 1979, 292 pp. ISBN: 0-07-017321-4. Cost: $24.80.

Unit construction costs are provided and detailed as to material and labor. Adjustment indexes are provided for over 50 trades and subtrades in numerous localities throughout the United States and Canada.

2337. Nineteen Seventy-Nine Dodge Construction Systems Costs. McGraw-Hill, 1979, 262 pp. ISBN: 0-07-017322-2. Cost: $38.80.

This book is excellent for estimating costs of construction and deals with such topics as the superstructure of buildings, exterior walls, partitions, and floors on grade.

2338. Nineteen Seventy-Nine Dodge Guide to Public Works and Heavy Construction Costs. MaGraw-Hill, 1979, 232 pp. ISBN: 0-07-017323-0. Cost: $24.80.

Covers over 6,000 unit costs on material, equipment rates, and labor. This is an excellent reference for estimating heavy construction costs.

2339. Niobium Alloys and Compounds. M. Neuberger; D.L. Grigsby; W.H. Veazie Jr. NTIS, 1972, 77 pp. Accession: A73-11878. Cost: $12.50.

Includes data tables on the composition, preparation, and temperature and field strength parameters of superconducting compounds in niobium alloys.

2340. NMR and the Periodic Table. Robin K. Harris; Brian E. Mann. Academic Press, Inc, 1978, 459 pp. LC: 78-52091. ISBN: 0-12-327650-0. Cost: $62.00.

Much of this data are presented in tabular format. Nuclear magnetic resonance spectroscopy data are provided for most of the chemical elements for which data were available.

2341. NMR Band Handbook. Herman A. Szymanski; Robert E. Yelin. IFI/Plenum Press, 1968, 432 pp. LC: 66-20321. ISBN: 0-306-65127-0. Cost: $75.00.

2342. NMR Data on Organic-Metal Carbonyl Complexes (1951-1971). P.W. Hickmott; Michael Cais; A. Modiano. Academic Press, 1977, 658 pp. LC: 68-17678. ISBN: 0-12-505347-9. Cost: $69.25. Series: *Annual Reports on NMR Spectroscopy*, Vol. 6C. E. F. Mooney (ed).

This is a tabular compilation of nuclear magnetic resonance spectroscopical data on approximately 3,000 organic-metal carbonyl complexes. Such complexes as the following are listed: cobalt complexes, iridium complexes, ruthenium complexes, osmium complexes, rhodium complexes, titanium complexes, vanadium complexes, niobium complexes, tantalum complexes, nickel complexes, paladium complexes, platinum complexes, and copper complexes.

2343. Nogyo Doboku Handbook (Handbook of Agricultural Engineering). Agricultural Engineering Society, 1957, 1,112 pp.

2344. Noise and Vibration Data. Trade and Technical Press. Renouf, 1976, 120 pp. ISBN: 0-85461-058-8. Cost: $19.00.

Various tables, charts, and technical measures are presented as a reference to noise and vibration information.

2345. Noise Control Handbook for Diesel-Powered Vehicles. R.J. Damkevala et al (Cambridge Collaborative, Inc). NTIS, 1974, 214 pp. Accession: PB-236382. Tech Report: DOT-TFC-OST-74-5.

2346. Noise Control: Handbook of Principles and Practices. David M. Lipscomb; Arthur C. Taylor. Van Nostrand Reinhold, 1978, 375 pp. LC: 77-20200. ISBN: 0-442-24811-3. Cost: $26.50.

This guide is designed primarily for writing effective environmental impact statements as they relate to noise control. Assistance is provided in explaining federal and local noise control legislation. Information is also provided on machine design as it relates to noise control and noise reduction. Such topics as aircraft noise, highway noise, rail traffic noise, construction noise, and basic measurement of sound are presented in this handbook.

2347. Nomenclature of Inorganic Chemistry. International Union of Pure and Applied Chemistry, 2d ed., 1971, 110 pp.; 36 pp. ISBN: 0-08-021999-3. Cost: $17.75.

This handbook is frequently referred to as IUPAC's *Red Book* and was originally published in the first edition in 1959. Basic instruction, methods, and procedures are provided for naming inorganic compounds. Such topics as cations derived by proton addition to molecular hydrides, condensed acids and their salts, ligands, coordination entities, boron hydrides, and polyanions are discussed. It should be noted that the use of a companion volume entitled *How to Name an Inorganic Substance* would be very helpful in understanding the *Red Book*.

2348. Nomenclature of Organic Chemistry, Sections A, B, C, D, E, F, and H. J. Rigaudy; S.P. Klesney (IUPAC). Pergamon Press, Inc, 4th ed., 1979, 559 pp. LC: 79-40358. ISBN: 0-08-022369. Cost: $82.00.

This reference tool provides information on recommended practices and procedures in naming organic compounds. The IUPAC refers to this book as the *Blue Book*. Nomenclature systems are provided for all types of organic compounds including acyclic hydrocarbons, monocyclic hydrocarbons, fused polycyclic hydrocarbons, bridged hydrocarbons, spiro hydrocarbons, hydrocarbon ring assembly, terpene hydrocarbons, and heterocyclic hydrocarbons. In addition, such organic compounds as halogens, aldehydes, carboxylic acids, organometallic compounds, organosilicon compounds and organoboron compounds are included in this nomenclature handbook.

2349. Non-Alcoholic Food Service Beverage Handbook. Marvin E. Thorner; R.J. Herzberg. AVI Publishing Company, Inc, 2d ed., 1979, 348 pp. ISBN: 0-87055-279-1. Cost: $28.00.

This handbook deals with processing, distribution, and storing of non-alcoholic beverages with emphasis primarily on soft drinks and coffee. In addition, such topics as beverage spoilage, dairy beverages, various types of juices, tea, cocoa beverages, and water are discussed.

2350. Non-Metallic Materials Handbook. Matin Marietta Corp (available from NTIS), 1974, 304 pp. Accession: N75-28229. Tech Report: NASA-CR-132673. Cost: $9.25.

Updated with N79-27307, Vol.1 *Epoxy Materials*, Report No. NASA-CR-3133, 1979, 249 pp. Covers thermochemical and other properties data for adhesives, electrical insulations, sealants, composite laminates, thermoinsulators, tapes, and inks.

2351. Nondestructive Testing Handbook. Robert Charles McMaster. Ronald Press Co, 1959, 2-Vol. set. LC: 59-14660. ISBN: 0-8260-5915-5. Cost: $36.00.

This handbook was prepared under the auspices of the Society of Nondestructive Testing, and includes information on the selection of the proper tests for specific applications and interpretation of test results. Such topics as X-ray and isotope gaging, ultrasonic testing, eddy current testing, and magnetic particle testing are discussed. Many tables, drawings, charts, and photographs are provided to help the user interpret the test data. An excellent index has been included with cross references to help the user gain access to the nondestructive testing information.

2352. Nonlinear Circuits Handbook; Designing with Analog Function Modules and IC's. Daniel H. Sheingold. Analog Devices, Inc, 1974, 502 pp. LC: 74-75329. ISBN: 0-916550-01-X. Cost: $5.95.

Covers such topics as time-function generation, logarithmic circuits, signal processing, discontinuous approximations and antilog applications.

2353. Normal Melting Points, Boiling Points and Critical Points of Monobasic Alkanoic Acids. Engineering Sciences Data Unit. NTIS, 1972, 6 pp. Accession: N75-29209. Tech Report: ESDU-72006. Cost: $12.00.

This work, sponsored by the Institute of Chemical Engineering, London provides boiling points, thermodynamic properties, and vapor pressures for monobasic alknanoic acids.

2354. Normal Table of Xenopus Laevis. P.D. Nieuwkoop; J. Faber. Elsevier, 2d ed., 1967 (rep. 1975), 252 pp. ISBN: 0-7204-4005-X. Cost: $74.25.

Information is provided for Xenopus Laevis, a South-African clawed toad. The Xenopus test is used for pregnancy determination.

2355. NRCA Roofing and Waterproofing Manual. National Roofing Contractors Association. National Roofing Contractors Association. Cost: $68.00.

This manual is available both as a hardbound volume or as a 3-ring binder. Covers recommended waterproofing practices by the National Roofing Contractors Association.

2356. NTIAC Handbook. Robert E. Englehardt. Nondestructive Testing Information Analysis Center, 1979, loose-leaf.

This service contains substantial information on nondestructive testing techniques and methods.

2357. Nuclear Data in Science and Technology. International Atomic Energy Agency. Unipub, 1974, 2 Vols. ISBN: Vol.1, 92-0-030073-1; Vol.2, 92-0-030173-8. Cost: $37.00 each.

2358. Nuclear Data Tables. U.S. Government Printing Office (NAS/NRC Nuclear Data Project), 1961, 4 Parts in 2 Vols., various paging.

Nuclear Data Tables and Atomic Data (Vol.1, 1965-1973; Vol. 2-5, 1969-1973) absorbed by *Atomic Data and Nuclear Data Tables*, 1973-.

2359. Nuclear Engineering Handbook. Harold Etherington. McGraw-Hill, 1958, 1,882 pp. (various paging). LC: 57-8000. ISBN: 0-07-019720-2. Cost: $55.00.

Although this handbook is somewhat dated, it still contains a considerable amount of information of value to nuclear engineers and scientists. Each section has been written by a recognized expert in his/her particular area of specialization. Reactor engineering receives a considerable amount of attention, including such areas of gas-cooled reactors, reactor control, reactor materials, and reactor physics. In addition, this handbook includes many mathematical tables, formulas, equations, and conversion tables. The nuclear engineer will also be interested in the information on isotopes, radiation, radiation shielding, heat transfer, fluid flow, and radiation damage to reactor materials. This hand-

book includes a comprehensive index as well as a very detailed contents listed at the beginning of each chapter.

2360. Nuclear Handbook. Otto Robert Frisch et al. Van Nostrand Reinhold, 1958. LC: 58-4687 (59-23015).

2361. Nuclear IEEE Standards. IEEE. John Wiley and Sons, 1978, Vol.1, 373 pp.; Vol.2, 413 pp. LC: Vol.1, 78-70785; Vol.2, 78-70587. ISBN: Vol.1, 0-471-05321-X; Vol.2, 0-471-05430-1. Cost: $32.50 per Vol.

This 2-volume set includes recommended tests, practices, procedures, and methods dealing with power systems, protection safety systems, standby power supplies, reliability, storage batteries, measuring and testing equipment, and control facilities and instrumentation.

2362. Nuclear Level Schemes A–45 through A–257 from Nuclear Data Sheets. D.J. Horen (Nuclear Data Group) (ed). Academic Press, 1973, 896 pp. LC: 73-14709. ISBN: 0-12-355650-3. Cost: $37.00.

2363. Nuclear Particles in Cancer Treatment. J.F. Fowler. Heyden, 1981, 188 pp. Cost: $28.00. Series: *Medical Physics Handbook*, 8.

This *Medical Physics Handbook* deals with the treatment of cancer patients by heavy-particle radiotherapy using nuclear particle accelerators.

2364. Nuclear Power Reactor Instrumentation Systems Handbook. U.S. Atomic Energy Commission. U.S. Government Printing Office, 1973.

2365. Nuclear Reactor Plant Data. Nuclear Engineering Division, Technical Data Committee, American Society of Mechanical Engineers, 1959, 2-Vol. set. LC: 59-3244.

Nuclear Reactor Plant Data consists of 2 volumes entitled as follows: Vol.1, *Power Reactors*, 2d ed.; Vol.2, *Research and Test Reactors*.

2366. Nuclear Tables. W. Kunz. Pergamon Press, 1958-68, 3 Vols.

2367. Nucleic Acid Sequence Handbook. Christian Gautier et al. Praeger, 1982, Vol.1, 320 pp.; Vol.2, 320 pp. LC: 81-19904. ISBN: Vol.1, 0-03-060626-8; Vol.2, 0-03-060627-6. Cost: $31.50 per Vol.

This 2-volume set contains nucleic acid sequences of approximately 150 continuous nucleotides. Such topics as single stranded SNA bacteriophages, single stranded RNA bacteriophages, double stranded DNA viruses, prokaryotes, eukaryotes, and mitochondria are discussed.

2368. Numerical Control Users' Handbook. W.H. Leslie. McGraw-Hill, 1970, 448 pp. LC: 78-115143. ISBN: 0-07-094216-1 (out of print).

Numerical Data and Functional Relationships in Science and Technology. *See* **Landolt-Boernstein Numerical Data and Functional Relationships in Science and Technology.**

2369. Numerical Values and Functions in Physics, Chemistry, Astronomy, Geophysics, and Engineering. Vol. 4 Engineering. Part 4B Thermodynamic Property Mixtures: Combustion, Heat Transfer. H. Hausen. Springer

Verlag, 1972, 798 pp. Accession: A73-25473. Cost: $152.20.

This is a German handbook that covers such topics as thermodynamics, combustion physics, and heat transfer. In addition, such topics as combustion products, adsorption, flame propagation, fuel combustion, and gas mixtures are discussed. Numerical data are presented for the thermodynamic properties of gases, vapors, liquids, and solids, and thermodynamic equilibrium of mixtures. Enthalpy and entropy of binary mixtures at evaporation and melting temperatures are also presented.

2370. Nurse's Drug Handbook. Suzanne Loebl; George Spratto; Andrew Wit; Estelle Heckheimer. John Wiley and Sons, 1980, 912 pp. LC: 76-54360. ISBN: 0-471-06092-5. Cost: $19.95.

Contains information on over 900 drugs. Includes table of poisons and antidotes.

2371. Nurses Handbook of Current Drugs. F.B. Gibberd; R.D. Tonkin. Interational Ideas, 3d ed., 1978. Cost: $12.00 paper.

Information on current drugs is presented.

2372. Nurses' Drug Reference. Joseph Albanese. McGraw-Hill, 2d ed., 1981, 1,184 pp. LC: 78-23554. ISBN: 0-07-000768-3 paper; 0-07-000767-5 cloth. Date of First Edition: 1979. Cost: $13.95 paper; $18.95 cloth.

Covers approximately 200 combination drugs, 800 drug monographs, and 400 over-the-counter drugs. Such information as brand names of each drug and their corresponding dosages are presented.

2373. Nurses' Guide to Diagnostic Procedures. Ruth M. French. McGraw-Hill, 5th ed., 420 pp. ISBN: 0-07-022146-4 hardcover; 0-07-022147-2 softcover. Cost: $10.95 hardcover; $7.95 softcover.

Medical laboratory testing and procedures are discussed along with patient preparation and the relationships between diagnosis and treatment.

2374. Nurses' Handbook of Fluid Balance. Norma M. Metheny; William D. Snively. Lippincott, 2d ed., 1974. Cost: $8.75.

2375. Nutritional Quality Index of Foods. R. Gaurth Hansen; Bonita W. Wyse; Ann W. Sorenson. AVI Publishing Co, Inc, 1979, 635 pp. ISBN: 0-87055-320-8. Cost: $25.50; $28.00 outside U.S. and Canada.

Nutrient profiles are provided for over 700 foods. Such information as nutrient density, nutrient ratios, calories, and food nutritional quality is presented.

2376. Observer's Book of Aircraft. William Green (comp). Warne, 1979. LC: 57-4425. ISBN: 0-684-16151-6. Cost: $3.95.

Similar to Jane's *All the World's Aircraft* in that it provides information on and illustrations of different types of aircraft.

2377. Observer's Handbook. John R. Percy (ed). Royal Astronomical Society of Canada, 1975.

This book is primarily concerned with telescopes and their associated equipment.

2378. Occurence, Diagnosis, and Sources of Hospital-Associated Infections. W.J. Fahlberg; D. Groschel. Marcel Dekker, 1978, 138 pp. ISBN: 0-8247-6724-1. Cost: $18.50. Series: *Handbook on Hospital-Associated Infections Series*, Vol. 1.

This publication provides information on diseases and infections that are hospital related.

2379. Oceanographic Handbook. U.S. Coast Guard. NTIS, 1968, 485 pp. Accession: PB-189281. Tech Report: CG-401.

2380. Ocular Therapeutics and Pharmacology. Philip Paul Ellis; D.L. Smith. Mosby (distributed in U.K. by Kimpton), 6th ed., 1981, 336 pp. LC: 81-9632. ISBN: 0-8016-1518-6. Cost: $29.95.

Formerly entitled *Handbook of Ocular Therapeutics and Pharmacology.*

2381. Official Methods of Analysis. Association of Official Analytical Chemists. Association of Official Analytical Chemists, 13th Ed, 1980, 1,038 pp.

This compendium contains methods of analysis for dairy products, beverages, disinfectants, oils and fats, air, water, drugs, and various hazardous substances.

2382. Oil and Gas Production Handbook. J.E. (Hank) Kastrop. Energy Communications, Inc; Petroleum Engineering Publishing Co, 1975, 404 pp. LC: 74-19928. Cost: $35.00.

Such information as water injection, gas wells, casing, tubing, oil recovery, logging, corrosion, gas processing, and maintenance of offshore wells are discussed in this handbook.

2383. Oil Economists' Handbook. Gilbert Jenkins. Applied Science Publishers, Ltd, 1977, 173 pp. ISBN: 0-85334-72-8. Cost: $66.00.

Includes world production of petroleum, seasonal ratios of petroleum costs, viscosity comparisons, and calorific values.

2384. Oil Spill Prevention and Removal Handbook—1974. Marshall Sittig. Noyes Data Corp, 1974, 466 pp. LC: 74-82351. ISBN: 0-8155-0543-4. Cost: $36.00.

2385. Oil/Gas Pipelining Handbook. Dean Hale. Petroleum Engineer Publishing Co, 3d ed., 1975, 386 pp. LC: 75-26144.

Construction of pipelines is presented in great detail in this handbook, particularly as it relates to oil and gas. A discussion of corrosion, leak detection, insulated oil pipelines, welding of pipes, and basic construction data are presented in this handbook. Coversion tables and charts are provided such as viscosity conversion charts, liquified natural gas conversion tables, and effect of pressure on oil viscosity data.

2386. OMNITAB. A Computer Program for Statistical and Numerical Analysis. National Bureau of Standards. NTIS, March, 1966. Cost: $9.00. Series: NBS Handbook 101.

2387. Operations Research Handbook: Standard Algorithms and Methods with Examples. Horst A. Eiselt; Helmut von Frajer. Walter de Gruyter, 1977, 398 pp. LC: 77-5572. ISBN: 3-11-007055-3. Cost: $32.90.

2388. Optical Properties of Minerals: A Determinative Table. Horace Winchell. Academic Press, Inc, 1965, 91 pp. ISBN: 0-12-759150-8. Cost: $21.00.

This is a concise guide to the optical properties of minerals.

Optical Properties of Optical Materials. *See McGraw-Hill/Cindas Data Series on Material Properties. Volume II-3*

2389. Optical Properties of Solids. Minko Balkanski. Elsevier-North Holland Publishing Co, 1980, 633 pp. LC: 79-16501. Cost: $122.50. Series: *Handbook on Semiconductors*, 2.

This handbook includes 9 major review articles by specialists in the field of optics of semiconductors.

2390. Optical Spectra of Transparent Rare Earth Compounds. S. Hüfner. Academic Press, Inc, 1978, 237 pp. ISBN: 0-12-360450-8. Cost: $36.50.

Such topics as the Europium Chalcogenides, rare earths in glasses, rare earth lasers, and magnetic interactions are discussed in this collection of data. In addition, such topics as the Jahn-Teller System, trivalent rare earth ions in a phonon field, and trivalent ions in the static crystal field are included.

2391. Optoelectronic D.A.T.A. Book. Derivation and Tabulation Associates, Inc. DATA, Inc, continuation issued semiannually, pages vary per issue. Cost: $54.50 per year by subscription.

2392. Optoelectronics Data Book for Design Engineers. Texas Instruments, Inc, Semiconductor Group, Engineering Staff, 1977. LC: 78-65638. ISBN: 0-89512-102-6. Cost: $3.50.

2393. Optometry Handbook. L. Rubin. Butterworths, 1975. ISBN: 0-407-98500-X. Cost: $17.95.

2394. Organic Chemistry of Bivalent Sulfur. E. Emmet Reid. Chemical Publishing Co, 1958-66, 6 Vols. (Vol. 4 out of print). Cost: $25.00 per Vol.

This 6-volume set contains over 20,000 references on organic chemistry of bivalent sulfur. Such topics as reactions of mercaptans, physical properties of sulfides, mustard gas, thioelastomers, disulfides, polysulfides, thiocarbonic acid, thiophosgene, and thiopyrimidine are discussed in this monumental series.

2395. Organic Electronic Spectral Data. John P. Phillips et al. Wiley-Interscience, 1960, multivolume set currently in 17 Vols., (Vols.1-4 out of print). LC: 60-16428. Cost: For price, consult publisher.

A collection of ultraviolet visible spectra of organic compounds in formula order. Includes over 300,000 spectra.

2396. Organicum; A Practical Handbook of Organic Chemistry. Trnaslated by B.J. Hazzard. Pergamon Press Inc, 1973, 768 pp. ISBN: 0-08-018964-4; 0-08-012789-4 h.

This translated edition presents information on the properties of organic chemicals.

2397. Ortho- and Pathomorphology of Human and Animal Cells in Drawings, Diagrams and Constructions. David Heinz. G. Fischer, 1978, 503 pp. ISBN: 3-437-30278-7. Cost: DM 138.

Contains over 203 plates with 981 separate diagrams.

2398. Orthodontic Diagnosis: Dental Practitioner Handbook No. 4. W.T.B. Houston. Year Book Medical Publishers, 2d ed., 1975, 88 pp. ISBN: 0-8151-4704-X. Cost: $9.95 paper.

2399. Orthopedic Roentgen Atlas. P.F. Matzen; H.K. Fleissner. Georg Thieme, 1970, 477 pp. ISBN: 3-13-459801-9. Cost: DM 198.

Provides drawings and photographs of radiological treatment of orthopedic disorders.

2400. Oscilloscope Handbook. Clyde N. Herrick. Reston Publishing Co, 1974, 243 pp. LC: 73-21056. ISBN: 0-87909-597-0. Cost: $18.95.

Practical applications of the oscilloscope are discussed in this work including testing of circuits, electronic troubleshooting, and signal tracing.

2401. OSHA Compliance Manual. Dan Peterson. McGraw-Hill, rev. ed., 1979, 241 pp. LC: 78-5060. ISBN: 0-07-049598-X. Cost: $27.50 spiral, paper.

Provides information on safety in industrial situations and covers such topics as record keeping, OSHA regulations and standards, and safety management.

2402. Outline of Oral Surgery: Part 1—Dental Practitioner Handbook No. 10. H.C. Killey et al. Year Book Medical Publishers, 1975, 196 pp. ISBN: 0-8151-5037-7. Cost: $14.50 paper.

2403. Oxide Handbook. G.V. Samsonov; Robert K. Johnson (trans). IFI/Plenum, 2d ed., 1981, 500 pp. ISBN: 0-306-65177-7. Cost: $75.00. Series: *IFI Data Base Library*.

Various properties are listed in this handbook, including optical, mechanical, electrical, nuclear, thermal and chemical. The main emphasis is the usage of oxides as refractory materials in metallurgical processes. Phase diagrams have been presented, and many bibliographic references are provided for further investigation. For related titles, please refer to *Metals Handbook*.

2404. Oxygen and Ozone—Gas Solubilities. R. Battino. ISBN: 0-08-023915-3.

See *Solubility Data Series*, Vol. 5.

2405. Paint Handbook. Guy E. Weismantel. McGraw-Hill, 1981, 640 pp., illustrated. LC: 80-12093. ISBN: 0-07-069061-8. Cost: $37.50.

Such topics as surface preparation, exterior wood finishes, masonry surfaces, marine finishes and coatings for steels and other types of metals are presented.

2406. Paints and Coatings Handbook. Abel Banov. McGraw-Hill, 2d ed., 1978, 444 pp. LC: 78-54102. ISBN: 0-07-003664-0. Cost: $32.50.

Provides information on different types of coatings for various types of surfaces.

2407. Parallel Tables of Slopes and Rises. C.K. Smoley; E.R. Smoley (rev); N.G. Smoley (rev). Smoley and Sons, Inc, rev. ed., 1968. ISBN: 0-911390-03-0. Cost: $12.00.

2408. Partial Grotrian Diagrams of Astrophysical Interest. C.E. Moore; P.W. Merrill (National Bureau of Standards). NTIS, 1968. Cost: $4.50. Series: NSRDS-NBS-23.

2409. Particle Atlas. Walter C. McCrone et al. Ann Arbor Science Publishers, 2d ed.; Vol.1, 1973; Vol.2, 1973; Vol.3, 1973; Vol.4, 1973; Vol.5, 1979; Vol.6, 1980, 6 Vols. ISBN: Vol.1, 0-250-40018-9; Vol.2, 0-250-40019-7; Vol.3, 0-250-40020-0; Vol.4, 0-250-40021-9; Vol.5, 0-250-40195-9; Vol.6, 0-250-40196-7. Cost: $90.00 per Vol; $540.00 set.

This multivolume set includes 6 volumes entitled as follows: Vol.1, *Instrumentation and Techniques*, Vol.2, *Light Microscopy Atlas*, Vol.3, *Electron Microscopy Atlas*, Vol.4, *Handbook for Analysts*, Vol.5, *Light Microscopy Techniques and Atlas*, Vol.6, *Electron-Optical Techniques and Atlas*. Includes 3,230 transmission and scanning electron micrographs of the identical 1,020 substances, making a total of 4,500 micrographs. These photomicrographs will help to identify such substances as fertilizers, insecticides, pigments, combustion products, industrial chemicals, minerals, explosives, and biologicals.

2410. Particle Size Enlargement. C.E. Capes. Elsevier-North Holland Publishing Co, 1980, 192 pp. Cost: $68.25.

Numerous tables, diagrams, and figures are provided on the topic of particle size enlargement. Such topics as agglomerate bonding, agitation methods, sintering, heat hardening, nodulizing, drying, and solidification are also discussed.

2411. Particle Size Measurement. Terence Allen. Chapman and Hall, Ltd, 3d ed., 1980, 454 pp. ISBN: 0-412-13490-X. Cost: $37.95.

2412. Particulate Pollutant Systems Study. Vol. 3. Handbook of Emission Properties. A.E. Vandergrift et al (Midwest Research Institute, Kansas City, MO). NTIS, 1971, 613 pp., Vol.3 of a 3-Vol. set. Accession: PB-203522. Tech Report: APTD-0745.

2413. Passive Solar Buildings: A Compilation of Data and Results. R.P. Stromberg. Sandia Laboratories, 1977. ISBN: 0-930978-040-7. Cost: $10.95.

Information has been compiled on solar structures.

2414. Patty's Industrial Hygiene and Toxicology. George D. Clayton; Florence E. Clayton. Wiley-Interscience, 3d rev. ed.; Vol.1, 1977; Vol.2, 1980; Vol.3, 1979, 3 Vols. LC: Vol.1, 77-17515; Vol.2, 78-27102. ISBN: Vol.1, 0-471-16046-6; Vol.2, 0-471-16042-3; Vol.3, 0-471-02698-0. Cost: Vol.1, $99.50; Vol.2, $95.00; Vol.3, $52.50.

This 3-volume set is entitled as follows: Vol.1, *General Principles*, Vol.2, *Toxicology*, Vol.3, *Theory and Rationale of Industrial Hygiene Practice*. Provides information on industrial hygiene and toxic substances.

2415. PCI Design Handbook; Precast and Prestressed Concrete. Prestressed Concrete Institute, 2d ed., 1978, 384 pp. ISBN: 0-937040-12-6. Cost: $42.00.

Includes load tables and information on flexure, shear, camber, thermal properties, acoustical properties, fire resistance, and various guide specifications.

2416. Pediatric Anesthesia Handbook. Richard M. Levin. Medical Examination Publishing Co, 1973. ISBN: 0-87488-637-6. Cost: $8.00.

2417. Pediatric Dosage Handbook. Harry C. Skirkey. American Pharmaceutical Association, 1973. LC: 73-161655. Cost: $2.00.

2418. Pediatric Drug Handbook. Harry C. Shirkey. Saunders, 1977. LC: 76-45966. ISBN: 0-7216-8247-2.

2419. Pediatric Neurology Handbook. J.T. Jabbour. Medical Examination Publishing Co, 1975. ISBN: 0-87488-636-8.

2420. Pediatric Nutrition Handbook. American Academy of Pediatrics. American Academy of Pediatrics, 1979, 472 pp. LC: 79-51786. Cost: $6.00.

Provides information on the nutrition of children. Numerous bibliographies, reference lists, and tables are included throughout the book.

2421. Pediatric Operative Dentistry: Dental Practitioner Handbook No. 21. D.B. Kennedy. Year Book Medical Publishers, 1976, 260 pp. ISBN: 0-8151-5013-X. Cost: $14.95.

2422. Pediatricians Handbook of Communication Disorders. Herold S. Lillywhite; Norton B. Young; Richard W. Olmstead. Lea & Febiger, 1970, 122 pp. LC: 74-123422. ISBN: 0-8121-0314-9. Cost: $5.75.

2423. Penguin Book of Tables. Penguin Books, Inc, 1974. ISBN: 0-14-080339-4. Cost: $0.95 paper.

This paperbound edition is primarily designed as an inexpensive reference for student usage.

2424. Peripheral Driver Data Book for Design Engineers. Texas Instruments, Inc, 1977, 118 pp.

Tabulated data are provided on integrated circuits.

2425. Permanent Magnet Design and Application Handbook. Lester R. Moskowitz. Cahners Books International, Inc, 1976, 385 pp. LC: 75-28109. ISBN: 0-8436-1800-0. Cost: $49.95.

Magnetic properties tables are presented in the appendices, along with demagneaization curves and physical properties tables. In addition, such information as an international index of permanent magnet materials, standards and specifications of permanent magnets, testing of permanent magnets, environmental effects, circuit effects, and leakage of permanent magnets are presented in this handbook.

2426. Permanent Magnet Handbook. Earl M. Underhill. Crucible Steel Co of America, 1957, loose-leaf format, various paging.

Includes such information as permanent magnet alloys, steels, cast Alnico, electromagnetism, and demagnetization.

2427. Permissible Dose from External Sources of Ionizating Radiations. National Bureau of Standards. NCRP Publications. Cost: $4.00. Series: NBS Handbook 59.

Available from NCRP Publications, PO Box 30175, Washington, DC 20014 as NCRP Report 39.

2428. Permits Handbook for Coal Development. Cooper Wayman; Gail Genasci. Colorado School of Mines Research Institute, 1981, 640 pp. LC: 80-22500. ISBN: 0-9180620-40-3. Cost: $35.00.

The issuing of permits is analyzed at different developmental levels from federal, state, to local. Six western states are primarily covered, namely Colorado, Wyoming, Utah, North Dakota, South Dakota, and Montana. Such topics as air pollution permits, water pollution permits, resource conservation, toxic substances, location of power plants, coal severence tax, and strip mining permits are discussed.

2429. Pesticide Formulations. W. Van Valkenburg. Marcel Dekker, Inc, 1973, 496 pp. ISBN: 0-8247-1695-7. Cost: $65.00.

Provides information on many chemicals used in various different types of pesticides.

2430. Pesticide Handbook—Entoma 1975-76. Donald Elisha Harding Frear. Entomological Society of America, 1975-76. LC: 52-44516. Cost: $7.50.

2431. Pesticide Manual; A World Compendium. Charles Worthing (ed). British Crop Protection Council, 6th ed., 1980, 655 pp. ISBN: 0-901436-44-5. Cost: $59.95.

Includes nomenclature, toxicology, and formulation of chemical compounds and microbial agents.

2432. Petroleum Exploration Handbook: A Practical Manual Summarizing the Application of Earth Sciences to Petroleum Explorations. G. Moody. McGraw-Hill, 1961, 860 pp. LC: 60-14046. ISBN: 0-07-042867-0. Cost: $65.00.

2433. Petroleum Measurement Tables. American Petroleum Institute, Measurement Coordination, 10 Vols; Vol.1, 678 pp.; Vol.2, 592 pp.; Vol.3, 563 pp.; Vol.4, 878 pp.; Vol.5, 812 pp.; Vol.6, 563 pp.; Vol.7, 958 pp.; Vol.8, 881 pp.; Vol.9, 587 pp.; Vol.10, 420 pp. LC: 80-68070. ISBN: 0-89364-021-2. Cost: $250.00; 20% less cost to ASTM members; Vol.1, $28.00; Vol.2, $28.00; Vol.3, $28.00; Vol.4, $32.00; Vol.5, $32.00; Vol.6, $28.00; Vol.7, $32.00; Vol.8, $32.00; Vol.9, $28.00; Vol.10 $25.00.

There is a substantial amount of density data that is presented in these computer-produced tables. In addition, thermal expansion coefficients are also listedl in these tables.

2434. Petroleum Processing Handbook. W.F. Bland; R.L. Davidson. McGraw-Hill, 1967, 1,102 pp. LC: 64-66366. ISBN: 0-07-005860-1. Cost: $56.50.

2435. Petroleum Production Handbook. Thomas C. Frick; R. William Taylor. McGraw-Hill, 1962, 2 Vols. (various paging). LC: 60-10601 (out of print).

This 2-volume handbook includes information on the production and recovery of both oil and gas, properties of crude oil and natural gas, liquid condensate, produced waters and reservoirs rocks. Production equipment is discussed, including all the necessary equipment to recover both oil and gas such as hydraulic pumps, and oil and gas separators. Data on phase diagrams are presented along with such information as well temperature, formation fracturing, and reservoir traps. Numerous mathematical tables are provided in addition to a well-prepared index to help retrieve data on petroleum technology.

2436. Petroleum Products Handbook. V. Guthrie. McGraw-Hill, 1960, 864 pp. LC: 58-13870. ISBN: 0-07-025295-5. Cost: $47.50.

2437. Petroleum Transportation Handbook. Harold Sill Bell. McGraw-Hill, 1963, various paging. LC: 63-12124.

2438. Pharmacopeia of the United States of America. Mack Publishing Co. Mack Publishing Co, 19th ed, 1975. Cost: $28.00.

2439. Phase Behavior in Binary and Multicomponent Systems at Elevated Pressures: *n*-Pentane and *Methane*-n-*Pentane.* V.M. Berry; B.H. Sage (National Bureau of Standards). U.S. Government Printing Office, 1970. SuDoc: C13.48:32. Cost: $1.15. Series: NSRDS-NBS-32.

2440. Phase Change Materials Handbook. A.D. Hale; M.J. Hoover; J.J. O'Neill. NTIS, 1971, 200 pp. Accession: N72-19956. Tech Report: NASA-CR-61363 LMSC/HREC-D225138.

This handbook includes information on the thermodynamic and heat transfer phenomena that are peculiar to phase-change materials design. Conventional temperature-control techniques are described in a space environment.

2441. Phase Diagrams for Ceramists. E.M. Levin et al. American Ceramic Society, 7th compilation, 1964, 601 pp. Cost: $35.00.

This seventh compilation includes over 2,064 phase diagrams. Some of the plates were revised and redrawn by E.F. Osborn and A. Muan and published as *Phase Equilibrium Diagrams of Oxide Systems*, American Ceramic Society.

2442. Phase Diagrams. J. Wisniak. Elsevier-North Holland Publishing Co, May, 1981, 2,101 pp.; 2-part set. ISBN: 0-444-41981-0. Cost: $319.50.

This is a literature source book covering the topic of phase diagrams between 1900 and 1980. The compounds are arranged in order of increasing carbon and hydrogen content. Binary compounds are listed first, followed by ternary, and quaternary systems. This 2-volume set includes compounds from A to H_2NiO_2 in Part A, and Part B includes compounds from H_2O to Z and the references.

2443. Phase Diagrams: Materials Science and Technology. Allen M. Alper. Academic Press, Inc, Vol.1, 1970; Vol.2, 1970; Vol.3, 1970; Vol.4, 1976; Vol.5, 1978, 5 vols; Vol.1, 375 pp.; Vol.2, 373 pp.; Vol.3, 343 pp.; Vol.4, 320 pp.; Vol.5, 352 pp. ISBN: Vol.1, 0-12-053201-8; Vol.2, 0-12-053202-6; Vol.3, 0-12-053203-4; Vol.4, 0-12-053204-2; Vol.5, 0-12-053205-0. Cost: Vol.1, $42.00; Vol.2, $42.00; Vol.3, $42.00; Vol.4, $40.00; Vol.5, $35.00.

This publication includes 5 volumes entitled as follows: Vol.1, *Theory, Principles, and Techniques of Phase Diagrams*; Vol.2, *The Use of Phase Diagrams in Metal, Refractory, Ceramic, and Cement Technology*; Vol.3, *The Use of Phase Diagrams in Electronic Materials and Glass Technology*; Vol.4, *The Use of Phase Diagrams in Technical Materials*; Vol.5, *Crystal Chemistry, Stoichiometry, Spinodal Decomposition, Properties of Inorganic Phases*. This 5-volume set includes such topics as phase equilibrium and magnetic oxide materials, phase diagrams in phosphor materials, phase equilibria of apatites, phase diagrams for sialons, phase diagrams of silicates, phase diagrams of rare earth borides, and superconductivity and phase diagrams.

2444. Phase Equilibria and Gas Mixtures Properties. American Institute of Chemical Engineers. American Institute of Chemical Engineers, 1968, 107 pp. Cost: $6.00 AIChe members; $20.00 others. Series: *AIChe Symposium Series*, S-88.

Information and papers are presented that include phase equilibria and properties of gas mixtures.

2445. Phase Equilibria and Related Properties. American Institute of Chemical Engineers. American Institute of Chemical Engineers, 1967, 133 pp. Cost: $7.50 AIChe members; $20.00 others. Series: *AIChe Symposium Series*, S-81.

Data are provided for the design of separation processes.

2446. Phase Transitions and Critical Phenomena. C. Domb; M. Green. Academic Press, Vol.1, 1972; Vol.2, 1972; Vol.3, 1974; Vol.5A, 1976; Vol.5B, 1976; Vol.6, 1976, 6 vols. ISBN: Vol.1, 0-012-220301-1; Vol.2, 0-12-220302-X; Vol.3, 0-12-220303-8; Vol.5A 0-12-220305-4; Vol.5B, 0-12-220351-8; Vol.6, 0-12-220306-2. Cost: Vol.1, $83.00; Vol.2, $84.50; Vol.3, $95.50; Vol.5A, $68.00; Vol.5B, $64.50; Vol.6, $84.50.

This represents a comprehensive compilation of phase diagrams.

2447. Photographic Atlas of the Mid-Atlantic Ridge Rift Valley. R.D. Ballard; J.G. Moore. Springer (available from Otto Harrassowitz), 1977, 114 pp. ISBN: 3-540-90247-3. Cost: $19.90.

Includes over 183 figures on the geologic formations of the Mid-Atlantic Ridge Rift Valley.

2448. Photographic Atlas of the Moon. Zdenek Kopal. University of Manchester, England, 1965, 277 pp. ISBN: 0-12-419368-4. Cost: $43.00.

Numerous lunar photographs are provided in this atlas.

2449. Photographic Dosimetry of X- and Gamma Rays. National Bureau of Standards. NTIS, Aug. 20, 1954. Accession: PB-248218. Cost: $3.50. Series: NBS Handbook 57.

2450. Photographic Lab Handbook. John S. Carroll. Prentice-Hall (also available from Amphoto—American Photogaphic Book Publishing Co, Inc), 5th ed., 1979, 800 pp. ISBN: 0-13-665422-3. Date of First Edition: 1970.

Covers such topics as photographic equipment, darkroom procedures, and photographic techniques.

2451. Photomicrographs of World Woods. Ann Miles. Department of the Environment, Building Research Establishment, H.M. Stationery Office, London, 1978, 233 pp. LC: 80-472148. ISBN: 0-1167-0754-2.

Numerous photomicrographs are provided for different types of woods which would be useful in wood identification and analysis.

2452. Photon Cross Sections, Attenuation Coefficients, and Energy Absorption Coefficients from 10 keV to 100 GeV. J.H. Hubbell (National Bureau of Standards). NTIS, 1969. Cost: $5.00. Series: NSRDS-NBS-29.

2453. Physical and Chemical Properties of JP-4 Fuel for 1975. L.C. Angelo; P.C. Baker. NTIS, 1975, 130 pp. Accession: N77-14273; AD A026135. Tech Report: AFAPL-TR-76-19. Cost: $3.50 microform; $12.50 hard copy.

An analysis of jet fuel JP-4 for 1975 has been provided. Distillation temperature data, thermal stability test data, Reid vapor pressure data, and American Petroleum Institute gravity values are provided.

2454. Physical and Geotechnical Properties of Soils. Joseph E. Bowles. McGraw-Hill, 1979, 478 pp. LC: 78-3790. ISBN: 0-07-006760-0. Cost: $23.50.

Such information as soil hydraulics, permeability, shrinkage, and capillarity are discussed in this book. Shear strength of soils, soil pressures, stability of slopes, and stress/strain characteristics of soils are also discussed in this book.

2455. Physical and Mathematical Tables. T.M. Yarwood; F. Castle. Macmillan, 1970.

2456. Physical Aspects of Irradiation. International Commission on Radiological Units and Measurements. National Bureau of Standards (available from U.S. Government Printing Office), March 31, 1964. SuDoc: C13.11:85. Cost: $1.65. Series: NBS Handbook 85.

Supersedes parts of Handbook 78; Handbooks 84 through 89 extend and largely replace Handbook 78.

2457. Physical Constants of Hycarbons C^1 to C^{10}. Commitee D-2 on Petroleum Products and Lubricants, American Society for Testing and Materials, 1971, 72 pp. LC: 79-170766. ISBN: 0-8031-0009-4. Cost: $4.00 paper. Series: *ASTM Data Series*, DS-4A.

A tabulation of physical constants of numerous hydrocarbons is presented.

2458. Physical Constants of Hydrocarbons. Gustav Egloff. American Chemical Society, Vol.1, 1939; Vol.2, 1940; Vol.3, 1946; Vol.4, 1947; Vol.5, 1953, 5 Vols, Vol.1, 418 pp.; Vol.2, 610 pp.; Vol.3, 675 pp.; Vol.4, 552 pp.; Vol.5, 536 pp. LC: 39-7977. Cost: Vol.1, $55.20; Vol.2, $83.20; Vol.3, $91.00; Vol.4, $71.30; Vol.5, $69.40. Series: *American Chemical Society Monograph Series*, No. 78.

This 5-volume set is entitled as follows: Vol.1, *Tabulated Data on Paraffins, Olefins, amd Acetylenes*; Vol.2, *Tabulated Data on Cyclanes, Cyclenes, Cyclynes, and other Alicyclic Hydrocarbons*; Vol.3, *Mononuclear Aromatic Hydrocarbons*; Vol.4, *Polynuclear Hydrocarbons*; Vol.5, *Tables of Revised Values Supplement and Revise Those in Volume 1*. This set includes physical constants of various types of hydrocarbons, including paraffins, olefins, acetylenes, cyclanes, cyclenes, cyclynes, and alicyclic hydrocarbons.

2459. Physical Metallurgy of the Rare Earth Metals. E.M. Savitskii; V.F. Terekhova. NTIS, 1975, 273 pp. Accession: A76-18450.

This Russian title presents data concerning the production, purification, and physico-chemical properties of rare earth metals. The electronic structure of rare earth metals is discussed with regard to their magnetic, nuclear, mechnical, optical, and chemical properties.

2460. Physical Principles of Audiology. Peter Haughton. Heyden and Son, Inc, 1980, 216 pp. Cost: $28.00. Series: *Medical Physics Handbook*, Vol. 3.

This publication includs over 29 tables of audiometric data. In addition, such topics as hearing aids, electrophysiology, physiology of the ear, middle ear function, and hearing disorders are discussed in this publication.

2461. Physical Properties Data Compilations Revelant to Energy Storage. George J. Janz. National Bureau of Standards, 1978, 244 pp. LC: 77-10824. Cost: $4.25. Series: NSRDS-NBS-61, Part 1.

This is a compilation of melting points and compositions of molten salts, and eutectic mixtures. Data have been compiled on mixtures melting in the range $-138°$ C to $2800°$ C.

2462. Physical Properties Data for Rock Salt. L.H. Gevantman. Superintendent of Documents, 1981, 282 pp. LC: 81-6589. SuDoc: C 13.44:167. Cost: $12.00. Series: NBS Monograph, 167.

This is a compendium of physical properties of rock salt, including numerous characteristics such as magnetic, optical, thermal, radiation, electrical, geological, mechanical, and physical properties. This data were primarily assembled to be used for nuclear waste storage, but it has wider applications.

2463. Physical Properties of Chemical Compounds. Robert R. Dreisbach. American Chemical Society, Vol.1, 1955; Vol.2, 1959; Vol.3, 1961, 3-Vol. set. ISBN: Vol.1, 0-8412-0016-5; Vol.2, 0-8412-0023-8; Vol.3, 0-8412-0030-0. Cost: Vol.1, $70.30; Vol.2, $37.25. Series: *Advances in Chemistry Series*, No. 22.

Covers numerous types of chemical compounds including alkanes, haloalkanes, alkenes, haloalkenes, dialefins, alkynes, and aliphatic compounds.

2464. Physical Properties of Hydrocarbons. Robert W. Gallant. Gulf Publishing Co, 1968, 2d ed., 1981, 2-Vol. set, Vol.1, 225 pp., Vol.2, 201 pp. LC: 68-9302. ISBN: Vol.1, 0-87201-688-9; Vol.2, 0-8720-690-0. Cost: $23.95 per Vol.

Covers such properties as vapor pressure, heat of vaporzation, heat capacity, density, vapor viscosity, liquid viscosity, surface tension, and thermal conductivity. Vol.1 covers the paraffinic hydrocarbons and halogenated hydrocarbons and Vol.2 covers the oxygenated hydrocarbons.

2465. Physical Properties of Inorganic Compounds in SI Units. A.L. Horvath. Hemisphere Publishing Co (also available from Crane, Russak), 1975, 466 pp. LC: 74-81574. ISBN: 8-8448-0523-8. Cost: $ 74.50.

This handbook lists many different physical properties of inorganic compounds including hydrogen sulphide, as well as many other relevant pure inorganic compounds. Various properties are listed including thermal conductivity, vapor pressure, refractive index, heat capacity, solubility in water, surface tension, and thermal expansion. Many other properties are listed, but are too numerous to mention. Bibliographic sources of information are provided so that the user can obtain additional information. Conversion tables and an excellent index, with numerous cross references, have been prepared. For related works, please refer to the *CRC Handbook of Chemistry and Physics*, the *JANAF Thermochemical Tables*, *Lange's Handbook of Chemistry*, and the *International Critical Tables*.

2466. Physical Properties of Liquid Crystalline Materials. W.H. De Jeu. Gordon and Breach, 1980, 140 pp. ISBN: 0-677-04040-7. Cost: $35.25.

This compilation provides data on the physical properties of liquid crystalline materials.

2467. Physical Properties of Martensite and Bainite. The Iron and Steel Institute; The Institute of Metals, 1965, 218 pp. ISBN: 0-900497-734. Cost: $12.00.

Metallurgical data are provided for bainite and martensite.

2468. Physical Properties of Materials for Engineers. Daniel D. Pollock. CRC Press, Inc, 1981, Vol.1, 224 pp.; Vol.2, 208 pp. ISBN: Vol.1, 0-8493-6201-6; Vol.2, 0-8493-6202-4. Cost: Vol.1, $69.95; Vol.2, $64.95.

This is currently a 2-volume set with a 3d volume planned for the future. Numerous types of properties are presented in this handbook, such as thermoelectric properties of alloys, dielectric properties, ferromagnetic properties of materials, heat capacity, thermal conductivity, and thermal expansion. In addition, such topics as the Schrodinger wave equation, band theory of solids, Fermi-Sommerfeld theory of metals, superconductivity, Seebeck effect, Peltier effect, and the Thomson effect are presented in this tabulation.

2469. Physical Properties of Plant and Animal Materials. N.N. Mohsenin. Gordon and Breach, 1970, 758 pp. ISBN: 0-677-02300-6. Cost: $101.25.

This recently reprinted title contains data on the physical properties of plants and animals.

Physical Properties of Rocks and Minerals. *See* **McGraw-Hill/CINDAS Data Series on Material Properties.**

2470. Physical Properties of Some Plutonium Ceramic Compounds—A Data Manual. B.J. Seddon. NTIS, 1968, 68 pp. Accession: N68-34000. Tech Report: TRG-1601/R/.

Various thermochemical and thermophysical, as well as mechanical properties have been presented for some plutonium compounds.

2471. Physical Properties: A Guide to the Physical, Thermodynamic and Transport Property Data of Industrially Important Chemical Compounds. Carl L. Yaws. McGraw-Hill, 1977, 239 pp. LC: 77-148-25. ISBN: 07-0099-712-8.

A wealth of property data are provided on numerous inorganic compounds as well as organic compounds. Graphs have been included and curves have been plotted to represent correlations. Equations and constants have been included that will be helpful in calculations dealing with thermodynamics, fluid flow, reaction engineering, heat transfer, process design, and mass transfer. Such chemicals as methanol, ethanol, propanol, butanol, ethylene, propylene, butylene oxides, xylenes, alkanes, olefins, and toluenes are listed in this compilation. In addition, sulfur oxides, halogens, carbon oxides, ammonia, hydrazine, helium, neon, and argon are also listed.

2472. Physician's Desk Reference for Ophthalmology. Medical Economics, 4th ed., 1975-76. LC: 73-640206 (first ed.). Date of First Edition: 1972.

2473. Physician's Desk Reference for Radiology and Nuclear Medicine. Medical Economics. Medical Economics, 1971-. LC: 77-649819.

2474. Physician's Desk Reference to Pharmaceutical Specialties and Biologicals (PDR). Medical Economics; J.M. Jones (ed. and publisher, 1947-52). Medical Economics, annual with supplements, 1947-. LC: Med 47-177.

NOTE: Commonly referred to as PDR.

2475. Physician's Handbook of Nutritional Science. Roger J. Williams. C.C. Thomas, 1975. ISBN: 0-398-03256-4. Cost: $9.75.

2476. Physician's Handbook on Orthomolecular Medicine. R.J. Williams; D.K. Kalita. Pergamon Press, 1977, 250 pp. LC: 77-8304. ISBN: 0-08-021533-5. Cost: $16.50.

2477. Physician's Handbook. Marcus A. Krupp et al. Lange Medical Publishers, 19th ed., 1979. LC: 41-9970. ISBN: 0-87041-024-5. Cost: $9.00.

2478. Physicians' Desk Reference for Nonprescription Drugs. Medical Economics Company, 1980, 232 pp. ISBN: 0-442-21312-3. Cost: $12.25.

This book is extensively cross-referenced. Several indexes, including product category index, manufacturers' index, brand name index, and the active ingredient index will be helpful in locating information. Drug interaction information is provided in the product information section, and color photographs of products are provided in the product identification section.

2479. Physico-Chemical Properties for Chemical Engineering. Kagaku Kogaku Kyokai; Society of Chemical Engineers. Maruzen Co, Ltd, 1977, 4 vols; Vol.1, 224 pp.; Vol.2, 228 pp.; Vol.3, 248 pp.; Vol.4, 260 pp.. Cost: Vol.1, $33.60; Vol.2, $37.10; Vol.3, $38.50; Vol.4, $38.50.

Such information as viscosity, thermal conductivity, solubility data, vapor-liquid equilibrium, heat capacity, vapor pressure, and heat of vaporization are included in this tabulation.

2480. Physics Data Series. Fachinformationszentrum. Fachinformationszentrum, Energie, Physik, Mathematik, 1976-.

This German series is a monumental effort to compile properties data from widely scattered literature. Numerous types of properties data are included in this set, including solubility, disassociation pressure of compounds, vapor pressure of volatile oxides, thermodynamic data, heat of transport, permeation of gases through metals, gas absorption, precipitation kinetics, and gas desorption kinetics. In some cases, bibliographies are listed, as well as graphs, charts, and diagrams. In some cases, the text will be in German, in English, or in both languages.

2481. Physics Demonstration Experiments. H.F. Mieners. John Wiley and Sons, 1970, 2 Vol set; 1,493 pp. LC: 69-14674. ISBN: Vol. 1, 0-471-7448-9; Vol. 2, 0-471-7449-7; 0-471-06759-8 Set. Cost: $67.50 set.

Over 1,100 experiments have been compiled in this listing which are examples of such subject areas as mechanics, wave motion, optics, nuclear physics, electricity, and heat. Over 2,200 photographs and line drawings have been included to assist in the explanation of these experiments.

2482. Physics Pocketbook. Hermann Ebert (ed). John Wiley and Sons, 1968, 575 pp. LC: 67-29717.

2483. Physiochemical Properties of Pesticides Handbook. A.A. Shamshurin; M.Z. Krimer. Khimiya, Moscow, U.S.S.R., 2d ed., 1976, 328 pp. Cost: Rub. 1.26.

This Russian title includes tabulated data on the physiochemical properties of various types of pesticides.

2484. Pictorial Handbook of Technical Devices. Paul Grafstein; Otto B. Schwarz. Chemical Publishing Co, 1971, 611 pp. ISBN: 0-8206-0234-5. Cost: $14.00.

This is a very fascinating volume covering many different types of machinery and devices with mechanical movements. The book is well illustrated with approximately 5,000 different diagrams, photographs, and sketches which will help the user understand these different devices. Various types of mechanisms and machinery are discussed from measuring devices and heating and refrigerating systems to cameras and power generation machinery. An index is provided for the user's convenience.

2485. Pipe Line Rules of Thumb Handbook. Gulf Publishing Co, 1978, 197 pp. LC: 76-52237. ISBN: 0-87201-698-6. Cost: $12.95.

Covers such topics as corrosion, pumps, natural gas, and pipeline construction.

2486. Pipefitters Handbook. Forrest R. Lindsey. Industrial Press, 3d ed., 1967, 464 pp. LC: 67-6326. ISBN: 0-8311-3019-9. Cost: $15.00.

Covers industrial pipebending and the fabrication of welding fittings.

2487. Pipefitting, Plumbing Handbook. Superintendent of Documents, 1966, 183 pp. SuDoc: D 209.14: P 66/3. Cost: $3.75.

Time data are provided for various different pipefitting and plumbing tasks.

2488. Piping Guide: A Compact Reference for the Design and Drafting of Industrial Piping Systems. David R. Sherwood. Syentek Book Company, rev. ed., 1976, 216 pp. ISBN: 0-914082-035. Cost: $33.00.

This guide contains numerous tables and charts of dimensions for different types of pipes, as well as flow resistances of fittings, connections, and valves. In addition, pressure drop data are provided, along with information on pumps, compressors, vents, and drains.

2489. Piping Handbook. Sabin Crocker. McGraw-Hill, 5th ed., 1967, 1,652 pp. ISBN: 0-07-013841-9. Cost: $49.50.

This handbook includes 19 authoritatively written articles on such subjects as thermal insulation, joining of commercial piping, corrosion of piping systems, cryogenic piping, fire protection for piping, hydraulic power transmission—piping, underground steam piping, sewerage systems piping, refrigeration piping. In addition, nuclear systems piping, piping for water supply, oil systems, flow of sludges and slurries, and other types of plumbing systems are discussed.

2490. Pit and Quarry Handbook and Buyers Guide for the Nonmetallic Minerals Industries. Pit and Quarry Publications, Inc, 69th ed, 1976-77, various paging. Cost: $50.00.

2491. Plane Coordinate Projection Tables, Special Publication Series. U.S. Coast and Geodetic Survey. Superintendent of Documents.

These plane coordinate projection tables are available for all states.

2492. Plane Strain Fracture Toughness (KIC) Data Handbook for Metals. William T. Matthews (Army Materiels and Mechanics Research Center, Watertowm, MA). NTIS, 19973, 92 pp. Accession: AD-773673. Tech Report: AMMRC-MS-73-6.

2493. Plant Classification. Lyman David Benson. D.C. Heath and Co, 2d ed., 1979. ISBN: 0-669-01489-3.

This guide provides clear and concise information on the classification of various plants.

2494. Plant Engineer's Desk Handbook. Cushing Phillips Jr. Prentice-Hall, Inc, 1979, 244 pp. ISBN: 0-13-680264-8. Cost: $34.95.

Provides information on management of manufacturing plants, ranging from manpower usage to utility usage.

2495. Plant Engineer's Handbook of Formulas, Charts, and Tables. Donald W. Moffat. Prentice-Hall, 1974. Cost: $24.95.

Includes information on different types of materials that would interest the plant engineer.

2496. Plant Engineering Handbook. William Staniar. McGraw-Hill, 2d ed., 1959, 2,391 pp. LC: 57-7244. ISBN: 07-060824-5. Date of First Edition: 1950. Cost: $64.50.

This handbook deals with the fundamental problems that are associated with plant engineering. For example, such subjects as reduction of vibrations, waste disposal, lubrication, corrosion problems, piping, and maintenance of boilers are discussed. In addition, such topics as welding, heating, refrigeration and air conditioning, and lubrication receive treatment. Industrial painting, power generation, power transmissions, and fire protection are further topics explored in this one-volume resource. For related work, please refer to such titles as *Management Handbook for Plant Engineers*, and *Maintenance Engineering Handbook*.

2497. Plant Health Handbook. Louis L. Pyenson. AVI Publishing Co, 1981, 241 pp. ISBN: 0-87055-377-1. Cost: $19.50.

Provides information on pesticides, lawn health, care of deciduous trees, evergreens, vegetable gardens, and shrubs.

2498. Plant Manager's and Safety and Pollution Manager's Emergency Manual. International Technical Information Institute, 450 pp. Cost: $70.00.

This emergency manual is a useful guide for such situations as fires, explosions, accidental emissions, toxic emissions, hazardous substances, rescue operations, and the management of accidents. Discusses such topics as avoiding damage to facilities and machinery, disposal of solid waste, and handling of power failures.

2499. Plant Manager's Handbook. Charles H. Becker. Prentice-Hall, 1974, 304 pp. ISBN: 0-13-680694-5. Cost: $24.95.

2500. Plastering: Practical Handbook. J.B. Taylor. International Ideas, Inc, 3d ed., 1980. ISBN: 0-7114-5588-0. Cost: $28.95.

This is a concise guide to the use of plasters and the practical art of plastering.

2501. Plastic and Swelling Properties of Illinois Coals. Orin Wainwright Rees; E.D. Pierron. Illinois State Geological Survey, 1954, 11 pp.

This is primarily a report on the results of the testing of Illinois coals.

Plastic Applications Handbook. *See* **Practical Guide to Plastics Applications.**

2502. Plastics 1980. International Plastics Selector, Inc, 2 Vols, Book A, 150 pp.; Book B, 1,400 pp. Cost: $72.00. Series: *Desk Top Data Bank.*

This 2-volume set includes as many as 23 different properties for 150 generic plastics. Bar charts are plotted for comparing property ranges of such properties as tensile strength and flexural modulus.

2503. Plastics Additives Handbook. Gaechter and Muller. Carl Hanser Verlag, 1982, approx. 600 pp. ISBN: 0-02-949430-3. Cost: $59.00.

Provides information on plastics additives for the manufacturing of plastics.

2504. Plastics and Their Properties. Hans Domininghaus. VDI-Verlag, Dusseldorf, Germany, 1976, 587 pp. Cost: DM 68.

Information is presented on the properties of plastics.

2505. Plastics Engineer's Data Book. Alan Birkett Glanvill. Industrial Press, 1971, 216 pp. LC: 73-10251. ISBN: 0-8311-1105-4. Cost: $17.00.

Covers injection moulding, thermostat moulding, extrusion processes, blow moulding, and machining.

2506. Plastics Engineering Handbook of the Society of the Plastics Industry, Inc. Joel Frados. Van Nostrand Reinhold, 4th ed., 1976, 1,000 pp. LC: 75-26508. ISBN: 0-442-22469-9. Date of First Edition: 1947. Cost: $42.50.

This handbook includes information on the various properties of plastics, different methods of molding plastics, different types of coatings that are put on plastics, and various tests that plastics undergo to provide performance information, including many tests of the American Society for Testing and Materials. Such topics as adhesive bonding, reinforced plastics, injection molding, extrusion, and cold stamping are discussed. A glossary is provided on the latest terms in the plastics industry, and the handbook includes an index and bibliographic references. For related titles, please refer to such handbooks as the *Polymer Handbook*; *The Plastics Industry Safety Handbook*; the *SPI Handbook of Technolgy and Engineering of Reinforced Plastics/Composites*; and the *Flammability Handbook for Plastics*.

2507. Plastics Extrusion Technology Handbook. Sidney Levy. Industrial Press, 1981, 300 pp. ISBN: 0-8311-1095-3. Cost: $27.50.

This book emphasizes the tools, equipment, and auxiliaries that are used in the process of plastics extrusion. In addition, such topics as foam extrusion, coextrusion, nonwire design, and plant layout are described in this handbook.

2508. Plastics for Electronics. International Plastics Selector, Inc, 1979, 937 pp. Cost: $29.45.

Provides information on over 2,000 plastics that are used in wire coating, encapsulating, circuit boards, switches, circuit breakers, insulation, and other types of electronic applications. Over 150 manufacturers are listed. Tables on flammability data are also presented.

2509. Plastics Industry Safety Handbook. D.V. Rosato. Society of the Plastics Industry (Cahners), 1973, 333 pp. LC: 72-91982. ISBN: 0-8436-1207-X. Cost: $15.95.

Covers fire protection materials handling, milling, laminating, coating, and casting.

2510. Plastics Manufacturing Handbook and Buyers Guide. Herman Publishing, annual, 250 pp. ISBN: 0-89047-011-1. Cost: $22.95.

Such topics as injection molding, extrusion systems, reinforced plastics systems, urethane systems, structural foam, thermoforming systems are discussed in this guide. In addition, such topics as chemicals used in the manufacturing of plastics are presented along with a discussion of welding and sealing systems, plastics additives, and compounding and mixing equipment. A directory of suppliers of plastics technology is also included in this guide.

2511. Plastics Materials Handbook. A.S. Athalye. Multi-Tech Publishing Co, 1980, 288 pp. Cost: $45.00.

This handbook provides 130 tables listing properties of plastics end-products. Such topics as machining plastics, adhesives for plastics, bonding of plastics to plastics, processing of plastics, and thermoplastics materials are discussed. Various materials such as acetal resins, acrylics, nylon, polystyrenes, PVC, polyvinylidene chloride, polyethylene, polypropylene, polycarbonate, polyester resins, polyurethane, and polytetrafluoroethylene are presented. Such thermosetting materials as epoxy resins, furan resins, ionomers, alkyd resins, amino resins, urea base compounds, and phenolic molding compounds are also discussed in this handbook.

2512. Plastics Mold Engineering Handbook. J. Harry DuBois; Wayne I. Pribble. Van Nostrand Reinhold, 3d ed., 1978, 653 pp. LC: 77-9911. ISBN: 0-442-22180-0. Cost: $39.50.

Design of molds is discussed in this handbook, and such topics as injection molding, sheet extrusion die design, structural foam, and extrusion-blow molds receive treatment.

2513. Plastics Product Design Engineering Handbook. Sidney Levy; J. Harry DuBois. Van Nostrand Reinhold, 1977, 342 pp. LC: 77-1044. ISBN: 0-442-24764-8. Cost: $28.95.

Areas covered include stress and strain behavior effects of fillers, structural design, and dynamic load response.

2514. Plastics Products Design Handbook. Part A: Materials and Components. Edward Miller. Marcel Dekker, Inc, 1981, 624 pp. ISBN: 0-82471339-7. Cost: $69.00. Series: *Mechanical Engineering Series*, Vol. 8.

This book was sponsored by the Society of Plastics Engineers and covers such information as reinforced plastics, flame retardants, vibration dampers, PVC piping, computer design, and mechanical properties of polymers. Part B of this handbook will

eventually be published and will emphasize processing and design of plastics.

2515. Plastics Tooling and Manufacturing Handbook: A Reference Book of the Use of Plastics as Engineering Materials for Tool and Workpiece Fabrication. American Society of Tool and Manufacturing Engineers. Prentice-Hall, 1965.

2516. Plastics. International Plastics Selector, Inc, 3 Vols; 2,800 pp. Cost: $69.95.

This 3-volume set is part of the *Desk-top Data Bank Series* and provides such information as chemical descriptions of plastics, bar charts, tensile strength, volume resistivity, dielectric strength, and flexural modulus. In addition, reinforced plastics, as well as flame retardant plastics are discussed.

2517. Platt's Oil Price Handbook and Oilmanac. 56th ed., 1979, 250 pp. Cost: $85.00.

Provides prices of gasoline and distillates for 48 U.S. cities. The prices are for 1979 and are listed by day, by week, and by month with averages, lows, and highs. In addition, European bulk prices and OPEC prices are included.

2518. Plumbers Handbook. Joseph P. Almond. Audel, 5th ed., 1979, 231 pp. LC: 79-64811. ISBN: 0-672-23339-8. Cost: $8.95.

This is an excellent guide for the installation of plumbing in residential dwellings and industrial buildings. Such topics as septic tanks, garbage disposals, grease traps, faucets, valves, cast iron soil pipe, and drainage systems are discussed in this handbook.

2519. Plumbing Estimating Handbook. Joseph J. Galeno. Van Nostrand Reinhold, 1976, 256 pp. LC: 76-57182. ISBN: 0-442-12157-1. Cost: $19.95.

Covers pricing techniques for standard building and site work plumbing services.

2520. Plutonium Handbook: A Guide to the Technology. Oswald J. Wick. Gordon and Breach (prepared under the auspices of the U.S. Atomic Energy Commission), 1967, 2-Vol. set. LC: 65-27512. ISBN: Vol.1, 0-677-01200-4; Vol.2, 0-677-01730-8. Cost: Vol.1, $64.00; Vol.2, $55.00.

Available from American Nuclear Society for $98.00 in one volume, 992 pp. Covers metallurgy, plutonium, properties of plutonium, chemical processing, and health and safety in radiation exposure.

2521. Pneumatic Data. Volume 2; Volume 3. Institute for Power Systems. Renouf, 1979, 2 Vols, Vol.2, 130 pp.; Vol.3, 100 pp. ISBN: Vol.2, 0-85461-012-X; Vol.3, 0-85461-069-3. Cost: Vol.2, $30.00; Vol.3, $21.00.

This handy collection of tables covers such topics as pressures, compressed air, pressure drop, discharge, flow rates, cylinder data, drop output, and conversion factors.

2522. Pneumatic Handbook. Trade and Technical Press, Ltd, 5th ed., 1978, 700 pp. ISBN: 0-85461-068-5. Cost: $90.00.

Covers compressed air, compressors, heat exchangers, air tools, vacuum pumps, pressure gauges and includes over 2,000 diagrams and tables.

2523. Pneumatic Technical Data. Trade and Technical Press, Ltd, multivolume set. ISBN: 0-85461-012-X. Cost: £3.60.

2524. Pocket Color Atlas of Dermatology. J. Kimmig; M. Janner; H. Goldschmidt. Georg Thieme, (distributed in the U.S. by Year Book Medical Publishers), 1975, 218 pp. ISBN: 3-13-511301-9. Cost: DM 28.

Various color photographs and figures for conditions related to skin disorders are provided.

2525. Pocketbook of Statistical Tables. Robert E. Odeh et al. Marcel Dekker, Inc, 1977, 184 pp. ISBN: 0-8247-6515-X. Cost: $8.75.

Includes information on normal distribution, chi-square dsitribution, Poisson distribution, and nonparametric tolerance limits.

2526. Polar Regions Atlas. Central Intelligence Agency. Superintendent of Documents, 1978, 66 pp. SuDoc: PrEx 3.10/4:P 75/2. Cost: $5.00.

Numerous illustrations in color and black-and-white are used to present information on such topics as sea ice, climate, permafrost, fisheries, mining, oil, gas, and transportation in the polar regions.

2527. Policy on Design of Urban Highways and Arterial Streets. American Association of State Highway and Transportation Officials, rev. ed., 1973, 740 pp. ISBN: 0-686-20949-4. Date of First Edition: 1957. Cost: 11.00.

2528. Policy on Geometric Design of Rural Highways. American Association of State Highway and Transportation Officials, 2d ed., 1965, 650 pp. Date of First Edition: 1954.

2529. Pollutant Removal Handbook. Marshall Sittig. Noyes Data Corp, 1973.

Although this book is somewhat dated, it provides a substantial amount of information on pollution removal, most of which has been obtained from patents. This book has been updated by a more recent book by Marshall Sittig entitled *How to Remove Pollutants and Toxic Materials from Air and Water; A Practical Guide*, 1977.

2530. Pollution Detection and Monitoring Handbook. Marshall Sittig. Noyes Data Corp, 1974, 401 pp. LC: 74-75905. ISBN: 0-8155-0529-9. Cost: $36.00.

Covers ambient quality standards, instrumentation, sampling, and emission standards.

2531. Pollution Engineering Practice Handbook. N. Cheremisinoff; Richard A. Young. Ann Arbor Science Publishers, Inc, 1975, 1,073 pp. LC: 74-14427. ISBN: 0-250-40075-8. Cost: $39.95.

This handbook covers such subjects as air pollution control, water pollution, noise pollution, solid waste disposal, and radioactive waste. Information is provided on odor control, wet scrubbers, filtration of waste water, cooling towers, corrosion, landfills, dumping of wastes in the ocean, and incineration. Many tables, charts, and figures are provided in the handbook, in addition to bibliographies of the literature, conversion factors, and a well-prepared index. For related works, please refer to such titles as *Environmental Engineers' Handbook*; *Handbook of Solid Waste Disposal*; *Handbook for the Operation and Maintenance of Air Pollution Control Equipment*; *Handbook of Industrial Noise Management*; *Handbook of Industrial Noise Control*; *Handbook of Environmental Civil Engineering*; *Betz Handbook of Industrial Water Conditioning*; the *CRC Handbook of Environmental Control*; *Air Pollution*; and *Air Pollution Engineering Manual*.

2532. Polymer Handbook. J. Brandrup; E.H. Immergut. Wiley-Interscience, 2d ed., 1975, 1,440 pp. (various paging). LC: 74-11381. ISBN: 0-471-09804-3. Date of First Edition: 1966. Cost: $47.95.

This handbook contains a wealth of information and tables on various properties of polymers such and heat capacity, refractive indices, rate of crystallization, and permeability coefficients. In addition, such data as relative viscosities, fractionation, and dipole moments of various polymers are included. Bibliographic references are presented so that the reader can consult the original articles from which the tables have been prepared. The major emphasis of this handbook is on experimental and theoretical polymer information. Physical constants are provided for several important polymers such as polyvinyl chloride and polyethylene. A subject index is included to help the user retrieve the informations/he is seeking. For related works, please refer to the *Plastics Engineering Handbook*.

2533. Polymer Science: A Materials Science Handbook. Aubrey Dennis Jenkins ed. North Holland/Elsevier, 1972, 2 Vols. ISBN: 0-7204-0245-X; 0-444-10355-4. Cost: $185.50.

2534. Popular Tube/Transistor Substitution Guide. TAB Books, 3d ed., 1971, 256 pp. LC: 74-105968. ISBN: 0-8306-9570-2. Cost: $7.95.

2535. Potentiometer Handbook; User's Guide to Cost-Effective Applications. Carl David Todd (Bourns, Inc). McGraw-Hill, 1975, 300 pp. LC: 75-20010. ISBN: 0-07-006690-6. Cost: $16.50.

2536. Powder Diffraction Data. Marlene Morris. Joint Committee on Powder Diffraction Standard, 1st ed., 1976, 2 Vols, 440 pp. LC: 76-151361. Cost: $150.00.

This 2-volume set is entited as follows: *Data Book* and *Search Manual*. The *Data Book* is composed of 949 images which represent 938 inorganic materials and 11 organic materials. Such information as *D* spacings and intensitites, composition, cell dimensions, space group, calculated density, optical data, isostructural relationship, and crystal systems are presented. The *Search Manual* helps gain access to the *Data Book* by such approaches as mineral name, organic formula, Hanawalt numerical section, and Fink numerical section.

2537. Power Cartridge Handbook. Naval Air Systems Command. NTIS, 4th ed., 1973, 326 pp. Accession: AD-775861 (Supersedes AD-660390). Tech Report: NAVAIR-7836.

2538. Power Semiconductor D.A.T.A. Book. Derivation and Tabulation Associates, Inc, continuation issued semiannually, pages vary per issue. Cost: $47.50 per year by supscription.

2539. Power Semiconductor Data Book of Design Engineers. Texas Instruments Inc, Components Group, Enginnering Staff, 816 pp. (various paging). Cost: $3.95.

2540. Power Supply Handbook. *73 Magazine* eds (available from TAB Books), 1979, 420 pp. LC: 78-31272. ISBN: 0-8306-9806-X. Cost: $15.95.

Such topics as DC to AC converters, battery chargers, high voltage power supplies, and sensing circuits are presented in this handbook.

2541. Practical Formulas for Hobby or Profit. Henry Goldschmiedt. Chemical Publishing Co, 1973, 512 pp. ISBN: 0-8206-0235-3. Cost: $17.00.

This guide presents 512 pages of practical formulas that are useful in the preparation of chemical products for the home experimenter and hobbyist.

2542. Practical Guide to Plastics Applications. Edward G. Crosby; Stephen N. Kochis. Cahners, 1972, 208 pp. LC: 77-156477. ISBN: 0-8436-1025-3. Cost: $14.95.

2543. Practical Handbook of Industrial Traffic Management. Richard C. Colton; Edmund S. Ward. Traffic Service Corp, 4th ed., 1965, 583 pp. LC: 65-27145.

2544. Practical Handbook of Low-Cost Electronic Test Equipment. Robert Genn Jr. Prentice-Hall, Inc (a Parker Publication), 1978, 224 pp. ISBN: 0-13-691071-8. Cost: $14.95.

This book contains numerous illustrations on how to use such test equipment as oscilloscopes and multimeters. In addition, this guide also presents information on how to construct low-cost test equipment.

2545. Practical Handbook of Solid State Troubleshooting. Robert Genn Jr. Parker Publishing Co, Inc. Cost: $14.95.

This handbook discusses such topics as troubleshooting integrated circuits, solid state t.v. circuits, and solid state regulated power supplies. In addition, such topics as servicing solid state radio receivers, solid state amplifiers, replacing bipolar transistors, and troubleshooting digital circuits are covered.

2546. Practical Inventor's Handbook. Orville Greene; Frank Durr. McGraw-Hill, 2d ed., 1979, 352 pp. LC: 78-26666. ISBN: 0-07-024320-4. Cost: $19.95.

This guide provides information on securing patents for inventions and describes major patents in several different areas, such as security equipment, tools, and kitchen utensils.

2547. Practical Oscilloscope Handbook. Rufus Turner; Howard Bierman; Paul Bierman. Hayden, 2d ed., 1981, 192 pp. ISBN: 0-8104-0851-1. Cost: $8.95.

This is basically a guide to the set-up display, and interpretation of waveforms. Such topics as voltage measurements, current measurements, phase measurements, audio amplifiers, receivers, and transmitters are discussed in this useful handbook.

2548. Practical Petroleum Engineering Handbook. Joseph Zaba; W.T. Doherty. Gulf Publishing Co, 5th ed., 1970, 949 pp. LC: 71-30368. ISBN: 0-87201-744-3. Cost: $39.95.

2549. Practical Semiconductor Data Book for Electronic Engineers and Technicians. John D. Lenk. Prentice-Hall, 1970, 260 pp. LC: 70-100085. ISBN: 0-13-693788-8. Cost: $15.95.

2550. Practicing Scientist's Handbook: A Guide for Physical and Terrestrial Scientists and Engineers. Alfred J. Moses. Van Nostrand Reinhold, 1977, 1,302 pp. LC: 77-5866. ISBN: 0-442-25584-5. Cost: $54.50.

The emphasis of this title is on property data. For example, it covers optical properties of materials, acoustic properties, magnetic properties, electrical properties, mechanical properties, and physical properties. Semiconductors, superconductors, glasses, polymers, organic, and inorganic properties are listed.

2551. Practitioner's Handbook of Ambulatory OB/GYN. J.D. Neeson; C.R. Stockdale. John Wiley and Sons, 1981, 408 pp. LC: 80-26151. ISBN: 0-471-05670-7. Cost: $18.95.

This handbook provides useful information in the area of maternity nursing.

2552. Pratt Guide: A Citizen's Handbook of Housing, Planning and Urban Renewal Procedures in New York City. R. Alpern. Pratt Institute, 1965.

Information on city planning for New York City with emphasis on urban renewal is provided.

2553. Preliminary Report on the Thermodynamic Properties of Lithium, Beryllium, Magnesium, Aluminum and Their Compounds with Hydrogen, Oxygen, Nitrogen, Fluorine, and Chlorine. T.B. Douglas; C.W. Beckett. NTIS, 1959, 305 pp. Accession: N79-75648 PB-287588. Tech Report: NBS-6484 TRS-2.

Such data as heat of formation and heat of vaporization and other thermodynamic properties are presented for lithium, beryllium, magnesium, and aluminum.

2554. Press Brake and Shear Handbook: A Basic Handbook on the Design, Selection, and Use of Press Brakes and Shears. Harold R. Daniels. Cahners, 2d ed., 1974, 184 pp. LC: 74-13643. ISBN: 0-8436-0815-3. Cost: $12.95.

Written as a companion volume to the *Mechanical Press Handbook*.

2555. Pressure Vessel Design Handbook. Henry H. Bednar. Van Nostrand Reinhold, 1981, 336 pp. ISBN: 0-442-25416-4. Cost: $26.50.

Design information is presented for pressure vessels, including the types of materials to be used, and design computations. Properties of typical pressure vessel constructioon materials are described.

2556. Pressure Vessel Handbook. Eugene F. Megyesy. Pressure Vessel Handbook Publishing, Inc, 4th ed., 1977, 415 pp. LC: 77-76432. ISBN: 0-914458-04-3. Date of First Edition: 1972. Cost: $24.50.

Covers such topics as construction, wind load, seismic load, deflection of towers, piping systems, and atmospheric tanks. Includes over 280 illustrations and 80 tables.

2557. Prestressed Concrete Designer's Handbook. P.W. Abeles et al. Cement and Concrete Association (available from Scholium International, Inc), 3d ed., 1981, 545 pp. ISBN: 0-7120-1227-2. Date of First Edition: 1962. Cost: $42.00.

This handbook includes many tables and charts that help to present the latest information on prestressed concrete. Such topics as fire resistance, load factors, deflection, and light weight concrete are discussed. Many examples of the uses of prestressed concrete are presented for different structures such as floors, roofs, bridges, and piles. Design information is provided in SI units and imperial units, and an index has been included for the convenience of the user. For related works, please refer to the following titles: *ACI Manual of Concrete Practice*, the *Handbook of Concrete Engineering*, *Prestressed Precast Concrete Design Handbook for Standard Products*, and the *Concrete Manual*.

Prestressed Concrete Institute. *See* **PCI.**

2558. Prestressed Precast Concrete Design Handbook for Standard Products. Concrete Materials, Inc, 1969, 628 pp.

Design information and specifications are provided in this handbook on concrete materials, products, and connections. Various tables are included such as load tables, as well as information on deflection curves and prestressed concrete properties. The terminology is well defined, and an index is provided for the user's convenience. The thumb index should make this handbook easy to use. For related titles, please refer to such handbooks as the *ACI Manual of Concrete Practice*; *Prestessed Concrete Designer's Manual*; the *Concrete Manual*; and the *Handbook of Concrete Engineering*.

2559. Preventive Dentistry: Dental Practitioner Handbook No. 22. John O. Forrest. Year Book Medical Publishers, 1976, 128 pp. ISBN: 0-8151-3266-2. Cost: $8.95 paper.

2560. Printed Circuits Handbook. C.F. Coombs Jr. McGraw-Hill, 2d ed., 1979, 704 pp. LC: 78-16800. ISBN: 0-07-012608-9. Cost: $32.50.

Covers repair of circuit boards, dry film, photo-resist, and cleanliness.

2561. Probability Charts for Decision Making. J.R. King. Industrial Press, 1971, 290 pp. ISBN: 0-8311-1023-6. Cost: $32.00.

Information is provided to assist managers in the interpretation of such probability charts and statistical data as logarithmic normal distribution, log-normal distribution, extreme value distribution, Weibull distribution data, reciprocal functions, gamma distribution, Chi-Square distribution, and binomial functions.

2562. Probability Tables for the Analysis of Extreme-Value Data. National Bureau of Standards. NTIS, 1953, 32 pp. Accession: PB-192337. Cost: $4.00. Series: NBS AMS22.

2563. Problems for the Numerical Analysis of the Future. National Bureau of Standards. NTIS, 1951, 21 pp. Accession: PB-251866. Cost: $3.50. Series: NBS AMS15.

2564. Procedure Handbook of Arc Welding. Lincoln Electric Co. Conquest Publications, 12th ed., 1973, 750 pp. Cost: $20.00.

Various methods and procedures are listed, including the welding of such materials as carbon steel and low alloy steel, cast iron and steel, aluminum and aluminum alloys, copper and copper alloys, and stainless steel.

2565. Process Engineer's Pocket Handbook. Carl Branan. Gulf Publishing Co, 1976, 136 pp. LC: 76-1680. ISBN: 0-87201-712-5.

Process engineering data are presented in this handbook. Such information as vacuum systems, pneumatic conveying, heat ex-

changers, air-cooled exchangers, compressors, pumps, and steam turbines is also included. In addition, information is presented on distillation, McCabe-Thiele diagram, absorption, hydrocarbon absorption, inorganic absorption, tank blending, and separators and accumulators such as gas scrubbers, liquid separators, and reflux drums.

2566. Process Instrumentation Manifolds: Their Selection and Use, A Handbook. John E. Hewson. Instrument Society of America, 1980, 320 pp. ISBN: 0-87664-447-7. Cost: $35.00.

This book includes over 300 drawings and photographs of process instrumentation manifolds. Such topics as using instruments to measure flow and static pressure, density, and level by differential pressure are covered.

2567. Process Instruments and Controls Handbook. D.M. Considine; S.D. Ross. McGraw-Hill, 2d ed., 1974, 1,344 pp. LC: 73-18261. ISBN: 0-07-012428-0. Date of First Edition: 1957. Cost: $49.50.

2568. Process Plant and Equipment Cost Estimation. O.R. Khabanda. Craftsman Book Co, 1979, 240 pp. Cost: $19.00.

Provides information on typical equipment and plant costs for over 100 chemicals. Selection of alternate materials and cost estimating are provided for construction of chemical processing plants.

2569. Process Plant Estimating, Evaluation and Control. Kenneth M. Guthrie. Craftsman Book Co, 1974, 608 pp. LC: 74-3106. Cost: $25.00.

Over 1,000 charts and diagrams are presented which help to describe material and equipment cost for refining and chemical process plants. In addition, pipe estimating data and computer cost systems are discussed.

2570. Product Safety and Liability: A Desk Reference. John Kolb; Steven S. Ross. McGraw-Hill, 1979, 832 pp. ISBN: 0-07-035380-8. Cost: $29.50.

Product safety is discussed in this publication, which includes numerous topics such as chemical products, mechanical products, electrical products, and their corresponding safety requirements and liability insurance.

2571. Production Handbook. Gordon B. Carson; Harold A. Bolz; Hewitt H. Young. Ronald Press Co, 3d ed., 1972, various paging. LC: 71-137775. ISBN: 0-8260-1820-3. Date of First Edition: 1944. Cost: $45.95.

Covers materials, management, inventory control, quality control, reliability, and operations research.

2572. Production Manager's Handbook of Formulas and Tables. L. Zeyher. Prentice-Hall, 1972. ISBN: 0-13-724427-4. Cost: $27.95.

2573. Production Processes: The Producibility Handbook. R.W. Bolz. Industrial Press, 5th ed., 1977 (rep. 1981), 1.089 pp. ISBN: 0-8311-1088-0. Cost: $45.00.

Covers metal removal methods, metal forming methods, metal working and forging methods, casting methods, finishing, fabricating, and molding.

2574. Programmes for Animation: Handbook for Animation Technicians. Brian Salt. Pergamon, 1978, 382 pp. LC: 78-40284. ISBN: 0-08-023153-5. Cost: $150.00.

Computer programs are included that deal with such topics as movements along a curve, circular pans, exponential movements, parabolic fairings, and movements with a constant acceleration.

2575. Promethium Isotopic Power Data Sheets. J.H. Jarrett; H.H. Van Tuyl. NTIS, 1970, 50 pp. Accession: N70-39027. Tech Report: BNWL-1309.

Chemical properties and mechanical properties are presented for promethium isotopes.

2576. Properties and Applications of Glass. H. Rawson. Elsevier, 1980, 318 pp. ISBN: 0-444-41922-5. Cost: $56.00. Series: *Glass Science and Technology*, 3..

The thermal expansion of glass, viscosity of glass, strength of glass, refractive index, reflective properties, and chemical durability, as well as absorption radiation by glasses, are presented in this handbook.

2577. Properties and Selection of Tool Materials. American Society for Metals, 1975, 314 pp. Cost: $24.00.

Materials are discussed in terms of their usage for production tools, dies, molds, and gauges. Numerous tables are provided to assist the user in selecting the proper tool material.

2578. Properties of Air and Combustion Products With Kerosine and Hydrogen Fuels. B. Banes; R.W. McIntyre; J.A. Sims. NTIS, 1967, various paging. Accession: N68-23629: Vol. 1.

This multivolume set includes thermodynamic properties data and chemical equilibrium composition for various types of mixtures.

2579. Properties of Air-Steam Mixtures Containing Small Amounts of Iodine. J.G. Knudsen. NTIS, 1970, 89 pp. Accession: N70-38439. Tech Report: BNWL-1326.

Thermodynamic properties of air-steam mixtures containing small amounts of iodine, representative of water-cooled reactor atmosphere, are included in this publication. Such values as Prandtl number, temperature effects, molecular weights, and pressure effects are provided.

2580. Properties of Aluminum and Aluminum Alloys. CINDAS, 1973, 802 pp. Cost: $10.00.

This is an extensive collection of evaluated physical properties data on aluminum and aluminum alloys.

Properties of Arsenides, Phosphides, Selenides, Sulfides, and Tellurides. *See* **McGraw-Hill/CINDAS Data Series on Materials Properties, Volume VIII-3.**

Properties of Borides, Carbides, Hydrides, Nitrides, and Silicides. *See* **McGraw-Hill/CINDAS Data Series on Material Properties, Volume VIII-2.**

2581. Properties of Building Materials. H.J. Eldridge. Construction Press, Ltd, 1974, 121 pp. ISBN: 0-904406-58-X. Cost: £3.25.

Many materials are listed including bricks, plaster, asphalt, concrete, sealants, and tile, etc. Originally published as part of the *Construction Industry Handbook*.

Properties of Building Materials. *See* **McGraw-Hill/CINDAS Data Series on Material Properties, Volume IX-3.**

Properties of Carbonates, Nitrates, Phosphates, Silicates, and Sulfates. *See* **McGraw-Hill/CINDAS Data Series on Material Properties, Volume VIII-4.**

Properties of Ceramics and Glasses. *See* **McGraw-Hill/CINDAS Data Series on Material Properties, Volume VI-7.**

Properties of Cermets. *See* **McGraw-Hill/CINDAS Data Series on Material Properties, Volume IV-6.**

2582. Properties of Combined Beam and Plate for High-Strength Steels and Titanium. S.F. Zemanek; N.S. Nappi. NTIS, 1977, 391 pp. Accession: N78-19278 AD-A048989. Tech Report: DTNSRDC-77-0069. Cost: $3.50 microform; $27.50 hard copy.

Structural design data are presented with tabulations of built-up T-beams. This report was prepared by the Naval Ship Research and Development Center at Bethesda, Maryland.

2583. Properties of Combustion Gases Systems—H$_2$-Air. W.G. Browne; D.L. Warlick. General Electric Co, 1962, 240 pp. Accession: N69-72610.

Thermodynamic properties are provided for hydrogen-air mixtures.

Properties of Commercial Graphites and Carbon-carbon Composites. *See* **McGraw/CINDAS Data Series on Material Properties, Volume VII-1.**

Properties of Commercial Refrigerants and Fluid Mixtures. *See* **McGraw-Hill/CINDAS Data Series on Material Properties, Volume V-2.**

Properties of Complex Oxides. *See* **McGraw-Hill/CINDAS Data Series on Material Properties, Volume VI-5.**

Properties of Composites (other than carbon-carbon composites). *See* **McGraw-Hill/CINDAS Data Series on Material Properties, Volume VII-2.**

Properties of Copper Alloys and Silver Alloys. *See* **McGraw-Hill/CINDAS Data Series on Material Properties, Volume IV-5.**

2584. Properties of Diamond. J.E. Field. Academic Press, 1979, 674 pp. ISBN: 0-12-255350-0. Cost: $75.00.

Numerous authoritative articles are presented on such topics as optical properties of diamonds, nuclear properties of diamonds, adsorbability of diamonds, strength of diamonds, and fracture of diamonds. Tables of properties are listed for the convenience of the user.

2585. Properties of Electrodeposited Metals and Alloys; A Handbook. William H. Safranek. American Elsevier Publishing Co (sponsored by American Electroplaters' Society, Inc), 1974, 517 pp. LC: 73-9214. ISBN: 0-444-00140-9. Cost: $26.00.

Properties of Electronic Oxides. *See* McGraw-Hill/CINDAS Data Series on Material Properties, Volume VI-2.

2586. Properties of Gases and Liquids. R.C. Reid; J.M. Prausnitz; T.K. Sherwood. McGraw Hill, 3d ed., 1977, 627 pp. LC: 76-42204. ISBN: 0-07-051790-8. Cost: $27.50.

This book is highly recommended for chemical engineering collections. Much of the material is in tabular format and a bibliography is included.

2587. Properties of Glass. George W. Morey. American Chemical Society, 2d ed., 1954. LC: 54-7946. Cost: $72.20.

Numerous properties of glass are listed, such as dielectric constant, optical properties, electrical conductivity, magnetic properties, thermal endurance, hardness, viscosity, composition, annealing, surface tension, heat capacity, conductivity, elasticity, and devitrification.

Properties of Halides (bromides, chlorides, fluorides, and iodides). *See* McGraw-Hill/CINDAS Data Series on Material Properties, Volume VIII-1.

Properties of Inorganic and Organic Fluids. *See* McGraw-Hill/CINDAS Data Series on Material Properties, Volume V-1.

Properties of Intermetallic Compounds. *See* McGraw-Hill/CINDAS Data Series on Material Properties, Volume VIII-5.

2588. Properties of Liquid and Solid Hydrogen. B.N. Eselson et al. NTIS, 1971, 127 pp. Accession: N72-27745. Tech Report: TT-70-50179. Cost: $3.00 hard copy.

This is a translation of a Russian title that was originally published in 1969. Data are given on specific heat, viscosity, and velocity of sound over wide ranges of pressures and temperatures for liquid and solid hydrogen.

Properties of Liquid Metal Elements. *See* McGraw-Hill/CINDAS Data Series on Material Properties, Volume III-4.

2589. Properties of Materials at Low Temperatures. V.J. Johnson. Pergamon Press, 1961. ISBN: 0-08-009617-4. Cost: $50.00.

2590. Properties of Materials. C. Vee Yong Chong. MacDonald and Evans, 1977, 318 pp. LC: 77-368446. Cost: $3.00. Series: *MacDonald and Evans Handbook Series*.

Provides information on the properties of materials that are used in the construction industry.

2591. Properties of Nonmetallic Fluid Elements. Y.S. Touloukian; C.Y. Ho. McGraw-Hill, 1981, 224 pp. LC: 80-18139. ISBN: 0-07-065033-0. Cost: $32.50. Series: *McGraw-Hill/CINDAS Data Series on Material Properties*, Vol. III-2.

Such properties as specific heat at constant pressure, thermal conductivity, viscosity, Prandtl number, density, dielectric constant, heat of fusion, vapor pressure, heat of vaporization, and melting point are presented for the nonmetallic fluid elements. For example, such elements as bromine, chlorine, deuterium, fluorine, helium-3, helium-4, hydrogen, iodine, krypton, neon, oxygen, ozone, radon, tritium, and xenon are listed in this compilation.

Properties of Nonstainless Alloy Steels, Carbon Steels, and Cast Irons. *See* McGraw-Hill/CINDAS Data Series on Material Properties, Volume IV-2.

2592. Properties of Ordinary Water-Substance in all its Phases. Noah Ernest Dorsey. American Chemical Society, 1940, 673 pp. LC: 40-6778. ISBN: 0-8412-0255-9. Cost: $48.50.

Includes water vapor, water, ice, and phase transition of water. This property data represents an update of the International Critical Tables.

Properties of Organic Compounds, Foods, and Biological Materials. *See* McGraw-Hill/CINDAS Data Series on Material Properties, Vol. IX-2.

Properties of Oxides Mixtures. *See* McGraw-Hill/CINDAS Data Series on Material Properties, Volume VI-6.

2593. Properties of Polymers. D.W. Van Krevelen. Elsevier-North Holland, 2d ed., 1976, 620 pp. ISBN: 0-444-41467-3. Cost: $119.50.

Transport properties of polymers, thermophysical properties of polymers, and numerous tables listing additional properties of polymers are presented in this book.

Properties of Polymers. *See* McGraw-Hill/CINDAS Data Series on Material Properties, Volume IX-1.

2594. Properties of Principle Cryogenics. Aerojet-General Corp. NTIS, 1965, 112 pp. Accession: N76-75869. Tech Report: AGC-9050-111-65.

Properties are listed for hydrogen, fluorine, nitrogen, and oxygen.

Properties of Rare-earth and Radioactive Elements. *See* McGraw-Hill/CINDAS Data Series on Material Properties, Volume III-6.

Properties of Rare-earth Oxides and Actinide Oxides. See *McGraw-Hill/CINDAS Data Series on Material Properties, Volume VI-1.*

2595. Properties of Refractory Materials. S.J. Burnett. NTIS, 1969, 326 pp. Accession: N69-78717. Tech Report: AERE-R-4657.

Mechanical properties and various graphs and charts are presented for refractory materials.

2596. Properties of Selected Ferrous Alloying Elements. Y.S. Touloukian; C.Y. Ho. McGraw-Hill, 1981, 288 pp. LC: 80-18072. ISBN: 0-07-065034-9. Cost: $37.50. Series: *McGraw-Hill/CINDAS Data Series on Material Properties*, Vol. III-1.

Such properties as thermal conductivity, specific heat, thermal linear expansion, thermal diffusivity, thermal emittance, thermal reflectance, thermal absorptance, thermal transmittance, viscosity, and electrical resistivity are listed for such elements as chromium, cobalt, iron, manganese, nickel, and vanadium.

Properties of Selected Nonferrous Alloying Elements and Precious Metals. *See* McGraw-Hill/CINDAS Data Series on Material Properties. Volume III-5.

Properties of Selected Nontransition-Metal Alloys. *See* McGraw-Hill/CINDAS Data Series on Material Properties. Volume IV-4.

Properties of Selected Nontransition-Metal Oxides. *See* McGraw-Hill/CINDAS Data Series on Material Properties. Volume VI-3.

Properties of Selected Refractory Elements. *See* McGraw-Hill/CINDAS Data Series on Material Properties. Volume III-3.

2597. Properties of Selected Rocket Propellants, Volume 1. M. George. NTIS, 1976, 261 pp. Accession: N76-73553. Tech Report: AD-450926 D2-11677.

Thermodynamic and chemical properties of beryllium, boranes, fluorine, and liquid oxygen are presented in tabular format.

Properties of Selected Semiconducting, Semimetallic, Nonmetallic Solid, and Other Elements. *See* McGraw-Hill/CINDAS Data Series on Material Properties. Volume III-7.

2598. Properties of Selected Superconductive Materials—1974 Supplement. B.W. Roberts (National Bureau of Standards). U.S. Government Printing Office, 1974. SuDoc: C13.46:825. Cost: $1.25. Series: NBS Tech. Note 825.

2599. Properties of Selected Superconductive Materials. B.W. Roberts (National Bureau of Standards). U.S. Government Printing Office, 1972. SuDoc: C13.46:724. Cost: $1.40. Series: NBS Tech. Note 724.

2600. Properties of Selected Superconductive Materials. B.W. Roberts. NTIS, 1978, 103 pp. Accession: PB-287013/7; N79-74281. Tech Report: NBS-TN-983; NBS-TN-825.

This report supersedes National Bureau of Standards-Technical Note-825. Provides mechanical properties data and cryogenic information for superconductors, superconducting magnets, and thin films.

Properties of Selected Transition-Metal Alloys. *See* McGraw-Hill/CINDAS Data Series on Material Properties. Volume IV-3.

Properties of Selected Transition-Metal Oxides and Oxides of Selected Nonmetallic Solid Elements. *See* McGraw-Hill/CINDAS Data Series on Material Properties. Volume VI-4.

2601. Properties of Silica. Robert B. Sosman. American Chemical Society, 1927, 860 pp. LC: 28-585. Cost: $105.00 library bound; $95.00 soft bound; $47.50 microfilm.

The structure of silica is discussed along with silica in magnetic fields, in electric fields, and in different phases and transformations.

Properties of Stainless Steels. *See* McGraw-Hill/CINDAS Data Series on Material Properties. Volume IV-1.

2602. Properties of Steel Weldments for Elevated Temperature Pressure Containment Applications. G.V. Smith. American Society of Mechanical Engineers, 1978, 205 pp. Cost: $30.00; $15.00 members.

Stress rupture properties are listed, in addition to thermal stress analysis and stress corrosion cracking behavior of various types of welds.

Properties of Systems. *See* McGraw-Hill/CINDAS Data Series on Material Properties. Volume VII-3.

2603. Properties of Water and Steam in SI Units: K J, BAR, 0-800° C. 0-1000 BAR. E. Schmidt. Springer-Verlag, 2d ed., 1979, 190 pp. ISBN: 0-387-09601. Cost: $40.20.

2604. Property Index to NSRDS Data Compilations, 1964-1972. David R. Lide Jr; Gertrude B. Sherwood; Charles H. Douglass Jr; Herman M. Weisman (National Bureau of Standards). U.S. Government Printing Office, 1975. SuDoc: C13.48:55. Cost: $0.85. Series: NSRDS-NBS-55.

2605. Protection Against Betatron-Synchrotron Radiations Up to 100 Million Electron Volts. National Bureau of Standards. U.S. Government Printing Office, Feb. 26, 1954. SuDoc: C13.11:55. Cost: $0.75. Series: Handbook 55.

2606. Protection Against Neutron Radiation Up to 30 MeV. National Bureau of Standards. NCRP Publications. Cost: $5.00. Series: NBS Handbook 63.

Available from NCRP Publications, PO Box 30175, Washington, DC 20014 as NCRP Report 38.

2607. Protection Against Radiation from Sealed Gamma Sources. National Bureau of Standards. NCRP Publications. Cost: $4.00. Series: Handbook 73 (Supercedes H54).

Available from NCRP Publications, PO Box 30175, Washington, DC 20014 as NCRP Report 40.

2608. Provisional Thermodynamic Functions for Para-Hydrogen. H.M. Roder; R.D. Goodwin. NTIS, 1961, 157 pp. Accession: PB-161631; N7570798. Tech Report: NBS-TN-130; NASA-CR-140919.

Thermodynamic properties of para-hydrogen are listed in tabular format.

2609. Provisional Values for the Thermodynamic Functions of Ethane. R.D. Goodwin. NTIS, 1974, 343 pp. Accession: N75-25754. Tech Report: NBSIR-74-398; COM-75-10130. Cost: $9.50.

This book was prepared by the National Bureau of Standards under partial sponsorship of the American Gas Association. Listed are thermodynamic properties, equations of state, specific heat, enthalpy, and entropy of ethane.

2610. Psychiatric Disorders in Old Age; A Handbook for the Clincial Team. Tony Whitehead. HM & M, 1979, 124. Cost: £3.75.

This handbook is designed for those people who are providing service to those elderly suffering from psychiatric illnesses.

2611. Pump Handbook. Igor J. Karassik et al. McGraw-Hill, 1976, 1,102 pp. LC: 75-22343. ISBN: 0-07-033301-7. Cost: $41.50.

This handbook, which has over 1,000 illustrations, contains information of different types of pumps including jet pumps, centrifugal pumps, rotary screw, steam, and power pumps. It contains articles written by experts in their particular specialization and includes such diverse subjects as noise and vibration in pumping operations, selection, operation, and installation of pumping systems, and various types of valves and controls used in pumping systems. Of interest to engineers will be the section on the usage of pumps in various applications and industries. This handbook contains references cited throughout the text and a considerable amount of tabulated data on pumps and pumping applications. For related works, please refer to such titles as the *ASPE 1971 Data Book*; *ASHRAE Handbook and Product Directory*; the *Piping Handbook*; and the various general mechanical engineering handbooks.

2612. Pump Selection—A Consulting Engineer's Manual. Rodger Walker. Ann Arbor Science Publishers, First ed., 1979, 128 pp. LC: 72-88891. ISBN: 0-250-40005-7. Date of First Edition: 1972. Cost: $12.50.

Includes information on pumping equipment, performance curves of various types of pumps, data sheets on maintenance of pumping equipment, and tables on properties of water, atmospheric pressures, comparison of engine fuels, and velocity.

2613. Pump Users' Handbook. British Pump Manufacturers' Association. Trade and Technical Press, Ltd, 250 pp. ISBN: 0-85461-070-7. Cost: $32.95.

Covers standards, materials, installations, operation, and maintenance of pumps. Includes physical properties of liquids and covers valves and pipelines.

2614. Pumping Data. Trade and Technical Press, Ltd, Vol.2, 1964; Vol.3, 1969, Vol.2, 140 pp. ISBN: Vol.1, 0-85461-014-6; Vol.2, 0-85461-015-4; Vol.3, 0-85461-033-2. Cost: Vol.1, £1.60; Vol.2, $16.00; Vol.3, $16.00.

Includes numerous formulas, charts, and tables dealing with pumps.

2615. Pumping Manual. Trade and Technical Press, Ltd, 6th ed., 1979, 800 pp. ISBN: 0-85461-063-4. Cost: $94.00.

Covers seals and packings, filtration, pipe-work systems, valves, power transmission equipment, and pipeline heating.

2616. Pyrolytic Graphite: Property Data. High Temperature Materials, Inc, 1962, various paging.

Property data are listed for pyrolytic graphite.

2617. Quality Assurance Reliability Handbook. Army Materiel Command. NTIS, 1968, 435 pp. Accession: AD-702936. Tech Report: AMC-PAM-702-3.

2618. Quality Control Handbook. J.M. Juran et al. McGraw-Hill, 3d ed., 1974, 1,780 ppp. LC: 74-2463. ISBN: 0-07-033175-8. Cost: $34.85.

Covers quality control for specific industries such as automotive, graphic arts, chemical, drug, metals, textiles, mechanical, and electrical components.

2619. Quality Technology Handbook. R.S. Sharpe; H.A. Cole; W.C. Heselwood. IPC Science and Technology Press, Ltd, 3d ed., 1978, 550 pp. ISBN: 0-902852-82-5. Cost: $46.00.

This handbook was launched by the Harwell Nondestructive Testing Centre primarily for the purpose of providing information on design and application of inspection procedures and manufacturing processes. Provides standards, safety, and patent information.

2620. Quanta: A Handbook of Concepts. P.W. Atkins. Oxford University Press, 1974, 309 pp. ISBN: 0-19-855494-X. Cost: $26.00.

A nonmathematical approach to the study of quantum mechanics.

2621. Quantitative Ultrastructural Data of Animal and Human Cells. David Heinz. Fischer Stuttgart, 1977, 495 pp. ISBN: 3-437-30275-2.

This German title provides 695 tables on ultrastructure data of animal and human cells.

2622. Quantum Radiation of Radioactive Nuclides: A Data Handbook. N.G. Gusev'; P.P. Dmitriev. Pergamon, 1979, 483 pp. LC: 79-40434. ISBN: 0-08-023058-X. Cost: $140.00.

Provides information on exposure rate constants of radioactive nuclides, and energies on gamma ray and KX-ray emission.

2623. Queueing Tables and Graphs. Frederick S. Hillier; Oliver S. Yu; David M. Avis et al. Elsevier-North Holland, 1981, 256 pp. ISBN: 0-444-00582-X. Cost: $44.00. Series: *Publications in Operations Research Series*, Vol. 3.

This book is arranged in tabular format and includes graphs for easy analysis. Such data as random arrivals, exponential service times, random arrivals and Erlang service times, and random arrivals and constant service times are listed in these tables.

2624. Quick Reference to Pediatric Emergencies. Delmer J. Pascoe. Lippincott, 1973, 421 pp. ISBN: 0-397-50305-9. Cost: $17.00.

2625. Quick Reference to Surgical Emergencies. Gerald W. Shaftan; Bernard Gardner. Lippincott, 1974, 664 pp. ISBN: 0-397-50307-5. Cost: $23.75.

2626. RAC Discrete Semiconductor Reliability Transistor/Diode Data. Roy C. Walker (Reliability Analysis Center). NTIS, 1976, 150 pp. Accession: AD-A031025. Tech Report: RAC-DSR-1. Cost: $50.00.

2627. Radar Cross Section Handbook. George T. Ruck et al. Plenum Press, 1970, 2 Vols., 949 pp. LC: 68-26774. ISBN: 0-306-30343-4. Cost: $115.00.

2628. Radar Handbook. Merrill I. Skolnik. McGraw-Hill, 1970, 1,000 pp (various pagings). LC: 69-13615. ISBN: 07-057908-3. Cost: $43.95.

2629. Radar Target Detection Handbook of Theory and Practice. Daniel P. Meyer; Herbert A. Mayer. Academic Press, Inc, 1973, 520 pp. ISBN: 0-12-492850-1. Cost: $55.50. Series: *Electrical Science Series*.

Technical information is provided on the theory and practice of radar. Both authors have had experience working at Sperry Rand Corporation.

2630. Radiation Chemistry of Ethanol: A Review of Data on Yields, Reaction Rate Parameters, and Spectral Properties of Transients. Gordon A. Freeman (National Bureau of Standards). U.S. Government Printing Office, 1974. SuDoc: C13.48:48. Cost: $0.80. Series: NSRDS-NBS-48.

2631. Radiation Chemistry of Gaseous Ammonia. D.B. Peterson (National Bureau of Standards). NTIS, 1974. Accession: COM 74-50175. Cost: $4.00. Series: NSRDS-NBS-44.

2632. Radiation Chemistry of Nitrous Oxide Gas. Primary Processes, Elementary Reactions, and Yields. G.R.A. Johnson (National Bureau of Standards). U.S. Government Printing Office, 1973. SuDoc: C13.48:45. Cost: $0.75. Series: NSRDS-NBS-45.

2633. Radiation Damage of Materials. G. Pluym; M.H. Van De Voorde. NTIS, 1966, 2 parts, Part 1, 131 pp.; Part 2, 62 pp. Accession: Part 1, N68-15832; Part 2, N68-15836. Tech Report: Part 1, MPS/INT-CO-66-25; Part 2, MPS/INT-CO-66-27.

This publication is entitled as follows: Part 1 *A Guide to the Use of Plastics Engineering Handbook*; Part 2 *A Guide to the Use of Elastomers*. Materials testing information is included, particularly as it relates to radiation effects on plastics. Properties data are supplied, including electrical properties, mechanical properties, radiation tolerance, and thermoplasticity data.

2634. Radiation Effects Design Handbook. Section 1—Semiconductor Diodes. D.J. Hamman; C.L. Hanks (Radiation Effects Information Center). NTIS, 1971, 41 pp. Accession: N71-28811. Tech Report: NASA-CR-1785.

2635. Radiation Effects Design Handbook. Section 2—Thermal-Control Coatings. N.J. Broadway. NTIS, 1971, 194 pp. Accession: N71-32280. Tech Report: NASA-CR-1786.

2636. Radiation Effects Design Handbook. Section 3—Electrical Insulating Materials and Capacitors. D.J. Hamman; C.L. Hanks. NTIS, 1971, 87 pp. Accession: N71-29776. Tech Report: NASA-CR-1787.

2637. Radiation Hardness of Radio-Technical Constructions: Handbook. V.K. Kniazev et al. NTIS, 1976, 568 pp. Accession: A77-24664.

Radiation-hardened materials are discussed, primarily for the purpose of protecting radio equipment from harmful doses of radiation. Such materials as liquid dielectrics, gaseous dielectrics, thermoplastics, glass-plastics, polymeric films, gas-filled plastics, ion exchange resins, membranes, ceramics, mica-based materials, laser single crystals, ferrites, graphite materials, single crystal dielectrics, and various metals and alloys are presented.

2638. Radiation Hygiene Handbook. McGraw-Hill, 1959, various paging.

2639. Radiation Protection by Shielding: Tables for the Calculation of Gamma Radiation Shielding. Paul-Friedrich Sauermann. Verlag Karl Thiemig, 192 pp. Cost: $9.50.

Tables are provided for protecting equipment against unnecessary radiation. Attenuation factors are given for concrete-lead-antimony alloys, and shielding glasses.

2640. Radiation Quantities and Units. National Bureau of Standards. ICRU Publications. Cost: $2.50. Series: NBS Handbook 84.

Available from ICRU Publications, PO Box 30165, Washington, DC 20014 as ICRU Report 19.

2641. Radio Amateur's Handbook. American Radio Relay League, Inc. American Radio Relay League, Inc, 56th ed., 1979. LC: 41-3345. ISBN: 0-87259-157-3. Date of First Edition: 1926. Cost: $15.75.

2642. Radio Amateur's Handbook. A. Frederick Collins; Robert Hertzberg rev. Thomas Y. Crowell Co, 14th ed., 1979. LC: 78-3303. ISBN: 0-690-01772-3. Cost: $9.95.

2643. Radio and Television Engineers' Reference Book. J.P. Hawker (Edward Molloy). Newnes, 4th ed., 1963. LC: 63-24354.

2644. Radio Control Handbook. Howard McEntee. TAB Books, 3d ed., 1971. LC: 78-105969. ISBN: 0-8306-1093-6. Cost: $9.95.

2645. Radio Electronic Laboratory Handbook. M.G. Scroggie. Butterworth Group, 9th ed., 1980, 592 pp. LC: 49-40026. ISBN: 0-408-00373-1. Date of First Edition: 1938 (*Radio Laboratory Handbook*). Cost: $53.95.

Covers choice and care of equipment, measurement of circuit parameters, and signal measurement. Emphasizes the trend towards displacing semiconductor devices by integrated circuits.

2646. Radio Engineering Handbook. K. Henney. McGraw-Hill, 5th ed., 1959, 1,800 pp. LC: 58-11174. ISBN: 0-07-028208-0. Cost: $49.50.

2647. Radio Handbook. William I. Orr. Editors and Engineers (Bobbs-Merrill Co, Inc), 21st ed., 1978, 78-64872. LC: 0-672-24034-3. Cost: $21.50.

2648. Radio Propagation Handbook. Peter N. Saveskie. TAB Books, 1980, 499 pp. LC: 79-25205. ISBN: 0-8306-9949-X. Cost: $17.95.

Information is presented on ionospheric and tropospheric propagation. In addition, information on the propagation of microwave, VHF, and UHF signals are discussed.

2649. Radio Spectrum Handbook. James M. Moore. Sams, 1970. LC: 71-118371. ISBN: 0-672-20772-9. Cost: $7.95.

2650. Radio, TV and Technical Reference Book. S.W. Amos. Newnes-Butterworths (distributed by Transatlantic Arts), 1977, approx. 1,000 pp. (various paging). ISBN: 0-408-00259-X. Cost: $60.00.

This book replaces Edward Molloy's *Radio and Television Engineer's Reference Book*, which was published in 4 editions between 1954 and 1963.

2651. Radio-Frequency Transmission Lines: Radio-Frequency Cables. I.E. Efimov; G.A. Ostankovich. NTIS, 2d rev. and enlarged ed., 1977, 408 pp. Accession: A78-36473.

This Russian title presents information on electromagnetic theory of transmission lines, and electrical and design characteristics of different types of cable, such as coaxial, symmetrical, and spiral cable. Cable testing data are also provided.

2652. Radioactive Waste Disposal in the Ocean. National Bureau of Standards. NCRP Publications. Cost: $2.00. Series: NBS Handbook 58.

Available from NCRP Publications, PO Box 30175, Washington, DC 20014 as NCRP Report 16.

2653. Radioactivity. National Bureau of Standards. National Bureau of Standards (available from U.S. Government Printing Office), Nov. 29, 1963. SuDoc: C13.11:86. Cost: $1.10. Series: NBS Handbook 86.

Supersedes parts of NBS Handbook 78; NBS Handbooks 84 through 89 extend and largely replace NBS Handbook 78.

2654. Radiobiological Dosimetry. National Bureau of Standards. National Bureau of Standards (available from U.S. Government Printing Office), April 30, 1963. SuDoc: C13.11:88. Cost: $0.65. Series: NBS Handbook 88.

Supersedes parts of NBS Handbook 78; NBS Handbooks 84 through 89 extend and largely replace NBS Handbook 78.

2655. Radiofrequency Radiation Dosimetry Handbook. C.H. Durney et al. NTIS, 3d ed., 1980, 139 pp. Accession: AD-A089-915/3. Tech Report: UTEC-80-057; SAM-TR-80-32. Cost: $12.50 paper; $3.50 microfilm.

Curves are included for synthetic aperture radar produced by near-field irradiation. Data are also included for SAR's produced by plane-wave irradiation.

2656. Radiological Health Handbook. Bureau of Radiological Health and The Training Institute, Environmental Control Administration, U.S. Department of Health, Education, and Welfare. U.S. Government Printing Office, rev. ed., 1970, 458 pp. Series: *Public Health Service Publication*, No. 2016.

2657. Radiological Monitoring Methods and Instruments. National Bureau of Standards. Superintendent of Documents, U.S. Government Printing Office, April 7, 1952. SuDoc: C13.11:51. Cost: $0.60. Series: NBS Handbook 51.

2658. Radiolysis of Methanol: Product Yields, Rate Constants and Spectroscopic Parameters of Intermediates. J.H. Baxendale; Peter Wardman (National Bureau of Standards). U.S. Government Printing Office, 1975. SuDoc: C13.48:54. Cost: $0.85. Series: NSRDS-NBS-54.

2659. Radiotherapy Treatment Planning. R.F. Mould. Heyden, 1981, 212 pp. Cost: $31.00. Series: *Medical Physics Handbook*, No. 7.

This book provides information on radiation isodose patterns; it would be a useful publication for those interested in teaching radiotherapy.

2660. Radiotron Designer's Handbook. Fritz Langford-Smith. Wireless Press for Amalgamated Wireless Valve Co, 4th ed., 1956, 1,498 pp.

2661. Radome Engineering Handbook; Design and Principles. J.D. Walton Jr. Marcel Dekker, 1970, 616 pp. LC: 74-131300. ISBN: 0-8247-1757-0. Cost: $65.00. Series: *Ceramics and Glass: Science and Technology Series*, Vol. 1.

2662. Rafter Length Manual. Benjamin Williams. Craftsman Book Co, 1979, 369 pp. ISBN: 0-910460-67-1. Cost: $9.00.

Roof framing information is provided and extensive rafter length tables are presented.

2663. Raising Laboratory Animals; A Handbook for Biological and Behavioral Research. James Silvan. Doubleday Publishing Co, 1966, 225 pp. LC: 66-17442.

2664. Raman/IR Atlas of Organic Compounds. B. Schrader; W. Meier. Verlag Chemie, 1st Issue (1st rep. of 1974 ed. 1978); 2d Issue (1st rep. of 1975 ed., 1978); 3d Issue 1977, 2 loose-leaf Vols. in 3 Issues, 1st issue, 314 pp.; 2nd issue, 344 pp.; 3rd issue, 432 pp. Cost: DM 795; $467. 70.

This publication includes numerous spectra, with 314 in the 1st issue, 301 in the 2nd issue, and 386 in the 3rd issue. Raman infrared and IR spectra are also presented for over 1,000 organic compounds.

2665. Rare Metals Handbook. A. Hampel. Reinhold (Krieger), 2d ed., 1971, 732 pp. LC: 61-10449. ISBN: 0-88275-024-0. Cost: $34.50.

2666. Rate Constants of Gas Phase Reactions, Reference Book. V.N. Kondratiev; L. Holslag trans; R. Fristrom trans. NTIS, Russian translation for the National Bureau of Standards, 1970, trans. 1972. Accession: COM 72-10014. Cost: $11.75.

2667. Raw Materials and Processing of Papermaking. H.F. Rance. Elsevier-North Holland, 1980, xiv + 289 pp. ISBN: 0-444-41778-8. Cost: $78.00. Series: *Handbook of Paper Science*, 1.

Provides information on interactions of cellulose and water, the preparation of pulp fibers, the process of forming the web of paper, and the beating process in papermaking.

2668. RCA Electro-Optics Handbook. RCA Corporation Commercial Engineering, 1974 (later editions may be available), 255 pp. ISBN: 0-913970-11-5. Cost: $4.95.

2669. Reaction Rate and Photochemical Data for Atmospheric Chemistry—1977. Robert F. Hampson Jr; David Garvin (Department of Commerce, Washington, D. C.). U.S. Government Printing Office, 1978, 106 pp. LC: 78-19758. SuDoc: C 13.10:513. S/N: 003-003-01924-1. Cost: $2.75.

Gas phase chemical reaction data are presented in tabular format. Reaction rate constants, photoabsorption cross sections, and quantum yields of photochemical processes are listed.

2670. Reaction Rate Compilation for H-O-N System.
G.S. Bahn. Gordon and Breach, 1968, 254 pp. LC: 68-20396. ISBN: 0-677-12570-2. Cost: $57.75.

A compilation of reaction rates is provided for hydrogen-oxygen-nitrogen systems.

2671. Reactivity of the Hydroxyl Radical in Aqueous Solutions. Leon M. Dorfman; Gerald E. Adams (National Bureau of Standards). NTIS, 1973. Accession: COM 73-50623. Cost: $4.50. Series: NSRDS-NBS-46.

2672. Reactor Handbook. Interscience Publishers, Inc, 2d ed, Vol.1, 1960; Vol.2, 1961; Vol.3a, 1962; Vol.3b, 1962; Vol.4, 1964, Vol.1, 1,223 pp.; Vol.2, 680 pp.; Vol.3a, 313 pp.; Vol.3b, 287 pp.; Vol.4, 857 pp. LC: 60-11027. ISBN: Vol.1, 0-470-71082-9;; Vol.2, 0-470-71115-9; Vol.3a, 0-470-71148-5; Vol.3b, 0-470-71150-5; Vol.4, 0-470-71152-3. Date of First Edition: 1955. Cost: Vol.1, $48.25; Vol.2, $31.00; Vo.3a, $17.00; Vol.3b, $14.25; Vol.4, $33.25.

This handbook was prepared under contract with the United States Atomic Energy Commission. It includes information on such subjects as neutron and gamma radiation attenuation by shielding, critical and nuclear data, nuclear fuels, radiation damage, binary alloy systems, and coolant materials. Information is also provided for such subjects as structural materials, such as light and heavy water. Generation of heat and shielding is discussed, as well as statics and dynamics of reactors. Indexes have been prepared for easy retrieval of the reactor data, and many bibliographic references have been cited.

2673. Reactor Handbook; Materials: General Properties. Argonne National Laboratory. McGraw-Hill, 1955, 610 pp.

2674. Reactor Physics Constants. U.S. Argonne National Laboratory, Reactor Physics Constants Center; Atomic Energy Commission. U.S. Government Printing Office, 2d ed., (ANL-5800), 850 pp.

2675. Reagent Chemicals. American Chemical Society, 6th ed., 1981, 612 pp. LC: 81-8111. ISBN: 0-8412-0560-4. Cost: $60.00.

This compilation provides data on reagent solutions, standard solutions, and solid reagent mixtures.

2676. Reagents for Organic Synthesis. Louis F. Fieser; Mary Fieser. John Wiley and Sons, 1980, 8 Vols. LC: 66-27894. ISBN: 0-471-08070-5. Cost: $265.00 per set.

This quick reference provides instant access to the preparation, properties, and uses of thousands of reagents.

2677. Reaktionen der Organischen Synthese. Cesare Ferri. Georg Thieme Verlag, 1978, 974 pp. ISBN: 3-13-487401-6. Cost: DM 295.

Includes approximately 2,700 organic chemistry reactions.

2678. Recommendations for the Disposal of Carbon-14 Wastes. National Bureau of Standards. U.S. Government Printing Office, October 26, 1953. SuDoc: C13.11:53. Cost: $0.40. Series: NBS Handbook 53.

2679. Recommendations for Waste Disposal of Phosphorus-32 and Iodine-131 for Medical Users. National Bureau of Standards. NCRP Publications. Cost: $2.00. Series: NBS Handbook 49.

Available from NCRP Publications, PO Box 30175, Washington, DC 20014 as NCRP Report 9.

2680. Recommended Reference Materials for Realization of Physicochemical Properties (Section Density). E.F.G. Herington. Pergamon Press, 1976. ISBN: 0-08-021017-1. Cost: $6.00.

2681. Recommended Values of the Thermophysical Properties of Eight Alloys, Major Constitutents and Their Oxides. Y.S. Touloukian. NTIS, 1966, 540 pp. Accession: B67-10062. Tech Report: NU-0095. Cost: $10.00.

Tabular and graphical data are provided on the thermophysical properties of basic alloys, their constituents and oxides.

2682. Red Book on Transportation of Hazardous Materials. Lawrence W. Bierlein. Cahners, 1977, 896 pp. LC: 76-44394. ISBN: 0-8436-1407-2. Cost: $65.00.

This book includes such data as vapor pressure curves for various flammable liquids, specifications for cargo tanks and numerous laws that pertain to the transportation of hazardous materials.

2683. Reference Book of Inorganic Chemistry. Wendall M. Latimer; Joel H. Hildebrand. Macmillan, 3d ed., 1964, 625 pp. LC: 66-68678.

2684. Reference Data for Acoustic Noise Control. W.L. Ghering. Ann Arbor Science Publishers Inc, 1978, 152 pp. LC: 78-62291. ISBN: 0-250-40257-2. Cost: $30.00.

Several special noise sources, such as fan noise, electric motor noise, compressor noise, gear noise, vent noise, hydraulic noise, and pump noise are discussed. In addition, such information as noise barriers, enclosures, hoods, and the effects of noise on people is presented. A listing of noise literature and references is provided in the appendix.

2685. Reference Data for Radio Engineers. ITT Staff. Howard W. Sams and Co, Inc, 6th ed., 1975, various paging. LC: 75-28960. ISBN: 0-672-21218-8. Date of First Edition: 1943. Cost: $30.00.

2686. Reference Guide to Practical Electronics. Robert G. Krieger. McGraw-Hill, 1981, 212 pp. LC: 80-14812. Cost: $7.50.

This is a compilation of approximately 100 useful electronics equations.

2687. Reference Manual for Program and Information Officials. Superintendent of Documents, 1978, 2 Vols, Vol.1, 69 pp.; Vol.2, 340 pp. SuDoc: Y 3.P 19:8 P 94/2. Cost: Vol 1, $2.50; Vol.2, $5.00.

This publication includes 2 volumes entitled as follows: Vol.1, *A Handbook for Managers*; Vol.2, *A Handbook for Technical Information Personnel*. Such topics as information resource management, computer administration, records management, and libraries are discussed.

Refractory Ceramics for Aerospace. *See* **Engineering Properties of Selected Ceramic Materials.**

2688. Refrigertion Processes: A Practical Handbook on the Physical Properties of Refrigerants and Their Appli-

cations. H.M. Meacock. Pergamon, 1979, 242 pp. ISBN: 0-08-024211-1. Cost: $30.00.

Approximately 30 refrigerants are presented with their corresponding data sheets, which list properties that will help to calculate temperature, pressure, volume, and enthalpy under various conditions. In addition, tables of saturation properties are listed, as well as pressure enthalpy charts.

2689. Registry of Mass Spectral Data. Einar Stenhagen; Sixten Abrahamsson. Wiley-Interscience, 1974, 3,358 pp., 4 vols. LC: 74-910. ISBN: 0-471-82115-2. Cost: $460.00.

This 4-volume set includes numerous tables presenting mass spectral data. Bibliographic references are provided.

2690. Registry of Toxic Effects of Chemical Substances. National Institute for Occupational Safety and Health. American Chemical Society, 1978. Cost: $19.50.

This compilation lists over 21,000 chemicals with their corresponding toxic dose information.

2691. Reinforced Concrete Designer's Handbook. Charles E. Reynolds; J.C. Steedman. Scholium International, Inc, 9th ed., 1980, 500 pp. Cost: $42.00; $36.00 paper.

Expanded in accordance with the requirements of CP 110. All material relates to limit-state design.

2692. Reinforced Masonry Engineering Handbook; Clay and Concrete Masonry. James E. Amrhein. Masonry Institute of America, 3d ed., 1978, 442 pp. LC: 73-185480. Cost: $21.75.

2693. Relay D.A.T.A. Book. Derivation and Tabulation Associates, Inc, continuation issued annually, pages vary per issue. Cost: $36.50 per year by subscription.

2694. Reliability Design Handbook. R.T. Anderson (Reliability Analysis Center). NTIS, 1976, 390 pp. Accession: AD-A024 601. Tech Report: RAC-RDH-376. Cost: $30.00.

2695. Reliability Handbook. Boris Anatol'evich Kozlov; I.A. Ushakov. Holt, Rinehart and Winston, 1970, 391 pp. LC: 79-85752. ISBN: 0-03-081417-0.

2696. Reliability Handbook. W.G. Ireson. McGraw-Hill, 1966, 702 pp. LC: 65-18747. ISBN: 0-07-03204-3. Cost: $55.00.

2697. Reliability Handbook. National Semiconductor Corp, 1979, 804 pp. Cost: $20.00.

Reliability engineering and quality assurance are discussed in connection with how electronic devices are manufactured and why they fail.

2698. Reliability/Design Handbook. Thermal Applications, Volumes 1-4. William Wallace Jr et al (Thermal Technology Laboratory, Buffalo, NY). NTIS, July 1973, Vol.1, 146 pp.; Vol.2, 304 pp.; Vol.3, 184 pp.; Vol.4, 238 pp. Accession: Vol.1, AD-A009013; Vol.2, AD-A009014; Vol.3, AD-A009015; Vol.4, AD-A009016. Tech Report: Vol.1, NAVELEX-0967-437-7010; Vol.2, NAVELEX-0967-437-7020; Vol.3, NAVELEX-0967-437-7030; Vol.4, NAVELEX-0967-437-7040. Cost: Vol.1, $6.00 paper, $2.25 microfiche; Vol.2, $9.75 paper, $2.25 microfiche;

Vol.3, $7.50 paper, $2.25 microfiche; Vol.4, $8.00 paper, $2.25 microfiche.

This 4-volume set was prepared to provide heat transfer data and electronic data for those Navy personnel involved with designing electronic equipment in accordance with the thermal sections of Naval specifications and standards for equipment. Such areas as forced air cooled and liquid cooled electronic equipment along with thermal design information are covered in this handbook. Information is also presented on the following: design of electronic equipment for operations at 150 to 350° C temperature, equipment installation, thermal evaluation of electronic equipment, standard hardware program (SHP) thermal design, special cooling techniques, and thermal design of vaporization cooled electronic equipment.

2699. Remodelers Handbook. Benjamin Williams. Craftsman Book Co, 1976, 416 pp. LC: 76-53565. ISBN: 0-910-460-21-3. Cost: $14.75.

Information is presented on repairing wood structures, painting surfaces, installing ventilation systems and gutters, and replacing windows and doors.

2700. Research and Report Handbook for Businesss, Industry, and Government. R. Moyer; E. Stevens; E. Switzer. John Wiley and Sons, 1980, 320 pp. LC: 80-18922. ISBN: 0-471-04257-9 cloth; 0-471-04258-7 paper. Cost: $14.95 cloth; $8.95 paper.

Information is presented on the written preparation of research results and the writing of technical reports. Information is also included on word processing.

2701. Reservoir Engineering Manual. Frank W. Cole. Gulf Publishing Co, 2d ed., 1969. ISBN: 0-87201-779-6. Cost: $22.95.

2702. Resonances in Electron Impact on Atoms and Diatomic Molecules. George J. Schulz (National Bureau of Standards). NTIS, 1973. Accession: AD 771200. Cost: $5.50. Series: NSRDS-NBS-50.

2703. Resource Recovery and Recycling Handbook of Industrial Wastes—1975. Marshall Sittig. Noyes Data Corp, 1975, 427 pp. LC: 75-22685. ISBN: 0-8155-0592-2. Cost: $36.00. Series: *Environmental Technology Handbook*, No. 3.

2704. Respiration and Circulation. Philip L. Altman; Dorothy S. Dittmer. Federation of American Societies for Experimental Biology, 1971, Vol.1, 930 pp. LC: 70-137563. ISBN: 0-913822-05-1. Cost: $30.00.

A compilation of 232 tables reporting data on such subjects as blood gases, capillaries, thorax, heart action, vascular system, blood vessels, and any topics pertinent to the respiratory or circulatory systems.

Retrieval Guide to Thermophysical Properties Research Literature. *See* **Thermophysical Properties Research Literature Retrieval Guide.**

2705. Review of Rate Constants of Selected Reactions of Interest in Re-Entry Flow Fields in the Atmosphere. M.H. Bortner (National Bureau of Standards). NTIS, 1969. Accession: AD-692231. Cost: $4.50. Series: NBS Tech. Note 484.

2706. Review of the Physical, Mechanical, and Irradiation Properties of U Sub 3 Si and U Sub 3 Si Sub 2. J. Bardsley. NTIS, 1968, 42 pp. Accession: N69-28086. Tech Report: AAEC/TM-487.

Such properties as electrical resistivity and thermal conductivity are listed for uranium-silicon alloys. In addition, phase diagrams are presented, along with information on corrosion and crystallography.

2707. RF Radiometer Handbook. G. Evans; C.W. McLeish. Artech, 1977, 152 pp. LC: 77-501. ISBN: 0-89006-055-X. Cost: $27.00.

Provides information for measuring radiation at short wavelengths.

Riegel's Industrial Chemistry. *See* **Handbook of Industrial Chemistry.**

2708. Rigid Urethane Foam Processing Handbook. Robert I. Lane. Technomic Publishing Co, Inc, 1974, 104 pp. LC: 74-78227. ISBN: 0-87762-130-6. Cost: $25.00.

The emphasis of this handbook is on processing equipment, material handling and evaluating processes for rigid urethane foam manufacturing.

2709. Ring Index: A List of Ring Systems Used in Organic Chemistry. A.M. Patterson. American Chemical Society, 2d ed., 1960, 1,425 pp, 1 Vol. with supplements. LC: A61-610.

2710. Road and Bridge Construction Handbook. Michael Lapinski. Van Nostrand Reinhold, 1978, 156 pp. LC: 78-1475. ISBN: 0-442-24681-1. Cost: $17.95.

Such topics as blasting, drilling, rock crushing, culvert installation, excavation, and compaction are discussed in this handbook.

2711. Rock-Forming Minerals. W.A. Deer. Wiley (Halsted Press), 1962, Vol.1, 333 pp.; Vol.2, 379 pp.; Vol.3, 270 pp.; Vol.4, 435 pp.; Vol.5, 372 pp. ISBN: 0-471-20518-4 (Vol. 1). Cost: Vols.1, 2, and 5,$29.50; Vol.3, $27.00; Vol.4, $32.50.

Vol.2a, 2d ed., 1979, LC: 78-40451, ISBN: 0-470-26455-1, $55.00.

2712. Rodale's Color Handbook to Garden Insects. Anna Carr. Rodale Press, Inc, 1979, 256 pp. LC: 79-4048. Cost: $12.95.

Color photographs are provided for identifying garden insects. Life cycle information, host plants, feeding habits, and natural controls are also discussed.

2713. Rodd's Chemistry of Carbon Compounds, A Modern Comprehensive Treatise. S. Coffey; E.H. Rodd (orig ed). Elsevier-North Holland, 2d ed., 1964-, 5 Vols. LC: 64-4605. ISBN: Vol.1, Part A, 0-444-40131-8. Cost: Consult publisher for price.

This multivolume set is not really a handbook but, as the title suggests, a comprehensive treatise on organic chemistry with exhaustive citations from the literature. The titles of the volumes are as follows: Vol.1, *Aliphatic Compounds*; Vol.2, *Alicyclic Compounds*; Vol.3, *Aromatic Compounds*; Vol.4, *Heterocyclic Compounds*; Vol. 5, *Miscellaneous General Index*.

2714. Roofers Handbook. Craftsman Book Co, 1976, 250 pp. LC: 76-5875. ISBN: 0-910460-17-5. Cost: $9.25.

This is a guide to the usage of wood and asphalt shingles. Information on roofing techniques and prevention of leaks is discussed in this handbook. Over 250 illustrations are provided.

2715. Royal Society Mathematical Tables. Royal Society of London. Cambridge University Press, Vol.1, 1950-.

The Royal Society is continuing a series of publications entitled *Mathematical Tables* that were originally started by the British Association for the Advancement of Science, 1931-1952.

2716. S-100 Bus Handbook. Dave Bursky. Hayden Book Co, 1980, 280 pp. ISBN: 0-8104-0897-X. Cost: $15.75.

The S-100 bus computer systems are discussed. Schematic drawings of circuit boards are provided. Such topics as computer memory, bulk-memory devices, and the CPU are also presented in this manual.

2717. Sadtler Handbook of Reference Spectra. Sadtler Research Laboratories, Inc, 1978, multivolume set. ISBN: 0-685-518442 set; Vol.1, 0-685-93054-8; Vol.2, 0-685-51845-0; Vol.3, 0-685-51846-9. Cost: $425.00 set; Vol.1, $165.00 Vol.2; $140.00 Vol.3, $140.00.

This 3-volume set is entitled as follows: Vol.1, *Infrared Spectra*; Vol.2, *Proton NMR Spectra*; Vol.3, *Ultraviolet Spectra*. This 3-volume handbook includes spectra on approximately 3,000 organic compounds. All spectra in these volumes are indexed by compound type, compound name, and the spectrum finder system.

2718. Sadtler Standard Spectra. Sadtler Research Laboratories, Inc.

The Sadtler Research Laboratories, Inc., publishes numerous publications which provide such information as nuclear magnetic resonance spectra, Carbon-13 NMR spectra, grating spectra, infrared absorption spectrograms, and ultraviolet standard spectra. Most of these services are loose-leaf so that they may be continuously updated. These publications are very useful in supplying information that helps to identify chemical compounds. Please refer to the publisher for prices.

2719. SAE Aerospace Applied Thermodynamics Manual. SAE Committee AC-9, Aircraft Environment Systems, Society of Automotive Engineers, Inc, 2d ed., 1969, 758 pp. LC: 74-13046 (1969); 60-13450 (1960). Date of First Edition: 1960.

NOTE: ARP 1168 (Aerospace Recommended Practices) &52 give dimensional, design/or performance recommendations intended as guides for standard engineering practices.

2720. SAE Handbook 1981. Society of Automotive Engineers, Inc, 1981, various paging. LC: 25-16527. ISBN: 0-89883-002-8. Date of First Edition: 1951. Cost: $75.00.

This 2-volume set is an annual publication of vehicle standards and recommended practices that have been approved by the Society of Automotive Engineers. The handbook includes standards on such subjects as brakes, wheels, transmissions, lighting, tires, trucks, and seat belts.

2721. Safe Handling of Bodies Containing Radioactive Isotopes. National Bureau of Standards. NCRP Publications. Cost: $4.00. Series: NBS Handbook 65.

Available from NCRP Publications, PO Box 30175, Washington, DC 20014 as NCRP Report 37.

2722. Safe Handling of Radioactive Materials. National Bureau of Standards. U.S. Government Printing Office, March 9, 1964. SuDoc: C13.11:92. Cost: $1.40. Series: NBS Handbook 92 (supersedes NBS Handbook 42).

2723. Safe-Load Tables for Solid Slabs. A.H. Allen. Scholium International Publishers, Inc, 1973, 56 pp. Cost: $6.50.

This handbook presents allowable loads for different spans and thicknesses of concrete.

2724. Safety Handbook for Science Teachers. K. Everett; E.W. Jenkins. London, Murray, 2d ed., 1977, 100 pp. Cost: £2.60.

This is a very useful book on safety in the laboratory. Such topics as handling hazardous materials are discussed.

2725. Safety Rules for the Installation and Maintenance of Electric Supply and Communication Lines. National Bureau of Standards. American National Standards Institute, Nov. 1, 1961. Cost: $6.50. Series: NBS Handbook 81 and its supplements.

Comprising Part 2, the definitions, and the grounding rules of the sixth edition of the National Electrical Safety Code. Supersedes NBS Handbook 32 and amends in part, Part 2, definitions and the grounding rules of NBS Handbooks 30 and 43. Available from American National Standards Institute, 1430 Broadway, New York, NY 10018 as ANSI C2.2-1976 (contained in ANSI C2-1981).

2726. Safety With Lasers and Other Optical Sources: A Comprehensive Handbook. David Sliney; Myron Wolbarsht. Plenum Publishers, 1980, 1,035 pp. ISBN: 0-306-40434-6. Cost: $49.50.

This is a very comprehensive collection of information on the usage of lasers and the various safety precautions that need to be considered.

2727. Safety: Electrical and Electronics Products. Media International Promotions, Inc. Media International Promotions, Inc, 400 pp. Cost: $100.00. Series: *World Standards Mutual Speedy Finder*, Vol. 5.

Tabular comparisons are presented on electrical safety standards for 6 countries and the International Electro-technical Commission Standards. Countries included are USA, West Germany, Canada, Australia, United Kingdom, and Japan. Numerous safety standards are included for such equipment as television receivers, audio receivers, radio receivers, household appliances, tape recorders, and amplifiers.

2728. Sanitary Landfill Design Handbook. George Noble. Technomic Publishing Co, Inc, 1976, 285 pp. LC: 76-1134. ISBN: 0-87762-100-4. Cost: $20.00.

Solid waste disposal is discussed in connection with landfills. Such topics as site planning, site preparation, and equipment selection are presented.

2729. Satellite Environment Handbook. Francis S. Johnson (ed). Stanford University Press, 2d ed., 1965, 193 pp. LC: 64-8894. ISBN: 0-8047-0090-7. Cost: $10.00.

2730. Saturn Base Heating Handbook. C.R. Mullen et al. NTIS, 1972, 506 pp. Accession: N72-30895. Tech Report: NASA-CR-61390 TD-050. Cost: $27.50.

This handbook contains a summary of flight test base heating data for the Saturn launch vehicles.

2731. Sausage and Small Goods Production: Practical Handbook on the Manufacture of Sausage and Other Meat-based Products. Frank Gerrard. International Ideas, Inc, 6th ed., 1976. ISBN: 0-7198-2587-3. Cost: $16.95.

This handbook provides practical information on the manufacturing of different types of sausages.

2732. Sawyer's Gas Turbine Engineering Handbook. John W. Sawyer. Gas Turbine Publications, Inc, 2d ed., 1972, 3 Vols. LC: 74-140403. Date of First Edition: 1966 (entitled *Gas Turbine Engineering Handbook*).

Includes such information as aerodynamic design of compressors, noise control, heat exchangers, and maintenance and troubleshooting of generator turbines.

2733. Scaffold Falsework Design. Murray Grant. Viewpoint Publicatons (distributed by Scholium International, Inc), 1978, 42 pp. ISBN: 0-7210-1063-6. Cost: $12.00.

The construction of scaffolding is discussed along with information on different loadings and capacities of various types of scaffold designs.

2734. Scheduling Handbook. J. O'Brien. McGraw-Hill, 1969, 736 pp. LC: 68-28149. ISBN: 0-07-047601-2. Cost: $55.00.

2735. Screw-Thread Standards for Federal Services, Part II. National Bureau of Standards. U.S. Government Printing Office, Nov. 16, 1959 (reprinted Dec. 1966 with corrections). SuDoc: C13.11:28/II. Cost: $2.00. Series: NBS Handbook 28.

Includes the changes to Part II listed in the 1963 Supplement and in the 1966 Amendment to H28.

2736. Screw-Thread Standards for Federal Services, Part III. National Bureau of Standards. U.S. Government Printing Office, Oct. 7, 1960, (reprinted Dec. 1966 with corrections). SuDoc: C13.11:28/III. Cost: $1.30. Series: Handbook 28.

Includes changes to Part III listed in the 1963 Supplement and in the 1966 Amendment to H28.

2737. Screw-Thread Standards for Federal Services. National Bureau of Standards. U.S. Government Printing Office, 1969; Part I, March 1970. SuDoc: C13.11:28.I. Cost: $3.80. Series: NBS Handbook 28.

Supersedes H28 (1957) Part I and that applicable to Part I in the 1963 Supplement to H28.

2738. Seal Users Handbook. R.M. Austin; B.S. Nau; D. Reddy. BHRA Fluid Engineering (distributed by Air Science Co), 2d ed., 1979, 220 pp. ISBN: 0-900983-90-6. Cost: $73.00.

The emphasis in this handbook is on seal design and usage. Information is also provided on seal failures.

2739. Search Manual for Selected Powder Diffraction Data for Metals and Alloys. JCPDS—International Centre for Diffraction Data, 1978, 729 pp. LC: 78-103462. Series: *JCPDS Publication SMMA-1-27.*

The *Search Manual* provides access information to the file on powder diffraction that has been prepared by the International Centre for Diffraction Data. The *Search Manual* is meant to be used as a companion volume with the *Data Book*, which contains the powder diffraction data.

2740. Seawater Corrosion Data. DECHEMA (available from MaGraw-Hill), 1981, 160 pp. ISBN: 0-07-016207-7. Cost: $57.50.

Over 100 different construction materials are covered in this compilation of data on seawater corrosion. Such information as corrosion tests, cathodic protection, and biofouling are included in this collection of data.

2741. Seawater Corrosion Handbook 1979. M.M. Schumacher. Noyes Data Corp, 1979, 494 pp. LC: 78-70745. ISBN: 0-8155-0736-4. Cost: $36.00.

This publication includes 194 tables and 126 figures on seawater corrosion. An annotated bibliography is provided. Such topics as corrosion of nonmetallic materials, effects of marine organisms, and corrosion at various depths of the ocean are discussed.

2742. Seed-Starter's Handbook. Nancy Bubel. Rodale Press, 1978, 363 pp. LC: 77-25332. ISBN: 0-87857-209-0. Cost: $12.95.

Information is provided on germination of seeds.

2743. Seidell's Solubilities: Inorganic and Metal Organic Compounds, Fourth Edition. W.F. Linke. American Chemical Society, 4th ed, Vol.1, 1958; Vol.2, 1965, 2 Vols, Vol.1, 1,486 pp.; Vol.2, 1,914 pp. ISBN: Vol.1, 0-8412-0097-1; Vol.2, 0-8412-0098-X. Cost: $42.00 Rev. Vol.

Evaluated data are provided on the solubilities of compounds in water and other solvents.

2744. Seizures, Epilepsy and Your Child: A Handbook for Parents, Techers and Epileptics of All Ages. Jorge C. Lagos. Harper and Row, 1974. LC: 74-5792. ISBN: 0-06-012504-7. Cost: $8.95.

2745. Selected Constants: Oxidation and Reduction Potentials of Inorganic Substances in Aqueous Solutions. G. Chariot. Pergamon Press,Inc, 1971, 78 pp. ISBN: 0-08-020836. Cost: $21.00. Series: *IUPAC Chemical Data Series*, Vol. 13.

This book was prepared under the auspices of the International Union of Pure and Applied Chemistry. Provides oxidation and reduction potentials for inorganic substances.

2746. Selected Electrical and Thermal Properties of Undoped Nickel Oxide. J.E. Keem; J.M. Honig. CINDAS/ Purdue University, 1978, 78 pp. Series: *Purdue University Report*, No. 52.

Evaluated data have been compiled providing information on the electrical and thermal properties of undoped nickel oxide.

2747. Selected Powder Diffraction Data for Metals and Alloys. JCPDS—International Centre for Diffraction Data, multivolume set. Cost: $490.00.

This 3-volume set includes a 2-volume *Data Book* and a 1-volume *Search Manual*. Each *Data Book* includes approximately 1,000 pages and the *Search Manual* has 725 pages. Powder diffraction data have been provided for metals, alloys, intermetallic compounds, carbides, hydrides, borides, and oxides. The *Search*

Manual is basically a key word in context index (KWIC) arranged alphabetically by chemical name. The *Search Manual* also includes a Hanawalt Numerical section and a Fink Numerical section.

2748. Selected Powder Diffraction Data for Minerals Supplment I. Peter Bayliss. JCPDS—International Centre for Diffraction Data, 1981.

The hardcover *Data Book* presents 1,100 powder diffraction patterns from data card sets 24-29, which includes descriptions for approximately 550 new mineral species. This *Supplement I* updated the *Selected Powder Diffraction Data for Minerals* file that was originally published in 1974 which represents a compilation of data card sets 1-23. The original 1974 work has approximately 2,300 powder diffraction patterns describing over 1,850 mineral species. The *1982 Supplement I Data Book* and the original *1974 Data Book* both have corresponding *Search Manuals* which help to provide access to the powder diffraction patterns.

2749. Selected Reference Material, U.S. Atomic Energy Program. U.S. Atomic Energy Commission. U.S. Government Printing Office, 1955, multivolume set.

This reference manual consists of 8 volumes entitled as follows: Vol.1, *Research Reactors*; Vol.2, *Reactor Handbook: Physics*; Vol.3, *Reactor Handbook: Engineering*; Vol.4, *Reactor Handbook: Materials*; Vol.5, *Neutron Cross-Sections*; Vol.6, *Chemical Processing and Equipment*; Vol.7, *Eight-Year Isotope Summary*; Vol.8, *Information Sources*.

2750. Selected Semiconductor Circuits Handbook. Seymour Schwartz. John Wiley and Sons, 1961.

2751. Selected Specific Rates of Reactions of the Solvated Electron in Alcohols. E. Watson, Jr; S. Roy (National Bureau of Standards). U.S. Government Printing Office, 1972. SuDoc: C13.48:42. Cost: $0.60. Series: NSRDS-NBS-42.

2752. Selected Specific Rates of Reactions of Transients from Water in Aqueous Solution, I. Hydrated Electron. A.B. Ross (National Bureau of Standards); M. Anbar; M. Bambenek. U.S. Government Printing Office, 1972. SuDoc: C13.48:43. Cost: $1.05. Series: NSRDS-NBS-43.

Supplement: *Selected Specific Rates of Reactions of Transients from Water in Aqueous Solution. Hydrated Electron, Supplemental Data.* A.B. Ross (National Bureau of Standards), U.S. Government Printing Office, 1975. Series No.: NSRDS-NBS-43, supplement; Superintendent of Documents No.: C13.48:43 supplement. $1.10.

2753. Selected Specific Rates of Reactions of Transients from Water in Aqueous Solution, II. Hydrogen Atom. Michael Anbar; Farhataziz; A.B. Ross (National Bureau of Standards). U.S. Government Printing Office, 1975. SuDoc: C13.48:51. Cost: $1.20. Series: NSRDS-NBS-51.

2754. Selected Specific Rates of Reactions of Transients from Water in Aqueous Solution, III. Hydroxyl Radical and Perhydroxyl Radical and Their Radical Ions. Farhataziz; Alberta B. Ross. U.S. Government Printing Office. SuDoc: C13.46:59. Series: NSRDS-NBS-59.

2755. Selected Table of Atomic Spectra, Atomic Energy Levels and Multiplet Tables, N IV, N V, N VI, N VII. Charlotte E. Moore (National Bureau of Standards). U.S.

Government Printing Office. SuDoc: C13.48:3/Sec. 4. Cost: $1.15. Series: NSRDS-NBS-3, Sec. 4.

2756. Selected Tables in Mathematics Statistics. H.L. Harter; D.B. Owens. American Mathematical Society, 1970-, 5 Vols. LC: 74-6283. ISBN: 0-8218-1905-4. Cost: Vol.1, $13.20; Vol.2, $16.40; Vol.3, $21.60; Vol.4, $18.00; Vol.5, $16.80.

2757. Selected Tables of Atomic Spectra, A: Atomic Energy Levels—Second Edition, B: Multiplet Tables O VI, O VII, O VIII: Data Derived from the Analyses of Optical Spectra. Superintendent of Documents. U.S. Government Printing Office, 1979, 36 pp. SuDoc: C 13.48:3/sec8. S/N: 003-003-01964-0. Cost: $1.60.

Atomic spectral data are provided in these selected tables.

2758. Selected Tables of Atomic Spectra, Atomic Energy Levels and Multiplet Tables, C I, C II, C III, C IV, C V, C VI. Charlotte E. Moore (National Bureau of Standards). U.S. Government Printing Office, 1970. SuDoc: C13.48:3/Sec. 3. Cost: $1.70. Series: NSRDS-NBS-3, Sec. 3.

2759. Selected Tables of Atomic Spectra, Atomic Energy Levels and Multiplet Tables, N I, N II, N III. Charlotte E. Moore (National Bureau of Standards). U.S. Government Printing Office, 1972. SuDoc: C13.48/3:Sec. 5. Cost: $1.80. Series: NSRDS-NBS-3, Sec. 5.

2760. Selected Tables of Atomic Spectra, Atomic Energy Levels and Multiplet Tables, O I. Charlotte E. Moore (National Bureau of Standards). U.S. Government Printing Office, 1976. SuDoc: C13.48:3/Sec. 7. Cost: $0.85. Series: NSRDS-NBS-3, Sec. 7.

2761. Selected Tables of Atomic Spectra, Atomic Energy Levels and Multiplet Tables, Si I. Charlotte E. Moore (National Bureau of Standards). NTIS, 1967. Cost: $3.50. Series: NSRDS-NBS-3/Sec. 2.

2762. Selected Tables of Atomic Spectra, Atomic Energy Levels and Multiplet Tables, Si II, Si III, Si IV. Charlotte E. Moore (National Bureau of Standards). U.S. Government Printing Office, 1965. SuDoc: C13.48:3/Sec. 1. Cost: $1.00. Series: NSRDS-NBS-3, Sec. 1.

2763. Selected Tables of Atomic Spectra, Atomic Energy Levels, and Multiplet Tables, H I, D, T. Charlotte E. Moore (National Bureau of Standards). U.S. Government Printing Office, 1972. SuDoc: C13.48:3/Sec. 6. Cost: $1. 15. Series: NSRDS-NBS-3, Sec. 6.

2764. Selected Thermodynamic Values and Phase Diagrams for Copper and Some of Its Binary Alloys. R. Hultgren; P.D. Desai. International Copper Research Association, 1973. Cost: $10.00.

2765. Selected Urologic Operations in Gynecology. Jamish, Palmrich, and Pecherstorfer. Walter de Gruyter, 1979, 73 pp. Cost: $58.00.

This publication includes an atlas of approximately 52 illustrations on urologic operations in gynecology.

2766. Selected Values for the Thermodynamic Properties of Metals and Alloys. Ralph Hultgren et al. John Wiley and Sons, 1963, 974 pp.

This project originally started at the University of California, Berkeley, and now participates in the National Bureau of Standards Reference Data Program.

2767. Selected Values of Chemical Thermodynamic Properties, Table for the First Thirty-Four Elements in the Standard Order of Arrangement. D.D. Wagman; W.H. Evans; V.B. Parker; I. Halow; S.M. Bailey; R.H. Schumm (National Bureau of Standards). U.S. Government Printing Office, 1968. SuDoc: C13.46:270-3. Cost: $3.25. Series: NBS Tech. Note 270-3.

2768. Selected Values of Chemical Thermodynamic Properties, Tables for Elements 35 through 53 in the Standard Order of Arrangement. D.D. Wagman; W.H. Evans; V.B. Parker; I. Halow; S.M. Bailey; R.H. Schumm (National Bureau of Standards). U.S. Government Printing Office, 1969. SuDoc: C13.46:270-4. Cost: $2.10. Series: NBS Tech. Note 270-4.

2769. Selected Values of Chemical Thermodynamic Properties, Tables for Elements 54 through 61 in the Standard Order of Arrangement. D.D. Wagman; W.H. Evans et al (National Bureau of Standards). U.S. Government Printing Office, 1971. SuDoc: C13.46:270-5. Cost: $0.95. Series: NBS Tech. Note 270-5.

2770. Selected Values of Chemical Thermodynamic Properties, Tables for the Alkaline Earth Elements (Elements 92 through 97 in the Standard Order of Arrangement). V.B. Parker; D.D. Wagman; W.H. Vans (National Bureau of Standards). U.S. Government Printing Office, 1971. SuDoc: C13.46:270-6. Cost: $1.90. Series: NBS Tech. Note 270-6.

2771. Selected Values of Chemical Thermodynamic Properties, Tables for the Lanthanide (Rare Earth) Elements (Elements 62 through 76 in the Standard Order of Arrangement). R.H. Schumm; D.D. Wagman; S. Bailey; W.H. Evans; V.B. Parker (National Bureau of Standards). U.S. Government Printing Office, 1973. SuDoc: C13. 46:270-7. Cost: $1.25. Series: NBS Tech. Note 270-7.

2772. Selected Values of Chemical Thermodynamic Properties. F.D. Rossini et al. U.S. Government Printing Office, 1952, 1,268 pp. Series: NBS Circular 500.

Selected Values of Chemical Thermodynamic Properties is updated by the following: U.S. National Bureau of Standards Technical Note No. 270-3: *Tables for the First 24 Elements in the Standard Order of Arrangement*, U.S. National Bureau of Standards Circular 500: Series No. 1.

2773. Selected Values of Electric Dipole Moments for Molecules in the Gas Phase. R.D. Nelson Jr; D.R. Lide Jr; A.A. Maryott (National Bureau of Standards). NTIS, 1967. Cost: $4.50. Series: NSRDS-NBS-10.

2774. Selected Values of Physical and Thermodynamic Properties of Hydrocarbons and Related Compounds. F.D. Rossini et al. Carnegie Press (Carnegie-Mellon University Press), 1953, 1,050 pp. LC: 53-9176.

2775. Selected Values of Properties of Chemical Compounds. B.J. Zwolinski. Texas A&M University, Thermodynamics Research Center, 1955-, 2 Vols.

NOTE: Originally sponsored by the Manufacturing Chemists Association. Data are provided on many inorganic and organic compounds including alkanols, phenols, xylenols, alkanoic acids, and nitrogen compounds. Such properties as Gibbs energy functions, vapor pressure, enthalpy, entropy, and heat capacity are tabulated.

2776. Selected Values of Properties of Hydrocarbons and Related Compounds. B.J. Zwolinski. American Petroleum Institute Research Project 44, Texas A&M University, Loose-leaf service, 6 Vols.

Many substances such as aliphatic and aromatic hydrocarbons, ketones, aldehydes, ethers, and alcohols are included. Such properties as refractive index, surface tension, Gibbs energy fuctions, vapor pressure, heat capacity, enthalpy, and entropy are tabulated.

2777. Selected Values of Properties of Hydrocarbons. Circular 461. National Bureau of Standards. U.S. Government Printing Office, 1947 (out of print), 483 pp.

2778. Selected Values of the Thermodynamic Properties of Binary Alloys. Ralph Hultgren et al. American Society for Metals, rev. ed., 1973, 1,435 pp. LC: 73-76588. Cost: $42.00.

NOTE: Originally published in 1963 by John Wiley as *The Selected Values of Thermodynamic Properties of Metals and Alloys.* Presents in tabular format thermodynamic data for binary alloys, including phase diagrams of each system.

2779. Selected Values of the Thermodynamic Properties of the Elements. Ralph Hultgren et al. American Society for Metals, 1973, 636 pp. LC: 73-76587. Cost: $32.00.

NOTE: Part of the National Bureau of Standards National Standards Reference Data Series. Data are presented in tabular format. All thermodynamic data for the elements are presented except for the effects of high pressures. Data have been critically evaluated and tested for consistency with thermodynamic laws.

2780. SEM/TEM Fractography Handbook. G.F. Pittinato et al (Metals and Ceramics Information Center). NTIS, 1975, 692 pp. Accession: AD-AQ20 793. Tech Report: AFML-TR-75-159; MCIC-HB-06. Cost: $70.00.

2781. Semiconductor Application Notes D.A.T.A. Book. Derivation and Tabulation Associates, Inc, continuation issued semiannually, pages vary per issue. Cost: $25.50 per year by subscription.

2782. Semiconductor Data Library, Series A. Motorola Semiconductor Products, Inc, 1974, multivolume set.

Data sheets and specification information are provided for semiconductors manufactured by Motorola. Such information as electrical, mechanical, thermal characteristics and maximum ratings for each semiconductor is provided. Various charts, diagrams, and schematics help to describe thermal response, thermal resistance, temperature coefficients, and other types of dynamic characteristics. Linear integrated circuits are also discussed in this comprehensive compilation.

2783. Semiconductor Diode and SCR D.A.T.A. Book. Derivation and Tabulation Associates, Inc, continuation issued semiannually, pages vary per issue. Cost: $48.50 per year by subscription.

2784. Semiconductor Diodes and Transistors Handbook. N.N. Goryunova. NTIS, translated from the Russian by the Army Foreign Science and Technology Center, 2d ed., 1974, 481 pp. Accession: AD-786811. Tech Report: FSTC-HT-23-814-74.

2785. Semiconductor Electronics: Properties of Materials. P.I. Baranskii; V.P. Klochkov; I.V. Potykevich. NTIS, 1975, 704 pp. Accession: A75-40148.

This Russian title provides such information as the magnetic, electrical, optical, galvanomagnetic, and thermomagnetic characteristics of semiconductors.

2786. Semiconductor Handbook. Robert D. Tomer. Sams, 2d ed., 1968, 287 pp. LC: 68-24769. Date of First Edition: 1961 (entitled *Industrial Transistor and Semiconductor Handbook.*).

2787. Semiconductor Reference Book 1976-1977. Association Internationale Pro Electron. A.E. Kluwer, Antwerp, 5th ed., 272 pp. Cost: $27.50.

This reference book contains over 6,500 semiconductors with PRO ELECTRON type number for users of semiconductor devices made in Western Europe.

2788. Semiconductors: Preparation, Crystal Growth, and Selected Properties. T.F. Connolly. Plenum Publishing Corp, 1972, 218 pp. ISBN: 0-306-68322-9. Cost: $65.00. Series: *Solid State Physics Literature Guides*, Vol. 2.

This publication is basically a bibliography on the topic of semiconductor preparation and crystal growth. The bibliography also covers articles on amorphous semiconductors and the ion implantation doping of semiconductors. This bibliography is based on the collection of documents at the Research Materials Information Center at Oak Ridge National Laboratory.

2789. Semimicro Qualitative Organic Analysis: Systematic Identification of Organic Compounds. Nicholas D. Cheronis et al. Wiley-Interscience, 3d ed., 1965, 1,065. LC: 64-25892. ISBN: 0-470-15315-6. Cost: $46.50.

2790. Sensory Integration. R. Bruce Masterton. Plenum Publishing Co, 1978, 567 pp. LC: 78-17238. ISBN: 0-306-35191-9. Cost: $39.50. Series: *Handbook of Behavioral Neurobiology*, Vol. 1.

This is a compilation of articles on such topics as the auditory systems, the visual system, olfaction, the vestibular system, and the gustatory system. These sensory systems are discussed in relationship to their interconnections to the brain and the nervous system.

2791. Separable Connector Design Handbook. F.A. Rathbun Jr. NTIS, 1974, 269 pp. Accession: N74-76953. Tech Report: NASA-CR-64944.

This manual was compiled at General Electric Company at the Advanced Technology Laboratories and includes such information as fluid transmission lines, propulsion system configurations, seals, leakage, and containment of liquid rocket propellants.

2792. Separation Procedures in Inorganic Analysis: A Practical Handbook. Roland S. Young. Halsted Press,

1979, approx. 400 pp. LC: 79-17725. ISBN: 0-470-26842-5. Cost: $65.95.

Provides information on the 57 common inorganic elements. Preparation and purification of elements and compounds in the inorganic chemical industry are discussed.

2793. Serial Number Reference Book for Metalworking Machinery. Industrial Machinery News, Book Publishing Division, 5th ed, 552 pp. Cost: $29.95.

Includes information on metalworking machinery such as manufacturer, year of manufacture, and serial number. U.S., European, and Asian manufacturers are listed, and many conversion tables and metric charts are provided.

2794. Setting Out Tables for Circular Curves. H. Osterloh; W. Weber. International Publications Service, 2d ed., 1975, 515 pp. (80 pp. text, 435 pp., tables). ISBN: 3-7625-0526-8. Cost: $22.50.

Setting out tables are provided for small building projects.

2795. Seven-Place Natural Trigonometrical Functions. Howard C. Ives. Wiley-Interscience, 1929, 222 pp. ISBN: 0-471-43098-6. Cost: $16.95.

2796. Sewer and Water-Main Design Tables (British and Metric). Tables of Flow in Sewers, Drains, and Water-Mains in British and Metric Units. Also Tables of Rainfall, Run-off, Repayment of Loans, etc. L.B. Escritt. Maclaren & Sons, Ltd., London, 1969, 260 pp. ISBN: 0-85334-035-8. Cost: Dfl.65.00 (approx. $18.00).

2797. Shands' Handbook of Orthopedic Surgery. Richard Beverly Raney; Robert H. Brashear. Mosby, 9th ed., 1978.

2798. Sheet Steel Handbook. Jack Blair. General Electric Co, Research and Development Center, 1st ed., 160 pp. Cost: $39.95.

Provides information on such topics as purchasing sheet steel, properties of sheet steel, material grades, finishes, tolerances, chemical composition, and coil sizes.

2799. Shielding for High-Energy Electron Accelerator Installations. National Bureau of Standards. NCRP Publications. Cost: $2.00. Series: NBS Handbook 97.

Available from NCRP Publications, PO Box 30175, Washington, DC 20014 as NCRP Report 31.

2800. Ships Instrumentation Handbook. O.C. Ledford (Federal Electric Corp). NTIS, 1971, 277 pp. Accession: AD-891518 (supersedes report dated April 1970, AD-874624). Tech Report: SAMTEC-TR-71-5.

2801. Shock and Vibration Handbook. Cyril M. Harris; Charles E. Crede. McGraw-Hill, 2d ed., 1976, 3 Vols., 1,344 pp. (various paging). LC: 76-55. ISBN: 0-07-026799-5. Date of First Edition: 1961. Cost: $84.50 set.

2802. Shock-Tube Thermochemistry Tables for High-Temperature Gases. T.E. Horton; W.A. Menard. NTIS, multivolume, Vol.1, 1969; Vol.4, 1970; Vol.5, 1971, Vol.1, 88 pp.; Vol.4, 65 pp.; Vol.5, 67 pp. Accession: Vol.1, N70-10726; Vol.4, N71-17899; Vol.5, N74-10864. Tech Report: Vol.1, NASA-CR-106660 JPL-TR-32-1048; Vol.4,

NASA-CR-116444 JPL-TR-32-1408; Vol.5, NASA-CR-135944 JPL-TR-32-1408.

Volumes 1, 4, and 5 of this set are entitled as follows: Vol.1, *Air*; Vol.4, *Nitrogen*; Vol.5, *Carbon Dioxide*. Lists such topics as thermodynamic properties, pressure gradients, and shock waves.

2803. Shorter Bergey's Manual of Determinative Bacteriology. John Holt. Williams and Wilkins, 1978. ISBN: 0-683-04105-3.

2804. Shortwave Listener's Handbook. Norman Fallon. Heyden, 2d ed., 1976, 144 pp. ISBN: 0-8104-5044-5. Cost: $5.30.

This handbook covers basic equipment, including all types of receivers and antennas. A listing by country and frequency of the major shortwave broadcast stations of the world is a key feature of this handbook.

2805. SI and Metrication Conversion Tables. Sapper Socrates. Butterworth Publishers, Inc, 1969, 28 pp. ISBN: 0-408-08473-1. Cost: $2.95.

Conversion tables are provided for the International System of Units and the metric system.

2806. SI Chemical Data. Gordon H. Aylward; Tristan J.V. Findlay. Wiley-Interscience, 2d ed., 1975, 136 pp. LC: 75-148002. ISBN: 0-471-03851-2. Cost: $8.50.

Includes information on such topics as properties of inorganic and organic compounds, physical and thermochemical data, latent heats, and dipole moments.

2807. SI Units in Engineering and Technology. S.H. Qasim. Pergamon Press, 1977, 47 pp. LC: 76-27670. ISBN: 0-08-021278-6. Cost: $6.50.

Covers conversion factors, properties of metals, units for applied mechanics, and electrical and magnetic units.

2808. Side Effects of Drugs Annual: A Worldwide Yearly Survey of New Data and Trends. M.N.G. Dukes. Elsevier, Annual 1, 1977; Annual 2, 1978; Annual 3, 1979; Annual 4, 1980, Annual 1, 420 pp.; Annual 3, 470 pp.; Annual 4, 376 pp. ISBN: Annual 1, 90-219-3038-2; Annual 2, 0-444-90023-3; Annual 3, 0-444-90072-1; Annual 4, 0-444-90130-2. Cost: Annual 1, $48.50; Annual 2, $53.75; Annual 3, $59.75; Annual 4, $63.50.

Annual 1, *SEDA, 1.* Annual 2, *SEDA, 2.* Annual 3, *SEDA, 3.* Annual 4, *SEDA, 4.* This annual publication provides information on the side effects of drugs.

2809. Silicon Data Sheets. M. Neuberger; S.J. Welles. NTIS, 1969, 270 pp. Accession: N70-22542; AD-698342. Tech Report: EPIC-DS-162.

Information has been supplied by Hughes Aircraft Corporation on optical, magnetic, thermal, and electronic properties of silicon. Energy band structure data are also presented.

2810. Silver Amalgam in Clinical Practice: Dental Practitioner Handbook No. 1. I.D. Gainsford. Year Book Medical Publishers, 2d ed., 1976. ISBN: 0-8151-3004-9. Cost: $9.95.

2811. Silver Azide, Cyanide, Cyanamides, Cyanate, Selenocyanate, and Thiocyanate—Solubilities of Solids.

M. Salomon. LC: 79-40362. ISBN: 0-08-022350-8 hard cover. Series: *Solubility Data Series*, Vol. 3.

2812. Silver Halides I—Solubilities of Solids. E.M. Woolley. ISBN: 0-08-023923-4. Series: *Solubility Data Series*, Vol. 7.

2813. Simplified Procedures for Water Examination. M.J. Taras. American Water Works Association, 1975, 158 pp. ISBN: 0-89867-070-5. Cost: $14.00. Series: *AWWA* No. M12.

Numerous tests are outlined in this book, including chemical examinations of water for such substances as ammonia nitrogen, carbon dioxide, chlorine, copper, fluorine, and manganese phosphates. In addition, testing methods are presented for odor, bacteria, and hardness.

2814. Simultaneous Linear Equations and the Determination of Eigenvalues. L.J. Paige; Olga Taussky (eds) (National Bureau of Standards). National Bureau of Standards (avail. from NTIS), 1953, 126 pp. Accession: AD-695953. Cost: $6.00. Series: NBS AMS29.

2815. Sines, Cosines, and Tangents, Ten Decimal Places with Ten Second Interval, 0°-6°. Coast and Geodetic Survey. NTIS, rev. of 1949 paper, 1971, 38 pp. Accession: COM-71-50369. SuDoc: C4.19:246. Series: *U.S.C.G.S. Special Publication*, No. 246.

Trigonometrical tables are presented up to 10 decimal places at 10 second intervals.

2816. Single Crystal Elastic Constants and Calculated Aggregate Properties: A Handbook. Gene Simmons; Herbert Wang. MIT Press, 2d ed., 1971, 370 pp. LC: 71-148848. ISBN: 0-262-19092-3. Cost: $20.00.

This handbook provides information on over 300 different metals and compounds, including such properties as elastic stiffness and elastic compliance as well as crystal density and temperature data.

2817. Six-Figure Tables of Trigonometric Functions. Leonid Sergeevich Khrenov; D.E. Brown (trans). Pergamon Press, 1965, 364 pp. LC: 63-18923. ISBN: 0-08-010101-1. Cost: $33.00. Series: *Mathematical Table Series*, Vol. 26.

2818. Six-Figure Trigonometrical Functions of Angles in Degrees and Minutes. C. Attwood. Pergamon Press, 5th ed., 1965. ISBN: 0-08-009893. Cost: $2.25. Series: *Practical Table Series*, Vol. 1.

2819. Skew Dimensions for Bridges, Grade Separations and Structures. Bartolome M. Ygay. Edwards Brothers, Inc, 1968, 359 pp. LC: 68-8797.

Tabular dimensions of sines, cosines, tangents, and secants of 660 given dimensions are provided in this manual. These dimensions will be helpful in calculating skew angles for detailing bridges and other structures.

2820. Small Animal Surgery: An Atlas of Operative Techniques. Wayne E. Wingfield; C.A. Rawlings. Saunders, 1979, 228 pp. LC: 78-57920. Cost: $28.00.

Methods and procedures are presented that will be useful for the veterinary surgeon.

2821. Small Computer Systems Handbook. Sol Libes. Hayden Book Co, 1978, 208 pp. ISBN: 0-8104-5678-8. Cost: $9.15.

Information is provided on the programing of small computers, as well as a discussion on digital logic and interconnecting components with microprocessors.

2822. Small Hydraulic Structures. D.B. Kraatz; S.K. Mahajan. Unipub, 1975-, 3 Vols. (3d Vol. in progress). ISBN: 0-685-54184-3. Cost: $47.50. Series: *Irrigation and Drainage Paper*, No. 26.

2823. Small Motor, Gearmotor and Control Handbook. Clay Bodine. Bodine Electric Co, 1979, 210 pp. LC: 78-57278. Cost: $3.50.

The emphasis of this handbook is on small electric motors.

2824. SME Mining Engineering Handbook. Arthur B. Cummins; Ivan A. Given. Society of Mining Engineers of AIME, 1973, 2-Vol. set. LC: 72-86922. ISBN: 0-89520-021-X. Cost: $45.00.

2825. Smithsonian Meterological Tables. Robert J. List (ed). Smithsonian Institution, 6th rev. ed., 1951 (rep. 1963), 527 pp. LC: 51-61623. Series: *Smithsonian Institution Publication 4014—Smithsonian Miscellaneous Collections*, V.114.

2826. Smithsonian Physical Tables. William Elmer Forsythe. Smithsonian Institution, 9th rev. ed., 1954 and 1969, 827 pp. LC: 54-60067. SuDoc: SI1.7:120. Series: Smithsonian Publication No. 4169—Smithsonian Miscellaneous Collections, Vol. 120.

This handbook consists chiefly of tables on physics. It includes bibliographical references and an index.

2827. Smoley's Four Combined Tables. C.K. Smoley; E.R. Smoley (rev); N.G. Smoley (rev). Smoley and Sons, Inc, rev. ed., 1974, 1 thumb-indexed Vol. ISBN: 0-911390-00-6. Cost: $26.00.

2828. Smoley's Three Combined Tables. C.K. Smoley; E.R. Smoley (rev); N.G. Smoley (rev). Smoley and Sons, rev. ed., 1974, 1 thumb-indexed Vol.. ISBN: 0-911390-01-4. Cost: $20.00.

Society of Automotive Engineers. *See* SAE

Society of Mining Engineers of AIME. *See* SME

Society of Photographic Scientists and Engineers. *See* SPSE

Society of the Plastics Industry. *See* SPI

2829. SOCMA Handbook, Commercial Organic Chemical Names. Synthetic Organic Chemical Manufacturer's Association. American Chemical Society, 1965, various paging.

2830. Sodium-NaK Engineering Handbook. O.J. Foust. Gordon and Breach, 1976, 5 Vols. LC: 70-129473. ISBN: Vol.1,0-677-03070-3; Vol.2, 0-677-03030-4; Vol.3, 0-677-03040-1; Vol.4, 0-677-03050-9; Vol.5, 0-677-03060-6. Cost: Vol.1, $73.25; Vol.2, $82.00; Vol.3, $73.75; Vol.4,

$73.75; Vol.5, $73.75; 5-Vol. Set $319.00. Series: *U.S. Atomic Energy Commission Monograph Series.*

The *Sodium-NaK Engineering Handbook* consists of 5 volumes with the following titles: Vol.1, *Sodium Chemistry and Physical Properties*; Vol.2, *Sodium Flow, Heat Transfer, Intermediate Heat Exchangers,and Steam Generators*; Vol.3, *Sodium Systems, Safety, Handling, and Instrumentation*; Vol.4, *Sodium Pumps, Valves, Piping, and Auxiliary Equipment*; Vol.5, *Sodium Purification, Materials, Heaters, Coolers, and Radiators.* This handbook woould be useful to designers of liquid metal fast breeder reactors.

2831. Soft-Soldering Handbook. C.J. Thwaites. International Tin Research Institute, 120 pp. Cost: £3.00. Series: *I.T.R.I. Publication*, No. 533.

Such topics as joint design, choice of flux, and solder alloy are discussed in connection with all types of soldered assemblies. In addition, testing methods, quality control, and safety are discussed in this publication.

2832. Solar Cell Array Design Handbook; the Principles and Technology of Photovoltaic Energy Conversion. Hans S. Rauschenbach. Van Nostrand Reinhold, 1980, 549 pp. LC: 79-9238. ISBN: 0-442-26842-4. Cost: $42.50.

This book was originally issued as 2 volumes in 1976 under NASA Report No. NASA-CR-149364. This substantially revised Van Nostrand edition provides information on such topics as terminals, connectors, space arrays, photovoltaic testing, design of solar cells, and various electrical, mechanical, and optical characteristics of solar arrays. Radiation shielding design and electromagnetic design are covered extensively in this handbook.

2833. Solar Energy Handbook. Jan F. Kreider; Frank Kreith. McGraw-Hill, 1981, 1,099 pp. LC: 79-22570. ISBN: 0-07-035474-X. Cost: $49.50.

Information is presented on the legal and economic aspects of solar energy, as well as its technical aspects, such as solar-thermal collection and conversion methods. In addition, low-temperature systems, high-temperature systems, and indirect solar conversion systems are discussed.

2834. Solar Energy Handbook; A Practical Engineering Approach to the Application of Solar Energy to the Needs of Man and Environment Including Sections on Terrestrial Cooling and Wind Power. Henry Clyde Landa et al. Ficoa/Seecoa, 5th ed., 1977. ISBN: 0-931974-03-8. Cost: $14.40.

2835. Solar Energy Technology Handbook. William C. Dickinson; Paul N. Cheremisinoff. Marcel Dekker, Inc, 1980, Part A, 912 pp.; Part B, 848 pp. LC: 80-1026. ISBN: Part A, 0-8247-6872-8; Part B, 0-8247-6927-9. Cost: $160.00 set; $85.00 Part A; $85.00 Part B. Series: *Energy, Power, and the Environment Series*, Vol. 6.

Part A is entitled *Engineering Fundamentals.* Part B is entitled *Applications, Systems Design, and Economics.* Such topics as photovoltaics, solar energy storage systems, solar-thermal collectors, bioconversion, and photosynthesis are discussed. In addition, numerous applications of solar technology are presented, such as heating for swimming pools, domestic water systems, space heating for buildings, distillation of seawater, irrigation pumping, electric power generation, and food dehydration processes.

2836. Solar Products Specifications Guide. Solar Vision, Inc, 1979-, loose-leaf service. ISBN: 0-686-65545-1. Cost: $165.00 per year.

This publication is a companion publication to the magazine, *Solar Age*, which is also published by Solar Vision. This guide provides information on over 400 different types of solar equipment from numerous U.S. manufacturers.

2837. Soldering Electrical Connections: A Handbook. NASA. Superintendent of Documents, rev. ed., 1979, 66 pp. SuDoc: NAS-1.21:5002/2. Date of First Edition: First ed., 1967. Cost: $2.10.

Information is provided on such topics as lacing of cable trunks, termination of shields by soldering, hand-soldering, and automatic machine soldering. Numerous photographs and detailed instructions are provided for different types of soldering techniques.

2838. Soldering Handbook; A Practical Manual for Industry and the Laboratory. B.M. Allen. Iliffe Books, Ltd, 1969, 120 pp. ISBN: 0-592-05752-8.

2839. Solubilities of Inorganic and Organic Compounds. Henry Stephen; T. Stephen. Pergamon Press (also Macmillan, New York), 1963, 3 Vols., 7,300 pp. LC: 79-40319. ISBN: 0-08-023599-9. Cost: $900.00.

This is a Russian translation of a publication originating in Moscow under the sponsorship of VINITI. Covers Binary systems as well as ternary and multi component systems of inorganic compounds. This 3-volume set has over 7,000 pages. (Volume three alone has over 5,500 tables) and an index of chemical component formulas and table numbers is included.

2840. Solubilities, Inorganic and Metal-Organic Compounds; A Compilation of Solubility Data from the Periodical Literature. William F. Linke; Atherton Seidell (ed) (Vols.1-3). American Chemical Society, 4th ed., Vol.1, 1958; Vol.2, 1965, Vol.1, 1,486 pp.; Vol.2, 1,914 pp. ISBN: Vol.1, 0-8412-0097-1; Vol.2, 0-8412-0098-X. Cost: $40.00 per Vol.

Provides solubility data of compounds in water and other solvents.

Solubility Data Series, Vol. 4, *See* **Argon—Gas Solubilities.**

2841. Solubility Data Series. A.S. Kertes (editor-in-chief). Pergamon Press, 1979, 18 Vols. (various paging). Cost: Vols. 1-8, $600.00 per set; $100.00 per Vol; Vols. 1-18, $1,200.00; Entire 80-100 Vol set, $10,000.00 prepaid.

This series is a monumental effort sponsored by the Union of Pure and Applied Chemistry and will eventually consist of approximately 100 volumes. It is projected that this set will be complete by 1988. Each volume will provide critically evaluated solubility data and recommended values. Data sheets will be provided with the corresponding source material for further investigation. All types of solubility data will be included in this series, such as solubility of gases in liquids, gases in solids, liquids in liquids, and solids in solids.

2842. Solubility in Inorganic Two-component Systems. M. Broul; J. Nyvlt; O. Soehnel. Elsevier-North Holland, 1980, 576 pp. ISBN: 0-444-99763-6. Cost: $90.25.

Solubility data are presented for approximately 500 inorganic substances in aqueous solutions.

2843. Solvents and Allied Substances Manual With Solubility Chart. Cyril Marsden. Elsevier-North Holland, 1954, 429 pp.

Solvents are listed in alphabetical sequence in dictionary format. Physical characteristics and properties, including boiling point, flash point, refractive index, dielectric constant, colorific values, specific heat, coefficient of cubic expansion, electrical conductivity, latent heat of fusion, viscosity, vapor pressure, and solubility are presented in this manual.

2844. Sound Recording Practice: a Handbook Compiled by the Association of Professional Recording Studios. John Borwick. Oxford University Press, 1976, 440 pp. ISBN: 0-19-311915-3. Cost: $35.00.

Covers such topics as acoustics, mixing consoles, equipment alignment, disc cutting, tape duplicating, television, and film.

2845. Source Book for Food Scientists. Herbert W. Ockerman. AVI Publishing Co, Inc, 1978, 926 pp. ISBN: 0-87055-228-7. Cost: $75.00.

Provides information on food composition and properties, such as dairy products, tangerine oil, and wheat products. The book is divided into 2 parts, and each part is arranged in alphabetical sequence for easy access to information.

2846. Source Book of Food Enzymology. Sigmund Schwimmer. AVI Publishing Co, Inc, 1981, 706 pp. ISBN: 0-87055-369-0. Cost: $79.50.

Enzyme action in foods is discussed in connection with a number of topics, such as flavors, food coloring, protein modification, and texture. Various types of foods are discussed in this handbook, including fish, meats, beverages, citrus fruits, vegetables, cheese, and bread.

2847. Source Book on Atomic Energy. S. Glasstone. Van Nostrand Reinhold, 3d ed., 1967, 883 pp. LC: 67-29947. ISBN: 0-442-02705-2. Cost: $16.95.

2848. Source Book on Brazing and Brazing Technology. Melvin M. Schwartz. ASM Press, Inc. Cost: $30.40 members; $38.00 nonmembers. Series: *ASM's Source Books Series*.

Such topics as brazing alloys, filler metals, brazing nonmetals to metals, and properties of brazed joints are discussed in this compilation.

2849. Source Book on Copper and Copper Alloys. ASM Press, Inc, 1979, 424 pp. Cost: $30.40 members; $38.00 nonmembers. Series: *ASM's Source Books Series*.

This is a compilation of authoritative articles on copper alloys and properties of copper. In addition, such topics as processing and fabrication of copper alloys are discussed.

2850. Source Book on Ductile Iron. A.H. Rauch. ASM Press, Inc, 1977, 400 pp. Cost: $30.40 members; $38.00 nonmembers. Series: *ASM's Source Books Series*.

Such topics as casting charactertistics of ductile iron, properties of ductile iron, heat treatment of ductile iron, quality control, and inspection of ductile iron are discussed in the compilation.

2851. Source Book on Electron Beam and Laser Welding. Melvin M. Schwartz. ASM Press, Inc, 1980, 400 pp. Cost: $30.40 members; $38.00 nonmembers. Series: *ASM's Source Books Series*.

This handbook covers such topics as the advantages of electron beam and laser welding when compared to arc welding.

2852. Source Book on Gear Design, Technology and Performance. Maurice A.H. Howes. ASM Press, Inc, 1979, 417 pp. Cost: $30.40 members; $38.00 nonmembers. Series: *ASM's Source Books Series*.

Such topics as stress analysis, lubrication, pitting, scoring, gear failure, and contact fatigue are discussed in this sourcebook.

2853. Source Book on Industrial Alloy and Engineering Data. ASM Press, Inc, 1978, 480 pp. Cost: $30.40 members; $38.00 nonmembers. Series: *ASM's Source Books Series*.

Property data on metals and alloys are presented, including properties of carbon and alloy steels, stainless steels, tool steels, wrought stainless steels, and cast irons. In addition, information is presented on aluminum alloys, copper alloys, refractory metals, and elevated temperature alloys.

2854. Source Book on Innovative Welding Processes. Melvin M. Schwartz. ASM Press, Inc, 1981, 364 pp. Cost: $30.40 members; $38.00 nonmembers. Series: *ASM's Source Books Series*.

High-frequency welding, diffusion welding, explosive welding, and friction welding are discussed in this sourcebook.

2855. Source Book on Maraging Steels. R.F. Decker. ASM Press, Inc, 1979, 400 pp. Cost: $30.40 members; $38.00 nonmembers. Series: *ASM's Source Books Series*.

Such topics as mechanical properties of maraging steels, heat treatment, casting, welding, stress corrosion cracking, and behavior of maraging steels are discussed. In addition, hydrometallurgy, plating and machining are also presented in this publication.

2856. Source Book on Materials for Elevated-Temperature Applications. Elihu E. Bradley. ASM Press, Inc, 1979, 408 pp. Cost: $30.40 members; $38.00 nonmembers. Series: *ASM's Source Books Series*.

Stainless steels, heat-resistant alloy castings, superalloys, and 12 percent carbon steels are presented in this sourcebook.

2857. Source Book on Materials Selection. Russell B. Gunia. American Society for Metals, 1977, 2 vols, Vol.1, 494 pp.; Vol.2, 477 pp. LC: 77-1347. ISBN: 0-87170-031-X. Cost: $38.00 per Vol.

A collection of articles on many different metals which discuss properties and performance. Also includes information on plastics, ceramics, composites, and coatings.

2858. Source Book on Nitriding. ASM Press, Inc, 1977, 328 pp. Cost: $30.40 members; $38.00 nonmembers. Series: *ASM's Source Books Series*.

Various different applications of nitriding are discussed in this handbook along with principles, processes, and the latest developments.

2859. Source Book on Powder Metallurgy. Samuel Bradbury. ASM Press, Inc, 1978, 480 pp. Cost: $30.40 members; $38.00 nonmembers. Series: *ASM's Source Books Series*.

Such topics as ferrous and nonferrous powders, sintering, powder mixing, machining, and properties of metal powders are discussed. In addition, applications, microstructure, and preparation of metal powders are considered.

2860. Source Book on Selection and Fabrication of Aluminum Alloys. ASM Press, Inc, 1978, 470 pp. Cost: $30.40 members; $38.00 nonmembers. Series: *ASM's Source Books Series.*

Numerous authoritative articles are presented on such topics as cold extrusion of aluminum, forging of aluminum, heat treating, welding, brazing, and soldering. In addition, aluminum and aluminum alloys are discussed in connection with forming, adhesive joining, cleaning, and finishing.

2861. Source Book on Wear Control Technology. David A. Rigney; W.A. Glaeser. ASM Press, Inc, 1978, 454 pp. Cost: $30.40 members; $38.00 nonmembers. Series: *ASM's Source Books Series.*

This collection of 47 authoritative articles presents such topics as abrasion, dry wear, erosion, fretting, coatings, case-hardening processes, and lubrication as they relate to wear control.

2862. Sourcebook of Electronic Circuits. John Markus. McGraw-Hill, 1968, 888 pp. LC: 67-15037. ISBN: 0-07-040443-7. Cost: $42.50.

2863. Sourcebook of Modern Transistor Circuits. Laurence G. Cowles. Prentice-Hall, 1976, 360 pp. ISBN: 0-013-823419-1. Cost: $19.95.

Contains over 300 transitor circuits.

2864. Sourcebook of Programmable Calculators. Texas Instruments, Inc. MaGraw-Hill, 1979, 320 pp. LC: 78-27823. ISBN: 0-07-063746-6. Cost: $14.50.

This book presents information on the many different applications of programmable calculators. Numerous equations and hundreds of worked out examples are provided in this publication.

2865. Sourcebook on Food and Nutrition. Marquis Who's Who, Inc, 1st ed, 1978, 498 pp. LC: 78-60134. ISBN: 0-8379-4501-1. Cost: $34.50.

Such topics as weight control and dieting, food-drug interactions, and malnutrition are included in this compilation.

2866. Soviet Aerospace Handbook. U.S. Department of the Air Force. U.S. Government Printing Office, 1978, 223 pp. LC: 78-20412. SuDoc: D 301.35:200-21. S/N: 008-070-00402-0. Cost: $3.00.

Information is provided on Soviet Aerospace Forces as well as Soviet Armed Forces. Military and aerospace resources are presented.

2867. Soviet Alloy Handbook. Metals and Ceramics Information Center. NTIS, 1st supplement, 1976, 208 pp. Accession: AD-A031 699. Tech Report: MCIC-HB-0551. Cost: $15.00.

See also report dated February, 1975, AD-A009 118, Technical Report No. MCIC-HB-05, $45.00, 362 pp.

2868. Space Groups and Lattice Complexes. Werner Fisher; Hans Burzlaff; Erwin Hellner; J.D.H. Donnay (Na-

tional Bureau of Standards). NTIS, 1973. Accession: COM 73-50582. Cost: $7.50. Series: NBS Monograph 134.

2869. Space Handbook: Astronautics and Its Applications. Rand Corp. Random House, 1959, 330 pp. LC: 59-10836.

2870. Space Materials Handbook. John R. Rittenhouse; John B. Singletary (Lockheed Missiles and Space Co). NTIS, 3d ed., 1968, 739 pp. Accession: AD-692353. Tech Report: AFML-TR-68-205.

Information is provided on mechanical, physical chemical properties, and characteristics are given for a wide variety of metallic and nonmetallic materials. Materials categories includes coverage of thermal control materials, optical materials, adhesives, and electronic compounds.

2871. Specifications and Criteria for Biochemical Compounds. National Research Council. NAS-NRC, 3d ed., 1972, 224 pp. ISBN: 0-309-01917-6. Cost: $21.25.

NOTE: NAS-NRC Publication No. 1344.

2872. Specifications and Tolerances for Field Standard Measuring Flasks. National Bureau of Standards. NTIS, Jan., 1971. Accession: COM 71-50065. Cost: $3.50. Series: NBS Handbook 105-2.

2873. Specifications and Tolerances for Field Standard Weights (NBS Class F). National Bureau of Standards. NTIS, rev., 1972. Accession: COM 72-50707. Cost: $3.50. Series: NBS Handbook 105-1.

2874. Specifications and Tolerances for Reference Standards and Field Standard Weights and Measures. National Bureau of Standards. U.S. Government Printing Office, May 1971. SuDoc: C13.11:105-3. Cost: $0.25. Series: NBS Handbook 105-3.

2875. Specifications for Adhesives. International Plastics Selector, Inc, 1979, 302 pp. Cost: $21.00 members; $25.95 nonmembers. Series: *Desk Top Data Bank Series.*

Specifications for adhesives, sealants, and resins are listed. Such topics as bond strengths, lap shear, and cure times are presented.

Specifications for Aluminum Structures. *See* **Aluminum Construction Manual.**

2876. Specifications for Dry Cells and Batteries. National Bureau of Standards. American National Standards Institute. SuDoc: C18.1-1972. Cost: $6.25. Series: NBS Handbook 71.

2877. Specifications for Plastics. International Plastics Selector, Inc, 1979, 350 pp. Series: *Desk Top Data Bank Series.*

Over 350 specifications dealing with thermosets, elastomeric resins, reinforced resins, and thermoplastics are discussed in this compilation.

2878. Specifications to Support Classification, Standards of Accuracy, and General Specifications of Geodetic Control Surveys. Federal Geodetic Control Committee. U.S. Department of Commerce, National Oceanic and Atmospheric Administration, National Geodetic Survey, 1975, 30 pp.

Information on conducting geodetic surveys is provided in this publication.

2879. Specifications, Tolerances, and Other Technical Requirements for Commercial Weighing and Measuring Devices. National Bureau of Standards. U.S. Government Printing Office, 4th ed., Nov. 1971 (supersedes Handbook 44, 3d ed.). SuDoc: C13.11:44. Cost: $7.90. Series: NBS Handbook 44.

2880. Spectral and Chemical Characterization of Organic Compounds: A Laboratory Handbook. W.J. Criddle; G.P. Ellis. John Wiley and Sons, 2d ed., 1980, 115 pp. LC: 80-40497. ISBN: 0-471-27813-0 cloth; 0-471-27812-0 paper. Cost: $39.00 cloth; $13.95 paper.

Includes revised melting point tables and information on the interpretation of spectra.

2881. Spectral Atlas of Nitrogen Dioxide—5530 to 6480 A. Donald K. Hsu; David L. Monts; Richard N. Zare. Academic Press, Inc, 1978, 646 pp. ISBN: 0-12-357950-3. Cost: $49.50.

This atlas is primarily a line atlas covering the region 5530 A to 6480 A and listing approximately 19,000 prominent lines. A selected bibliography is provided as well as indexes for the convenience of the user.

2882. Spectral Data and Physical Constants of Alkaloids. J. Holubek. Heyden, 1965-73, 5 Vols. ISBN: 0-85501-070-3. Cost: $400.00 set.

This is a collection of 1,000 data cards in 5 volumes, including an index for 1965-1973. Ultraviolet and infrared spectral data are provided for approximately 1,000 alkaloid compounds.

2883. Spectral Data of Natural Products, Volume l. K. Yamaguchi. Elsevier-North Holland, 1970, 765 pp. ISBN: 0-444-40841-X. Cost: $134.25.

A substantial amount of spectral data are provided for the identification of natural products.

2884. Spectrometric Identification of Organic Compounds. R.M. Silverstein; G.C. Bassler; T.C. Morrill. John Wiley and Sons, 4th ed., 1981, 704 pp. est. LC: 80-20548. ISBN: 0-471-02990-4. Cost: $23.95.

Mass spectra, infrared spectra, ultraviolet spectra, and nuclear magnetic resonance data are provided for numerous organic compounds.

2885. Spectroscopic Data Relative to Diatomic Molecules. Tables Internationales de Constantes, Paris. Pergamon Press, 1971, 530 pp. ISBN: 0-08-016546-X. Cost: $123.00.

International Table of Selected Constants, Vol. 17.

2886. Spectroscopic Data. S.N. Suchard. IFI/Plenum Press, Vol.1, 1975; Vol.2, 1976, Vol.1, 1,235 pp; Vol.2, 585 pp. LC: 74-34288. ISBN: Vol.1, 0-306-68311-3; Vol.2, 0-306-68312-1. Cost: Vol.1, $150.00; Vol.2, $75.00.

Vol. 1 is entitled *Heteronuclear Diatomic Molecules*. Vol. 2 is entitled *Homonuclear Diatomic Molecules*.

Spencer-Meade Cane Sugar Handbook. *See* **Cane Sugar Handbook.**

2887. SPI Handbook of Technology and Engineering of Reinforced Plastics/Composites. J. Gilbert Mohr et al. Van Nostrand Reinhold, 2d ed., 1973, 406 pp. ISBN: 0-442-25448-2. Cost: $32.50.

Covers curing processes, thermoset molding processes, reinforced thermoplastics, properties of materials, tool and product design, as well as processing temperature and equipment requirements.

SPI Plastics Engineering Handbook. *See* **Plastics Engineering Handbook.**

2888. Spillman's Handbook of Conversion Factors. James R. Spillman. Dorrance and Company, 1977, 224 pp. LC: 76-151874. Cost: $8.95.

Numerous conversion factors are included in this handy publication.

2889. Spray Drying Handbook, 3rd Edition. K. Masters; G. Godwin. Halsted Press, 3d ed., 1979, 687 pp. LC: 79-40732. ISBN: 0-470-26549-3. Cost: $113.95.

Such topics as atomization, thermal efficiency, heat balance, mass balance, spray mixing, and spray flow are discussed. Numerous applications are presented for spray during equipment in the chemical industry, food industry, and the pharmaceutical industry.

2890. Spring Designer's Handbook. Harold Carlson. Marcel Dekker, Inc, 1978, 368 pp. ISBN: 0-8247-6623-7. Cost: $29.50. Series: *Mechanical Engineering Series*, Vol. 1.

Spring materials are presented with details on the manufacturing of springs. Such topics as corrosion, heat treatment, tolerance, and design stresses are discussed in connection with springs. Various metals, such as high carbon steels, chromium-vanadium steels, hot-rolled bars, stainless steel, and nickel based alloys are discussed in connection with spring manufacture.

2891. SPSE Handbook of Photographic Science and Engineering. Woodlief Thomas Jr. John Wiley and Sons, 1973, 1,416 pp. LC: 72-10168. ISBN: 0-471-81880-1. Cost: $72.00.

2892. Stability and Control Handbook for Compound Helicopters. E.K. Garay; E. Kisielowski. NTIS, 1971, 625 pp. Accession: AD-722250. Tech Report: DCR-314 USA ABLABS-TR-70-67.

Stability of Metal-Ion Complexes. *See* **Critical Stability Constants.**

2893. Stahschlüssel (Key to Steel). Verlay Stahlschlüssel Verlag, 11th ed., 1977, 460 pp. Cost: $36.00.

This publication is presented in English, French, and German. Includes over 40,000 brands and designations of steels. German standards are listed and other countries' standards are cross-listed.

2894. Stainless Steels. International Plastics Selector, Inc. International Plastic Selector, Inc, 1st ed.

This is a compilation of thousands of commercially available stainless steels, with the corresponding manufacturers, commercial names, and standard specification numbers.

2895. Stair Builders Handbook. T.W. Love. Craftsman Book Co, 1974, 416 pp. LC: 74-4298. ISBN: 0-910460-07-8. Cost: $8.00.

This is a collection of tables with over 3,500 rise and run combinations for every stairway between a 3 foot and 12 foot rise.

2896. Standard Handbook for Civil Engineers. Frederick S. Merritt. McGraw-Hill Book Co., 2nd ed, 1976, 1,344 pp (various pagings). LC: 75-25850. ISBN: 0-07-041510-2. Date of First Edition: 1968. Cost: $46.50.

This handbook provides a very comprehensive collection of information in the field of civil engineering including such areas as surveying, transportation, water, and sanitary engineering. Materials used in construction are discussed, including concrete, laminates, asphalts, polymers, and metallics. Although this handbook is oriented towards practical applications, some theory is prsented such as stress-strain relations, bending moment diagrams, axial and bending loads on beams, continuous loads on beams and frames, torsion, and thin-shell structures. Information on heavy equipment and excavation is presented along with blasting and the use of explosives. In addition, such subjects areas as bridges, highways, tunnels, sanitary landfills, and specification writing are discussed. An index is provided, and references are made to additional sources of information throughout the text.

2897. Standard Handbook for Electrical Engineers. Donald G. Fink; John M. Carroll. McGraw-Hill, 11th ed., 1978, 2,448 pp (various paging). LC: 56-6964. ISBN: 0-07-020974-X. Date of First Edition: 1907. Cost: $54.50.

Covers design data and methods relating to all components and systems employed in generation, transmission, control, conversion, and distribution of electricity. Includes sub-station design, relay protection, grounding, gas-insulated substations, power electronics. Also covers alternate sources and converters of power in reference to solar, geothermal, windpower, and nuclear fusion. For related works, please refer to the *Electrical Engineer's Reference Books*, the *Electronics Engineers' Handbook*, the *Basic Electronic Instrument Handbook*, the *Handbook of Electronic Packaging*, the *Handbook of Materials and Processes for Electronics*, the *Handbook of Modern Electronic Data; Industrial Electronics Handbook*, and the *Handbook of Semiconductor Electronics*. Also refer to the *Handbook for Electronics Engineering Technicians; Electronics Handbook*, and *Electronic Conversions, Symbols, and Formulas*.

Standard Handbook for Mechanical Engineers. *See* **Mark's Standard Handbook for Mechanical Engineers.**

2898. Standard Handbook of Engineering Calculations. T.G. Hicks. McGraw-Hill, 1972, 1,216 pp. LC: 73-130674. ISBN: 0-07-028734-1. Cost: $32.50.

Includes 5,000 calculation procedures, some with worked-out numerical examples, for the 12 major branches of engineering.

2899. Standard Handbook of Fastening and Joining. Robert O Parmley. McGraw-Hill, 1977, 704 pp. (various paging). LC: 75-45399. ISBN: 0-07-048511-9. Cost: $31.50.

This handbook covers a wide range of fastening techniques from welding to adhesive bonding. Fasteners for piping systems, riveting, and threaded products are also discussed. An index and general tables are included.

2900. Standard Handbook of Lubrication Engineering. James J. O'Connor; John Boyd. McGraw-Hill, 1968, 1,000 pp. (various paging). LC: 64-16489. ISBN: 0-07-047605-5. Cost: $46.50.

2901. Standard Handbook of Textiles. A.J. Hall. Halsted Press, 8th ed., 1975. ISBN: 0-470-34297-8. Cost: $26.95.

2902. Standard Methods for the Analysis for Oils, Fats, and Derivatives. C. Paquot. Pergamon Press, Inc, 6th ed., 1979, 140 pp. LC: 78-40305. ISBN: 0-08-022379-6-h. Cost: $25.00.

Standard methods are described for the determination of density of fats and oils in the liquid state. Such topics as the analysis of fruits and seeds, and the oil extracted, are discussed.

2903. Standard Methods for the Examination of Water and Waste Water. American Public Health Association; American Water Works Association; Water Pollution Control Federation. American Public Health Association Publications, 14th ed., 1976, 1,193 pp. LC: 55-1979 (rev). ISBN: 0-87553-078-8. Date of First Edition: 1917. Cost: $3.50.

2904. Standard Methods of Chemical Analysis. N. Howell Furman; F.J. Welcher. Van Nostrand Reinhold (Krieger), 6th ed., 1962-66, 3-Vol set. LC: Vol.1, 74-23465. ISBN: Vol.1, 0-88275-254-5. Cost: Vol.1, $48.50; Vol.2A, $54.75; Vol.2B $53.75; Vol.3A, $39.50; Vol.3B, $47.50.

Standard Methods of Chemical Analysis consists of 3 volumes entitled as follows: Vol.1, *The Elements*; Vol.2, *Industrial and Natural Products and Non-Instrumental Methods*; Vol.3, *Instrumental Methods*.

2905. Standard Plant Operators' Manual: A Basic Reference on the Operation and Maintenance of Energy Systems Equipment, Complete With License Examination Questions and Answers. Stephen Michael Elonka. McGraw-Hill, 3d ed., 1980, 416 pp. LC: 79-22089. ISBN: 0-07-019298-7. Cost: $22.95.

Such topics as steam equipment, steam turbines, diesel engines, air conditioning, gas burners, oil burners, pumps, weldings, and accident prevention are discussed in this manual.

2906. Standard Specfications for Transportation Materials and Methods of Sampling and Testing. American Association of State Highway and Transportation Officials. American Association of State Highway and Transportation Officials, 12th ed., 1978, 2-Vol set. ISBN: 0-686-20936. Cost: $33.00.

Part 1 is entitled *Specifications*, and Part 2 is entitled *Methods of Sampling and Testing*.

2907. Standards and Practices for Instrumentation. Instrument Society of America, 6th ed., 1980, 1,056 pp. ISBN: 0-87664-450-7. Cost: $120.00.

This is a guide to instrument design, testing, and maintenace. Numerous standards are listed for U.S. organizations as well as international organizations. Such topics as standards on control valves, electrical safety, transducers, and strain gages are among the many types of standards listed in this compilation.

2908. Standards Cross-Reference List. La Donna Thompson et al. MTS Systems Corp, 2d ed., 1977, 126 pp. Cost: $10.00 prepaid.

This is a listing of engineering standards adopted by 2 or more organizations with cross-references for ease of location and identification.

2909. State Plane Coordinate Systems (A Manual for Surveyors). H.C. Mitchell; L.G. Simmons. U.S. Department of Commerce, Coast and Geodetic Survey, rev. 1974, 62 pp. Date of First Edition: 1945. Series: *Special Publication*, No. 235.

Detailed information is presented on state plane coordinate systems. This is an excellent guide for those people who perform control surveys or employ the state coordinate system.

2910. Static Methods of Testing Reinforced Plastics. Y.M. Tarnopolskiy; T.Y. Kintsis. NTIS, 1976, 286 pp. Accession: N76-21588. Tech Report: NASA-TT-F-16669. Cost: $9.25 hard copy.

Detailed tables are presented on such topics as properties of composite materials, and property characteristics of reinforced plastics, elastic properties, strength properties, structural properties. Tensile testing, compression testing, bending testing, shear testing, and crushing testing data are recorded and tabulated.

2911. Statistical Tables for Biological, Agricultural, and Medical Research. Ronald A. Fisher; Frank Yates. Hafner, 6th rev. ed., 1974, 146 pp. Cost: $7.95.

2912. Statistical Tables for Science, Engineering, Management and Business Studies. John Murdoch. Wiley(Halsted), 1977. ISBN: 0-470-62515-5. Cost: $1.95.

2913. Statistical Theory of Extreme Values and Some Practical Applications. E.J. Gumbel (National Bureau of Standards). NTIS, 1954, 51 pp. Accession: PB-175818. Cost: $4.50. Series: NBS AMS33.

2914. Statistics Tables. H.R. Neave. Allen and Unwin, Inc, 1978, 80 pp. Cost: $3.95.

The tables in this compilation represent standard statistical techniques; it also includes tables on nonparametric methods, quality control, correlation, and analysis of variance. Numerous mathematical tables are also included for the convenience of the user.

2915. Steam and Air Tables in SI Units: Including Data for Other Substances and a Removable Fold-Out Mollier Chart for Steam. Thomas F. Irvine Jr; James P. Hartnett. Hemisphere Publishing Co, 1976, 136 pp. LC: 75-34007. ISBN: 0-89116-004-3. Cost: $9.95.

Steam Charts. *See* **ASME Steam Charts.**

2916. Steam Tables, 1964: Physical Properties of Water and Steam 0-800° C. 0-1000 Bars. Great Britain National Engineering Laboratory. H.M.S.O., Edinburg, 1964, 152 pp.

2917. Steam Tables; Thermodynamic Properties of Water, Including Vapor, Liquid, and Solid Phases. Joseph H. Keenan et al. Wiley, 2d ed., 1978. LC: 77-28321. ISBN: 0-471-04210-2. Cost: $20.95.

Extends the range of the original Keenan and Keyes Tables to 1300&So C and 1000 bars pressure. The authors have independently correlated all new experimental data on the thermodynamic transport properties of water.

2918. Steel and Its Heat Treatment: Bofors Handbook. K.E. Thelning. Butterworth and Co, 1975, 584 pp. ISBN: 0-408-70651-1. Cost: $47.50.

2919. Steel Buying Handbooks. General Electric Co, 3-Vol set. Cost: $35.95 per Vol.

The 3 volumes of this set are entitled as follows: Vol.1, *Sheet Steel Handbook*; Vol.2, *Strip Steel Handbook*; Vol.3, *Plate Steel Handbook*. These handbooks are designed for those who specify and purchase steel.

2920. Steel Castings Handbook. Peter F. Wieser. Steel Founders' Society of America, 5th ed., 1980, 456 pp. Cost: $35.00.

2921. Steel Designers' Manual. Constructional Steel Research and Development Organization; C.S. Gray et al (ed, Vols 1 and 2). John Wiley and Sons (Halsted Press), 4th ed., 1972, 1,089 pp. LC: 75-19073. ISBN: 0-470-16865-X. Date of First Edition: 1955. Cost: $51.95.

This handbook provides basic information on the design of steel frame buildings. It contains information on such subjects as the deflection of compound girders, frame structures, beams, steel piling, welding principles and applications, fire resistant construction, and cantilevers. The information provided is primarily in metric terms, and many references are made to standards sponsored by the British Standards Institution. Steel designers will find the bibliographies and the index a useful means of finding additional information and retrieving the data from this handbook. Several examples of steel design problems are included, in addition to background information on basic principles of design to help the user understand several different approaches to the same problem. For related works, please refer to *Metals Handbook; ASME Handbook; Thermophysical Properties of Matter; Woldman's Engineering Alloys; Corrosion Guide; Handbook of Aluminum; Metals Reference Book; Fatigue Design Handbook; Handbook of Binary Phase Diagrams; CRC Handbook of Materials Science*, and the *Aerospace Structural Metals Handbook*. Handbook of Materials Science, *and the* Aerospace Structural Metals Handbook.

2922. Steel Diaphragm Roof Decks: A Design Guide With Tables for Engineers and Architects. E.R. Bryan; J.M. Davies. Halsted Press, 1981, 96 pp. ISBN: 0-470-27158-2. Cost: $24.95.

Tabulated values are presented for strength and stiffness of steel roof decks.

2923. Steel Wire Handbook. Allan B. Dove. The Wire Assocation, Inc, Vol.1, 1965; Vol.2, 1969, Vol.1, 318 pp.; Vol.2, 389 pp.

2924. Stereogram Book of Rocks, Minerals, and Gems. David Techter. Hubbard Press, 1970, 64 pp. LC: 71-23931.

This book includes primarily stereoscopic photographs of various rocks and minerals.

2925. Sterotaxic Atlas of the Rat Brain. Louis J. Pellegrino; Ann S. Pellegrino; Anna J. Cushman. Plenum Publishing Corp, 2d ed., 1979, 260 pp. ISBN: 0-306-40269-6. Cost: $22.50.

This atlas includes illustrations with the brainstem and cerebellum of rats. This is a useful guide for the sterotaxic placement of electrodes, cannulae, and lesions.

2926. Stock Drive Products Handbook of Commercial Drive Components. Ferdinand Freudenstein. Stock Drive, 1971, 520 pp. ISBN: 0-686-01078-7. Cost: $1.75 paper.

2927. Storm Data, 1970-1974 and Storm Data, 1975-1979. James A. Ruffner; Frank E. Blair. Gale Research Co, 1981, 2 Vols. Cost: $85.00 per Vol.

This publication was originally issued in monthly parts by the Environmental Data Service of the National Oceanic and Atmospheric Administration or its predecessor organization. It is now compiled into 2 volumes, '70-'74 and '75-'79, and represents a compilation of the monthly reports of storm activity recorded by the U.S. Weather Bureau. The data are arranged by state, territory, and day of month. Meteorological data and an account of damage and personal injury are listed in this publication.

2928. Stratigraphical Atlas of Fossil Foraminifera. D.G. Jenkins; J.W. Murray. Halsted Press, 1981, 250 pp. ISBN: 0-470-27191-4. Cost: $89.95. Series: *British Micropalaeontological Society Series.*

This atlas describes approximately 570 individual species of foraminifera in the Phanerozoic rocks.

2929. Strength of Materials: Handbook for Calculations and Design. P.G. Korolev. NTIS, 1974, 288 pp. Accession: A75-43475.

This handbook, written in Russian, emphasizes such topics as stress analysis, structural design, elastodynamics, trusses, and beams.

2930. Strength of Turbine Disks. V.P. Rabinovich. NTIS, 1968. Accession: N69-27554; AD-683273. Tech Report: FTD-HT-23-494-68.

This translated Russian title deals with the topic of axial flow turbine wheel strength.

2931. Strength Properties of Timber. Building Research Establishment. Construction Press, Ltd, 1974, 248 pp. ISBN: 0-904406-25-3. Cost: $24.50.

Different types of timber are discussed including redwoods, plywoods, laminated timbers, etc.

2932. Stress Analysis Manual. Gene E. Maddux et al (Technology, Inc). NTIS, 1969, 576 pp. Accession: AD-759199. Tech Report: TI-219-69-24; AFFDL-TR-69-42.

2933. Stress Analysis of Cracks Handbook. Hiroshi Tada. Del Research Corp, Hellertown, PA, 1978, loose-leaf.

Various formulas and stress analysis information is provided on different types of crack problems. Such topics as crack propagation, fracture mechanics, plane strain fracture, Green's function, orthotropic effects, anisotropic effects, dynamic effects, plasticity analysis, and J-integral methods are discussed in this tabulation.

2934. Stress Concentration Factors: Charts and Relations Useful in Making Strength Calculations for Machine Parts and Structural Elements. Rudolph E. Peterson. Wiley-Interscience, 1974, 317 pp. LC: 73-9834. ISBN: 0-471-68329-9. Cost: $27.95.

2935. Stress Distribution Tables for Soil Under Concentrated Loads. A.R. Jumikis. Rutgers University, College of Engineering, 1969.

2936. Structural Alloys Handbook. Mechanical Properties Data Center. Mechanical Properties Data Center, 1978, 2-Vol set with annual supplements, loose-leaf service. Cost: $150.00 plus $40.00 for annual supplement service.

Published as a companion volume to the *Aerospace Structural Metals Handbook.* This new handbook was designed to cover alloys used in construction, machine tool, heavy equipment, automotive, and general manufacturing industries. Covers wrought steel, high-strength low-alloy steel, cast iron, wrought aluminum, wrought stainless steel, and cast magnesium.

2937. Structural and Construction Design Manual. James M. Gere; Helmut Krawinkler. Equipment Guide-Book Co, 1977, various paging. Cost: $49.95.

Numerous charts and tables are provided for designing structural members and determining allowable loads. In addition, such topics as wood design, design of formwork for concrete, structural analysis, structural steel design, and reinforced concrete design are also discussed in this handbook.

2938. Structural Engineering Handbook. Edwin H. Gaylord Jr; Charles N. Gaylord. McGraw-Hill Book Co, 2d ed., 1979, 1,248 pp. LC: 78-25705. ISBN: 0-07-023123-0. Cost: $46.50.

Structural Metals Handbook. *See* **Aerospace Structural Metals Handbook.**

2939. Structural Properties of Plastics. I.V. Kragelskii et al. NTIS, translated from the Russian by Foreign Technology Division, Wright Patterson Air Force Base, 1970, 234 pp. Accession: AD-719780. Tech Report: FTD-8C-23-65-70.

2940. Structural Steel Designers' Handbook. F.S. Merritt. McGraw-Hill Book Co, 1972, 1,000 pp. LC: 73-159313. ISBN: 0-07-041507-2. Cost: $41.50.

2941. Student Engineer's Companion. J. Carvill. Butterworth, 1980, 133 pp. ISBN: 0-408-00438-X. Cost: $12.50.

This is a collection of descriptions of approximately 800 mechanical engineering components and constructions. Numerous drawings and illustrations are provided.

2942. Student's Engineering Manual. George A. Hawkins. McGraw-Hill, 1968, 584 pp. LC: 68-11608. ISBN: 0-07-02730-3. Cost: $12.50.

2943. Subsidence Engineers' Handbook. National Coal Board, 1975, 109 pp. Date of First Edition: 1965.

Information is provided on ground movement, horizontal strains, surface damage, mining damage, and case histories as they relate to subsidence problems.

2944. Sunrise and Sunset Tables for Key Cities and Weather Stations in the U.S. Gale Research. Gale Research, 1977, 369 pp. LC: 76-24796. ISBN: 0-8103-0464-3. Cost: $48.00.

This tabulation provides the hour and minute of sunrise and sunset for over 300 cities for the remaining years of the twentieth century.

2945. Superconductive Materials and Some of Their Properties. B.W. Roberts. NTIS, 1966, 80 pp. Accession: N70-71766. Tech Report: NBS-TN-408.

This National Bureau of Standards publication provides information on the electrical properties, composition properties, critical temperature, crystallography, and metallography of superconductors.

2946. Supplement to the TTL Data Book for Design Engineers. Texas Instruments, Inc, 2d ed., 1977, 56 pp.

Information is provided on integrated circuits and electronic circuit design.

2947. Suppliers Material Specification Service. General Electric Co. General Electric Co, 4 sets, various paging. Cost: $895.00 all 4 Sets; $250.00 Set 1; $195.00 Set 2; $300.00 Set 3; $195.00 Set 4.

The 4 sets are entitled as follows: Set 1, *Nonmetallic Materials, Chemical Materials and Test Methods*; Set 2, *Ferrous Metallic Materials and Test Methods*; Set 3, Nonferrous Metallic Materials and Test Methods; *Set 4, Machine Parts and Finishes.* Chemical properties and composition of over 12,000 raw materials, specifications for over 10 million machine parts, and 1,000 finishes are provided in this compilation. This information is available on microfiche.

2948. Surface Coatings for Savings in Engineering: Seminar Handbook. Welding Institute, 2d ed., 1974, 50 pp.

Different types of coating materials are discussed in this handbook.

2949. Surface Ionization. E. Ya. Zandberg; N.L. Ionov. NTIS, Russian translation for the National Bureau of Standards, 1969, translated 1971. Accession: TT70-50148. Cost: $10.50.

2950. Survey of the Heat Treatment and Mechanical Properties of Uranium/Low Titanium Alloys. G.P.R. McCarthy. ERDA Depository Libraries, 1975, 28 pp. Accession: N76-21323. Tech Report: AWRE-0-16. Cost: $4.50 hard copy.

Tables and data are presented on heat treatment and mechanical properties of titanium alloys and uranium alloys.

2951. Survey of the Properties of the Hydrogen Isotopes Below Their Critical Temperatures. H.M. Roder et al. Superintendent of Documents, 1973, 121 pp. Accession: N74-10664. Tech Report: NBS-TN-641. SuDoc: C12. 46:641. Cost: $1.25 hard copy.

This survey presents thermodynamic, thermal, electrical, radiative, transport, optical, and mechanical properties for hydrogen isotopes below 40° K.

2952. Survey of Thermodynamic Properties of the Compounds of the Elements CHNOPS: Progress Report, 1 March 30-June 1968. E.S. Domalski; G.T. Furukawa; M.L. Reilly. NTIS, 1968, 51 pp. Accession: N68-33427. Tech Report: NASA-CR-96468; NBS-9883.

Information is provided on the thermodynamic properties for hydrogen compounds, nitrogen compounds, oxygen compounds, phosphorus compounds, and sulfur compounds. Heat of combustion and heat formation data are included for these organic compounds.

2953. Surveying Tables. Lefax Publishing Co, loose-leaf binding. ISBN: 0-685-14168-3. Cost: $3.00. Series: *Lefax Data Books*, No. 649.

2954. Switchgear and Control Handbook. Robert Smeaton. McGraw-Hill, 1976, 1,056 pp. LC: 76-17925. ISBN: 0-07-058439-7. Cost: $37.50.

This handbook includes information on electrical switchgear, electric controllers, and automatic control. Information on AC and DC switchgear is provided along with information on emergency power supply during power outages, and load and safety considerations. Different types of control circuit devices are discussed including timers and relays as well as the latest information on seismic requirements with regard to the earthquake precautions in connection with switchgear installation.

2955. Synchro Engineering Handbook. A.R. Upson; J.H. Batchelor. Hutchinon & Co, 1965, 235 pp.

2956. Synfuels Handbook: Including the Yellow Pages of Synfuels. Roseanne Schwaderer. *Coal Week*, McGraw-Hill Publishing Co, 1980, irregular pagination. LC: 80-65731. ISBN: 0-07-6066-86. Cost: $57.00.

Such topics as coal gasification, coal liquefaction, oil shale, pyrolysis, hydrogenation, solvent extraction, and surface retorting are discussed in this handbook.

2957. Synthetic Fuels Data Handbook. Thomas A. Hendrickson. Cameron Engineers, 1978, 418 pp. Cost: $95.00.

Includes over 450 figures and tables on such topics as oil shale, oil sands, and coal. Various properties and processes are extensively listed.

Synthetic Organic Chemical Manufacturer's Association. *See* SOCMA.

2958. System of Mineralogy. James Dwight Dana; Edward Salibury Dana. Wiley-Interscience, 7th ed., 1944-62, Vol.1, 834 pp.; Vol.2, 1,124 pp.; Vol.3, 334 pp. ISBN: Vol.1, 0-471-19239-2; Vol.2, 0-471-19272-4; Vol.3, 0-471-19287-2. Cost: Vol.1, $46.50; Vol.2, $39.95; Vol.3, $23.95.

This 3-volume set has the following titles: Vol.1, *Elements, Sulfides, Sulfosalts, Oxides*; Vol.2, *Halides, Nitrates, Borates, Carbonates, Sulfates, Phosphates, Arsenates, Tungstates, Molybdates, etc.*; Vol.3, *Silica Minerals.*

2959. System of Ophthalmology. Sir Stewart Duke-Elder. Mosby/Henry Kimpton, 1958-, 15 Vols. ISBN: Vol.1, 0-8016-8282-7. Cost: Consult publisher for price.

NOTE: This monumental work is more like a treatise than a handbook, and it covers the total field of ophthalmology including such topics as ocular motility, diseases of the eyes, glaucoma, and injuries to the eye.

2960. Systematic Energy Conservation Management Guide. American Management Assns, Energy Management Div, loose-leaf, 11 Vols. Cost: $325.00.

This 11-volume, loose-leaf service provides information on how to conduct an energy audit and how to monitor an energy conservation program. Such topics as the conservation of energy in heating, ventilation and air conditioning systems, lighting systems, steam distribution systems, boilers, motors, electric motors, compressors, furnaces, dryers, and compressors are discussed in this set of volumes.

2961. Systems Engineering Handbook. R.E. Machol; W.P. Tanner; S.N. Alexander. McGraw-Hill, 1965, 1,084 pp. (varous paging). LC: 64-19214. ISBN: 0-07-039371-0. Cost: $47.50.

2962. Table and Charts of Equilibrium Normal-Shock Properties for Hydrogen-Helium Mixtures with Velocities to 70 km/sec. C.G. Miller; S.E. Wilder. NTIS, rev. ed., 1976, 4 Vols. Accession: Vol.1, N77-18234; Vol.2, N77-18235; Vol.3, N77-18236; Vol.4, N77-18237. Tech Report: NASA-SP-3085. Cost: $18.50 per Vol., hard copy; $3.50 per Vol., microform.

The 4 volumes of this set are entitled as follows: Vol.1, *0.95 H₂O .05 He*; Vol.2, *0.90 H₂O .10 He*; Vol.3, *0.85 H₂O .15 He*; Vol.4, *0.75 H₂O .25 He*. The properties in this compilation include temperature, pressure, density, speed of sound, molecular-weight ratio, entropy, isentropic exponent, species mole fractions, and velocity for hydrogen-helium mixtures.

2963. Table for Converting pH to Hydrogen Ion Concentration [H+] over the Range 5-9. Vincent Fiorica. Department of Transporation, Federal Aviation Administration, 1968, 30 pp. SuDoc: TD4.210:68-23.

This publication provides information on hydrogen ion concentration in blood.

2964. Table of Arc Tan x. National Bureau of Standards. NTIS, 1942, 169 pp. Accession: PB-178415. Cost: $6.75. Series: NBS.

NOTE: Reissued as AMS26.

2965. Table of Arctangents of Rational Numbers. J. Todd (National Bureau of Standards). NTIS, 1951, 105 pp. Accession: AD-702411. Cost: $5.50. Series: NBS AMS11.

2966. Table of Associated Legendre Functions. National Bureau of Standards. Columbia University Press, 1945, 306 pp. Series: CUP6.

2967. Table of Bessel Clifford Functions of Orders Zero and One. National Bureau of Standards. NTIS, 1953, 72 pp. Accession: AD-695952. Cost: $4.50. Series: NBS AMS28.

2968. Table of Circular and Hyperbolic Tangents and Cotangents for Radian Arguments. National Bureau of Standards. Columbia Univesity Press (out of print), 1943, 412 pp. Series: CUP3.

2969. Table of Coefficients for Obtaining the First Derivative without Differences. H.E. Salzer (National Bureau of Standards). NTIS, 1948, 20 pp. Accession: PB-194385. Cost: $3.50. Series: NBS AMS2.

2970. Table of Hyperbolic Sines and Cosines, x=2 to x=10. National Bureau of Standards. NTIS, 1955, 81 pp. Accession: COM 74-10639. Cost: $4.00. Series: NBS AMS45.

NOTE: This is an extension of AMS36.

2971. Table of Integrals, Series, and Products. I.M. Ryzhik; I.S. Gradshteyn; Alan Jeffrey (trans). Academic Press, 5th ed., 1980, 1,248 pp. ISBN: 0-12-294760-6. Cost: $21.50.

Covers vector field theory, algebraic inequalities, integral inequalities, ordinary differential equation, Fourier and Laplace transforms, matrices, and determinants.

2972. Table of Integrals. Ralph G. Hudson; Joseph Lipka. Wiley-Interscience, 1971, 25 pp. ISBN: 0-471-43065-X. Cost: $6.50.

2973. Table of Isotopes. C. Michael Lederer et al. John Wiley and Sons, 7th ed., 1978, 1,600 pp. LC: 78-14938. ISBN: 0-471-04179-3. Cost: $50.50.

2974. Table of Laplace Transforms. George E. Roberts; Hyman Kaufman. Saunders, 1966, 367 pp. LC: 65-23098.

2975. Table of Laser Lines in Gases and Vapors. R. Beck; W. Englisch; K. Gürs. Otto Harrassowitz, 202 pp. LC: 77-26659. ISBN: 3-540-08603-X. Cost: $28.40. Series: *Springer Series in Optical Sciences*, 2.

Tabulated data are presented on laser lines in gases and vapors.

2976. Table of Meta-Stable Transitions for Use in Mass Spectrometry. John Herbert Beynon et al. Elsevier Publishing Co, 1965. LC: 65-25201.

2977. Table of Miscellaneous Thermodynamic Properties for Various Substances with Emphasis on the Critical Properties. D. Rathmann; J. Bauer; P.A. Thompson. NTIS, 1978, 81 pp. Accession: N79-28040. Tech Report: MPIS-6. Cost: $9.50 hard copy; $3.50 microfilm.

Presented in this compilation are such properties as molecular weight, normal boiling point in Kelvin, critical temperature in Kelvin, triple pressure in bars, critical pressure in bars, critical specific volume in cm/g, and triple temperature in Kelvin.

2978. Table of Natural Logarithms for Arguments Between Zero and Five to Sixteen Decimal Places. National Bureau of Standards. NTIS, 1953, 501 pp. Accession: COM 74-11112. Cost: $12.75. Series: AMS31.

NOTE: A reissue of NBS MT10—*See* NBS AMS53 for x–5 (.0001)10.

2979. Table of Natural Logarithms for Arguments from Five to Ten to Sixteen Decimal Places. National Bureau of Standards. NTIS, 1958, 506 pp. Accession: PB-186430. Cost: $12.75. Series: NBS AMS53.

NOTE: A reissue of NBS MT12.

2980. Table of Natural Logarithms. National Bureau of Standards. NTIS, 1941, 4 Vols., (Vols. 1 and 2 out of print). Accession: Vol.III, COM 73-10499; Vol.IV, PB-186430. Cost: Vol.III, $12.75; Vol.IV, $12.75. Series: Vol.I, NBS MT7; Vol.II, NBS MT9; Vol.III, NBS MT10; Vol.IV, NBS MT12.

NOTE: Vol.I, Values may be obtained by using Vol.III. Vol.II, Values may be obtained by using Vol.IV. Vol.III, reissued as NBS AMS31. Vol.IV, reissued as NBS AMS53.

2981. Table of Neutron Activation Constants. Franz Baumgartner. Verlag Karl Thiemig, 2d ed., 1967, 47 pp. LC: 72-409029. ISBN: 3-521-06007-1. Cost: $3.00.

Tables are presented on neutron activation constants.

2982. Table of Reciprocals of the Integers from 100,000 through 200,009. National Bureau of Standards. Columbia University Press (out of print), 1943, 204 pp. Series: CUP1.

2983. Table of Salvo Kill Probabilities for Square Targets. National Bureau of Standards. NTIS, 1955, 33 pp. Accession: AD-698954. Cost: $4.00. Series: NBS AMS44.

2984. Table of Secants and Cosecants to Nine Significant Figures at Hundredths of a Degree. National Bureau of Standards. NTIS, 1954, 46 pp. Accession: PB-186428. Cost: $4.00. Series: NBS AMS40.

2985. Table of Series and Products. Eldon R. Hansen. Prentice-Hall, 1975, 523 pp. LC: 74-6346. ISBN: 0-13-881938-6. Cost: $74.00.

This reference source is an exhaustive collection of tables of mathematical series and infinite products. It includes such information as trigonometric series, logarithmic functions, orthogonal polynomials, and many other mathematical functions. An index is provided as well as a bibliography of the literature for further investigation. For related works, please refer to such titles as the *Handbook of Mathematical Tables and Formulas*; *Eight-place Tables of Trigonometric Functions for Every Second of Arc*; *CRC Handbook of Tables for Probability and Statistics*; *CRC Standard Mathematical Tables*; *CRC Handbook of Tables for Mathematics*; *Engineering Mathematics Handbook*; and the various mathematics series of the U.S. National Bureau of Standards.

2986. Table of Sine and Cosine Integrals for Arguments from 10 to 100. National Bureau of Standards. NTIS, 1954, 187 pp. Accession: COM 73-10499. Cost: $7.50. Series: NBS AMS32.

NOTE: A reissue of NBS MT13.

2987. Table of Sine and Cosine Integrals for Arguments from 10 to 100. National Bureau of Standards. NTIS, 1942, 185 pp. Accession: COM 73-10499. Cost: $7.50. Series: NBS MT13.

NOTE: Reissued as NBS AMS32.

2988. Table of Specific Gamma Ray Constants. Dieter Nachtigall. Verlag Karl Thiemig, 1969, 112 pp. ISBN: 3-521-06051-9. Cost: $5.00.

Gamma ray constants are presented for approximately 700 gamma-emitting nuclides.

2989. Table of the Bessel Functions $J^0(z)$ and $J^1(z)$ for Complex Arguments. National Bureau of Standards. Columbia University Press, 1943, 403 pp. Series: CUP2.

2990. Table of the Exponential Integral for Complex Arguments. National Bureau of Standards. NTIS, 1958, 634 pp. Accession: PB-248467. Cost: $16.25. Series: NBS AMS51.

2991. Table of the First Ten Powers of the Integers from 1 to 1000. National Bureau of Standards. National Bureau of Standards, 1939, 80 pp. (out of print). Series: NBS MT1.

NOTE: Value of N^k, $k = 1(1)10$, $n = 2(1)999$, exact or 10S are included in Table 3.1 of AMS55.

2992. Table of the Gamma Function for Complex Arguments. National Bureau of Standards. U.S. Government Printing Office, 1954, 105 pp. SuDoc: C13.32:34. Cost: $4.70. Series: NBS AMS34.

2993. Table of the Incomplete Elliptic Integral of the Third Kind. R.G. Selfridge; J.E. Maxfield. Dover, 1958, 805 pp. ISBN: 0-486-60501-9. Cost: $10.00.

2994. Tables and Charts of Equilibrium Thermodynamic Properties of Ammonia for Temperatures from 500 to 50,000 K. A.L. Simmonds; C.G. Miller; J.E. Nealy. NTIS, 1976, 259 pp. Accession: N76-30097. Tech Report: NASA-SP-3099. Cost: $9.00 hard copy.

Properties for ammonia are tabulated, including pressure, temperature, enthalpy, speed of sound, density, entropy, molecular-weight ratio, specific heat at constant pressure, specific heat at constant volume, isentropic exponents, and species mole fractions.

2995. Tables and Charts of Equilibrium Thermodynamic Properties of Carbon Dioxide for Temperatures from 100 K to 25,000 K. C.G. Miller III; S.E. Wilder. NTIS, 1976, 493 pp. Accession: N76-32272. Tech Report: NASA-SP-3097. Cost: $11.50 hardcopy.

Properties for pure carbon dioxide are presented in tabulated and graphical format. Properties include temperature, pressure, enthalpy, specific heat at constant pressure, molecular-weight ratio, specific heat at constant volume, isentropic exponent, and species mole fractions.

2996. Tables and Factors and Formulas for Computing Respiratory Exchange and Biological Transformations of Energy. Thorne M. Carpenter. Carnegie Institute (distributed by Academic Press), 4th ed., 1964. ISBN: 0-87279-311-7. Cost: $4.50.

2997. Tables and Formulae for Spherical Functions. M.I. Zhurina; L.N. Karmazina; E.L. Albasiny (trans). Pergamon Press, 1966, 107 pp. LC: 66-18241 (GB). ISBN: 0-08-011538-1. Cost: $14.50. Series: *Pergamon Mathematical Tables Series*, Vol. 40.

2998. Tables and Formulas for Fixed End Moments of Members of Constant Moment of Inertia and for Simply Supported Beams. Paul Rogers. Ungar, 2d ed., 1965, 101 pp. LC: 65-28016. ISBN: 0-8044-4850-7. Cost: $8.50.

2999. Tables and Monograms of Hydrochemical Analysis. I. Yu Sokolov. Consultants Bureau, 1960, 85 pp. LC: 60-13952. ISBN: 0-306-10644-2. Cost: $20.00.

3000. Tables and Operating Data for Electroplaters. Hans Benninghoff. International Publications Service, 1976, 414 pp. LC: 75-326615. ISBN: 0-901994-70-7. Cost: $50.00.

3001. Tables for Computing Elevations and Topographic Levelling. Leonid Sergeevich Khrenov; D.E. Brown (trans). Pergamon Press, 1964, 199 pp. LC: 63-18924. Series: *Pergamon Mathematical Tables Series*, Vol. 31.

3002. Tables for Continuous Beams, Vol. 3: Beams Over Three Spans, 11-12-13. E. Brandt. Adler's Foreign Books, Inc, 1972, 360 pp. ISBN: 3-7625-0485-7. Cost: $62.80.

Explanatory notes are provided in French, German, and English. Deals primarily with the subject of continuous girders with constant moment of inertia frequently encountered in reinforced concrete design.

3003. Tables for Conversion of X-Ray Diffraction Angles to Interplanar Spacing. H. Swanson (National Bureau of Standards). NTIS, 1951, 159 pp. Accession: PB-251960. Cost: $6.75. Series: NBS AMS10.

3004. Tables for Estimating Median Fatigue Limits. American Society for Testing and Materials, 1981, 176 pp. ISBN: 04-731000-30. Cost: $15.00. Series: *American Society for Testing and Materials. Special Technical Publication*, No. 731.

Over 100 tables are presented on median fatigue limits. Strength distribution tables are provided in this tabulation.

3005. Tables for Limit State Design of Singly Reinforced Rectangular Beams and Slabs. A.H. Allen. Scholium International, Inc, 1973, 32 pp. Cost: $3.50.

Tables are presented using steel with different characteristic strengths and concrete of different grades.

3006. Tables for Microscopic Identification of Ore Minerals. W. Uytenbogaardt; E.A. Burke. Elsevier Science, 2d ed., 1971. ISBN: 0-444-40876-2. Cost: $51.25.

3007. Tables for Normal Tolerance Limits, Sampling Plans, and Screening. Robert E. Odeh; D.B. Owen. Marcel Dekker, Inc, 1980, 328 pp. LC: 79-27905. ISBN: 0-8247-6944-9. Series: *Statistics: Textbooks and Monographs Series*, Vol. 32.

Statistical tables are presented for normal distribution, bivariate normal distribution, and univariate normal procedures.

3008. Tables for Numerical Integration of Functions with Logarithmic and Power Singularities. V.I. Krylov. Halsted Press, 1972. LC: 72-181529. ISBN: 0-470-50902-3. Cost: $11.75.

3009. Tables for Rapid Sub-Frame Analysis. S. Rajendran. Singapore University Press, 1979, 377 pp. Cost: $30.00.

Tabulated data are presented for rapid sub-frame analysis.

3010. Tables for Rocket and Comet Orbits. S. Herrick (National Bureau of Standards). NTIS, 1953, 100 pp. Accession: PB-251870. Cost: $5.50. Series: NBS AMS20.

3011. Tables for Texture Analysis of Cubic Crystals. J. Hansen; J. Pospiech; K. Lücke. Springer-Verlag New York, Inc, 1978, 300 pp. ISBN: 0-387-08689-7; 3-540-08689-7 (Springer). Cost: $99.20.

Twenty-two figures and 11 tables are presented on texture analysis of cubic crystals.

3012. Tables for the Analysis of Beta Spectra. National Bureau of Standards. NTIS, 1952, 61 pp. Accession: PB-251865. Cost: $4.50. Series: NBS AMS13.

3013. Tables for the Analysis of Plates, Slabs, and Diaphragms. R. Bares. Bauverlag, Weisbaden, 3d ed., 1979, 676 pp. ISBN: 3-7625-0177-2. Cost: $78.80.

Covers bending of isotropic plates with small deflections, bending of orthotropic plates, stability of plates, natural frequency of plates, and plane state of stress in isotropic diaphrams.

3014. Tables for the Computation of Toroidal Shells. L.N. Osipova; S.A. Tumarkin; Morris D. Friedman (trans). Noordhoff International Publishing, 1965, 126 pp. LC: 65-6182.

3015. Tables for the Design and Analysis of Stiffened Steel Plates. N.W. Murray; G. Thierauf. Heyden and Son, Inc, 1981. Cost: $46.00.

Approximately 4,000 different examples of stiffened steel plates are presented with their corresponding section properties for stress analysis and buckling properties for stability analysis. These tables will be useful in calculating loads for stiffened steel plates.

3016. Tables for the Energy and Photon Distribution in Equilibrium Radiation Spectra. P.A. Apanasevich; V.S. Aizenshtadt; Prasenjit Basu (trans). Macmillan (also Pergamon), 1965, 250 pp. LC: 63-19955; 65-5752. Series: *Mathematical Table Series*, Vol. 35.

3017. Tables for the Internal Forces of Pierced Shear-Walls Subject to Lateral Loads. Rosman, 1966, 93 pp. Cost: 18,60 DM. Series: *Bauingenieur-Praxis*, Vol. 66.

This publication is presented in German and English. Twelve tables and 19 illustrations are provided on shear-walls subject to lateral loads.

3018. Tables for the Numerical Solution of Boundary Value Problems for the Theory of Harmonic Functions. L.V. Kantorovich et al; Alexis N. Obolensky (trans). Ungar, 1963. LC: 63-12908. ISBN: 0-8044-4497-8. Cost: $22.50.

3019. Tables for the Rigid Asymmetric Rotor: Transformation Coefficients from Symmetric to Asymmetric Bases and Expectation Values and P^2_z, P^4_z, P^6_z. R.H. Schwendeman (National Bureau of Standards). U.S. Government Printing Office, 1968. SuDoc: C13.48:12. Cost: $1.45. Series: NSRDS-NBS-12.

3020. Tables for Use in High Resolution Mass Spectrometry. R. Binks; J.S. Littler; R.L. Cleaver. Heyden and Sons, Inc, 1970, 180 pp. + 32 pp. supplement. ISBN: 0-85501-026-6. Cost: $97.00.

This publication is presented in English, German and French. These tables will be useful in determining the masses of ions. Exact masses and C13 ratios of different combinations of carbon, nitrogen, and fluorine atoms are listed.

3021. Tables of 10^x. (Antilogarithms to the Base 10). National Bureau of Standards. NTIS, 1953, 543 pp. Accession: AD-694116. Cost: $13.00. Series: NBS AMS27.

3022. Tables of All Primitive Roots of Odd Primes Less Than 1,000. Roger Osborn. University of Texas Press, 1961, 70 pp. LC: 61-10046. ISBN: 0-292-73394-1. Cost: $6.00.

3023. Tables of Angular Scattering Functions for Heterodisperse Systems of Spheres. Wilfried Heller et al. Wayne State University Press, 1969, 1,288 pp. LC: 68-57471. ISBN: 0-8143-1364-7. Cost: $50.00.

3024. Tables of Angular Spheroidal Wave Functions. A.L. Van Buren; B.J. King; R.V. Baier. Naval Research Laboratory, 1975, 2 Vols. LC: 75-15170.

These 2 volumes are actually 2 of a series of 8 volumes that tabulate values of angular and radial spheroidal wave functions.

3025. Tables of Antenna Characteristics. Ronald W.P. King. IFI/Plenum Press, 1971, 385 pp. LC: 74-157425. ISBN: 0-306-65154-8. Cost: $45.00.

3026. Tables of Arcsin x. National Bureau of Standards. Columbia University Press (out of print), 1945, 124 pp. Series: CUP5.

3027. Tables of Arctan x. National Bureau of Standards. NTIS, 1953, 170 pp. Accession: PB-178415. Cost: $6.75. Series: NBS AMS26.

NOTE: A reissue of MT16.

3028. Tables of Bessel Functions of Fractional Order. National Bureau of Standards. Columbia University Press, Vol.1, 1948; Vol.2, 1949, Vol.1, 413 pp.; Vol.2, 365 pp. Series: Vol.1, CUP10; Vol.2, CUP11.

3029. Tables of Bessel Transforms. F. Oberhettinger. Springer-Verlag, 1972, 289 pp. LC: 72-88727. ISBN: 0-387-05997-0. Cost: $12.40.

3030. Tables of Bimolecular Gas Reactions. A.F. Trotman-Dickenson; G.S. Milne (National Bureau of Standards). NTIS, 1967. Cost: $6.00. Series: NSRDS-NBS-9.

3031. Tables of Chebyshev Polynomials $S_n(x)$ and $C_n(x)$. National Bureau of Standards. NTIS, 1952, 161 pp. Accession: COM-73-10498. Cost: $6.75. Series: NBS AMS9.

3032. Tables of Chemical Kinetics, Homogenous Reactions. U.S. Bureau of Standards. U.S. Government Printing Office, 1951, 731 pp. Series: *U.S. National Bureau of Standards Circular*, No. 510. Also includes *NBS Monograph*, 34, Vol.1 and 2, 1961-1964.

3033. Tables of Circular and Hyperbolic Sines and Cosines for Radian Arguments. National Bureau of Standards. NTIS, 1953, 407 pp. Accession: PB-251871. Cost: $11.00. Series: NBS AMS36.

3034. Tables of Circular and Hyperbolic Sines and Cosines for Radian Arguments. National Bureau of Standards. NTIS, 1939, 405 pp. Accession: PB-251871. Cost: $11.00. Series: NBS MT3.

NOTE: Reissued as AMS36.

3035. Tables of Co-efficients for the Analysis of Triple Angular Correlations of Gamma-Rays from Aligned Nuclei. Geoffrey Kaye et al. Pergamon Press, 1968, 218 pp. LC: 67-28102. ISBN: 0-08-12260-4. Cost: $36.00.

3036. Tables of Coefficients for the Numerical Calculation of Laplace Transforms. H.E. Salzer (NBS). NTIS, 1953, 36 pp. Accession: PB-175817. Cost: $4.00. Series: NBS AMS30.

3037. Tables of Collision Integrals and Second Virial Coefficients for the (m,6,8) Intermolecular Potential Function. Max Klein; H.J.M. Hanley; Francis J. Smith; Paul Holland (National Bureau of Standards). U.S. Government Printing Office, 1974. SuDoc: C13.48:47. Cost: $2.25. Series: NSRDS-NBS-47.

3038. Tables of Coulomb Wave Functions (Whittaker Functions). A.V. Lukianov et al; D.E. Brown (trans). Pergamon Press, 1965, 221 pp. LC: 63-20585. Series: *Mathematical Table Series*, Vol. 24.

3039. Tables of Coulomb Wave Functions. Volume I. National Bureau of Standards. NTIS, 1952, 141 pp. Accession: PB-251868. Cost: $6.00. Series: NBS AMS17.

3040. Tables of Critical-Flow Functions and Thermodynamic Properties for Methane and Computational Procedures for both Methane and Natural Gas. Robert C. Johnson. National Aeronautics & Space Administration, 1972, 171 pp. Tech Report: NASA SP-3074.

3041. Tables of Dimensions, Indices, and Branching Rules for Representations of Simple Lie Algebras. W.G. McKay; J. Patera. Marcel Dekker, 1981, 336 pp. LC: 80-27663. ISBN: 0-8247-1227-7. Cost: $39.75. Series: *Lecture Notes in Pure and Applied Mathematics Series*, Vol. 69.

This tabulation includes a table of branching rules, a table of dimensions, and second and fourth indices of irreducible dimensions.

3042. Tables of Elliptical Integrals. Vasilii Mikailovich Belikov et al; Prasenjit Basu (trans). Pergamon Press, 1965, 647 pp. LC: 64-21221; 65-5883. Series: *Mathematical Table Series*, Vol. 37.

3043. Tables of Equilibrium Thermodynamic Properties of Argon. H.S. Brahinsky; C.A. Neel. NTIS, Vol.1, Jan 1966-Jan 1968; Vol.2, Jan 1966-Jan 1968; Vol.3, Jan 1966-Jan 1968, 3-Vol. set, Vol.1, 127 pp.; Vol.2, 215 pp.; Vol.3, 398 pp. Accession: Vol.1, N69-29412, AD-684177; Vol.2, N69-29413, AD-684178; Vol.3, N69-29414, AD-684532. Tech Report: AEDC-TR-69-19.

This 3-volume set is entitled: Vol.1, *Constant Temperature Final Report*; Vol.2, *Constant Pressure Final Report*; Vol.3, *Constant Entropy Final Report*. This publication lists thermodynamic properties of argon including entropy, enthalpy, density, temperature, and pressure.

3044. Tables of Equilibrium Thermodynamic Properties of Nitrogen. H.S. Brahinsky; C.A. Neel. NTIS, Vol.1, Jan 1966-Jan 1969; Vol.4, Jan 1966-Jan 1969, Vol.1, 155 pp.; Vol.2, 213 pp.; Vol.3, 457 pp.; Vol.4, 155 pp. Accession: Vol.1, N70-13708, AD-693134; Vol.2, N70-13709, AD-

692712; Vol.3, N70-13710, AD-692713; Vol.4, N70-11285, AD-692172. Tech Report: AEDC-TR-69-126.

Thermodynamic properties of nitrogen are listed including such data as compressibility, acoustic velocity, gas density, gas dynamics, thermodynamic equilibrium, temperature data, specific heat, speed of sound data, entropy, and pressure.

3045. Tables of Experimental Dipole Moments. A.L. McClellan. W.H. Freeman, 1963, 713 pp. LC: 63-14844. ISBN: 0-7167-0122-7. Cost: $28.75.

3046. Tables of Fractional Powers. National Bureau of Standards. Columbia University Press (out of print), 1946, 488 pp. Series: CUP7.

3047. Tables of Functions and Zeros of Functions. Collected Short Tables of the Computational Laboratory. National Bureau of Standards. NTIS, 1954, 211 pp. Accession: PB-251872. Cost: $7.75. Series: NBS AMS37.

3048. Tables of Functions with Formulae and Curves. Eugene Jahnke; Fritz Emde. Dover, 4th ed., 1945, 379 pp. ISBN: 0-486-60133-1. Cost: $5.00 paper.

3049. Tables of Gamma Ray Energies for Activation Analysis. Christoph Meixner. Verlag Karl Thiemig, 1970, 256 pp. LC: 78-493179. ISBN: 3-521-06055-1. Cost: $10.50.

3050. Tables of Generalized Airy Functions for the Asymptotic Solution of Differential Equations. Liubov' Nikolaevna Nosova; S.A. Tumarkin; D.E. Brown (trans). Pergamon Press, 1965, 89 pp. LC: 63-19332. Cost: $12.00. Series: *Pergamon Mathematical Tables Series*, Vol.3.

3051. Tables of Indefinite Integrals. G. Petit-Bois. Dover, 1906, 150 pp. ISBN: 0-486-60225-7. Cost: $4.50.

3052. Tables of Integral Transforms. Arthur Erdelyi. McGraw-Hill, 1954, 2-Vol. set. LC: 54-6214. ISBN: Vol.1, 0-07-019549-8; Vol.2, 0-07-019550-1. Cost: Vol.1, $32.50; Vol.2, $39.50.

3053. Tables of Integrals and Other Mathematical Data. Herbert B. Dwight. Macmillan, 4th ed., 1961, 335 pp. LC: 61-6419. ISBN: 0-02-331170-3. Cost: $16.95.

3054. Tables of Irreducible Representations of Space Groups and Co-Representations of Magnetic Space Groups. S.C. Miller; W.F. Love. Pruett Publishing Co, 1967. LC: 67-30015. ISBN: 0-87108-134-2. Cost: $54.95.

3055. Tables of Lagrange Coefficients for Cubic Interpolation. Mieczyslaw Warmus. Panstwowe Wydawnictwo, Computing Centre of the Polish Academy of Sciences, 1965, 500 pp. LC: 67-50272 (Pol.). Series: *Mathematical Tables*, Vol. 1.

3056. Tables of Lagrange Coefficients for Quadratic Interpolations. Mieczyslaw Warmus. Panstwowe Wydawnictwo, Computing Centre of the Polish Academy of Science, 1966, 502 pp. LC: 67-6199 (Pol.). Series: *Mathematical Tables*, Vol. 2.

3057. Tables of Lagrangian Coefficients for Sexagesimal Interpolation. National Bureau of Standards. NTIS, 1954, 157 pp. Accession: PB-251103. Cost: $6.75. Series: NBS AMS35.

3058. Tables of Lagrangian Interpolation Coefficients. National Bureau of Standards. Columbia University Press (out of print), 1944 (2d printing, 1948), 394 pp. Series: CUP4.

3059. Tables of Laguerre Polynomials and Functions. V.S. Aizenshtadt et al; Prasenjit Basu (trans). Pergamon Press, 1966, 149 pp. LC: 66-14492 (G.B.).

3060. Tables of Laplace Transforms. F. Oberhettinger; L. Badii. Springer-Verlag, 1973, 428 pp. LC: 73-81328. ISBN: 0-387-06350-1. Cost: $34.90.

3061. Tables of Laplace Transforms. William D. Day. Transatlantic Arts, Inc, 1967, 36 pp. ISBN: 0-685-20639-4. Cost: $1.50.

3062. Tables of Laplace, Heaviside, Fourier and Z Transforms. Martin Healey. Halsted Press (also Chambers), 1972, 78 pp. LC: 67-92791 (G.B.). ISBN: 0-470-36663-X. Cost: $3.95.

3063. Tables of Light Scattering Functions for Spherical Particles. William J. Pangonis et al. Wayne State University Press, 1957, 116 pp. LC: 56-12604. ISBN: 0-8143-1076-1. Cost: $9.95.

3064. Tables of Lommel's Functions of Two Pure Imaginary Variables. L.S. Bark; P.I. Kuznetsov; D.E. Brown (trans). Pergamon Press, 1965, 263 pp. LC: 63-18940. Series: *Mathematical Table Series*.

3065. Tables of Mathematical Functions, Vol. 3. Harold T. Davis; Vera Fisher. Trinity University Press, 1962, multivolume set. LC: 63-5764. ISBN: 0-911536-17-5. Cost: $8.75.

3066. Tables of Mellin Transforms. F. Oberhettinger. Springer-Verlag, 1975, 275 pp. ISBN: 3-540-6942-9; also 0-387-06942-9. Cost: $19.30.

3067. Tables of Molecular Vibrational Frequencies, Consolidated Volume I. T. Shimanouchi (National Bureau of Standards). U.S. Government Printing Office, 1972. SuDoc: C13.48:39. Cost: $5.10. Series: NSRDS-NBS-39.

3068. Tables of Molecular Vibrational Frequencies, Part 1. T. Shimanouchi (National Bureau of Standards). NTIS, 1967. Series: NSRDS-NBS-6.

NOTE: Superseded by NSRDS-NBS-39 Part 2 of this title—NSRDS-NBS-11.

3069. Tables of Molecular Vibrational Frequencies, Part 2. T. Shimanouchi (National Bureau of Standards). NTIS, 1967. Series: NSRDS-NBS-11.

NOTE: Superseded by NSRDS-NBS-39 Part 1 of this title—NSRDS-NBS-6.

3070. Tables of Molecular Vibrational Frequencies, Part 3. T. Shimanouchi. U.S. Government Printing Office,

1968, 46 pp. Accession: N68-34705. Tech Report: NSRDS-NBS-17. Cost: $ 0.30.

This National Bureau of Standards publication includes information on molecular spectroscopy, and the vibrational frequencies for computation of ideal gas thermodynamic properties. Superseded by NSRDS-NBS-39.

3071. Tables of n and Γ (n + ½) for the First Thousand Values of n. H.E. Salzer (National Bureau of Standards). NTIS, 1951, 10 pp. Accession: PB-175967. Cost: $10.00. Series: NBS AMS23.

3072. Tables of Normal Probability Functions. National Bureau of Standards. NTIS, 1953, 344 pp. Accession: PB-175967. Cost: $10.00. Series: NBS AMS23.

3073. Tables of Normalized Associated Legendre Polynomials. S.L. Belousov. Pergamon Press, 1962, 379 pp. LC: 62-17650. ISBN: 0-08-009723-5. Cost: $35.00. Series: *Pergamon Mathematical Tables Series*, Vol. 18.

3074. Tables of Nuclear Quadrupole Resonance Frequencies. I.P. Biryukov. Halsted Press, 1969. LC: 70-15351. ISBN: 0-470-07549-X. Cost: $17.50.

3075. Tables of Ordinary and Extraordinary Refractive Indices, Group Refractive Indices, and -H-SUB-OX-F Curves for Standard Ionospheric Layer Models. Walter Becker. Springer-Verlag, 1960. ISBN: 0-387-02580-4. Cost: $7.10.

3076. Tables of Osculatory Interpolation Coefficients. H.E. Salzer (National Bureau of Standards). NTIS, 1959, 34 pp. Accession: PB-190608. Cost: $4.00. Series: NBS AMS56.

3077. Tables of Percentage Composition of Organic Compounds. Heinrich Gysel. International Publications Service, 2d ed., 1969, 1,310 pp. LC: 72-474600. ISBN: 3-7643-0157-0. Cost: $105.00.

3078. Tables of Physical and Chemical Constants. G.W. Kaye et al. Wiley (Longman), 14th ed., 1973, 320 pp. LC: 73-85205. ISBN: 0-582-46326-2. Cost: $19.95.

3079. Tables of Powers of Complex Numbers. H.E. Salzer (National Bureau of Standards). NTIS, 1950, 44 pp. Accession: PB-251864. Cost: $4.00. Series: NBS AMS8.

3080. Tables of Probability Functions. National Bureau of Standards. NTIS, Vol.1, 1941; Vol.2, 1942 (2d ed., 1948), Vol.1, 302 pp.; Vol.2, 344 pp. Accession: Vol.1, PB-176521; Vol.2, PB-175967. Cost: Vol.1, $9.75; Vol.2, $10.00. Series: Vol.1, NBS MT8; Vol.2, NBS MT14.

NOTE: Vol.1 reissued as NBS AMS41. Vol.2 reissued as NBS AMS23.

3081. Tables of Racah Coefficients. A.F. Nikiforov; Prasenjit Basu (trans). Pergamon Press, 1965, 319 pp. LC: 63-25775. Series: *Mathematical Tables Series*, Vol. 36.

3082. Tables of Scattering Functions for Heterodisperse Systems. Arthur F. Stevenson; Wilfried Heller. Wayne State Univerity Press (Ambassador), 1961, 214 pp. LC: 61-8315. ISBN: 0-8143-1135-0. Cost: $9.95.

3083. Tables of Scattering Functions for Spherical Particles. National Bureau of Standards. NTIS, 1949, 119 pp. Accession: PB-251962. Cost: $5.50. Series: NBS AMS4.

3084. Tables of Similar Solutions to the Equations of Momentum, Heat and Mass Transfer in Laminar Boundary Layer Flow. E. Elzy; R.M. Sisson. Oregon State University Engineering Experiment Station, 1967, 333 pp.

3085. Tables of Sin 0 and Sin 2 0 for Values of 0 From 2° to 87°. H.A. Plettinger. Gordon and Breach, 1965, 46 pp. ISBN: 0-677-01100-8. Cost: $25.25.

Useful trigonometric tables are presented in this manual.

3086. Tables of Sine, Cosine, and Exponential Integrals. National Bureau of Standards. National Bureau of Standards, 1940, Vol.1, 444 pp.; Vol.2, 225 pp. (out of print). Series: Vol.1, NBS MT5; Vol.2, NBS MT6.

3087. Tables of Sines and Cosines for Radian Arguments. National Bureau of Standards. NTIS, 1955, 278 pp. Cost: $9.25. Series: NBS AMS43.

NOTE: A reissue of MT4.

3088. Tables of Sines and Cosines for Radian Arguments. National Bureau of Standards. National Bureau of Standards (avail. from NTIS), 1940, 275 pp. Accession: PB-176127. Cost: $9.25. Series: NBS MT4.

NOTE: Reissued as NBS AMS43.

3089. Tables of Sines and Cosines to Fifteen Decimal Places at Hundredths of a Degree. National Bureau of Standards. NTIS, 1949, 95 pp. Accession: COM-73-10501. Cost: $5.00. Series: NBS AMS5.

3090. Tables of Spectral Lines of Neutral and Ionized Atoms. A.R. Striganov; N.S. Sventitskii. IFI/Plenum, 1968, 899 pp. LC: 68-28091. ISBN: 0-306-65139-4. Cost: $75.00.

3091. Tables of Spectral Lines. A.N. Zaidel et al. IFI/Plenum, 3d ed., 1970, 782 pp. LC: 70-120028. ISBN: 0-306-65151-3. Cost: $75.00.

3092. Tables of Spherical Bessel Functions. National Bureau of Standards. Columbia University Press, Vol.1, 1947; Vol.2, 1947, Vol.1, 375 pp.; Vol.2, 328 pp. Series: Vol.1, CUP8; Vol.2, CUP9.

3093. Tables of Squares, Cubes, Square Roots, Cube Roots, and Reciprocals of All Integers Up To 12,500. Peter Barlow; L.J. Cormie (trans). Methuen, Inc, 4th ed., 1965, 258 pp.

3094. Tables of Standard Electrode Potentials. Guilio Milazzo; S. Caroli; V.K. Sharma. Wiley-Interscience, 1978, 419 pp. LC: 77-8111. ISBN: 0-471-99534-7. Cost: $76.25.

Tables include all electrode-systems and are arranged according to the periodic table. Includes bibliographic references.

3095. Tables of Summable Series and Integrals Involving Bessel Functions. Albert D. Wheelon. Holden-Day, 1968. LC: 67-13849. ISBN: 0-8162-9512-3. Cost: $14.95.

3096. Tables of the Anger and Lommel-Weber Functions. Gary D. Bernard; Akira Ishimaru. University of Washington Press, 1962. LC: 62-17144. ISBN: 0-295-73956-8. Cost: $10.00.

3097. Tables of the Bessel Function $Y_0(z)$ and $Y_1(z)$ for Complex Arguments. National Bureau of Standards. Columbia University Press, 1950, 427 pp. Series: CUP12.

3098. Tables of the Bessel Functions $Y_0(x)$, $Y_1(x)$, $K_0(x)$, $KI1(x)$, $\leq$ x $\leq$. National Bureau of Standards. NTIS, 1952, 60 pp. Cost: $4.50. Series: NBS AMS25.

NOTE: A reissue of NBS AMS1 with minor revisions in the introduction.

3099. Tables of the Binomial Probability Distribution. National Bureau of Standards. NTIS, 1950, 387 pp. Accession: PB-251862. Cost: $10.75. Series: NBS AMS6.

3100. Tables of the Bivariate Normal Distribution Function and Related Functions. National Bureau of Standards. NTIS, 1959, 258 pp. Accession: PB-176520. Cost: $9.00. Series: NBS AMS50.

3101. Tables of the Characteristic Functions of the Eclipse and Related Delta-Functions for Solution of Light Curves of Eclipsing Binary Systems. Masatoshi Kutamura. University Park Press, University of Tokyo Press, 1968, 341 pp. LC: 67-30318 (Jap.). ISBN: 0-8391-0001-9. Cost: $60.00.

3102. Tables of the Clebsch-Gordon Coefficients. Institute of Atomic Energy. Science Press, Peking, 1965.

3103. Tables of the Confluent Hypergeometric Function F (n/2, 1/2;x) and Related Functions. National Bureau of Standards. NTIS, 1949, 73 pp. Accession: PB-251959. Cost: $4.50. Series: NBS AMS3.

3104. Tables of the Descending Exponential, x = 2.5 to x = 10. National Bureau of Standards. NTIS, 1955, 76 pp. Accession: PB-186429. Cost: $5.00. Series: NBS AMS46.

NOTE: This is an extension of AMS14.

3105. Tables of the Error Function and its Derivative. National Bureau of Standards. NTIS, 1954, 302 pp. Accession: PB-176521. Cost: $9.75. Series: NBS AMS41.

NOTE: A reissue of NBS MT8.

3106. Tables of the Exponential Function e^x. National Bureau of Standards. U.S. Government Printing Office, 1951, 537 pp. Cost: $9.00. Series: NBS AMS14.

NOTE: A reissue of NBS MT2. For extension of this table, *see* NBS AMS46.

3107. Tables of the Exponential Function e^x. National Bureau of Standards. U.S. Government Printing Office, 1939, 535 pp. SuDoc: C13.32:14. Cost: $9.00. Series: NBS MT2.

NOTE: Reissued as NBS AMS14.

3108. Tables of the F-E Related Distribution Algorithms. K. Mardia; P.J. Zemroch. Academic Press, 1979, 256 pp. LC: 77-74368. ISBN: 0-12-471140-5. Cost: $34.00.

This book is intended for the practicing statistitian, a comprehensive library of Algol 60 algorithms for the calculation of percentage points and probability integrals is included. Also included are 5-point Lagrangian coefficients for interpolation between tabulated percentage points, as well as the gamma function, failure indicators, and listings of algorithms.

3109. Tables of the Fabry Factors G and V, and of Magnetic Field Homogeneity, for Thick Cylindrical Coils. M.W. Garrett. NTIS, 1969, 58 pp. Accession: N69-27223. Tech Report: ORNL-4281.

This is a tabular collection of properties of cylindrical coils with 2 dimensions of current density.

3110. Tables of the Fractional Functions for the Planck Radiation Law. Marianus Czerny; A. Walther. Springer-Verlag, 1961. ISBN: 0-387-02642-8. Cost: $17.30.

3111. Tables of the Function (Mathematical Function Given) in the Complex Domain. Konstantin Andrianovich Karpov; D.E. Brown (trans). Pergamon Press, 1965, 519 pp. LC: 63-19541. Series: *Mathematical Tables Series*, Vol. 37.

3112. Tables of the Hypergeometric Probability Distribution. Gerald J. Lieberman; Donald B. Owen. Stanford University Press, 1961, 726 pp. LC: 61-6879. ISBN: 0-8047-0057-5. Cost: $25.00.

3113. Tables of the Incomplete Beta-Function. Karl Pearson. Cambridge, 2d ed., 1968, 505 pp. LC: 68-11285 (G. B.).

3114. Tables of the Legendre Functions. Mariia Ivanovna Zhurina; L.N. Karmazina; D.E. Brown (trans). Pergamon Press, 1964-65, 2-Vol. set. LC: 63-18939. Series: *Mathematical Tables Series*, Vol. 22, 38.

3115. Tables of the Moments of Inertia and Section Moduli of Ordinary Angles, Channels, and Bulb Angles with Certain Plate Combinations. National Bureau of Standards. National Bureau of Standards, 1941, 197 pp. (out of print). Series: NBS MT11.

3116. Tables of the Normal Probability Integral, the Normal Density, and Its Normalized Derivatives. Nikolai Vasilevich Smirnov; D.E. Brown (trans). Pergamon Press, 1965, 125 pp. LC: 63-18936. Series: *Mathematical Tables Series*, Vol. 32.

3117. Tables of the Principal Unitary Representations of Fedorov Groups. Dmitrii Kontantinovich Faddeev; Prasenjit Basu (trans). Pergamon Press, 1964, 155 pp. LC: 63-18934. Series: *Mathematical Tables Series*, Vol. 34.

3118. Tables of the Thermodynamic Properties of Air and the Exhaust Gas from a Turbine Engine. J.M. Pelton; K.L. Hannah. NTIS, 1976, 167 pp. Accession: N76-29393, AD-A021954. Tech Report: ARO-ETF-TR-75-93; AFDC-TR-76-16.

Tabular data are presented on the thermodynamic properties of air and exhaust gas from a turbine engine. Air enthalpy, entropy, and specific heat at a constant pressure are among the properties presented.

3119. Tables of the Velocity of Sound in Sea Water. L.S. Bark et al; D.E. Brown (trans). Pergamon Press, 1964, 180 pp. LC: 63-18921. Series: *Mathematical Tables Series*, Vol. 21.

3120. Tables of Thermal Properties of Gases. J. Hilsenrath et al. NTIS, 1955, 499 pp. Accession: N78-79496; PB-280020. Tech Report: NBS-CIRC-564.

This National Bureau of Standards publication presents data on the thermodynamic properties of argon, carbon dioxide, carbon, and carbon monoxide.

3121. Tables of Thermodynamic Data. International Atomic Energy Agency. Unipub, 1964. ISBN: 92-0-145264-0. Cost: $4.00 paper copy. Series: *IAEA Technical Reports*, No. 38.

3122. Tables of Thermodynamic Functions of a Substance at a High Concentration of Energy. N.N. Kalitkin; L.V. Kuzmina. NTIS, 1978, 50 pp. Accession: N79-71057. Tech Report: UCRL-TRANS-11373.

This Russian translation presents information on quantum mechanics and thermodynamic properties of a substance with a high concentration of energy.

3123. Tables of Thermodynamic Properties of Water and Steam. M.P. Vukalovich. NTIS, 1968, 38 pp. Accession: N68-87766. Tech Report: RTS-4198.

This is the English translation of a Soviet title of tabular data on the thermodynamic properties of water and steam. The Soviet title was originally published in 1957.

3124. Tables of Thomson Functions and Their First Derivatives. L.N. Nosova. Pergamon Press, 1961, 422 pp. LC: 61-12445. ISBN: 0-08-009518-6. Cost: $35.00. Series: *Pergamon Mathematical Tables Series*, Vol. 13.

3125. Tables of Transformation Brackets for Nuclear Shell-Model Calculations. T.A. Brody; M. Moshinsky. Gordon and Breach, 2d ed., 1967, 175 pp. LC: 67-23633. ISBN: 0-677-01320-5. Cost: $49.50.

3126. Tables of Trigonometric Functions for the Numerical Computation of Electron Density in Crystals. I.M. Kuntsevich. Halsted Press (International Scholarly Book Service), 1972, 218 pp. ISBN: 0-470-51090-0. Cost: $19.50.

3127. Tables of Two-Associate-Class Partially Balanced Designs. W.H. Clatworthy; J.M. Cameron (contrib) J.A. Speckman (National Bureau of Standards). U.S. Government Printing Office, 1973, 327 pp. SuDoc: C13.32:63. Cost: $3.45. Series: NBS AMS63.

3128. Tables of Wavenumbers for the Calibration of Infra-Red Spectrometers. A.R.H. Cole. Pergamon Press, 2d ed., 1977, 219 pp. LC: 76-28298. ISBN: 0-08-021247-6. Date of First Edition: 1961. Cost: $52.00.

This work is sponsored by the International Union of Pure and Applied Chemistry, Commision on Molecular Structure and Spectroscopy and is part of the *Chemical Data Series*, Vol.9. This work includes tables for low as well as high resolution spectrometers. Also has information on the refractive index of air. The range of calibration data has been extended to cover 4,350cm⁻¹ to 1cm⁻¹.

3129. Tables on the Thermophysical Properties of Liquids and Gases in SI Units; in Normal and Dissociated States. N.B. Vargaftik. Hemisphere Publishing Co (also available from Halsted Press), 2d ed., 1975, 772 pp. LC: 75-14260. ISBN: 0-470-90310-4. Cost: $75.00. Series: *Advances in Thermal Engineering Series*.

Thermophysical properties are presented for a number of gases including hydrogen, lithium, nitrogen, argon, and steam. The book is divided into tabulations of pure substances in Part 1 and mixtures of substances in Part 2.

3130. Tables Relating to Mathieu Functions. National Bureau of Standards. U.S. Government Printing Office, 1967, 311 pp. SuDoc: C13.32:59. Cost: $6.80. Series: NBS AMS59.

NOTE: A reissue with additions of CUP13.

3131. Tables to Facilitate Sequential t-Tests. National Bureau of Standards. NTIS, 1951, 82 pp. Accession: PB-251863. Cost: $5.00. Series: NBS AMS7.

3132. Tabulation of Data on Semiconductor Amplifiers and Oscillators at Microwave Frequencies. C.P. Marsden. NTIS, 1971, 48 pp. Accession: N72-13208. Tech Report: NBS-TN-597.

The chemical characteristics of semiconductor microwave amplifiers and oscillators are presented in this tabulation.

3133. Tabulation of Infrared Spectral Data. David Dolphin; Alexander E. Wick. Wiley-Interscience, 1977, 549 pp. LC: 76-48994. ISBN: 0-471-21780-8. Cost: $24.95.

3134. Tacheometric Tables for the Metric User. D.T.F. Munsey. International Ideas, Inc, 1971. ISBN: 0-211-39336-5. Cost: $14.95.

These tables provide conversion factors for measuring the revolutions per minute of a rotating shaft.

3135. Technical Communicator's Handbook of Technology Transfer. Hyman Olken. Olken Publications, 1980, 144 pp. ISBN: 0-934818-01-0. Cost: $12.50.

Information is provided on communication techniques, technical writing, and various types of technical presentations.

3136. Technical Data on Fuel. J.W. Rose; J.R. Cooper. British National Committee, World Energy Conference (distributed by Scottish Academic Press), 7th ed., 1976, 343 pp. ISBN: 0-7073-0129-7. Cost: £30 ($75.00).

NOTE: Also available from Wiley/Halsted, LC: 77-24872, ISBN: 0-470-99239-5, $74.95. Includes 280 tables and 140 diagrams. Covers such properties as thermoconductivity, viscosity, heat capacity, enthalpy of gases, ignition energies gross calorific value, molar enthalphies of combustion, vapor pressures of hydrocarbons, density, melting point, and coefficient of expansion.

3137. Technical Handbook for Facilities Engineering and Construction Manual. U.S. of Health, Education, and Welfare Department. U.S. Government Printing Office, earliest verified date: Nov. 2, 1970 (continually updated), loose-leaf service, multivolume set, issued with perforations. SuDoc: HE 1.108/2: Part No..

3138. Technical Handbook. Royal Netherlands Aircraft Factories. NTIS, 1966, 85 pp. Accession: N68-80193. Tech Report: TH-3-011-E.

This handbook, originally entitled *Technisch Handboek*, was prepared by the Royal Netherlands Aircraft Factories, Amsterdam. The publication deals with fatigue materials, load calculations, reinforced shells, and mechanical properties of metals.

3139. Technician's Handbook of Plastics. Peter A. Grandilli. Van Nostrand Reinhold, 1981, 272 pp. LC: 80-19379. ISBN: 0-442-23870-3. Cost: $19.95.

Such information as plastics applications in electronics, vacuum encapsulation, thermoset labs, graphite fibers, Kelvar fibers, laminates, and resins are presented.

3140. Techniques in Pedology; A Handbook for Environmental and Resource Studies. R.T. Smith; K. Atkinson. Scientific Books Ltd (distributed by Merrimack Book Service), 1975, 213 pp. ISBN: 0-236-30939-0. Cost: $15.95.

Such topics as soil sampling, aerial photographs for soil surveying, chemical properties of soil, and analysis of soils are discussed in this handbook.

3141. Technology Mathematics Handbook; Definitions, Formulas, Graphs, Systems of Units, Procedures, Conversion Tables, Numerical Tables. Jan J. Tuma. McGraw-Hill, 1975, 370 pp. LC: 74-26962. ISBN: 0-07-065431-X. Cost: $19.50.

3142. Telecommunication Transmission Handbook, 2nd Edition. Roger L. Freeman. Wiley-Interscience, 2d ed., 1981, 688 pp. LC: 81-7499. ISBN: 0-471-08029-2. Date of First Edition: 1975. Cost: $36.50.

Such topics as speech telephony, facsimile transmission, video transmission, and telegraphic data are presented in this handbook. In addition, such subjects as cable carrier, channel coding, earth station engineering, scatter, HF radio, and subscriber loop design are presented in this handbook, as well as fiber optic communications, digital radio, and transmission troubleshooting.

3143. Telescope Handbook and Star Atlas. Neale E. Howard. Crowell, 2d ed., 1975, 226 pp. LC: 75-6601. ISBN: 0-690-00686-1. Cost: $15.95.

3144. Television Engineering Handbook. Donald G. Fink. McGraw-Hill, 1957.

3145. Temperature Measurement Handbook and Catalog. Omega Engineering, Inc, 1981, 360 pp. Cost: No cost.

This handbook-catalog provides specification data for over 10,000 measurement and control products.

3146. Temperature Measurements in Engineering. Henry Dean Baker et al. Omega Enginering, 1975, 2 Vols.

This 2-volume set provides information on methods of temperature measurements and the various instruments associated with temperature measurments.

3147. Ten-Decimal Tables of the Logarithms of Complex Numbers and for the Transformation from Cartesian to Polar Coordinates. L.A. Liusternik; D.E. Brown trans. Pergamon Press, 1965, 110 pp. LC: 63-18926. Series: *Mathematical Tables Series*, Vol. 33.

3148. Terms of Contract Handbook. British Mechanical Engineering Confederation.. Bookstax. Cost: $40.00.

3149. Textile Handbook. American Home Economics Association, 4th ed., 1970.

The United States Dispensatory. *See* U.S. Dispensatory and Physicians Pharmacology.

3150. Theoretical Mean Activity Coefficients of Strong Electrolytes in Aqueous Solutions from 0 to 100° C. Walter J. Hamer (National Bureau of Standards). Office of Standard Reference Data, 1968. Cost: $6.10. Series: NSRDS-NBS-24.

3151. Theoretical Steam Rate Tables. American Society of Mechanical Engineers, 1969. LC: 75-88047. ISBN: 0-685-06532-4. Cost: $15.00.

3152. Therapist's Handbook: Treatment Methods of Mental Disorders. Benjamin B. Wolman. Van Nostrand Reinhold, 1976, 556 pp. LC: 75-28356. ISBN: 0-442-29570-7. Cost: $24.50.

3153. Thermal Accommodation and Adsorption Coefficients of Gases. Y.S. Touloukian; C.Y. Ho (eds); S.C. Saxena; R.K. Joshi (authors). McGraw-Hill, 1st ed., 1981, 448 pp. LC: 80-17822. ISBN: 0-07-065031-4. Cost: $42.50. Series: *McGraw-Hill/CINDAS Data Series on Material Properties*, Vol. II-I.

This publication presents data on gas-solid systems, heat transfer information at different pressures, along with the adsorption characteristics of various gases on specific solid surfaces.

3154. Thermal Conductivity and Emittance of Solid UO_2. R. Brandt; G. Haufler; G. Neuer; Y.S. Touloukian (trans ed). Purdue Research Center, 160 pp. Cost: $39.50.

This English translation of a critical review originally written in German, covers 231 references that deal with uranium dioxide. Such topics as stoichiometry, composition, porosity, irradiation, zone melting, and annealing are discussed in this review.

3155. Thermal Conductivity of Carbon Dioxide Gas and Liquid. NTIS. ESDU, Springfield, VA, 1976, 26 pp. ISBN: 0-85679-167-9. Accession: N77-21200. Tech Report: ESDU-76030.

This data, compiled by the Engineering Sciences Data Unit in England, covers such topics as equations and values of thermal conductivity of carbon dioxide for temperatures between 240° K and 1,500° K and for pressures from 1 to 1,000 bar.

3156. Thermal Conductivity of Liquid Aliphatic Alcohols. ESDU, 1975, 1,499 pp. Accession: N76-21295. Tech Report: ESDU-75024. Cost: $170.50.

The Engineering Sciences Data Unit has compiled information on the thermal conductivity of liquid aliphatic alcohols. The range covered is from melting point to approximately .9 of the critical temperature.

3157. Thermal Conductivity of Selected Materials, Part 2. C.Y. Ho; R.W. Powell; P.E. Liley (National Bureau of Standards). NTIS, 1968. Cost: $6.75. Series: NSRDS-NBS-16, Part 2.

3158. Thermal Conductivity of Selected Materials. R.W. Powell; C.Y. Ho; P.E. Liley (National Bureau of Standards). NTIS, 1966. Accession: PB-189698. Cost: $7.50. Series: NSRDS-NBS-8, Part 1.

3159. Thermal Constants of Compounds. V.P. Glushko. U.S.S.R. Academy of Sciences (available from NTIS), 1966, 32 pp. Accession: N68-15751; PB-175626. Tech Report: TT-67-62363.

Thermodynamic properties of inorganic and organic compounds are listed, along with free energy data, Gibbs' free energy, heat balance, thermochemistry data, and phase transformation.

3160. Thermal Constants of Materials. V.P. Glushko. Viniti, Moscow (available from NTIS), Vol.6, 1973; Vol.7, 1974, multivolume set, Vol.6, 465 pp.; Vol.7, 344 pp. Accession: Vol.6, A74-46629; Vol.7, A75-43698.

This multivolume Russian tabulation provides such information as phase transformations, thermodynamic properties, thermophysical properties, ionization potentials, crystal structure, enthalpy, Gibbs' free energy, and transition temperature. Bibliographic references are provided for original source material.

3161. Thermal Expansion Properties of Aerospace Materials. E.F. Green. NTIS, 1969. Accession: B69-10055. Tech Report: M-FS-18335.

Charts and tables are presented on thermal expansion properties of aerospace materials, including data on elevated and cryogenic temperatures.

3162. Thermal Insulation Handbook. William C. Turner; John F. Malloy. McGraw-Hill, 1981, 624 pp. ISBN: 0-07-039805-4 (McGraw-Hill); 0-88275-510-2 (Krieger). Cost: $59.50.

Such topics as vapor-retarders, weather-barriers, properties of thermal insulation, and fundamentals of heat transfer are discussed in this handbook.

3163. Thermal Linear Expansion of Nine Selected AISI Stainless Steels. P.D. Desai; C.Y. Ho. CINDAS/Purdue University, 1978, 51 pp. Series: *Purdue University Report*, No. 51.

Provides evaluated data on thermal linear expansion of stainless steels.

3164. Thermal Properties of Aqueous Uni-Univalent Electrolytes. V.B. Parker (National Bureau of Standards). U.S. Government Printing Office, 1965. SuDoc: C13.48:2. Cost: $1.10. Series: NSRDS-NBS-2.

3165. Thermal Properties of the Gases H_2, CH_4, H_2O, N_2, CO, and CO_2 in the Range 0 C Less Than or Equal to Theta Less Than or Equal to 1000 C 1 Bar Less Than or Equal to Rho Less Than or Equal to 50 Bar. R. Harth. NTIS, 1974, 108 pp. Accession: N75-19065. Tech Report: JUL-1085-RB.

This is a collection of thermal properties of hydrogen, methane steam, nitrogen, carbon monoxide, and carbon dioxide. This publication is written in German, with an English summary.

Thermal Radiative Properties of Coatings. *See* McGraw-Hill/CINDAS Data Series on Material Properties. Volume II-4.

Thermal Radiative Properties of Solids. *See* McGraw-Hill/CINDAS Data Series on Material Properties. Volume I-5.

3166. Thermochemical Properties of Inorganic Substances. I. Barin; O. Knacke. Springer-Verlag, 1973; Supplement I, 1977, 1 Vol. plus supplement. LC: 72-95058. ISBN: 0-387-06053-7; Supplement I, 3-540-08031-7. Cost: $97.50; Supplement I, $100.30.

3167. Thermochemistry for Steelmaking. J.F. Elliott et al. Addison-Wesley (Vol. 2 currently available from Pergamon Press), Vol.1, 1960; Vol.2, 1963, Vol.1, 296 pp.; Vol.2, 550 pp. ISBN: Vol.2, 0-08-010050-3. Cost: Vol.2, $37.50 (Pergamon).

Vol.2 is entitled *Thermodynamics and Transport Properties*. and includes phase diagrams. This work was partially sponsored by the American Iron and Steel Institute, MIT, and the U.S National Bureau of Standards. It discusses compounds and elements involved in the steelmaking process and their related thermodynamic properties such as Gibbs energy functions, heat capacity, enthalpy, entropy, vapor pressure, and many more.

3168. Thermocouple Reference Data Expressed on the Kelvin Scale. R.L. Powell; W.J. Hall; C.H. Hyink Jr. NTIS, 1971, 186 pp. Accession: N75-78110. Tech Report: NBS-10-724.

This National Bureau of Standards publication lists data on thermocouples.

3169. Thermodynamic and Related Properties of Parahydrogen from the Triple Point to 100 K at Pressures to 340 Atm. H.M. Roder; L.A. Weber; R.D. Goodwin. NTIS, 1963, 104 pp. Accession: N77-77408. Tech Report: NASA-CR-152799; NBS-7987.

This National Bureau of Standards publication provides information on the thermodynamic properties of parahydrogen, as well as data on the isochoric process and the isothermal process.

3170. Thermodynamic and Related Properties of Parahydrogen from the Triple Point to 100 K at Pressures to 340 Atmospheres. H.M. Roder; L.A. Weber; R.D. Goodwin (National Bureau of Standards). NTIS, 1965. Accession: N65-32001. Cost: $5.50. Series: NBS Monograph 94.

3171. Thermodynamic and Thermophysical Properties of Combustion Products. V.P. Glushko et al (trans). International Scholarly Book Service, 1974-75, multivolume set. ISBN: Vol.1, 0-7065-1471-8. Cost: Vol.1, $32.50; Vol.2, $42.50; Vol.3, $42.00.

This title consists of several volumes entitled as follows: Vol.1, *Computation Methods*; Vol.2, *Oxygen-Based Propellants*; and Vol.3, *Oxygen and Air-Based Propellants*. Vol.4, *Nitrogen Tetroxide-Based Propellants*. Includes information on reaction kinetics, vapor phases, transport properties, chemical composition, enthalpy, density, and other properties. Such fuels as nitrogen tetroxide dimethylhydrazines, kerosene, oxygen fluorides and chlorine fluorides.

3172. Thermodynamic and Thermophysical Properties of Combustion Products. Vol. 1, Computation Methods. V.E. Alemasov; A.F. Dregalin; A.P. Tishin; V.A. Khudyakov; V.P. Glushko (ed). NTIS, Russian translation for the National Bureau of Standards, 1971, trans. 1974. Accession: TT74-50019. Cost: $12.00.

3173. Thermodynamic and Thermophysical Properties of Combustion Products. Vol. 2, Oxygen-Based Propellants. V.E. Alemasŏv; A.F. Dregalin; V.A. Khudyakov; V.N. Kostin; V.P. Glushko (ed). NTIS, Russian translation for the National Bureau of Standards, 1972, trans. 1975. Accession: TT74-50032. Cost: $12.75.

3174. Thermodynamic and Thermophysical Properties of Combustion Products. Vol. 3, Oxygen- and Air-Based Propellants. V.E. Alemasov; A.F. Dregalin; A.P. Tishin; V.A. Khudyakov; V.N. Kostin; V.P. Glushko (ed). NTIS, Russian translation for the National Bureau of Standards, 1973, trans. 1975. Accession: TT75-50007. Cost: $16.25.

3175. Thermodynamic and Thermophysical Properties of Helium. N.V. Tsederberg; V.N. Popov; N.A. Morozova; A.F. Alyab'ev (ed). NTIS, Russian translation for the National Bureau of Standards, 1969, trans. 1971. Accession: TT70-50096. Cost: $9.00.

3176. Thermodynamic and Thermophysical Properties of Helium. N.V. Tsederberg. NTIS, 1971, 260 pp. Accession: N71-38756. Tech Report: TT70-50096.

Tables and diagrams on the thermodynamic properties of helium are provided in this Russian translation.

3177. Thermodynamic and Transport Properties of Ethylene and Propylene. I.A. Neduzhii et al. NTIS, Russian translation for the National Bureau of Standards, 1971, trans. 1972. Accession: COM 75-11276. Cost: $7.75. Series: NBSIR 75-763.

3178. Thermodynamic and Transport Properties of Fluids, SI Units. Y.R. Mayhew; G.F.C. Rogers. Basil Blackwell (distributed by International Scholarly Book Services), 1976, 20 pp. ISBN: 0-631-96400-2. Cost: $1.50.

A collection of 11 tables of thermodynamic data and transport properties of fluids.

3179. Thermodynamic and Transport Properties of Gaseous Tetrafluoromethane in Chemical Equilibrium. J.L. Hunt; L.R. Boney. NTIS, 1973, 107 pp. Accession: N73-27805. Tech Report: NASA-TN-D-7181; L-8483. Cost: $3.00 hard copy.

Pressure-enthalpy diagrams are presented and equilibrium, thermodynamic, and transport property data are tabulated.

3180. Thermodynamic and Transport Properties of Liquid Sodium. M.E. Durham. NTIS, 1974, 39 pp. Accession: N75-11054. Tech Report: RD/B/M-2479-REV; CFR/THWP/P-(72)28-REV. Cost: $5.00 hard copy.

Graphical and tabular data are presented for saturated liquid sodium. Thermodynamic and transport properties can be determined as a function of temperature.

3181. Thermodynamic and Transport Properties of Organic Salts. International Union of Pure and Applied Chemistry. Pergamon, 1980, 370 pp. LC: 80-40689. ISBN: 0-08-022378-8. Cost: $105.00. Series: *Chemical Data Series*, No. 28.

Tables on transport properties of pure molten salts with organic ions are provided in this tabulation.

Thermodynamic and Transport Properties of Steam. *See* **ASME Steam Tables.**

3182. Thermodynamic Characteristics of Helium—Formula and Tables. R. Drut. NTIS, 1969, 62 pp. Accession: N69-34214. Tech Report: CEA-R-3791.

Various properties such as specific mass, compressiblity coefficient, specific heats, dynamic viscosity, thermal conductivity, enthalpy, and entropy of helium are presented in tabular format in this compilation. This publication is written in French, with an English summary.

3183. Thermodynamic Constants of Substances. V.P. Glushko. USSR Academy of Sciences, All-Union Institute of Scientific and Technological Information, Moscow (available in the United States from Victor Kamkin, Inc., Bookstore, Washington, D.C.), Part I, 1965, Handbook in 10 parts.

3184. Thermodynamic Data for Ferric Sulfate and Indium Sulfate. L.B. Pankratz; W.W. Weller. Bureau of Mines, Albany, OR, 1969, 9 pp. Accession: N69-30807. Tech Report: BM-RI-7280.

High temperature data and low temperature test data are presented on ferric sulfate and indium sulfate.

3185. Thermodynamic Data for Inorganic Sulphides, Selenides, and Tellurides. K.C. Mills. Halsted Press (also Butterworths), 1974. LC: 74-19506. ISBN: 0-470-60655-X; 0-408-70537-X (Butterworths). Cost: $65.00.

Covers such properties as heat capacity, heat of formation, entropy, vapor pressure and disassociation pressure.

3186. Thermodynamic Data for Silver Chloride and Silver Bromide. L.B. Pankratz. Bureau of Mines, Albany, OR, 1970, 15 pp. Accession: N70-40910. Tech Report: BMRI-7430.

High temperature testing data and heat measurement data are provided for silver chlorides and silver bromides.

3187. Thermodynamic Data for Waste Incineration. American Society of Mechanical Engineers, 1979, 160 pp. LC: 78-71787. ISBN: 0-685-95761-6. Cost: $30.00.

Numerous materials are listed in alphabetical sequence with their corresponding thermodynamic properties. Such properties as specific heat, heat of fusion, heat of vaporization, heat of combustion, heat of explosion, explosion temperature, and vapor pressure are provided for these materials. These tables were originally prepared by the Chemical Thermodyanmics Data Center of the National Bureau of Standards.

3188. Thermodynamic Data on Oxygen and Nitrogen. J. Brewer. NTIS, 1961, 171 pp. Accession: N76-73566. Tech Report: AD-275728; ASD-TR-61-625.

Thermodynamic property data are presented for nitrogen and oxygen.

3189. Thermodynamic Derivatives for Water and Steam. Solomon L. Rivkin; Aleksey A. Aleksandrov; Elena A. Kremenevskaya. John Wiley and Sons, 1978, 264 pp. LC: 78-5789. ISBN: 0-470-26363-6. Cost: $25.00.

This is a tabular and graphical representation of the thermodynamic derivatives defining the properties of steam. Comparisons are made with such measurements as specific heats, Joule-

Thomson coefficients, and velocity of sound. In addition, analytical expressions such as the Gibbs free enthalpy, and Helmholtz free energy have been used.

3190. Thermodynamic Functions of Gasses. F. Din (ed). Butterworth and Co., Ltd, Vol.1 and 2, 1956; Vol.3, 1961, Vol.1, 175 pp.; Vol.2, 201 pp.; Vol.3, 218 pp.

The 3 volumes are entitled as follows: Vol.1, *Ammonia, Carbon Dioxide, Carbon Monoxide*; Vol.2, *Air, Acetylene, Ethylene, Propane and Argon*; Vol.3, *Methane, Nitrogen, Ethane*.

3191. Thermodynamic Properties and Reduced Correlations for Gases. Lawrence N. Canjar; Francis S. Manning. Gulf Publishing Co, 1967, 212 pp. LC: 66-30022. ISBN: 0-87201-867-9. Cost: $19.95.

3192. Thermodynamic Properties in SI; Graphs, Tables, and Computational Equations for Forty Substances. William C. Reynolds. Department of Mechanical Engineering, Stanford University, 1979, 167 pp. LC: 79-84863. ISBN: 0-917606-05-01. Cost: $4.50.

Thermodynamic data are presented for a wide variety of working fluids, including refrigerants, air, ammonia, water, hexane, hydrogen, oxygen, propane, and propyl alcohol. A psychrometric chart is included with this publication.

3193. Thermodynamic Properties of Ammonia as an Ideal Gas. L. Haar (National Bureau of Standards). U.S. Government Printing Office, 1968. SuDoc: C13.48:19. Cost: $0.65. Series: NSRDS-NBS-19.

3194. Thermodynamic Properties of Argon from the Triple Point to 300 K at Pressures to 1000 Atmospheres. A.L. Grossman; J.G. Hust; R.D. McCarty. Superintendent of Documents, 1969, 154 pp. Accession: N69-22221. Tech Report: NASA-CR-1000532; NSRDS-NBS-27. SuDoc: C13.48:27. Cost: $1.25.

Data on argon are presented such as specific heat, entropy, equations of state, density, enthalpy, and compressibility.

3195. Thermodynamic Properties of Benzene and Toluene. J.F. Counsell; I.J. Lawrenson; E.B. Lees. NTIS, 1976, 16 pp. Accession: N77-31001. Tech Report: NPL-CHEM-52. Cost: $5.00 hard copy; $3.50 microfilm.

Methods for calculating the thermodynamic properties of benzene and toluene are presented. Such values as temperature, volume, and pressure are presented in this publication.

3196. Thermodynamic Properties of Benzene. Engineering Sciences Data Unit, Sponsored by the Institute of Chemical Engineers (available from NTIS), 1973, 39 pp. Accession: N75-29200. Tech Report: ESDU-73009. Cost: $122.50 hard copy.

Such data as specific heat, pressure gradients, ethalpy, entropy, and temperature gradients are presented for benzene.

3197. Thermodynamic Properties of Chemical Substances. V.P. Glushko et al (eds). USSR Academy of Sciences, 1962, Vol.1, 1,164 pp.; Vol.2, 916 pp., 2 Vols.

3198. Thermodynamic Properties of Combustion Gases. Jerry D. Pearson; Robert C. Fellinger. Iowa State University Press, 1966. ISBN: 0-8138-1701-3. Cost: $7.50.

3199. Thermodynamic Properties of Compressed Gaseous and Liquid Fluorine. R. Prydz; C.C. Straty. NTIS, 1970, 199 pp. Accession: N71-19869. Tech Report: AD-716286; NBS-TN-392. Cost: $1.50.

This National Bureau of Standards publication presents such thermodynamic properties as vapor pressure and pressure-volume-temperature data for compressed gaseous and liquid fluorine.

3200. Thermodynamic Properties of Copper and Its Inorganic Compunds. E.G. King; Alla D. Mak; L.B. Pankratz. International Copper Research Association, 1973. Cost: $10.00.

3201. Thermodynamic Properties of Gaseous Ge_2, Ge_4, GeF_2 and GeF. P.A.G. Ohare. NTIS, 1968, 18 pp. Accession: N69-38386. Tech Report: ANL-7523.

This Argonne National Laboratory publication presents thermodynamic properties of gaseous germanium, germanium fluoride, germanium difluoride, and germanium tetrafluoride.

3202. Thermodynamic Properties of Helium 3 Helium 4 Solutions with Applications to the Helium 3 Helium 4 Dilution Refrigerator. R. Radebaugh. U.S. Government Printing Office, 1967, 145 pp. Accession: N68-35807. Tech Report: NBS-TN-362. Cost: $0.70.

This National Bureau of Standards publication presents thermodynamic data on helium 3 helium 4 solutions at cryogenic temperatures.

3203. Thermodynamic Properties of Helium from 6 to 540 Degrees R Between 10 and 1500 Psia. D.B. Mann. NTIS, 1962, 43 pp. Accession: N69-74308; PB-182435. Tech Report: NBS-154A; R-264.

This National Bureau of Standards publication presents thermodynamic properties of helium, including information on density, pressure distribution, and temperature control.

3204. Thermodynamic Properties of Helium, Nitrogen, and Helium-Nitrogen Mixtures from 240 Degrees R to 950 Degrees R for Pressures Between 14.696 and 3000 Psia. R.E. Wood. NTIS, 1968, 371 pp. Accession: N68-36342. Tech Report: BM-RI-7190.

Equations of state data and Joule-Thomson effect information are presented along with other thermodynamic properties of helium, nitrogen, and helium-nitrogen mixtures.

3205. Thermodynamic Properties of Hydrogen-Helium Plasmas. H.F. Nelson. NTIS, 1971, 125 pp. Accession: N72-14938. Tech Report: NASA-CR-1861.

This collection of data was compiled as a result of research performed at the University of Missouri under NASA contract. Thermodynamic Pproperties of hydrogen and helium plasmas are presented for temperatures from 10,000 to 100,000° K.

3206. Thermodynamic Properties of Individual Substances. G.A. Bergman et al. NTIS, 1967-, multivolume set, Vol.1, Part 1, 940 pp.; Vol.1, Part 3, 626 pp.; Vol.2, 868 pp. Accession: Vol.1, Part 1, N68-10297; TT-67-63067; AD-659659; Vol.1, Part 3, N68-11073; TT-67-63056; AD-659679; Vol.2, N68-10260; TT-67-63062; AD-659660. Tech Report: Vol.1, Part 1 FTD-HT-66-251; Vol.1, Part 3 FTD-HT-66-251; Vol.2, FTD-HT-66-251.

Some of the volumes of this publication are entitled as follows: Vol.1, Part 1, *Calculation of the Thermodynamic Properties*; Vol.1, Part 3, *Calculation of the Thermodynamic Properties*; Vol.2, *Calculation of the Thermodynamic Properties*. This handbook, originally published in Russian in 1962, was translated into English in 1967. This multivolume set includes calculation methods and reference material for thermodynamic properties of numerous substances, such as lead compounds, mercury compounds, potassium compounds, sodium compounds, and lithium compounds.

3207. Thermodynamic Properties of Isopropyl Alcohol. Engineering Sciences Data Unit, London (available from NTIS), 1976, 41 pp. ISBN: 0-85679-148-2. Accession: N77-72248. Tech Report: ESDU-76012. Cost: $434.50.

This publication was prepared by the Engineering Sciences Data Unit and includes such thermodynamic properties of isopropyl alcohol as enthalpy, entropy, and vapor pressure.

3208. Thermodynamic Properties of Nitrogen Gas Derived From Measurements of Sound Speed. Ben Younglove; R.D. McCarty. NTIS, 1979, 53 pp. Tech Report: NASA-RP-1051.

This report has been prepared by NASA and includes information on the thermodynamic properties of nitrogen gas.

3209. Thermodynamic Properties of Organic Compounds. George J. Janz. Academic Press, rev. ed., 1967. ISBN: 0-12-380451-5. Cost: $37.50. Series: *Physical Chemical Series*, Vol. 6.

3210. Thermodynamic Properties of Oxygen. R.B. Stewart. NTIS, 1966, 217 pp. Accession: N69-74834.

This collection of data was compiled by the Department of Chemical Engineering at the University of Iowa as part of a Ph.D. dissertation. Includes such information as equations of state, data on ambient temperature and various thermodynamic properties of oxygen.

3211. Thermodynamic Properties of Saline Water. B.M. Fabuss. NTIS, 1965, 67 pp. Accession: N76-72305. Tech Report: R/D-PR-136.

This report was prepared by Monsanto Research Corporation. Includes information on the thermodynamic properties of seawater.

3212. Thermodynamic Properties of Silicates. P.J. Spencer. NTIS, 1973, 39 pp. Accession: N73-24938. Tech Report: NPL-CHEM-21. Cost: $4.00.

Tabulated data are presented on solid and molten silicates.

3213. Thermodynamic Properties of Solid Mercury at Temperature Intervals of from 0 Deg K to Melting Point at Normal Pressure. M.P. Vukalovich. NTIS, 1969, 33 pp. Accession: N69-34198. Tech Report: NASA-CR-103884.

This NASA report is an English translation of a Russian document and includes thermodynamic properties of solid mercury.

3214. Thermodynamic Properties of Some Chalcogen Fluorides. P.A.G. Ohare. NTIS, 1968, 95 pp. Accession: N69-24735. Tech Report: ANL-7315.

This compilation was prepared by the Argonne National Laboratories and includes data on the thermodynamic properties of chalcogen fluorides.

Thermodynamic Properties of Steam. *See* **ASME Steam Charts.**

3215. Thermodynamic Properties of the Coexisting Phases and Thermochemical Properties of the NaCl Component in Boiling NaCl Solutions. J.L. Haas Jr. NTIS, 1976, 76 pp. Accession: N77-85577. Tech Report: BULL-1421-B; LC-76-608097.

This U.S. Geological Survey Bulletin provides information on the thermodynamic properties of sodium chlorides.

3216. Thermodynamic Properties of the Elements. Daniel R. Stull; Gerard C. Sinke. American Chemical Society, 1956, 234 pp. LC: 57-1340. ISBN: 0-8412-0019-X. Cost: $21.00. Series: *Advances in Chemistry Series*, No. 18.

Covers tabulated values of heat capacity, heat content, entropy, and free energy function of solid, liquid, and gas states.

3217. Thermodynamic Properties of Toluene. Engineering Sciences Data Unit, London (available from NTIS), 1974, 49 pp. Accession: N75-29296. Tech Report: ESDU-74024. Cost: $218.50 hard copy.

Evaluated data are provided on the thermodynamic properties of toluene, including numerous tables and a temperature-entropy chart.

3218. Thermodynamic Stability Constants of Organic/Metal-Ion Complexes. E.D. Fultz. NTIS, 1967, 63 pp. Accession: N68-14552. Tech Report: UCRL-50200.

This Lawrence Radiation Laboratory publication provides tables of thermodynamic stability constants of organic metal-ion complexes.

3219. Thermodynamic Tables and Charts. Lefax Publishing Co. Lefax Publishing Co, loose-leaf binding. ISBN: 0-685-14171-6. Cost: $3.00. Series: *Lefax Data Books*, No. 634.

3220. Thermodynamics of the Polystyrene-Hydrocarbon Vapor System. F.J. Krieger. NTIS, 1973, 43 pp. Accession: N74-33617; AD-779905. Tech Report: RDA-TR-206-DNA; DNA-3254F.

Thermodynamic data are provided for polystyrene for a range of temperatures up to 6,000° K and pressures up to 1,000 atm.

3221. Thermodynamics—Data and Correlations. AIChe, 1974, 140 pp. Cost: $7.50 members; $20.00 nonmembers.

Experimental data are tabulated for the thermodynamics of such systems as aqueous solutions containing electrolytes, solar energy conversion systems, and data on solid fossil fuels.

3222. Thermodynmic Properties of Water to 1000°C. and 10,000 Bars. C. Wayne Burnham et al. Geological Society of America, 1969, 98 pp. LC: 73-96715. ISBN: 0-8137-2132-6. Accession: N72-70738; PB-203815. Cost: $3.00. Series: Special Paper No. 132.

Includes information on thermal expansion, free energy, density, and enthalpy and entropy.

3223. Thermoluminescence Dosimetry. A.F. McKinlay. Heyden, 1981, 180 pp. ISBN: 0-9960020-4-9. Cost: $28.00. Series: *Medical Physics Handbook*, 5.

Provides information on the accurate measurement of exposure and dosage of ionizing radiations.

3224. Thermophysical and Heat Transfer Properties of Alkali Metals. P.Y. Achener et al. NTIS, 1968, 259 pp. Accession: N68-31745. Tech Report: AGN-8195.

This Aerojet-General Nucleonics publication provides data on thermophysical properties and heat transfer information on alkali metals.

3225. Thermophysical Properties at High Temperatures Measured by Direct Heating Methods; Technical Report, 1 Jul. 1969 - 31 Oct. 1970. R.E. Taylor. NTIS, 1970, 69 pp. Accession: N71-32949; AD-724592. Tech Report: AFML-TR-70-286.

Thermophysical properties data are tabulated for tantalum alloys. Such data as electrical resistivity, equations of state, and information on tungsten alloys are also presented.

3226. Thermophysical Properties of a Type 308 Stainless Steel Weld. L.M. Greene et al. NTIS, 1975, 51 pp. Accession: N75-33405. Tech Report: Y-1967. Cost: $5.45 hard copy.

This publication was prepared at the Y12 plant at Oak Ridge and includes such properties as thermal expansion, thermal diffusivity, specific heat, and thermal conductivity for type 304-308 stainless steel weldments. This type of stainless steel is used in the liquid-metal breeder reactor.

3227. Thermophysical Properties of Air and Air Components. A.A. Vasserman. NTIS, 1971, 393 pp. Accession: N71-38755. Tech Report: TT70-50095. Cost: $6.00 hard copy; $0.95 microfilm.

This is an English translation of a Russian work that was originally published in 1966. This publication includes a Mollier diagram, equations of state, and other thermophysical data on air, argon, nitrogen, and oxygen.

3228. Thermophysical Properties of Freon-22. A.V. Kletskii. NTIS, 1971, 78 pp. Accession: N71-37695. Tech Report: TT70-50178.

This is an English translation of a Russian publication that was originally published in 1970. Substantial information is tabulated on the thermodynamic properties of freon-22.

3229. Thermophysical Properties of Gaseous and Liquid Methane. V.A. Zagoruchenko; A.M. Zhuravlev. NTIS, Russian translation for the National Bureau of Standards, 1969, trans. 1970. Accession: TT70-50097. Cost: $9.00.

3230. Thermophysical Properties of Gases and Liquids, No. 1. V.A. Rabinovich. NTIS, Russian translation for the National Bureau of Standards, 1968, trans. 1970. Accession: TT69-55091. Cost: $7.75.

NOTE: Other volumes have English title *Thermophysical Properties of Matter and Substances.*

3231. Thermophysical Properties of Helium-4 from 2 to 1500 K with Pressures to 1000 Atmospheres. R.D. McCarty. NTIS, 1972, 161 pp. Accession: N73-15953. Tech Report: NBS-TN-631; NASA-CR-130325. SuDoc: C13.46:631. Cost: $ 0.95 microfilm; $1.25 Superintendent of Documents.

This National Bureau of Standards publication provides tables of thermophysical properties of helium-4.

3232. Thermophysical Properties of Helium-4 from 4 to 3000 R with Pressures to 15000 Psia. R.D. McCarty. NTIS, 1972, 147 pp. Accession: N73-11966. Tech Report: NASA-CR-128445; NBS-TN-622. SuDoc: C13.46:622. Cost: $1.25 Superintendent of Documents.

Critically evaluated proptery data have been compiled on the thermophysical properties of helium-4.

3233. Thermophysical Properties of High Temperature Solid Materials. Thermophysical Properties Research Center, Purdue University; Y.S. Touloukian (ed). Macmillan, 1967, multivolume set, 6 Vols. in 9 parts. LC: 67-15295.

The volumes of this collection are entitled: Vol.1, *Elements*; Vol.2 in 2 parts, *Nonferrous Alloys*; Vol.3, *Ferrous Alloys*;, Vol.4 in 2 parts, *Oxides and their Solutions and Mixtures*; Vol.5, *Nonoxides and their Solutions and Mixtures*; Vol.6 in 2 parts, *Intermetallics, Cermets, Polymers and Composite Systems.*

3234. Thermophysical Properties of Liquid Air and its Components. A.A. Vasserman; V.A. Rabinovich. NTIS, Russian translation for the National Bureau of Standards, 1968, trans. 1970. Accession: TT69-55092. Cost: $9.75.

3235. Thermophysical Properties of Matter and Substances, Volume 2. V.A. Rabinovich (ed). NTIS, Russian translation for the National Bureau of Standards, 1970, trans. 1974. Accession: TT72-52001. Cost: $10.75.

NOTE: English title, Vol. 1, *Thermophysical Properties of Gases and Liquids, No. 1.*

3236. Thermophysical Properties of Matter and Substances, Volume 3. V.A. Rabinovich (ed). NTIS, Russian translation for the National Bureau of Standards, 1971, trans. 1975. Accession: TT73-52009. Cost: $10.25.

3237. Thermophysical Properties of Matter and Substances, Volume 4. V.A. Rabinovich (ed). NTIS, Russian translation for the National Bureau of Standards, 1972, trans. 1975. Accession: TT73-52029. Cost: $10.25.

Thermophysical Properties of Matter and Substances. See *Thermophysical Properties of Gases and Liquids, No. 1.*

3238. Thermophysical Properties of Matter. Y.S. Touloukian. IFI/Plenum, 1970-76, 13 Vols., approx. 16,389 pp. LC: 73-129616. ISBN: 0-306-67020-8 (Set). Cost: $795.00.

This 13-volume set is a massive collection of data on such materials as metallic elements and alloys, nonmetallic liquids and gases, coatings, and nonmetallic solids. Many thermophysical properties of these materials are covered including thermal conductivity, specific heat, thermal radiative properties, thermal diffusivity, viscosity, and thermal expansion. Experimental methods are given for the derivation of these properties, and each presentation is well documented with references to data sources and further information. Each material is well indexed and cross-indexed from volume to volume to help the user retrieve the data desired. For related works, please refer to *Metals Handbook, CRC Handbook of Materials Science,* and *Handbook of Heat Transfer.*

3239. Thermophysical Properties of Methane. R.D. Goodwin. NTIS, 1972, 12 pp. Accession: A73-15050.

This National Bureau of Standards publication provides information on the methods used to prepare tables on the thermophysical properties of methane. Bibliographic references are provided for the main sources of pressure-volume-temperature data.

3240. Thermophysical Properties of Oxygen From the Freezing Liquid Line to 600 R for Pressures to 5000 Psia. R.D. McCarty; L.A. Weber. NTIS, 1971, 190 pp. Accession: N71-34089. Tech Report: NASA-CR-121739; NBS-TN-384. Cost: $1.50.

Such properties as viscosity, specific heat, refractivity, enthalpy, entropy, temperature, and pressure data are provided for oxygen from the freezing liquid line to 600 R.

3241. Thermophysical Properties of Parahydrogen From the Freezing Liquid Line to 5000 R for Pressures to 1000 Psia. R.D. McCarty; L.A. Weber. NTIS, 1972, 169 pp. Accession: N72-28917. Tech Report: NASA-CR-127701; NBS-TN-617. Cost: $1.50.

Numerous tables on the thermophysical properties of parahydrogen are provided, including such properties as entropy, internal energy, enthalpy, speed of sound, density, volume, specific heat, diffusivity, viscosity, thermal conductivity, and dielectric constant. This publication was prepared by the National Bureau of Standards Cryogenics Division in Boulder, Colorado.

3242. Thermophysical Properties of Refrigerants. American Society of Heating, Refrigerating, and Air-Conditioning Engineers, Inc, 1972, 237 pp.

Numerous properties are provided for approximately 40 different refrigerants. Pressure-enthalpy diagrams are provided, along with such data as properties of superheated vapor, properties of liquid and saturated vapor, and temperature-entropy diagrams. In addition, data on viscosity, thermal conductivity, and specific heat are provided.

3243. Thermophysical Properties of Selected Aerospace Structural Materials. Y.S. Touloukian; C.Y. Ho. Hemisphere Publishing Corp, 1977, 2 Vol. set, Vol.1, 1,060 pp.; Vol.2, 240 pp. ISBN: Vol.1, 0-89116-056-6; Vol.2, 0-89116-057-4. Cost: $250.00 per set.

Part One is entitled *Thermal Radiative Properties* and Part Two is entitled *Thermophysical Properties of Seven Materials*.

3244. Thermophysical Properties Research Literature Retrieval Guide. Y.S. Touloukian. IFI/Plenum, Supplement II, 1480 pp. LC: Supplement I, 60-14426; Supplement II, 79-16324; Set, 67-31831. ISBN: Supplement I, 0-306-7200-6; Supplement II, 0-306-67210-3; Set, 0-306-65125-4. Cost: $275.00; Supplement II, $395.00 set.

NOTE: First published under the title *Retrieval Guide to Thermophysical Properties Research Literature* Supplement I, a 6-volume set published in 1973, covers the years 1964-70. Each volume is entitled as follows: Vol.1, *Elements and Inorganic Compounds*; Vol.2, *Organic Compounds and Polymeric Materials*; Vol.3, *Alloys, Intermetallic Compounds and Cermets*; Vol.4, *Oxide Mixtures and Minerals*; Vol.5, *Mixtures and Solutions*; Vol.6, *Coatings, Systems, and Composites.* Supplement II, also a 6-volume set, published in 1979 covers the years 1971-79. Each volume is entitled as follows: Vol.1, *Elements and Inorganic Compounds*; Vol.2, *Organic Compounds and Polymeric Materials*; Vol.3, Alloys, Intermetallic Compounds, and

Cermets; *Vol.4,* Oxide Mixtures and Minerals; *Vol.5, Mixtures and Solutions*; Vol.6, *Coatings, Systems, and Composites.*

3245. Thermophysical, Electrical, and Optical Properties of Selected Metal-Nonmetal Transition Materials: A Comprehensive Bibliography with Typical Data. Y.S. Touloukian; C.Y. Ho; J.F. Chaney. CINDAS/Purdue University, 1978, 154 pp. Series: *CINDAS/Purdue University Report*, No. 50.

This publication provides a listing of thermophysical, electrical, and optical properties of materials with their corresponding bibliographic citations where the results of the experimental research were originally reported.

3246. Thorium Ceramics Data Manual. C.E. Curtis; S. Peterson. NTIS, 1970, Vol.1, 74 pp.; Vol.2, 13 pp. Accession: Vol.1, N71-15063; Vol.2, N71-20594. Tech Report: Vol.1, ORNL-4503-VOL-1; Vol.2, ORNL-4503-VOL-2.

Vol.1 is entitled *Oxides*; Vol.2 is entitled *Nitrides*. This 2-volume set includes information on ceramic nuclear fuels and provides physical, chemical, and mechanical property data of thorium ceramics.

3247. Threshold Limit Values for Chemical Substances and Physical Agents in the Workroom Environment With Intended Changes for 1977. American Conference of Governmental Industrial Hygienists. Cost: $1.50.

This publication may be purchased from the Secretary-Treasurer, PO Box 1937, Cincinnati, Ohio, 45201. This publication provides threshold limit values for chemicals in the work place.

3248. Thyristor D.A.T.A. Book. Derivation and Tabulation Associates, Inc. Derivation and Tabulation Associates, Inc, continuation issued semiannually, pages vary per issue. Cost: $31.50 per year by subscription.

3249. Timber and Design Manual. Laminated Timber Institute of Canada. Laminated Timber Institute of Canada, 1972, 458 pp.

3250. Timber Construction Manual; A Manual for Architects, Engineers, Contractors, Laminators and Fabricators Concerned with Engineered Timber Buildings and Other Structures. American Institute of Timber Construction. John Wiley and Sons, 2d ed., 1974, 816 pp., various paging. LC: 73-11311. ISBN: 0-471-02549-6. Date of First Edition: 1966. Cost: $22.95.

NOTE: This thumb-indexed work covers physical and mechanical properties of wood, laminates, adhesives, and fasteners.

3251. Timber Design and Construction Handbook. Timber Engineering Company. McGraw-Hill, 1956, 622 pp. LC: 56-10879. ISBN: 0-07-064606-6. Cost: $55.00.

3252. Time-Saver Standards for Architectural Design Data. J.H. Callender. McGraw-Hill, 5th ed., 1974, 1,042 pp. LC: 73-17383. ISBN: 0-07-009647-3. Cost: $41.50.

Companion volume to: *Time-Saver Standards for Building Types.*

3253. Time-Saver Standards for Building Types. J.E. DeChiara; J.H. Callender. McGraw-Hill, 1973, 1,024 pp. LC: 73-6663. ISBN: 0-07-016218-7. Cost: $49.50.

3254. Titanium Alloys Handbook. R.A. Wood; R.J. Favor. Ohio Metals and Ceramics Information Center (also NTIS), 3d ed., 1972, 644 pp., (various paging). Accession: AD-758335. Tech Report: MCIC-HB-02. Date of First Edition: 1965. Cost: $35.00 microfiche; $35.00 paper copy.

3255. Tool and Design and Tool Engineering. John G. Jergens, 1955, 506 pp. Date of First Edition: 1942.

Primary emphasis of this manual is on tool and machine design. Information is provided on the preparation of drawings for jigs and fixtures.

3256. Tool and Manufacturing Engineers Handbook: A Reference Work for Manufacturing Engineers. Daniel B. Dallas. McGraw-Hill, 3d ed., 1976, 2,950 pp., (various paging). LC: 76-4892. ISBN: 0-07-059558-5. Date of First Edition: 1949. Cost: $65.50.

This handbook includes information on machining metals, nondestructive testing, powder metallurgy, lubrication, and surface preparation. Properties of metals are included along with all the necessary information that is required in the area of metalworking.

3257. Tool Steels. International Plastics Selector, Inc. Series: *Desk-top Data Bank Series.*

Provides information on commercially available steel products. Commercial names are listed as well as standard specification numbers, such as the AISI number, the UNS number, the JIS number, or the DIN number. In addition, hundreds of manufacturers are listed.

3258. Topographic Positions of the Measurement Points in Electro-Acupuncture. Reinhold Voll. Medizinisch Literarische, 1977-78, 4 Vols. + 1 supplement; 1st supplement 50 pp. ISBN: Vol.1, 3-88136-042-5; Vol.2, 3-88136-053-0; Vol.3, 3-88136-049-2; Vol.4, 3-88136-060-3; 1st Supplement, 3-88136-063-8. Cost: Vol.1, DM 120; Vol.2, DM 120; Vol.3, DM 120; Vol.4, DM 120; 1st Supplement DM 60.

The volumes and supplement of this publication are presented as follows: Vol.1, Illustrated; Vol.2, Textural; Vol.3, Illustrated; Vol.4, Textural and Illustrated. This 4-volume set plus supplement provides textural and illustrated information on the topographic positions of the measurement points for electro-acupuncture. This set has been translated from Germany by Hartwig Schuldt.

3259. Toxic and Hallucinogenic Mushroom Poisoning: A Handbook for Physicians and Mushroom Hunters. Gary Lincoff; D.H. Mitchel MD. Van Nostrand Reinhold, 200 pp. ISBN: 0-442-24580-7. Cost: $14.95.

3260. Toxic and Hazardous Industrial Chemicals Safety Manual for Handling and Disposal with Toxicity and Hazard Data. International Technical Information Institute, Japan, 580 pp. Cost: $75.00 plus $12.00 handling charge.

Contains information on over 700 industrial chemicals, including toxicity, leakage, fire, spills, and disposal. Includes chemical name, formula, uses, flash point, ignition point, vapor pressure, vapor density, and explosion limits.

3261. Toxic Substances Control Act Chemical Substances Inventory: Supplement I. U.S. Environmental Protection Agency, 1979.

The initial inventory listed 43,278 chemical substances produced in or imported into the United States since January 1, 1975. Supplement 1 lists an additional 3,000 substances. This inventory lists such information as the CAS registry number and preferred names of the chemical substances. The Environmental Protection Agency has prepared indexes to gain access to this information by substance name, molecular formula, and an index of chemical substances of unknown or variable composition.

3262. Toxic Substances Control Sourcebook. Center for Compliance Information. Aspen Systems Corp, 1978, 700 pp. LC: 77-84943. Cost: $59.50.

Legal compliance information is provided for toxic substances. Summaries of laws, bills, and acts dealing with toxic substances are provided in this sourcebook. Information on such substances as arsenic, asbestos, benzene, polybrominated biphenyls, lead, and vinyl chloride are discussed in this book.

3263. TPRC Data Books. Thermophysical Properties Research Center. Purdue Universty.

NOTE: *See Thermophysical Properties of Matter* and *Thermophysical Properties Research Literature Retrieval Guide.*

3264. Trace Substances and Health: A Handbook, Part 1. P.M. Newberne. Dekker, 1976, 448 pp. ISBN: 0-8247-6341-6. Cost: $49.75.

3265. Trade Designations of Plastics and Related Materials (Revised). Joan B. Titus (Plastics Technical Evaluation Center). NTIS, supersedes report dated Dec. 1965, AD-481788, 1970, 176 pp. Accession: AD-715401.

This handbook includes a listing of approximately 5,100 trademarks and brand names of different types of plastics and polymers. Included in the listing is the name of the manufacturer along with a description of the plastic.

3266. Traffic Control Devices Handbook: An Operating Guide. U.S. Department of Transportation, Federal Highway Administration, 1975.

Information is provided on the design and dimension of traffic control devices, such as traffic lights and signs.

3267. Traffic Control Systems Handbook. U.S. Department of Transportation, Federal Highway Administration. U.S. Government Printing Office, 1976, 650 pp. SuDoc: TD2.36: 76-10. S/N: 050-001-00114-4. Cost: $12.00.

3268. Traffic Engineering Handbook. John E. Baerwald. The Institute of Traffic Engireers, Washington, DC, 3d ed., 1965, 770 pp. LC: 65-17560. Date of First Edition: 1941.

This handbook covers such topics as parking, speed regulations, roadway lighting, geometric design of roadways, and highway capacity. Traffic signs and markings, accidents, the pedestrian and the driver are also covered in this handbook.

3269. Transformer and Inductor Design Handbook. William T. McLyman. Marcel Dekker, 1978, 464 pp. ISBN: 0-8247-6801-9. Cost: $39.75. Series: *Electrical Engineering and Electronics Series,* 7.

Such topics as transformer design, static inductors, magnetic material selection, inductor design, 3-phase transformer design, and toroidal power core selection are discussed in this handbook.

3270. Transistor and Diode Data Book for Design Engineers. Texas Instruments Inc, 1973, 1,248 pp. Cost: $7.95.

3271. Transistor D.A.T.A. Book. Derivation and Tabulation Associates, Inc. Derivation and Tabulation Associates, Inc, continuation issued semiannually, pages vary per issue. Cost: $42.50 per year.

3272. Transistor Specifications Manual. Howard Sams and Co, Inc, 9th ed., 1978. LC: 78-57208. ISBN: 0-672-21516-0. Cost: $8.95.

3273. Transistor Substitution Handbook. Sams, 17th ed., 1978. LC: 78-54456. ISBN: 0-672-21515-2. Cost: $5.95.

3274. Transition Metal Oxides, Crystal Chemistry, Phase Transition, and Related Aspects. C.N.R. Rao; G.V. Subba Rao (National Bureau of Standards). U.S. Government Printing Office, 1974. SuDoc: C13.48:49. Cost: $1.70. Series: NSRDS-NBS-49.

Transport Properties of Fluids: Thermal Conductivity, Viscosity, and Diffusion Coefficient. *See* McGraw-Hill/CINDAS Data Series on Material Properties, Volume I-1.

3275. Transport Properties of Mixed Plasmas—He-N$_2$, Ar-N$_2$, and Xe-N$_2$ Plasmas at One Atmosphere, Between 5000 K and 35,000 K. M. Capitelli; E. Ficocelli. NTIS, 1970, 55 pp. Accession: N71-23082.

Such properties as viscosity, thermal conductivity, and collision parameters are presented in this compilation for nitrogen plasma, argon plasma, and helium plasma.

3276. Transport Properties of Selected Elements and Compounds in the Gaseous State. P.E. Liley. NTIS, 1972, 78 pp. Accession: N73-23101; AD-757528. Tech Report: TPRC-20; AFOSR-73-0478TR.

Such properties as viscosity, thermal conductivity, and diffusion coefficients are listed for 38 binary mixtures and 9 pure substances. Temperature range is from 100 to 3,000° K. Helium, hydrogen, deuterium, fluorine, hydrogen fluoride, and deuterium fluoride are among the elements and compounds listed. This tabulation was prepared at the Thermophysaical Properties Research Center at Purdue University.

Transport Properties of Solids: Thermal Conductivity, Electrical Resistivity, and Thermoelectric Properties. *See* McGraw-Hill/CINDAS Data Series on Material Properties, Volume I-2.

3277. Transportation and Traffic Engineering Handbook. Wolfgang S. Homburger; Louis E. Keefer; William R. McGrath. Prentice-Hall, Inc, 2d ed., 1982, 883 pp. LC: 82-340. ISBN: 0-13-930362-6. Cost: $75.00.

This new edition has been prepared under the sponsorship of the Institute of Traffic Engineers. Such topics as pipelines, air transportation, water transportation, railroad engineering, legal liability, and the economics of transport are discussed in this updated edition.

3278. Transportation Energy Conservation Data Book: Edition 4. G. Kulp et al. NTIS, 4th ed., 1980, 397 pp. Tech Report: ORNL-5654. Cost: $3.50 microform; $27.50 paper copy.

This is a statistical compendium compiled and published by Oak Ridge National Laboratories for the U.S. Department of Energy. Statistics on energy characteristics are presented, including energy prices, energy use, energy use by fuel source, and energy use by transportation mode. All modes of transportation are listed, including highway, rail, air, marine, and pipeline.

3279. Traverse Tables. Lefax Publishing Co, loose-leaf binding. ISBN: 0-685-14174-8. Cost: $3.00. Series: *Lefax Data Books*, No. 38-50.

3280. Trends in Properties of Unleaded Gasolines. J.N. Bowden. NTIS, 1975, 37 pp. Accession: N75-24968. Tech Report: AD-A008407.

This report represents a sampling of 72 types of unleaded gasoline between the years of 1973 and 1974. Each sample was analyzed for hydrocarbon composition. The unleaded samples were compared to data on leaded fuels.

3281. Tribology Handbook. M.J. Neale. Butterworths, (also Halsted Press), 1973, various paging. LC: 72-13394. ISBN: 0-408-00082-1; 0-470-63081-7 (Halsted). Cost: $45.00.

3282. Troubleshooters' Handbook for Mechanical Systems. R.H. Emerick. McGraw-Hill, 1969, 448 pp. LC: 68-28413. ISBN: 0-07-019314-2. Cost: $45.00.

3283. Trucking Permit Guide: Private, Contract, Common, Exempt. Harold C. Nelson. J.J. Keller Associates, rev. ed., 1979, loose-leaf binder. LC: 75-16944. ISBN: 0-93467-00-0. Cost: $95.00.

This loose-leaf service is a compilation of U.S. federal and state, as well as Canadian and Mexican regulatory requirements for common carrier permits and certificates. Fuel use tax, mileage tax, highway use tax, and weight permit information is also included in this compilation.

3284. Trucking Safety Guide; Driver, Vehicle, Cargo, Highway. Terence J. Quirk. J.J. Keller, 1979, loose-leaf binder. LC: 74-3865. ISBN: 0-934674-03-5. Cost: $95.00.

Information is provided on the transporation of hazardous materials; OSHA standards, with respect to trucking; and inspection and maintenance of vehicles.

3285. TTL Data Book for Design Engineers. Texas Instruments, Inc, Components Group, Engineering Staff, 2d ed., 1977, 832 pp. Cost: $4.95.

3286. Tube Caddy: Tube Substitution Guide. H.A. Middleton. Hayden, annual, 1960-.

3287. Tube Substitution Handbook. Sams, 20th ed., 1977. LC: 76-42880. ISBN: 0-672-21405-9. Cost: $2.95 paper copy.

3288. Tunnel Engineering Handbook. John O. Bickel; Thomas Kuesel. Van Nostrand Reinhold, 1982, 640 pp. ISBN: 0-442-28127-7. Cost: $52.50.

Information is provided on the design and construction of tunnels. Such topics as ecological and environmental considerations are discussed in addition to new tunneling methods and new tools that are used in tunnel construction.

3289. Turf Managers' Handbook. William Hugh Daniel. Harcourt, Brace, Jovanovich, 1979. LC: 78-71794. Cost: $18.95.

A wealth of information is provided on the topic of turf management. Data are provided on different types of grasses, fertilizer application, and watering methods.

3290. TV Field and Bench Servicer's Handbook. John Spillane. TAB Books, 1978, 308 pp. LC: 78-27297. ISBN: 0-8306-9847-7. Cost: $9.95.

Numerous troubleshooting charts are provided in this handbook to analyze television receiver systems. Such topics as high voltage circuits, sweep circuits, picture tubes, IC modules, low voltage power supply, and convergence are discussed.

3291. Two-Six (2-6) Semiconducting Compounds Data Tables. M. Neuberger. NTIS, 1969, 168 pp. Accession: N70-22589; AD-698341.

This compilation was prepared by Hughes Aircraft Corporation, and provides data tables on properties of group 2A and group 6A binary semiconducting compounds.

3292. Two-Six (II-VI) Ternary Compounds Data Tables. M. Neuberger. NTIS, 1972, 103 pp. Accession: N72-27799; AD-739359. Tech Report: EPIC-S-15; LC-71-189484.

Property data are included for groups II-VI ternary semiconductor compounds. Such properties as crystallographic, magnetic, optical, thermal, mechanical, and physical properties are listed in this tabulation.

3293. U.S. Dispensatory and Physicians Pharmacology. (Title varies). Arthur Osol; Robertson Pratt. Lippincott, 27th ed., 1973, 1,292 pp. LC: 73-2673. ISBN: 0-397-55901-1. Cost: $30.00.

3294. UFAW Handbook on the Care and Management of Laboratory Animals. Universities Federation for Animal Welfare, London, 5th ed., 1976, 648 pp. ISBN: 0-443-01404-3.

The care and management of laboratory animals is extensively covered in this publication. Information is provided on numerous types of laboratory animals, such as guinea pigs, rats, mice, bats, and cats.

3295. UITP Handbook of Urban Transport 1979. International Union of Public Transport, Brussels, Belgium, 1979, 2 Vols., 1,120 pp. Cost: $110.00.

This handbook includes statistical information on transportation systems in over 600 cities of the world. Seventy-four pages of tables provide such data as comparison of passenger boarding rates for surface equipment, rapid transit data, and information on metropolitan railways.

3296. Ultimate Load Design of Reinforced Concrete—A Practical Handbook. M. Nadim Hassoun, 1977, 192 pp. Cost: $27.50.

Tables and charts are provided for calculating load design of reinforced concrete. Numerous examples of analyzing load design for reinforced concrete members are provided in addition to equations for the analysis and design of beams and columns.

3297. Ultimate Strength Handbook. American Concrete Institute, 2nd ed, 1973. Date of First Edition: 1968.

3298. Ultrasonics John Woodcock. Heyden (Adam Hilger), 1979, 176 pp. ISBN: 0-85274-506-0. Cost: $29.50. Series: *Medical Physics Handbooks*, Vol. 1.

Such topics as acoustical holography, ultrasonic imaging systems, the Doppler effect, and pulse-echo methods are discussed in this handbook.

3299. Ultrastructure of the Mammalian Cell: An Atlas. R.V. Krstic. Springer-Verlag, 1979, 380 pp. ISBN: 0-387-09583-7; 3-540-09583-7. Cost: $31.00.

This atlas provides approximately 176 illustrations representing the ultrastructure of the mammalian cell.

3300. Ultraviolet Muiltiplet Table; Section 1. The Spectra of Hydrogen through Vanadium; Section 2. The Spectra of Chromium through Niobium. Charlotte E. Moore (National Bureau of Standards). NTIS, 1950, 1952. Accession: PB-252093/AS. Cost: $7.75. Series: NBS Circular 488, Secs. 1 & 2.

3301. Ultraviolet Multiplet Table, Section 3. The Spectra of Molybdenum through Lanthanum and Hafnium through Radium; Section 4. Finding List for Spectra of the Elements Hydrogen to Niobium (Z=1 to 41); Section 5. Finding List for Spectra of the Elements Molybdenum to Lanthanum (Z=72 to 88). Charlotte E. Moore (National Bureau of Standards). NTIS, 1962. Accession: PB-25094/AS. Cost: $7.75. Series: NBS Circular 488, Sections 3, 4, 5.

3302. Underground Mining Methods Handbook. W.A. Hustrulid. American Institute of Mining, Metallurgical, and Petroleum Engineers, 1982, 754 pp. LC: 80-70416. ISBN: 0-89520-049-X. Cost: $120.00.

This is a very comprehensive loose-leaf service written by many distinguished experts. This publication covers such topics as mine ventilation, foundation design, caving methods, longwall mining, shortwall mining, open-pit mining, underground mining, and blasting.

3303. Underwater Acoustics Handbook II. Vernon M. Albers. Pennsylvania State University Press, 2d ed., 1965, 356 pp. LC: 64-15069. ISBN: 0-271-73106-0. Cost: $20.00.

3304. Underwater Handbook: A Guide to Physiology and Performance for the Engineer. Charles W. Shilling. Plenum Publishing Co, 1976, 912 pp. LC: 76-7433. ISBN: 0-306-30843-6. Cost: $75.00.

3305. Unified Numbering System for Metals and Alloys. Society of Automotive Engineers. Society of Automotive Engineers, 2d ed., 1977, 288 pp. ISBN: 0-89883-395-7; 0-686-50148-9. Cost: $49.00. Series: *ASTM* DS-56A; *SAE* HS 1086.

This publication contains over 3,500 metals and alloys with their corresponding designations and specifications. Each listing provides a unified or master number, chemical composition, a description of the metal or alloy, and the different numbers and specifications by which it is known. Carbon alloys, stainless steels, tool steels, castings, aluminum, copper, nickel, cobalt alloys, superalloys, and numerous other ferrous and nonferrous metals and alloys are included.

3306. Unit Conversions and Formulas Manual. Nicholas P. Cheremisinoff; Paul N. Cheremisinoff. Ann Arbor Science, 1980, 171 pp. LC: 79-55140. ISBN: 0-250-40331-5. Cost: $8.95.

This is a very useful collection of formulas and tables on such topics as electrical engineering, general chemistry, hydraulics, heat transfer, statics, dynamics, and fan laws.

3307. United States Cotton Handbook. International and National Cotton Council of America, 1972.

3308. United States Energy Atlas. David J. Cuff; William J. Young. Free Press (division of Macmillan Publishing Co), 1980, 416 pp. ISBN: 0-02-691250-3. Cost: $75.00.

This is an excellent graphic analysis of energy resources in the United States. The atlas includes over 50 graphs, charts, and tables, approximately 200 maps, and 150 illustrations. Such topics as coal, synthetic crude oil, nuclear fuels, geothermal heat, hydroelectric power, wind power, solar energy, and ocean thermal energy are discussed in this atlas.

3309. United States Pharmacopeia Dispensing Information. United States Pharmacopeial Convention Dist. Dept, 1981. Cost: $18.75 (including updates).

Provides dosage information, preparation, packaging, storage, and labeling information for drugs. This compilation is a cross-indexed listing of drugs both by generic and brand names. Side effect information is also listed as well as drug interaction information.

3310. United States Standard Atmosphere, 1976. National Oceanic and Atmospheric Administration. U.S. Government Printing Office, 1976, 227 pp. Date of First Edition: 1962. Cost: $6.20.

This handbook includes tables on the properties of the upper atmosphere such as composition, temperature, pressure, and density.

3311. Universal Tables for Magnetic Fields of Filamentary and Distributed Circular Currents. Philip J. Hart. American Elsevier, 1967, 489 pp. LC: 67-24195. ISBN: 0-444-00019-4. Cost: $25.00.

3312. Uranium Prospecting Handbook. S.H.U. Bowie; M. Davis; D. Ostle. Institute of Mining, England, 1972 (rep. ed., 1980), 346 pp. ISBN: 0-900488-15-8. Cost: $95.00.

This handbook includes several case histories of uranium exploration in Australia, Canada, and Scotland. In addition, such topics as airborne gamma-ray survey techniques, borehole logging techniques for uranium exploration, and numerous new methods for uranium prospecting are presented in this handbook.

3313. Urodynamics—The Mechanics and Hydrodynamics of the Lower Urinary Tract. D.J. Griffiths. Adam Hilger Ltd, 1980. ISBN: 0-85274-507-9. Cost: £11.95. Series: *Medical Physics Handbooks*, 4.

Such topics as urethral obstruction, the bladder, continence, incontinence, and basic information on the mechanics and hydrodynamics of the lower urinary tract are discussed in this publication.

3314. USAF Bioenvironmental Noise Data Handbook. Wright Patterson Air Force Base, Aerospace Medical Research Laboratory (available from NTIS), 1975, multivolume set. Accession: Vol.1, AD-031865. Tech Report:

Vol.1, AMRL-TR-75-50. Cost: $5.00 hard copy; $3.50 microform.

Noise testing data have been compiled in this multivolume series that has been produced by the Medical Research Laboratory at Wright Patterson Air Force Base. Measured data defining the bioacoustic environment at flight crew locations inside aircraft during normal flight operations are provided in these tabulations.

3315. User's Handbook of D/A and A/D Converters. Eugene R. Hnatek. John Wiley and Sons, 1976, 472 pp. LC: 75-14341. ISBN: 0-471-40109-9. Cost: $32.50.

Covers digital-to-analog converters, analog-to-digital converters, multiplexers, and sample-and-hold circuits.

3316. User's Handbook of Integrated Circuits. Eugene R. Hnatek. Wiley-Interscience, 1973, 449 pp. LC: 72-13596. ISBN: 0-471-40110-2. Cost: $45.00.

3317. User's Handbook of Semiconductor Memories. Eugene R. Hnatek. Wiley-Interscience, 1977, 688 pp. LC: 77-362. ISBN: 0-471-40112-9. Cost: $36.95.

Discusses shift registers, read-only memories (ROM), random-access memories (RAM), programmable logic arrays (PLA), and content-addressable memories (CAM).

3318. User's Practical Selection Handbook for Optimum Plastics, Rubbers, and Adhesvies. International Technical Information Institute, Japan, 550 pp. Cost: $60.00 plus $12.00 handling charge.

3319. Using the Telescope: A Handbook for Astronomers. J.H. Robinson. Halsted Press, 1979, 112 pp. LC: 78-10759. ISBN: 0-470-26514-0. Cost: $12.95.

Such topics as maintenance of the telescope and observational techniques are discussed in this handbook.

3320. Utility Vehicle Design Handbook. Society of Automotive Engineers, Inc, 1981, 472 pp. ISBN: 0-89883-008-7. Cost: $32.00. Series: *SAE*, AE-8.

Such topics as body design, hydraulic systems, winches, aerial devices, and materials used in the construction of utility vehicles are discussed.

3321. UTU Handbook of Transportation in America. C. Luna. Popular Library, Inc, 1971.

This publication provides information on urban transportation with data on the role of the federal government in funding and regulating rail transportation and highway transportation.

3322. UV-Atlas of Organic Compounds. Verlag Chemie, 1966-71, 5 Vols., loose-leaf set. LC: 66-21542. ISBN: 0-306-68300-8. Cost: 675 DM.

Ultraviolet absorption spectra are provided for approximately 1,000 organic compounds in this 5-volume set. The spectra are supplemented by tables showing the effects of substituents and solvents. A cummulative formula index is provided as well as explanatory text in both English and German.

3323. Vanadium and Its Alloys. V.V. Baron; Iu. V. Efimov; E.M. Savitskii. NTIS, 1979, 260 pp. Accession: A70-30627.

Tables and phase diagrams are provided for vanadium and vanadium alloys. In addition, information on chemical composition, mechanical properties, and physical properties of vanadium and vanadium alloys are included.

3324. Vanderbilt Rubber Handbook. Robert O. Babbit. R. T. Vanderbilt Co, 1978.

Covers such commercial elastomers as buytl rubber, neoprenes, polyacrylic rubber silicon elastomers, thermoplastic elastomers, ethylene acrylic elastomers, chlorinated polyethylenes, and fluor-elastomers. Also covers styrenes, butadiene rubbers, synthetic polyisoprene, and nitrite rubber. Various properties of rubber are tabulated and processing methods and equipment are discussed.

3325. Vapor-Liquid Equilibrium Data Bibliography. I. Wichterle; J. Linek; E. Hala. Elsevier-North Holland, Vol.1, 1973; Supplement I, 1976; Supplement II, 1979, 3 Vols, Vol.1, 1,053 pp.; Supplement I, 334 pp.; Supplement II, 286 pp. ISBN: Vol.1, 0-444-41161-5; Supplement I, 0-444-41464-9; Supplement II, 0-444-41822-9. Cost: Vol.1, $102.50; Supplement I, $70.75; Supplement II, $65.75.

This 3-volume bibliography is useful in identifying articles dealing with the chemical problems of distillation and rectification in the chemical industry. The original work includes 4,800 references that were reported between 1900 and December, 1972. The first supplement includes 1,000 references on articles that were written between January, 1973 through December, 1975. The second supplement includes articles between January, 1976 until December, 1978. This computerized compilation lists the substances by the Hill notation system that is used in the *Chemical Abstracts Formula Index.*

3326. Vapor-Liquid Equilibrium Data Collection. DE-CHEMA, Vol.1, Part 1, 1977; Vol.1, Part 1A, in preparation; Vol.1, Part 2A, 1977; Vol.1, Part 2B, 1978; Vol.1, Part 2C, in preparation; Vol.1, Part 3-4, 1979; Vol.1, Part 5, in preparation; Vol.1, Part 6A, 1980; Vol.1, Part 6B, 1980; Vol.1, Part 7, in preparation; Vol.1, Part 8, in preparation, multivolume set, Vol.1, Part 1, 750 pp.; Vol.1, Part 2A, 752 pp.; Vol.1, Part 2B, 576 pp.; Vol.1 Part 6A, 687 pp.; Vol.1, Part 6B, 506 pp. ISBN: Vol.1, Part 1, 3-921567-01-7; Vol.1, Part 2A, 3-921567-09-2; Vol.1, Part 2B, 3-921567-12-2; Vol.1, Part 3-4, 3-921567-14-9; Vol.1, Part 6A, 3-921567-30-0; Vol.1, Part 6B, 3-921567-31-9. Cost: Vol.1, Part 6A, $111.00; Vol.1, Part 6B, $98.00. Series: *Chemistry Data Series*, Vol. 1.

This multivolume set is entitled as follows: Vol.1, Part 1, *Aqueous-Organic Systems*; Vol.1, Part 1A, *Aqueous Organic Systems Supplement I*; Vol.1, Part 2A *Organic-Hydroxy Compounds: Alcohols*; Vol.1, Part 2B, *Organic-Hydroxy Compounds: Alcohols and Phenols*; Vol.1, Part 2C, *Organic-Hydroxy Compounds Supplement I*; Vol.1, Part 3-4, *Aldehydes and Ketones: Ethers*; Vol.1, Part 5, *Esters and Carboxylic Acids*; Vol.1, Part 6A, *Aliphatic Hydrocarbons*; Vol.1, Part 6B, *Aliphatic Hydrocarbons*; Vol.1, Part 7, *Aromatic Hydrocarbons*; Vol.1, Part 8, *Halogen, Nitrogen, Sulphur, and Other Compounds*. This multivolume set provides a compilation of thermodynamic data for various binary and multicomponent mixtures. Boiling points and vapor-equilibrium curves are included as function of composition at selected temperatures. This data collection is restricted mainly to mixtures of organic compounds, but does include water.

3327. Vapour Liquid Equilibrium Data. Ju-chin Chu et al. University Microfilms International, 1956, 762 pp. Cost: $91.50.

University Microfilms order No.: EC1-OP25876. A substantial amount of vapor-liquid equilibrium data are provided in this compilation.

3328. Vapour Pressures and Critical Points of Pure Substances. 6: C3 to C15 Aliphatic Ketones. Engineering Sciences Data Unit, London, 1975, 29 pp. Accession: N76-21296. Tech Report: ESDU-75025. Cost: $290.50.

Thermophysical property data are presented for aliphatic ketones. Saturated vapor pressures are provided in both tabular and graphical form for aliphatic ketones.

3329. Vapour Pressures of Pure Substances up to their Critical Points. 1: C1 to C8 Alkanes. Engineering Sciences Data Unit, London (available from NTIS), 1972, 35 pp. Accession: N75-29210. Tech Report: ESDU-72028. Cost: $98.50.

Evaluated vapor pressure data for liquid hydrocarbons are presented in both tabular and graphical format.

3330. Vapour Pressures of Pure Substances up to their Critical Points. 2: C2 to C6 Alkenes (Monoolefines). Engineering Sciences Data Unit, London (available from NTIS), 1973, 42 pp. Accession: N75-29201. Tech Report: ESDU-73008. Cost: $170.50 hard copy.

Evaluated vapor pressure data are provided for alkenes (monoolefines).

3331. Vapour Pressures of Pure Substances: Selected Values of the Temperature Dependence of the Vapour Pressures of Some Pure Substances in the Normal and Low Pressure Region. Tomas Boublik; Yojtech Fried; Eduard Hala. Elsevier Scientific Publishing Co, 1973, 626 pp. LC: 72-97420. ISBN: 0-444-41097-X. Cost: $150.00.

3332. Vapour Pressuress and Critical Points of Liquids. 9: C2 to C11 Aliphatic Ethers and Three Aromatic Ethers. Engineering Sciences Data Unit, London. NTIS (distributed by ESDU, Springfield, VA), 1976, 29 pp. ISBN: 0-85679-161-X. Accession: N77-20176. Tech Report: ESDU-76024.

Vapor pressures and critical points are provided for aliphatic ethers and 3 aromatic ethers.

3333. Variable Regions of Immunoglobin Chains: Tabulations and Analyses of Amino Acid Sequences. Elvin A. Kabat; Tai-te Wu; Howard Bilotsky. Bolt, Beranek, and Newman, 1976, 130 pp.

Data are presented on amino acide sequences.

3334. Variable Star Observer's Handbook. John Stephen Glasby. Norton & Co, Inc, 1st American ed., 1971, 213 pp. LC: 72-175943. ISBN: 0-283-48470-5.

3335. Variables of State and Characteristics for Isentropic Discharge Phenomena of Water, Starting with Saturation. H. Baudisch. NTIS, 1968, 187 pp. Accession: N68-24136. Tech Report: DLR-FB-68-11; DVL-694.

This publication is presented in German with an English summary. Thermodynamic values for isentropic discharge values of water are tabulated.

3336. Vegetable Growing Handbook. Walter E. Splittstoesser. AVI Publishing Co, 1979, 400 pp. ISBN: 0-87055-319-4. Cost: $12.50.

Such information as pest control, plant nutrition, soils, fertilizers, harvesting, and storing of vegetables is provided in this handbook.

3337. Vehicle Electrical Equipment Handbook. Yu Borovskikh. NTIS, translated from the Russian by the U.S. Army Foreign Science and Technology Center, 1974, 81 pp. Accession: AD-786833. Tech Report: FSTC-HT-23-1692-73.

3338. Vehicle Sizes and Weight Manual: Legal Limitations, Oversize and Overweight Movements. Harold C. Nelson et al. J.J. Keller, rev. ed., 1979, loose-leaf, 586 pp. LC: 74-31863. ISBN: 0-934671-21-3. Cost: $69.00.

This loose-leaf service provides information on federal regulations, state regulations, and Canadian regulations on weight limitations and overweight vehicles.

3339. Venomous Snakes of the World: A Checklist. Keith A. Harding; Kenneth R.G. Welch. Pergamon, 1980, 188 pp. ISBN: 0-08-025495. Cost: $44.00.

This is a compilation of different types of venomous snakes throughout the world.

3340. Vertical Stress Tables for Uniformly Distributed Loads on Soil. Alfreds R. Junikis. Rutgers University Bureau of Engineering Research, 1971. Cost: $12.50. Series: *Engineering Research Publication*, No. 52.

This set of tables contains vertical stress influence values for the elastic medium, namely soil, from pressures of differently-shaped bearing areas on the surface of the soil. Such differently-shaped bearing areas as squares, rectangles, finite and infinitely long strips, circles, and trapezoidal loading of strips are given to the fifth decimal place. This publication includes graphs that accompany the tables.

3341. Veterinary Handbook for Cattlemen. Jackson Will Baily. Springer, 5th ed., 1980, 590 pp. LC: 79-16235. ISBN: 0-8261-0285-9. Cost: $22.50.

This handbook provides practical veterinary information for those dealing with cattle.

3342. VHF Handbook for Radio Amateurs. Herbert S. Brier; William R. Orr. Radio Publications, Inc, 1974, 336 pp. LC: 74-75450. Cost: $5.95.

3343. Vibration and Acoustic Measurement Handbook. Michael P. Blake; William S. Mitchell. Hayden, 1972, 656 pp. ISBN: 0-8104-9195-8. Cost: $34.50.

This handbook is useful in detecting, analyzing, and measuring vibrations. Shock and impact measurement techniques are discussed.

3344. Vibration Isolation: Use and Characterizations. John C. Snowdon. Superintendent of Documents, 1979, 119 pp. LC: 79-600062. SuDoc: C13.11:128. Cost: $4.00. Series: *National Bureau of Standards Handbooks*, No. 128.

Information is provided on anti-vibration mountings for the control of noise and vibration. Numerous literature references are provided.

3345. Video User's Handbook. Peter Utz. Prentice-Hall, 1979, 368 pp. LC: 79-379. ISBN: 0-13-941823-7. Cost: $19.95.

Such information as proper lighting for both black and white and color cameras, special effects, editing, video and audio tape, and the design and use of graphics for television production is among the topics discussed in this handbook. Information is provided for operating, maintaining, and troubleshooting video equipment.

3346. Virial Coefficients of Pure Gases and Mixtures: A Critical Compilation. J.H. Dymond; E.B. Smith. Oxford University Press, 1980, 518 pp. LC: 79-40667. ISBN: 0-19-855361-7. Cost: $69.00. Series: *Oxford Science Research Papers*, 2.

This compilation provides thermodynamic properties of numerous gases and mixtures.

3347. Viscoelastic Properties of Polymers. J.D. Ferry. John Wiley and Sons, 3rd ed, 1980, 641 pp. LC: 79-28666. ISBN: 0-471-04894-1. Cost: $51.00.

A discussion of viscoelastic properties of polymers is presented in this publication. Stress-strain relationships are discussed in connection with deformation of polymers.

3348. Viscosity Index Tables Calculated from Kinematic Viscosity. American Society for Testing and Materials, 1965. Cost: $23.00. Series: ASTM No. 05-039010-12.

3349. Viscosity Index Tables for Celsius Temperatures. American Society for Testing and Materials, 1975, 950 pp. LC: 75-10096. ISBN: 0-686-52070-X. Series: *ASTM Data Series Publication*, DS 39B.

Kinematic viscosity values are given for 40° C and 100° C. These tables will be useful in obtaining the viscosity index of petroleum products and lubricants.

3350. Viscosity Tables for Kinematic Viscosity Conversions and Viscosity Index Calculations. American Society for Testing and Materials, 1972, 35 pp. ISBN: 0-8031-0136-8. Cost: $5.00. Series: *American Society for Testing and Materials Special Technical Publication*, No. 43C.

This set of viscosity tables is meant to be used with ASTM viscosity testing methods, D 2161 and D 2270.

3351. VNR Metric Handbook of Architectural Standards. P. Tutt; D. Adler. Van Nostrand Reinhold Co, 1979, 450 pp. ISBN: 0-442-25189-0. Cost: $24.95.

Design criteria for over 25 different types of buildings are provided in this publication.

3352. VNR Metric Handbook. Leslie Fairweather; Jan Sliwa. Van Nostrand Reinhold (originally published in England), 1972, 206 pp. ISBN: 0-442-22364-1. Cost: $7.95.

3353. Voltage Regulator Handbook. John D. Spencer; Dale E. Pippenger. Texas Instruments, Inc, 1977, 198 pp. LC: 77-87869. ISBN: 0-89512-101-8. Cost: $4.40.

Such topics as heat-sink design, input filter design, and different regulator design considerations are presented in this handbook. In addition, numerous individual data sheets are provided on different types of regulators.

3354. Waste Recycling and Pollution Control Handbook. A.V. Bridwater; C.J. Mumford. Van Nostrand Reinhold, 1979, 706 pp. LC: 79-27240. ISBN: 0-442-21937-7. Cost: $35.00. Series: *Environmental Engineering Series.*

Numerous examples are provided for the management and control of liquid, solid, and gaseous waste. Such topics as pollution for combustion processes, metals recovery, noise control, and the economics of waste treatment processes are discussed.

3355. Water and Water Pollution Handbook. Leonard L. Ciaccio. Marcel Dekker, 1971-73, Vol.1, 478 pp.; Vol.2, 365 pp.; Vol.3, 528 pp.; Vol.4, 648 pp.; 4 vols. ISBN: Vol.1, 0-8247-1104-1; Vol.2, 0-8247-1116-5; Vol.3, 0-8247-1117-3; Vol.4, 0-8247-1118-1. Cost: $58.50 per Vol.

Water Engineers Handbook. *See* **Water Surfaces Handbook.**

3356. Water Quality and Treatment: A Handbook of Public Water Supplies. American Water Works Association. McGraw Hill, 3d ed., 1971, 640 pp. LC: 71-116657. ISBN: 0-07-001539-2. Cost: $34.50.

3357. Water Quality Criteria Data Book. U.S. Environmental Protection Agency. U.S. Government Printing Office, 1970-72, 4 Vols.

This compilation has 4 volumes entitled as follows: Vol.1, *Organic Chemical Pollution of Freshwater*; Vol.2, *Inorganic Chemical Pollution of Freshwater*; Vol.3, *Effects of Chemicals on Aquatic Life*; Vol.4, *An Investigation Into Recreational Water Quality.*

3358. Water Quality Handbook. H.J. Mark; J. Mattson. Marcel Dekker, 1981, 496 pp. ISBN: 0-8247-1334-6. Series: *Pollution Engineering and Technology Series*, Vol. 18.

This is a new series published by Marcel Dekker. Vol.1 is entitled *Water Quality Measurement: The Modern Analytical Techniques*. Several other volumes are planned for the future.

3359. Water Surfaces Handbook 1976-1977. Derek Eddowes et al. Fuel and Metallurgical Journals, Ltd, 44th year of publication, 1976-77, 544 pp. ISBN: 0-90199490-1. Cost: £7.5.

3360. Water Treatment Data: A Handbook for Chemists and Engineers in Industry. William M.T. Boby; George Stefan Solt. Hutchinson, 1965.

3361. Water Treatment Handbook. Jacques Bechaux (coord); Donald F. Long (trans and ed). Halsted Press, 5th ed., 1979, 1,186 pp. LC: 79-97503. ISBN: 0-470-26749-6. Cost: $81.85.

Although this is listed as the 5th edition, it should be clarified that this is the 5th edition in the English language. The original French has undergone 18 editions since 1950. Such topics as waste water, sludge, hydraulics, domestic sewage, drinking water, and various types of treatment methods are discussed in this handbook. In addition such topics as flocculant settling, filtration methods, flotation methods, ion exchange methods, aerobic treatment, anaerobic treatment, chemical precipation, and reverse osmosis are just a few of the numerous topics that are presented in this fine handbook.

3362. Water-Based Paint Formulations 1975. Ernest W. Flick. Noyes Data Corp, 1975, 396 pp. ISBN: 0-8155-0571-X. Cost: $28.00.

Approximately 350 formulas are provided in this book on both interior and exterior water-based paints. In addition, physical constants are provided such as viscosity, total solids, test results, and weight per gallon.

3363. Waveguide Handbook. N. Marcuvitz. McGraw-Hill, 1951, 428 pp.

Covers such topics as coaxial waveguides, circular waveguides, rectangular waveguides, elliptical waveguides, radio waveguides, spherical waveguides, microwave networks, arrays, and antennas.

3364. Wavelength Tables with Intensities and Arc, Spark, or Discharge Tube of More than 100,000 Spectrum Lines. MIT Spectroscopy Laboratory (available from MIT Press), 1969 ed., 1970, 429 pp. LC: 73-95288. ISBN: 0-262-08002-8. Cost: $28.00.

3365. Wavelengths of X-Ray Emission Lines and Absorption Edges. Y. Cauchois; C. Senemaud. Pergamon Press, 1978, 340 pp. LC: 78-40419. ISBN: 0-08-022448-2. Cost: $150.00. Series: *International Tables of Selected Constants*, Vol. 18.

This compilation includes selected experimental values such as wavelengths of X-ray emission lines and absorption edges. In addition, various notations, abbreviations, and conversion factors are included along with a bibliography of approximately 750 literature references.

3366. Waverly Handbook. S.G. Symons. Waverly Oil Works Company, Pittsburgh, PA, 9th ed., 1937, 902 pp.

Although this is a rather dated handbook, it includes useful information on lubricating oils, vegetable oils, animal oils, fuel oils, crude oil, and gasoline. In addition, numerous formulas are listed in the back of this handbook, including formulas for bleaching, cement coatings, cleaners, disinfectants, inks, insecticides, lacquer, soaps, and various sprays.

3367. Wear Control Handbook. M. Peterson; W. Winer. American Society of Mechanical Engineers, 1980, 1,300 + pp. Cost: $75.00.

Such topics as cavitation, particle erosion, wear-resistant materials, hard surfacing, bushings, gears, couplings, chains, and brushes are discussed in this compilation. This handbook also includes aproximately 550 figures and 200 tables.

3368. Weather Almanac: A Reference Guide to Weather and Climate of the U.S. and Its Key Cities. James A. Ruffner; Frank E. Bair. Gale Research, 2d ed., 1977. LC: 73-9342. ISBN: 0-8103-1043-0. Cost: $35.00.

3369. Weather Atlas of the United States. U.S. Environmental Data Service. Gale Research Company, 1975, 262 pp. LC: 74-11931. ISBN: 0-8103-1048-1. Cost: $44.00.

This weather atlas includes approximately 271 maps and 15 statistical tables. Such information as precipitation, barometric pressures, relative humidity, solar radiation, heating degree days, and evaporation data are presented in this atlas.

3370. Weather Data Handbook: For HVAC and Cooling Equipment Design. Yale Adams; Ecodyne Corp. Mc-

Graw-Hill, 1980, 384. LC: 79-27370. ISBN: 0-07-018960-9. Cost: $32.50.

Contains such information as summer wet bulb design values, summer dry bulb design values, wind data, and combinations of wet bulb temperatures, wind speed, and relative humidity. This book will be very useful in the design of atomospheric heat exchangers.

3371. Weather Handbook: A Summary of Weather Statistics for Principal Cities Throughout the United States and Around the World. H. McKinley Conway Jr; Linda L. Liston. Conway Research, Inc, 1974. LC: 79-54253. ISBN: 0-910436-00-2. Cost: $25.00.

3372. Weather of U.S. Cities. National Oceanic and Atmospheric Administration. Gale Research Company, 1981, 2 Vols., 1,169 pp. LC: 80-22694. ISBN: 0-8103-1034-1. Cost: $75.00 set.

This 2-volume set is entitled: Vol.1, *City Reports, Alabama-Missouri*; Vol.2, *City Reports, Montana-Wyoming*. Weather data are provided for approximately 293 cities in the United States and island territories. Such information as the range of temperature in each city, rainfall, snowfall, and other special weather features is provided in this book.

3373. Webb Society Deep-Sky Observer's Handbook, Volume 3: Open and Globular Clusters. Kenneth Glyn Johns. Enslow Pubs, Inc, 1980, 206 pp. LC: 78-31260. ISBN: 0-89490-034-X. Cost: $8.95 paper. Series: *Webb Society Deep-Sky Observer's Handbooks Series*, Vol. 3.

This handbook provides the amateur astronomer with a guide to deep-sky observations.

3374. Webster's Medical Office Handbook. Anne H. Soukhanov. Merriam, 1979, 596 pp. LC: 78-26235. ISBN: 0-87779-035-3. Cost: $10.95.

Provides information on health insurance and medical law, and offers numerous answers to clerical and administrative questions. This handbok is designed primarily for the medical office assistant.

3375. Welding Handbook. American Welding Society. American Welding Society, 7th ed., 1976, multivolume set (only Vol.1 currently available), 373 pp., additional volumes planned. Cost: Vol.1, $15.00.

3376. Wellington Sears Handbook of Industrial Textiles. Ernest R. Kaswell. Wellington Sears Co (subs of West Point-Pepperill, Inc), 1963, 757 pp. LC: 63-14165.

3377. Westcott's Plant Disease Handbook. R. Kenneth Horst. Van Nostrand Reinhold, 4th ed., 1979, 832 pp. LC: 78-15312. ISBN: 0-442-23543-7. Cost: $34.50.

Covers numerous plants and the different types of diseases that attack these plants. A description of chemicals available for controlling fungi, viruses, and bacertia is provided.

3378. Western Fertilizer Handbook. California Fertilizer Association. Interstate, 6th ed, 1980, 252 pp. ISBN: 0-8134-2122-5. Cost: $5.50.

Data are provided on different types of fertilizers used for plant nutrition.

3379. Wharfbuilding Handbook. U.S. Government Printing Office, 1963 (rep. 1977), 143 pp. SuDoc: D 209.14:W 55/963. Cost: $3.00.

Construction information is provided on wharfbuilidng along with time data for different types of jobs associated with the construction of wharfs.

3380. Whole Pediatrician Catalog: A Compedium of Clues to Diagnoses and Management. Julia A. McMillan. Saunders, 1979, 495 pp. LC: 76-27060. Cost: $16.95.

This book can be used both as a handy reference or as a textbook for the clinician. Provides information on both common pediatric problems as well as more esoteric pediatric conditions.

3381. Windpower: A Handbook on Wind Energy Conversion Systems. V. Daniel Hunt. Van Nostrad Reinhold, 1981, 610 pp. LC: 80-12581. ISBN: 0-442-27389-4. Cost: $39.50.

Such topics as wind characteristics, applied aerodynamics, wind towers, energy conversion, and energy storage are discussed in this handbook. This book contains a good list of references and a compilation of equipment sources.

3382. Woldman's Engineering Alloys. Robert C. Gibbons. American Society for Metals, 6th ed., 1979, 1,832 pp. LC: 79-20379. ISBN: 0-89170-086-7. Cost: $68.00. Series: *ASM Technical Books Series*.

Previous editions of this compendium were entitled *Engineering Alloys* and were edited by Norman Emme Woldman. This compendium includes more than 40,000 alloys with their corresponding composition and chemical and mechanical properties. Approximately 9,000 new alloys have been added since the fifth edition. This compilation has a listing of obsolete alloys as well as a listing of alloy manufacturers.

3383. Wood Handbook: Wood as Engineering Material. U.S. Department of Agriculture, Forest Products Laboratory, Forest Service. U.S. Government Printing Office, 1974, 432 pp. LC: 73-600335. SuDoc: A1.76:72/974. Cost: $10.00. Series: *Agriculture Handbook*, No. 72.

Covers properties of wood and wood-based products and the principles of how wood is dried, fastened, finished, and preserved.

3384. Wool Handbook. Werner Von Bergen. John Wiley and Sons, 3d ed., 1963-70, 2-Vol. set. LC: 63-11600. ISBN: Vol.1, 0-471-91014-7 (out of print); Vol.2, Part 1, 0-471-91015-5; Vol.2, Part 2, 0-471-91016-3. Cost: $72.50 per Vol.

3385. Work Items for Construction Estimating. Craftsman Book Co, 208 pp. Cost: $20.00.

This is a computer-generated printout of estimating and scheduling data of numerous different heavy construction projects, for example, personhour requirements are included, as well as work scheduling and equipment requirements. Over 10,000 individual data records have been cataloged to produce this publication.

3386. Working Drawing Handbook: A Guide for Architects and Builders. Robert C. McHugh, 166 pp. Cost: $12.95.

Such topics as layout drawing, drafting standards, time budgeting, and project planning are discussed in this handbook.

3387. World Energy Book: An A-Z, Atlas and Statistical Source Book. David Crabbe; Richard McBride. Nichols Publishing Co, 1978, 259 pp. LC: 78-50805. ISBN: 0-89397-032-8. Cost: $27.50.

This publication has approximately 40 pages of tables, diagrams, and charts, and 35 world maps which help to locate the world's energy resources. This publication is laid out in the form of an encyclopedic dictionary and provides approximately 1,500 alphabetically arranged entries on world energy topics. Such topics as oil shale, solid fuels, wind, lightning, fission energy, methane, and magnetohydrodynamics are discussed in this compilation.

3388. World Forestry Atlas. Claus Wiebecke. Verlag Paul Parey, 1978. Cost: DM 2188.

Economic statistics are listed in this publication on forestry and the forest industry. Seventy-one maps are provided.

3389. World Guide to Battery-Powered Road Transportation. Jeffery Christian; Gary G. Reibsamen. McGraw-Hill, 1980, 350 pp. LC: 79-5272. ISBN: 0-07-010790-4. Cost: $49.50.

Over 100 electric vehicles are illustrated in this title. Approximately 65 manufacturers for 14 different countries have provided technical specifications for their vehicles, in this compilation.

3390. World List of Mammalian Species. G.B. Corbet; J.E. Hill. Cornell University Press, 1980, 226 pp. LC: 79-53396. ISBN: 0-8014-1260-9. Cost: $35.00.

This is a comprehensive compilation of mammalian species throughout the world.

3391. World Metric Standards for Enginering. Knut O. Kverneland. Industrial Press, Inc, 1978, 760 pp. LC: 77-25875. ISBN: 0-8311-1113-5. Cost: $47.50.

This is a compilation of metric standards for converting mechanical design projects and manufacturing processes into international metric standards. Numerous computer-produced tables are provided with dimension and tolerance information in the metric system. Numerous cross-references for steel designations are provided from one country to another. Such topics as screw threads, fasteners, mechanical power transmission systems, fluid power systems, electrical components, metal-cutting tools, valves, bearings, and steel material data are covered in these metric tables.

3392. World Ocean Atlas. S.G. Gorschkov. Pergamon Press, Vol.1, 1976; Vol.2, 1979, multivolume set, Vol.1, 340 pp.; Vol.2, 352 pp. LC: 78-40616. ISBN: 0-08-020181-4 set; Vol.1, 0-08-021144-5; Vol.2, 0-08-021953-5. Cost: Vol.1, $330.00; Vol.2, $300.00.

The first 2 volumes of this set are entitled: Vol.1, *Pacific Ocean*; Vol.2, *Atlantic and Indian Ocean*. This atlas was originally published in the Soviet Union and includes such information as the physical elements of the ocean to a depth of 5,000 meters and the physical elements of the atmosphere up to 18 kilometers. Such topics as hydrology, climate, hydrochemistry, and navigation charts are presented in this multivolume set.

3393. World Radio TV Handbook 1977. Jans Frost (ed). Watson-Guptill Publications, 35th ed., 1981, 560 pp. ISBN: 0-8230-5907-9. Cost: $16.50.

3394. World Standards Mutual Speedy Finder. Media International Promotions, Inc, Vol.1, 1976; Vol.2, 1976; Vol.3, 1976; Vol.4, 1976; Vol.5, 1976, multivolume set, Vol.1, 850 pp.; Vol.2, 1,150 pp.; Vol.3, 1,100 pp.; Vol.4, 800 pp.; Vol.5, 400 pp.; Vol.6, 450 pp. Cost: Vol.1, $95.00; Vol.2, $95.00; Vol.3, $95.00; Vol.4, $95.00; Vol.5, $100.00; Vol.6, $100.00.

The volumes of this set are entitled: Vol.1, *Chemicals*; Vol.2, *Electrical and Electronics*; Vol.3, *Machinery*; Vol.4, *Materials—Iron and Steel, Nonferrous Metals, Paper and Pulp Fibers and Textile Ceramics*; Vol.5, *Safety: Electrical and Electronic Products*; Vol.6, *Steel*. This is a series of books on world standards covering such topics as electronics, machinery, materials, chemicals, and steel. Each standard is cross-referenced from 5 different countries including the United States, the United Kingdom, France, West Germany, and Japan.

3395. World Weather Records. U.S. Weather Service. U.S. Government Printing Office, Vol.4, 1959-.

Earlier volumes issued by Smithsonian Institution: to 1920; 1927, 1,199 pp. (Smithsonian Miscellaneous Collections, V. 79), Publication 2913. 1921-30; 1934, 616 pp. (Smithsonian Miscellaneous Collections, V. 90), Publication 3218. 1931-40; 1947, 646 pp. (Smithsonian Miscellaneous Collections, V. 105), Publication 3803.

3396. Worldwide Compilation of Published Multicomponent-Analyses of Ferromanganese Concretions. J.M. Monget; J.W. Murray; J. Mascle. NTIS, 1976, 182 pp. Accession: PB-263 389. Cost: $3.50 microfilm; $15.50 hard copy.

Geochemical data are provided for manganese, iron, aluminum, silica, calcium, titanium, cobalt, nickel, copper, rubidium, tin, tellurium, lead, lanthium, and numerous other elements.

3397. Worldwide Guide to Equivalent Irons and Steels. American Society for Metals. American Society for Metals, 1979, 575 pp. LC: 79-24216. ISBN: 0-87170-088-3. Cost: $92.00. Series: *ASM Engineering Handbook*.

Information is provided on specifications and designations of equivalent irons and steels produced in numerous countries, such as the United States, France, Germany, the United Kingdom, Canada, Japan, Mexico, Sweden, and other nations of the world. Such data as chemical composition and mechanical properties are provided for such types of steel as bar steel, wire, sheet steel, castings, et cetera. Such properties as tensile strength, yield strength, and elongation data are provided.

3398. Worldwide Guide to Equivalent Nonferrous Metals and Alloys. American Society for Metals. American Society for Metals, 1980, 626 pp. Cost: $92.00. Series: *ASM Engineering Handbook*.

This is a companion volume to the *Worldwide Guide to Equivalent Irons and Steels* that was published in 1979. It, too, presents specifications and designations of equivalent nonferrous metals and alloys for numerous nations, such as the United States, Germany, France, Japan, the United Kingdom, Australia, Canada, Italy, Sweden, et cetera. Chemical composition data and mechanical properties of different forms of nonferrous metals and alloys are provided.

3399. X-Ray Diffraction Tables. J.H. Fang; F. Donald Bloss. Southern Illinois University Press, 1966, various paging. LC: 66-21919. ISBN: 0-8093-0211-X. Cost: $15.00.

3400. X-Ray Emission Line and Absorption Wavelengths and Two-Theta Tables. ASTM, 1970, 306 pp. ISBN: 0-8031-0069-8. Cost: $54.00 hard copy. Series: *ASTM Data Series*, DS 37A.

This is a tabulation of X-ray emission lines and their corresponding wavelengths up to 160 angstroms.

3401. X-Ray Emission Wavelengths and Kev Tables for Nondiffractive Analysis. G.G. Johnson Jr; W. White. American Society for Testing and Materials, 1970, 38 pp. LC: 71-121001. Series: *ASTM Data Series*, DS 46.

This *American Society for Testing and Materials Data Series* publication provides X-ray emission wavelengths for nondiffractive analysis.

3402. X-Ray Wavelengths and X-Ray Atomic Energy Levels. J.A. Bearden (National Bureau of Standards). NTIS, 1967. Cost: $4.50. Series: NSRDS-NBS-14.

3403. Yields of Free Ions Formed in Liquids by Radiation. A.O. Allen (National Bureau of Standards). U.S. Government Printing Office, 1976. SuDoc: C13.46:57. Cost: $0.55. Series: NSRDS-NBS-57.

Author/Editor Index

Subject Index

Publisher's Index

Academic Press
111 5th Ave
New York, NY 10003

Addison-Wesley Publishing
Co Inc
Reading, MA 01867

Adler's Foreign Books, Inc
152 5th Ave
New York, NY 10010

Agricultural Engineering
Society
c/o Tokyo Daigku Nogakuba
1-1-1
Yayoi, Bunk-yo-ku, Tokyo,
Japan

Alcan Aluminum Corp
100 Erieview Plaza
Cleveland, OH 44101

Allegheny Press
522 East St
California, PA 15819

Allen and Unwin, Inc
9 Winchester Terr
Winchester, ME 01890

Alliance of American
Insurers
20 N Wacker Dr
Chicago, IL 60606

Allyn & Bacon, Inc
470 Atlantic Ave
Boston, MA 02210

Aluminum Association Inc
750 3rd Ave
New York, NY 10017

American Ceramic Society,
Inc
65 Ceramic Dr
Columbus, OH 43214

American Chain Association
160 Meredith Dr
Englewood, FL 33533

American Chemical Society
1155 16th St NW
Washington, DC 20036

American Concrete Institute
PO Box 1950, Redford
Station
Detroit, MI 48219

American Council on
Education
1 Dupont Circle NW
Washington, DC 20036

American Electromedics
Corp
145 Palisades
Dobbs Ferry, NY 10522

American Foundrymen's
Society
Golf and Wolf Rds
Des Plaines, IL 60016

American Gear
Manufacturers Association
1940 N Fort Myer Dr
Arlington, VA 22209

American Home Economics
Association
2010 Massachusetts Ave,
NW
Washington, DC 20036

American Hospital
Association
840 N Lakeshore Dr
Chicago, IL 60411

American Industrial Hygiene
Association
66 S Miller Rd
Akron, OH 44313

American Institute of
Chemical Engineers
345 E 47th St
New York, NY 10017

American Institute of Steel
Construction
1221 Ave of the Americas
New York, NY 20036

American Management
Assoc
135 W 50th St
New York, NY 10020

American National
Standards Institute
1430 Broadway
New York, NY 10018

American Petroleum
Institute
1801 K St, NW
Washington, DC 20006

American Petroleum
Institute, Measurement
Coordination
2101 L Street N.W.
Washington, D. C. 20037

American Pharmaceutical
Association
2215 Constitution Ave, NW
Washington, DC 20036

American Phytopathological
Society
3340 Pilot Knob Rd
Saint Paul, MN 55121

American Public Health
Association Publications
1015 18th St, NW
Washington, DC 20036

American Public Works
Association
1313 E 60th St
Chicago, IL 60637

American Radio Relay
League, Inc
225 Main St
Newington, CT 06111

American Society for Metals
9275 Kinsman Rd
Metals Park, OH 44073

American Society of Brewing
Chemists
3340 Pilot Knob Rd
Saint Paul, MN 55121

American Society of Heating,
Refrigerating and Air
Conditioning Engineers,
Inc
United Engineering Center,
345 E 47th St
New York, NY 10017

American Society of Hospital
Pharmacists
4630 Montgomery Ave
Washington, DC 20014

American Society of
Mechanical Engineers
United Engineering Center,
345 E 4th St
New York, NY 10017

American Society of
Photogrammetry
105 N Virginia Ave
Falls Church, VA 22046

American Society of
Plumbing Engineers
15233 Ventura Blvd, Suite
616
Sherman Oaks, CA 91403

American Trucking
Associations
1616 P St NW
Washington, DC 20036

American Water Works
Association
6666 W Quincey Ave
Denver, CO 80235

American Welding Society,
Inc
2501 W 7th St
Miami, FL 33125

Anaconda Company
Greenwich Office Park 3
Greenwich, CT 06830

Analog Devices, Inc
PO Box 280
Norwood, MA 08016

Appleton-Century-Croft
292 Madison Ave
New York, NY 10017

Applied Science Publishers
Ltd.
Ripple Road, Barking
Essex IG11 0SA, England

Arco Publishing Co
219 Park Ave S
New York, NY 10003

Arizona State University
Tempe, AZ 85281

ARMCO Drainage and
Metal Products Inc
Middletown, OH 45042

Artech House, Inc
610 Washington St
Dedham, MA 02026

Asbestos Textile Institue
131 N York Ave, PO Box
471
Willow Grove, PA 19090

ASI Publications Inc
127 Madison Ave
New York, NY 10016

Asia Publishing House
440 Park Ave S
New York, NY 10016

ASM Press, Inc
56 E 16th St
New York, NY 10003

Aspen Systems Corp
Dept TL, 20010 Century
Blvd
Germantown, MD 20767

Asphalt Institute
Asphalt Inst Bldg
College Park, MD 20740

Association of Official
Analytical Chemists
1111 N 19th Ave
Arlington, VA 22209

Audel
4300 W 62nd St
Indianapolis, IN 46268

AVI Publishing Co, Inc
PO Box 831
Westport, CT 06880

Bailey Bros. and Swinfen
Warner House, Folkestone
Kent, England

Ballantine Books, Inc
1201 E 50th St
New York, NY 10022

Barnes and Noble (division
of Harper and Row)
Orders to: Harper and Row,
Keystone Industrial Park
Scranton, PA 18512

Battelle, Columbus
Laboratories, Mechanical
Properties Data Center
505 King Ave
Columbus, OH 43201

Bauverlag GmBH
P.O. Box 1468
D-6200 Wiesbaden 1, West
Germany

Beekman Publishers, Inc
38 Hicks St
Brooklyn Heights, NY 11201

Benjamin Co, Inc
485 Madison Ave
New York, NY 10022

Birkhauser Boston, Inc
380 Green St
Cambridge, MA 02139

Bobbs-Merrill
Box 558, 430 W. 62nd St
Indianapolis, IN 46206

Bodine Electric Co
2500 W Bradley Pl
Chicago, IL 60618

Books for Industry
777 3d Ave
New York, NY 10017

Boston Technical Press
116 Austin St
Boston, MA 02319

British Astronomical
Association
Burlington House, Picadilly
London, England W1V ONL

British Book Centre
153 E 78th St
New York, NY 10021

British Construction Steel
Work Assoc
Silvertown House, 1 Vincent
Square
London, England SW1P 2PJ

British Crop Protection
Council
c/o Clacks Farm, Borceley,
Ombersley
Droitwich, Worcester,
England WR9 0HX

British Railways Board
222 Marleybone Rd
London, England NW1 6JJ

Bureau of National Affairs,
Inc
1231 25th St., N.W.
Washington, DC 20037

Burgess Publishing Co
7108 Ohms Lane
Minneapolis, MN 55435

Butane-Propane News, Inc
PO Box 1408
Arcadia, CA 91006

Butterworths Publishing Inc
10 Tower Office Park
Woburn, MA 01801

Cahners Books International,
Inc
221 Columbus Ave
Boston, MA 02116

Cambridge University Press
22 E 57th St
New York, NY 10022

Cameron Engineers
1315 S Clarkson St
Denver, CO 80210

Canadian Institute of Steel
Construction
201 Consumer Rd, Suite 300
Willowdale, ON M2J 4G8

Canfield Press (division of
Harper-Row)
1700 Montgomery St
San Francisco, CA 94111

Carnegie Press (Carnegie-
Mellon University Press)
Scaife Hall
Pittsburgh, PA 15213

Cast Iron Pipe Research
Association
1301 W 22nd St, Suite 509
Oak Brook, IL 60521

Center for Urban Policy
Research
Rutgers Univ, Bldg 4501-
Kimer Campus
New Brunswick, NJ 08902

Chapman and Hall, Ltd
11 New Fetter Ln
London EC4P 4EE England

Chelsea Publishing Co
159 E Tremont Ave
Bronx, NY 10453

Chemical Publishing
Company, Inc
155 W 19th St
New York, NY 10011

Chilton Book Co
201 King of Prussia Rd
Radnor, PA 19089

Churchill Livingstone
19 W 44th St
New York, NY 10036

CIBA Pharmaceutical Co
Saw Mill River Rd
Ardsley, NY 10502

Climax Molybdenum Co
1 Greenwich Plaza
Greenwich, CT 06830

Coblentz Society
Perkin Elmer Corp, 761
Main Ave
Norwalk, CT 06856

Colorado School of Mines
Research Institute
P. O. Box 112
Golden, CO 80401

Compressed Air and Gas
Institute
1621 Euclid Ave
Cleveland, OH 44115

CompuSoft
8643 Navajo Rd, Suite B
San Diego, CA 92119

Comstock Editions Inc
3030 Bridgeway
Sausilito, CA 94645

Concrete Materials, Inc
638 Riverside
Owatonna, MN 55060

Concrete Reinforcing Steel
Institute
180 N LaSalle St
Chicago, IL 60601

Conquest Publications
PO Box 11965
Winston-Salem, NC 27106

Construction Press, Ltd
Lunesdale House
Hornby, Lancaster, England
LA2 8NB

Conway Research, Inc
Peachtree Air Terminal, 1954
Airport Rd
Atlanta, GA 30341

Cornell Maritime
Box 456
Centreville, MD 21617

Corona Publishing Co. Ltd
10-46-4 Sergoko Bunkyo-Ku
Tokyo, Japan

Craftsman Book Co
542 Stevens Ave
Solana Beach, CA 92075

CRC Press
2000 NW 24th St
Boca Raton, FL 33431

Crucible Steel Co of America
4 Gateway Center
Pittsburgh, PA 15222

D. Reidel Publishing Co
P.O. Box 17, 3300 AA
   Dordrecht
Holland

D.C. Heath and Co
Distribution Center, 2700
   Richardt Ave
Indianapolis, IN 46219

Datel-Intersil, Inc
Attn: Marketing Department,
   11 Cabot Blvd
Mansfield, MA 02048

DECHEMA
Frankfurt/Main
Federal Republic of
   Germany

Dorrance and Company
Cricket Terrace Center
Ardmore, PA 19003

Dos Reals Publishing
2490 Channing Way
Berkeley, CA 94704

Doubleday Publishing Co
245 Park Ave
New York, NY 10017

Dover Publications
180 Varick St
New York, NY 10014

Dufour Editions, Inc
Chester Springs, PA 19425

Duxbury Press (division of
   Wadsworth Inc)
6 Bound Brook Court
North Scituate, MA 02060

E. and F.N. Spon Ltd
11 Newfetter Ln
London, EC4 4EE (available
   from Methunen Inc., 733
   Third Ave, New York, NY
   10017)

East-West Center Press
   (University of Hawaii
   Press)
2840 Kolowalu St
Honolulu, HI 96822

Editions Eyrolles
61 Blvd Saint-German
Paris

Edwards Brothers,Inc
2500 S State St
Ann Arbor, MI 48104

Elsevier-North Holland
   Publishing Co
52 Vanderbilt Ave
New York, NY 10017

Engineering Sciences Data
   Unit
251/9 Regent St
London W1R 7AD, England

Enslow Publishers
Bloy St and Ramsey Ave
Hillside, NJ 07205

Entomological Society of
   America
4603 Calvert Rd, Box A-J
College Park, MD 20740

Environment Information
   Center, Inc
292 Madison Ave
New York, NY 10017

Equipment Guide-Book Co
PO Box 10113FX
Palo Alto, CA 94303

ESDU
Springfield, VA 22161

ESDU Marketing Ltd
251/9 Regent St
London W1R 7AD England

F and J Publishing Corp
Agoura Rd, Suite 232
Westlake Village, CA 91361

Fachinformationszentrum,
   Energie, Physik,
   Mathematik
GmbH,
   Kernforschungszentrum, D
   7514 Eggenstein-
   Leopoldshafen 2
Federal Republic of
   Germany

Factory Mutual Engineering
1151 Boston-Providence
   Turnpike
Norwood, MA 02062

Fairmont Press, Inc (division
   of Van Nostrand Reinhold
P.O. Box 14227
Atlanta, GA 30324

Federation of American
   Societies for Experimental
   Biology
9650 Rockville Pike
Bethesda, MD 20014

Film Instruction Co of
   America
2901 S Wentworth Ave
Milwaukee, WI 53207

Food and Drug
   Administration
Rockville, MD

Forging Industry Association
1121 Illuminating Bldg, 55
   Public Square
Cleveland, OH 44113

Franklin Watts, Inc
730 5th Ave
New York, NY 10019

Frederick Research Corp
2601 University Blvd W
Wheaton Br, MD 20902

Freeman, Cooper and Co
1736 Stockton St
San Francisco, CA 94133

Fuel and Metallurgical
   Journals, Ltd
Sales Promotion Mgr, 17/19
   John Adams St
London, England WC2N
   6JH

Futura Publishers
Box 330, 295 Main St
Mt Kisco, NY 10549

Garland Press
2000 N 24th St
Boca Raton, FL 33431

Garland STPM Press
136 Madison Ave
New York, NY 10016

Gas Processors Suppliers
   Asociation
1812 1st Place
Tulsa, OK 74103

Gas Turbine Publications,
   Inc
80 Lincoln Ave
Stamford, CT 06904

Gembooks
PO Box 808
Mentone, CA 92359

Gemological Institute of
   America
1600 Stewart St
Santa Monica, CA 90404

General Electric Co
Business Growth Services
Schenectady, NY 12305

Geological Society of
   America
3300 Penrose Pl
Boulder, CO 80301

Geron-X
PO Box 1108
Los Altos, CA 94022

Glencoe Press
17337 Ventura Blvd
Encino, CA 91316

Goodyear Publishing Co.,
   Inc
1640 5th St
Santa Monica, CA 90401

Goodyear Tire and Rubber
   Co
Industrial Products Div
Akron, OH 44316

Gordon and Breach, Science
   Publishers, Inc
1 Park Ave
New York, NY 10016

Government Institutes, Inc
4733 Bethesda Ave
Bethesda, MD 20014

Greenwood Press
88 Post Rd W
Westport, CT 06881

Grune & Stratton
111 5th Ave
New York, NY 10003

Gulf Publishing Co
Box 2068
Houston, TX 77001

Halsted Press (division of
   John Wiley and Sons, Inc)
605 3rd Ave
New York, NY 10016

Harper and Row
10 E 53rd St
New York, NY 10022

Harvard University Press
79 Garden St
Cambridge, MA 02138

Hayden Book Co
50 Essex St
Rochelle Park, NY 07662

Heinman
Montan-Und
  Wirtschaftsverlag. GmBH,
  Lang Str. 13
Frankfurt AM Main, Federal
  Republic of Germany

Hemisphere Publishing Corp
19 W 44th St
New York, NY 10036

Her Majesty's Stationary
  Office
Atlantic House, Holburn
  Viaduct
London, England NW4 3XX

Herman Publishing, Inc
45 Newbury St
Boston, MA 02116

Heyden and Sons Inc
247 S 41st St
Philadelphia, PA 19104

Holden-Day Inc
500 Sansome St
San Francisco, CA 94111

Holt, Rinehart and Winston
383 Madison Ave
New York, NY 10017

Howard W. Sams and Co,
  Inc
4300 W. 62nd St, PO Box
  7092
Indianapolis, IN 46206

Humana Press
PO Box 2148
Clifton, NJ 07015

Hydraulic Institute
712 Lakewood Center N, 146
  Detroit Ave
Cleveland, OH 44107

I.P.C. Science and
  Technology Press Ltd
PO Box 63, Westbury House,
  Bury St
Guilford, Surrey, GU2 5BH
  England

IIT Research Institute
10 W 35 St
Chicago, IL 60616

Iliffe Science and Technology
  Publishers, Ltd
Iliffe House, High Street
Guilford, Surrey, England

Illinois State Geological
  Survey
Urbana, IL 61801

Illuminating Engineering
  Society
345 E 64th St
New York, NY 10017

Industrial Machinery News,
  Book Publishing Division
29510 Southfield Rd
Southfield, MI 48037

Industrial Press
Bldg 424, Raritan Center
Edison, NJ 08817

Ingersoll-Rand Co
1200 Chestnut Ridge Rd
Woodcliff Lake, NJ 07675

Institute for Management,
  Inc
Old Saybrook, CT 06475

Institute for Materials
  Research, National Bureau
  of Standards
Washington, DC 20234

Institute of Fracture and
  Solid Mechanics
Lehigh University
Bethlehem, PA 18015

Institute of Laboratory
  Animal Resources
2101 Constitution Ave
Washington, DC 20418

Institute of Science and
  Technology, Industrial
  Development Div, Univ of
  Michigan
2200 Bonistel Blvd
Ann Arbor, MI 48105

Institute of Welding
Abington Hall
Abington, Cambridge CB1
  6AL, England

Instrument Society of
  America
400 Stanwix St
Pittsburgh, PA 15222

Interchange, Inc
PO Box 16012 B
St Louis Park, MN 55416

International and National
  Cotton Council of America
PO Box 12285, 1918 North
  Pkwy
Memphis, TN 38112

International Conference of
  Building Officials
5360 S Workman Mill Rd
Whittier, CA 90601

International Copper
  Research Association
708 3d Ave
New York, NY 10017

International Ideas Inc
1627 Spruce St
Philadelphia, PA 19103

International Plastics
  Selector, Inc
2251 San Diego Ave, Suite
  A216
San Diego, CA 92110

International Publishing
  Service
114 E 32nd St
New York, NY 10016

International Technical
  Information Institute,
  Japan
Toranomon Tachikawa Bldg,
  1-6-5 Nishi-Shimbashi
Minato-Ku, Tokyo, Japan

International
  Telecommunications
  Union
Place des Nations
1121 Geneva, 20, Switzerland

International Tin Research
  Institute
Fraser Rd
Perivale Greenford
  Middlesex 4B6 7 4
  England

Interstate Printers &
  Publishers, Inc
19-27 N Jackson St
Danville, IL 61832

Iowa State University Press
S State Ave, 112C Press
  Office
Ames, IA 50010

Iron and Steel Institute
1 Carlton House Terrace
London, England SW1

Iron Castings Society
Cast Metals Federation Bldg,
  2601 Center Ridge Rd
Rocky River, OH 44116

Israel Program for Scientific
  Translations
14 Shammai St, PO Box
  7145
Jerusalem, Israel

Izdate'stvo Metallurgiia
2-j Obdenskij Per. 14
Moscow G-34, USSR

J.J. Keller and Associates
145 W Wisconsin Ave
Neenah, WI 54965

John Wiley and Sons, Inc
605 3rd Ave
New York, NY 10016

Johns Hopkins University
  Chemical Propulsion
  Information Agency
Baltimore, MD 20205

Kendall-Hunt
2460 Kerper Dr
Dubuque, IA 52001

Key Books
Dist by: Associated
  Booksellers, 147 McKinley
  Ave
Bridgeport, CT 06606

Knapp Press, (div of Knapp
  Communications)
5900 Wilshire Blvd
Los Angeles, CA 90036

Kynoch Press
PO Box 216
Witton, Birmingham,
  England B6 7BA

Lake Publishing Co
PO Box 1595
Beverly Hills, CA 90213

Laminated Timber Institute
  of Canada
4916 Elliott St
Delta, BC V4S 2Y1

Lange Medical Publishers
Drawer L
Los Altos, CA 94022

Lea and Febiger
600 S Washington Sq
Philadelphia, PA 19106

Lefax Publishing Co
2867 E Allegheny Ave
Philadelphia, PA 19134

Lehi Publishing Co
303 Gretna Green Way
Los Angeles, CA 90049

Lexington Books (division of
  D.C. Heath and Co)
125 Spring St
Lexington, MA 02173

Lippincott
Washington Sq
Philadelphia, PA 19015

Little, Brown & Co
34 Beacon St
Boston, MA 02106

Longman Inc
19 W 44th St, Suite 1012
New York, NY 10036

Los Alamos Scientific
Laboratory
Los Alamos, CA 93440

Mack Publications
20th & Easton Sts
Easton, PA 18042

Manufacturing Chemists'
Association
1825 Connecticut Ave, NW
Washington, DC 20009

Marcel Dekker Inc
270 Madison Ave
New York, NY 10016

Marcell Dekker, Inc
270 Madison Ave
New York, NY 10016

Marquis Who's Who, Inc
200 E Ohio St
Chicago, IL 60611

Maruzen Co, Ltd
P.O. Box 5050
Tokyo International 100-31
Japan

Masson Publishing USA
14 E 60th St
New York, NY 10022

Matheson Gas Products
Box E
Lyndhurst, NJ 07071

McGraw-Hill
1221 Ave of the Americas
New York, NY 18042

McKnight and McKnight
211 Prospect Rd
Bloomington, IL 60701

Media International
Promotions Inc
114 E 42nd St
New York, NY 10016

Media International
Promotions, Inc
114 E 42nd St, PO Box 292
New York, NY 10016

Medical Examination
Publishing, Inc
969 Stewart Ave
Garden City, NY 11530

Merrow Publishing Co, Ltd
Meadowfield House
Pontefield, Newcastle-upon-
Tyne NE20 950

Metals and Ceramics
Information Center
PO Box 8125
Columbus, OH 43201

Metals and Plastics
Publications, Inc
1 University Dr
Hackensack, NJ 07601

Methuen, Inc
572 5th Ave
New York, NY 10036

Miller Freeman Publishing,
Inc
500 Howard St
San Francisco, CA 94105

MIT Press
28 Carleton St
Cambridge, MA 02142

Morgan-Grampian Book
Publishers, Ltd
20 Calderwood St
Woolwich, London, England
SE18 6QH

MTS Systems Corp
Box 24012
Minneapolis, MN 55424

Multi-Tech Publishing Co
15, Yogesh, Hingwala Ln
Ghatkopar (East), Bombay
400 077

NACE Publications
2400 West Loop S
Houston, TX 77027

NAS-NRC Committee on
Medical and Biological
Effects of Environmental
Pollution (available from
NAS-NRC)
Washington D.C.

National Academy Press
2101 Constitution Ave NW
Washington, DC 20418

National Association of
Corrosion Engineers
2400 West Loop S
Houston, TX 77027

National Audubon Society
950 3rd Ave
New York, NY 10022

National Bureau of
Standards
Washington, DC 20234

National Council of
Architectural Registration
Boards
1735 New York Ave NW,
Suite 700
Washington, DC 20006

National Fire Protection
Assocation
Batterymarch Park
Quincy, ME 02269

National Foremen's Inst
24 Rope Ferry Rd
Waterford, CT 06385

National Geodetic Survey,
U.S. Dept. of Commerce
NOAA
Rockville, MD 20850

National Roofing
Contractors Association
8600 Bryn Mawr Avenue
Chicago, Illinois 60631

National Safety Council
444 N Michigan Ave
Chicago, IL 60611

National Semiconductor
Corp
2900 Semiconductor Dr
Santa Clara, CA 95051

NCRP Publications
PO Box 30175
Washington, DC 20014

Nichols Publishing Company
Box 96
New York, NY 10024

Noordhoff International
Publishing
Schuttersveld 9, PO Box 26
Leyden, The Netherlands

North Atlantic Treaty
Organization: Advisory
Group for Aerospace
Research and Development
b-B-1110 Brussells, Belgium

Northern Miner Press
Limited
7 Labatt Ave
Toronto, Canada M5A 3P2

Norton & Co, Inc
500 5th Ave
New York, NY 10036

Noyes Data Corp
Mill Rd at Grand Ave
Park Ridge, NJ 07656

NTIS
5285 Port Royal Rd
Springfield, VA 22161

Ohio State University Press
Hitchcock Hall, Room 316,
2070 Niel Ave.
Columbus, OH 43210

Olken Publications
2830 Kennedy St
Livermore, CA 94550

Omega Enginering
1 Omega Dr
Stamford, CT 06907

Oregon State University
Engineering Experiment
Station
Corvallis, OR 97331

Oriental Publishing Co
PO Box 5115
Honolulu, HI 96814

Otto Harrassowitz
6200 Wiesbaden, POB 29 29
Federal Republic of
Germany

Oxford University Press
200 Madison Ave
New York, NY 10016

Pachart Publishing House
PO Box 35549
Tucson, AZ 85740

Pacific Coast Publishers
4085 Campbell Ave
Menlo Park, CA 94025

Palmerton Publications
461 8th Ave
New York, NY 10001

Parker Publishing Co, Inc
West Nyack, NY 10994

Pasadena Technology Press
3543 E California Blvd
Pasadena, CA 91107

Penguin Books, Inc
625 Madison Ave
New York, NY 10022

Pennsylvania State University
Press
215 Wagner Bldg
University Park, PA 16802

Pergamon Press Inc
Maxwell House, Fairview
Park
Elmsford, NY 10523

Pest Control Technology
2803 Bridge Ave
Cleveland, OH 44113

Petrocelli (division of
McGraw-Hill)
1221 Avenue of the Americas
New York, NY 10020

Petroleum Engineer
Publishing Co
Box 159
Dallas, TX 75221

Philosophical Library, Inc
15 E 40th St
New York, NY 10016

Physical Electronics
Industries
6509 Flying Cloud Dr
Hopkins, MN 55343

Pit and Quarry Publications,
Inc
105 W Adams
Chicago, IL 60603

Plenum Publishing Corp
227 W 17th St
New York, NY 10011

Popular Library, Inc
1515 Broadway
New York, NY 10036

Portland Cement Association
Old Orchard Rd
Skokie, IL 60077

Prakken Publications, Inc
416 Longshore Dr
Ann Arbor, MI 48107

Prentice-Hall, Inc
Box 500
Englewood Cliffs, NJ 07632

Presses Polytechniques
Romandes
1015 Lausanne
Switzerland

Pressure Vessel Handbook
Publishing, Inc
PO Box 35365
Tulsa, OK 74135

Prestressed Concrete
Institute
20 N Wacker Dr
Chicago, IL 60606

Pruett Publishing Co
3245 Prarie Ave
Boulder, CO 80301

Public Document
Distribution Center
Dept 19
Pueblo, CO 81009

Publishing Sciences Group
545 Great Rd
Littleton, MA 01720

R.C. Krieger Publishing Co
Inc
Box 542
Huntington, NY 11743

R.S.*Means Co
100 Construction Plaza
Kingston, MA 02364

Railway Systems and
Management Association,
Commodity Safety System
Box 330
Ocean City, NJ 08226

Rand Corp
1700 Main St
Santa Monica, CA 90406

Raven Press
1440 Ave of the Americas
New York, NY 10036

RCA Corporation
Commercial Engineering
30 Rockefeller Plaza
New York, NY 10020

Reston (division of Prentice-
Hall)
11480 Sunset Rd
Reston, VA 22090

Rodale Press, Inc
33 E Minor St
Emmaus, PA 18049

Ross Books
Box 3404
Berkeley, CA 94704

Royal Astronomical Society
of Canada
124 Merton St
Toronto, ON M4S 2Z2

Rutgers University, College
of Engineering
New Brunswick, NJ 08903

Sadtler Research
Laboratories, Inc
3316 Spring Garden St
Philadelphia, PA 19104

Saunders Press
W Washington Sq
Philadelphia, PA 19105

Schaevitz Engineering
Pennsauken, NJ 08110

Scholium International Inc
130-30, 31st Ave
Flushing, NY 11354

Scientific Information
Consultants Ltd.
661 Finchley Rd
London NW2 2HN England

Scientific Publishing Co
40 Walton St
Manchester, England M4
4JP

Smithsonian Institution
Arts and Industries Bldg,
Room 2280
Washington, DC 20560

Smoley and Sons, Inc
PO Box 14
Chautauqua, NY 14722

Society of Automotive
Engineers
400 Commonwealth Dr
Warrendale, PA 15096

Solar Vision, Inc
Church Hill
Harrisville, NH 03450

Southern Illinois University
Press
Box 3697
Carbondale, IL 62901

SP Medical and Scientific
Books (division of
Spectrum Publications Inc)
175-20 Wexford Terrace
Jamaica, NY 11432

Span, E and FN, Ltd
11 Fetter Lane
London, EC4 4EE England

Springer-Verlag
175 5th Ave
New York, NY 10010

Stanford Research Institute
333 Ravenswood Ave
Menlo Park, CA 94025

Stanford University Press
Stanford, CA 94305

Stark Research Corp
Cedarburg, WI 53021

State of California, Dept of
Food and Agriculture
Div of Plant Industry Lab
Services-Entomology
Sacramento, CA

Stein and Day
122 E 42nd St, Suite 3602
New York, NY 10017

Stock Drive Products
55 S Denton Ave
New Hyde Park, NY 11040

Stratton Intercontinental
Medical Book Co
381 Park Ave S
New York, NY 10016

Structures Publishing Co
PO Box 423
Farmington, MI 48024

Superintendent of
Documents, U.S.
Government Printing
Office
Washington, D.C. 20402

TAB Books
Blue Ridge Summit, PA
17214

Technomic Publishing Co.,
Inc
265 Post Road W
Westport, CT 06880

Texas Instruments, Inc,
Semiconductor Group,
Engineering Staff
PO Box 5012, MS16
Dallas, TX 75222

Textile Book Service
1447 E 2d St, PO Box 907
Plainfield, NJ 07061

Thermodynamics Research
Center
Texas A & M University
College Station, TX 77480

Thomas Publishing Co
301 27 E Lawrence Ave
Springfield, IL 62717

Thomas Y. Crowell (subs of
Harper-Row)
10 E 53rd St
New York, NY 1002

Trade and Technical Press,
Ltd
Crown House
Monden, Surrey, SM4 5EW,
England

Trans Tech Publications
16 Bearskin Neck
Rockport, MD 01966

Transatlantic Arts, Inc
88 Bridge St
Central Islip, NY 11722

Transportation Research
  Board
National Research Council,
  2101 Constitution Ave,
  NW
Washington, DC 20418

Tri-Ocean Books
62 Townsend St
San Francisco, CA 94107

Trinity University Press
715 Stadium Dr
San Antonio, TX 78284

U.S. Environmental
  Protection Agency,
  Analytical Quality Control
  Laboratory
Cincinnati, OH 45268

UITP (Union Internationale
  des Transports Publics)
Avenue de L'Uruguay 19
Brussels, Belgium

Ungar Publishing Co
250 Park Ave S
New York, NY 10013

Unie-Bell Plastic Pipe
  Association
2655 Villa Creek Drive, Suite
  164
Dallas, TX 75234

UNIPUB (division of R R
  Bowker)
345 Park Ave S
New York, NY 10010

United States Gypsum Co
101 E Wacker Dr
Chicago, IL 60606

United States Pharmacopeial
  Convention Dist. Dept
20th and Northhampton Sts
Easton, PA 18042

United States Steel Corp
600 Grant St
Pittsburgh, PA 15230

Univelt
Box 28310
San Diego, CA 92128

University of Arizona
Tucson, AZ

University of California Press
2223 Fulton St
Berkeley, CA 94720

University of Chicago Press
Orders to: 11030 S Langley
  Ave
Chicago, IL 60628

University of Miami Press
Drawer 9088
Coral Gables, FL 33124

University of Michigan Press
615 E University
Ann Arbor, MI 48106

University of Texas Press
Box 7819, University Station
Austin, TX 78712

University of Toronto Press
Orders to: 33 E Tupper St
Buffalo, NY 14208

University of Washington
  Press
Seattle, WA 98105

Urban and Schwarzenberg
7 E Redwood St
Baltimore, MD 21202

Van Nostrand Reinhold Co
  (division of Litton
  Educational Publishing,
  Inc)
135 W 50th St
New York, NY 10020

Verlag Chemie International
175 5th Ave
New York, NY 10010

Verlag Parey Berlin
Lindenstrasse 44-47
B-1000 Berlin, Federal
  Republic of Germany

Verlag Paul Parey
Spitalerstrasse 12, IV
Hamburg 1, West Germany

Viewpoint Publications
Wexham Springs, Slough,
  England

W. Canning and Co
PO Box 288, Great Hampton
  St
Birmingham, England B18
  6AS

W.B. Saunders Co
W Washington Sq
Philadelphia, PA 19105

Walter De Gruyter, Inc
3 Westchester Plaza
Elmsford, NY 10523

Water Information Center
7 High St
Huntington, NY 11743

Watson-Guptill Publications
1515 Broadway
New York, NY 10036

Wayne State University Press
5980 Cass Ave
Detroit, MI 48202

Wellington Sears Co (subs of
  West Point-Pepperill, Inc)
400 W 10th St
West Point, GA 31833

Western Reserve Press Inc
PO Box 675
Ashtabula, OH 44004

Wiley-Interscience
605 3d Ave
New York, NY 10016

Willard Grant Press
20 Providence St
Boston, MA 02116

William C. Brown Co,
  Publishers
2460 Kerper Blvd
Dubuque, IA 52001

Williams and Wilkins
428 E Preston St
Baltimore, MD 21202

Wireless Press for
  Amalgamated Wireless
  Valve Co
47 York St
Sydney, Australia

World Publications
PO Box 366
Mountain View, CA 94042

Wright-Patterson Air Force
  Base
Dayton, OH 45433

Yale University Press
302 Temple St
New Haven, CT 06511

Year Book Medical
  Publishers, Inc
35 E Wacker Dr
Chicago, IL 60601

Yorke Medical Books
666 5th Ave
New York, NY 10019